MANAGEMENT OF HAZARDOUS AND TOXIC WASTES IN THE PROCESS INDUSTRIES

MANAGEMENT OF HAZARDOUS AND TOXIC WASTES IN THE PROCESS INDUSTRIES

Edited by

S. T. KOLACZKOWSKI

and

B. D. CRITTENDEN

School of Chemical Engineering
University of Bath, UK

ELSEVIER APPLIED SCIENCE
LONDON and NEW YORK

ELSEVIER APPLIED SCIENCE PUBLISHERS LTD
Crown House, Linton Road, Barking, Essex IG11 8JU, England

Sole Distributor in the USA and Canada
ELSEVIER SCIENCE PUBLISHING CO., INC.
52 Vanderbilt Avenue, New York, NY 10017, USA

WITH 140 TABLES AND 189 ILLUSTRATIONS

British Library Cataloguing in Publication Data

Management of hazardous and toxic wastes in the
process industries.
1. Factory and trade waste   2. Hazardous waste
treatment facilities—Management
I. Kolaczkowski, S. T.   II. Crittenden, B. D.
628.5′4   TD897.6

Library of Congress Cataloging-in-Publication Data

Management of hazardous and toxic wastes in the process
industries.

Papers presented at the International Congress on
Recent Advances in the Management of Hazardous and Toxic
Wastes in the Process Industries held in Vienna,
Austria in Mar. 1987.
    Bibliography: p.
    1. Hazardous wastes—Management—Congresses.
2. Factory and trade waste—Congresses.   3. Sewage—
Purification—Congresses.   I. Kolaczkowski, S. T.
II. Crittenden, B. D. (Barry D.)   III. International
Congress on Recent Advances in the Management of
Hazardous and Toxic Wastes in the Process Industries
(1987: Vienna, Austria)
TD811.5.M34   1987   628′.4   87-24568

ISBN 1-85166-163-8

Printed in Great Britain by Galliard (Printers) Ltd, Great Yarmouth

PREFACE

The increasing concern being shown worldwide for protection of the environment and the concomitant spread of ever more restrictive legislation, have led to rapid advances being made in the management of hazardous and toxic wastes. Despite such advances, regrettable incidents which damage the environment still occur, sometimes on a dramatic scale. Nominating 1987 to be the **European Year of the Environment** has helped to focus the mind on the need to research and develop new techniques and processes which will be successful both environmentally and commercially. Each year brings new difficult wastes and new ways of handling them.

The aim of the International Congress on Recent Advances in the Management of Hazardous and Toxic Wastes in the Process Industries was to promote a timely exchange of information between the industries which are responsible for the generation, handling, recovery, treatment and disposal of difficult wastes, the industries which can provide cost effective technologies for ensuring environmental protection and the academic and industrial researchers whose efforts today will provide the foundations for tomorrow's technologies.

The Congress held in Vienna in March 1987, certainly provided a forum within Europe for a lively exchange of ideas between academics and industrialists. The texts of all the keynote and session papers, published in full in this book, reveal that attention was focussed on a broad range of industries which included oil refining, petrochemical, agrochemical, pharmaceutical, metallurgical and water treatment processes. Waste management practices which relate solely to the nuclear industry were specifically excluded from the Congress.

This book will act as a valuable reference for professionals active in the field of hazardous waste management, for industrialists and academics engaged in research and teaching, and for those exploring business opportunities in this rapidly developing multi-disciplinary subject.

There is much overlap between the themes of the Congress and therefore to unify the subject matter there are in Chapter One, eight state-of-the-art overviews and case studies written by invited experts. The session papers are subsequently presented in Chapters 2 to 6.

Chapter 1 Invited Overviews and Case Studies

Chapter 2 Waste Management Methodologies and Practices

Chapter 3 Physical Treatment Methods

Chapter 4 Biological Treatment Methods

Chapter 5 Chemical Treatment Methods

Chapter 6 Thermal Treatment Methods

Appendix Overview of waste management practices in major industrial groups.

Many of the papers reflect that industrial hazardous waste management practices have indeed undergone significant changes arising out of stringent and comprehensive environmental regulations, the increasing cost of disposal facilities and the increasing awareness of environmental problems which are caused by dubious or marginal disposal practices. The minimisation of waste and the anticipation of future environmental liabilities, which are key factors in the United States approach, are likely to spread worldwide. As more and more countries begin to prohibit the disposal of untreated wastes directly onto the land so the technology associated with the management of hazardous and toxic wastes will itself become more process orientated. The ideal of achieving zero waste is considered to be impracticable as industrialised countries adopt new chemicals and processing routes in order to remain competitive.

The rapid changes taking place within both the process industries and the waste management industries will require frequent exchanges between academics and industrialists. With this in mind the organisers, Congress Team International (UK) Ltd, plan to hold a second meeting on this theme in three years' time.

As technical advisers to the Vienna Congress, we wish to express our thanks to the Keynote and session speakers for their contributions and to Congress Team International (UK) Ltd for making this event possible.

Stan Kolaczkowski
Barry Crittenden

School of Chemical Engineering
University of Bath
Bath, United Kingdom, BA2 7AY

vii

Contents

Chapter 3

PHYSICAL TREATMENT METHODS

Chapter 4

BIOLOGICAL TREATMENT METHODS

Chapter 5

CHEMICAL TREATMENT METHODS

Chapter 6

THERMAL TREATMENT METHODS

Chapter 1

INVITED OVERVIEWS AND CASE STUDIES

OVERVIEW OF WASTE MANAGEMENT PRACTICES
IN MAJOR INDUSTRIAL GROUPS

Amir M. Metry, Ph.D., P.E.
Robert J. Schoenberger, Ph.D., P.E.
Michael H. Corbin, P.E.
Roy F. Weston, Inc.
Weston Way
West Chester, PA 19380

ABSTRACT

Industrial Hazardous waste management practices have undergone significant changes which have been driven by stringent and comprehensive environmental regulations, increasing cost at disposal facilities, and increasing awareness of environmental problems caused by marginal disposal practices. The traditional methods for managing industrial waste are being phased out or are undergoing significant upgrading/modification. Industry is working to minimize and anticipate potential future environmental liabilities. This paper will address current and future trends in waste management including several of the major industry sectors.

Hazardous waste management practices for major industries have undergone significant changes within the last 5-year period. These changes have been driven by enactment of more stringent and comprehensive environmental regulations, increasing cost at off-site disposal facilities, closing of commercial facilities and fewer available options, and increasing awareness of environmental problems caused by current and past disposal practices. It is expected that this trend of major changes in the waste management field will continue for the next ten years or longer. The traditional methods for managing industrial waste are being phased out or are undergoing significant upgrading/modification in response to the more stringent regulations and public pressure. Industrial firms are also working to minimize and anticipate potential future liabilities due to environmental and public health concerns related to hazardous waste management practices.

Several of the major industry sectors, confronted with significant changes in the management of their hazardous wastes include: petroleum refining, petrochemicals, pharmaceuticals and electronics.

In the petroleum refining industry, wastes typically fall into both the inorganic and organic categories. Inorganic residues include spent catalysts from process reactor units. Past disposal practices for these spent catalysts typically utilized uncontrolled on-site land disposal. Current trends are for off-site controlled landfill or recovery/regeneration of the catalysts. Other inorganic residues include spent caustic, lime sludge, boiler ash, and filtering material.

Organic residues from petroleum refining have typically been treated by discharging to the process sewer which historically has treated the wastewaters through oil separation and/or an effluent settling pond system. Major upgrading of the wastewater treatment systems have been undertaken by the refining industry in response to clean water laws and now include biological treatment, physical/chemical treatment and enhanced oil separation techniques. Organic sludges such as API separator sludge are generated from the wastewater system and these have typically been land-disposed or land-applied. Other organic sludges include tank bottoms from tank cleaning operations. Current trends in managing of these organic residues are by either thermal destruction or land application for biological degradation. The wastewater treatment ponds are being phased out or retrofitted with liner systems for the protection of groundwater quality.

The petrochemical industry is also undergoing major changes with respect to hazardous waste management. The types of waste generated by the petrochemical manufacturer vary widely from plant to plant and are highly dependent upon the type of process and plant production. A portion of the wastes have been handled via the process sewer and a wastewater system consisting of oil/water separation, and settling of solids in a pond/lagoon system. Major upgrading of wastewater treatment facilities is underway to incorporate biological treatment in addition to physical/chemical treatment processes such as filtration and carbon adsorption. Residues are generated from the wastewater treatment system along with the plant process units that are typically produced in the form of residues and sludges. Historically, these have been handled via land disposal; however, the current trend is toward thermal destruction, biological degradation, and stabilization. Thermal destruction processes involve incinerators capable of handling both liquid and solid materials and typically incorporate heat recovery units. Chlorinated and non-chlorinated solvents are major waste streams in the petro-chemical industry and the current trends for managing these materials involve thermal destruction or solvent recovery. Figure 1 depicts the typical waste management options for handling spent solvents and organic solutions.

The major waste stream from the pharmaceuticals industry is typically wastewater from production operations. This has historically been handled either through discharge to a municipal sewer or through an on-site facility. Current practices and trends involve installation of pretreatment facilities prior to sewer discharge and upgrading of the existing treatment facilities. This upgrading involves the construction of physical/chemical treatment modules or biological treatment. Other pharmaceutical waste materials are generated by laboratory chemicals from research, organic sludges from wastewater treatment or fermentation steps, spent solvents, off-spec products, and other biological materials from research. Current management trends for these materials include thermal destruction, solvent recovery, destruction via

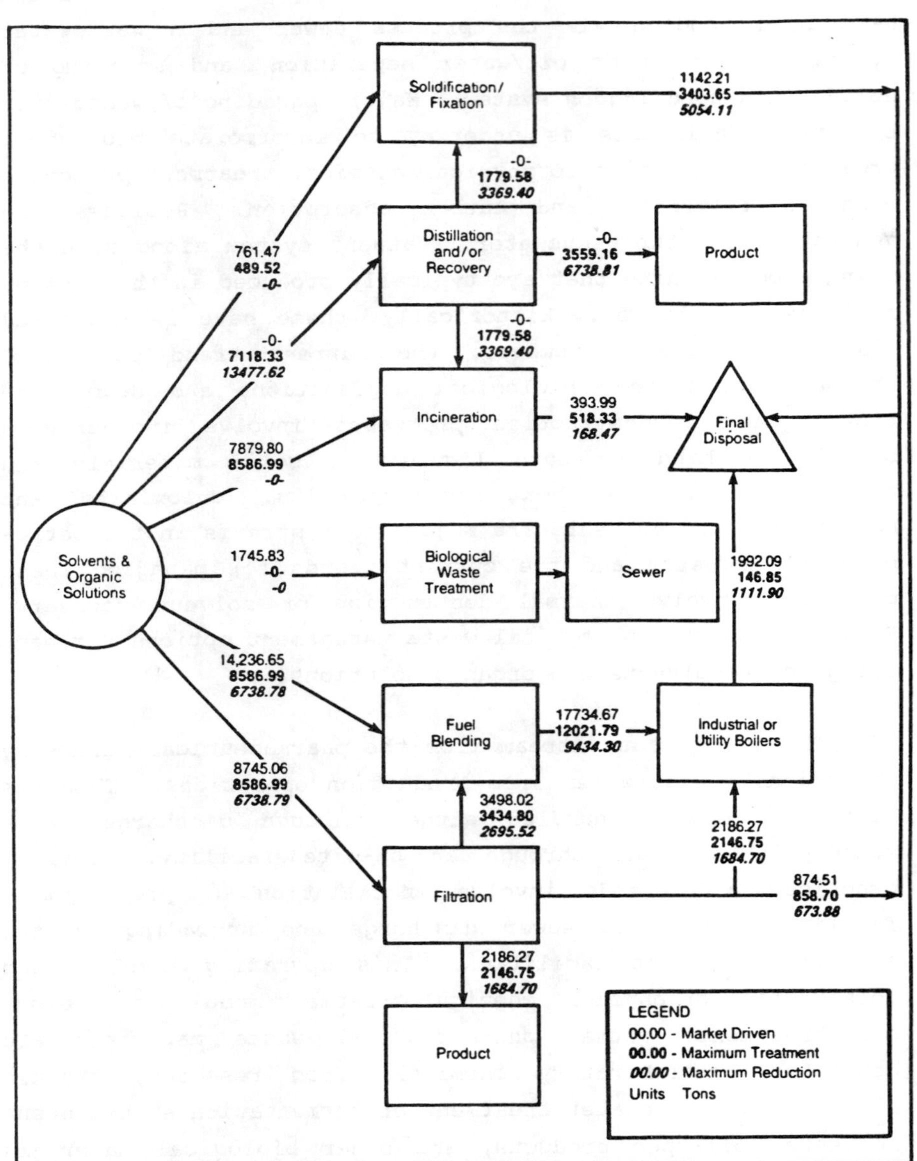

WASTE MANAGEMENT MICRONETWORK

SOLVENTS AND ORGANIC SOLUTIONS

Figure 1

incineration, and waste reduction. Figure 2 depicts a rotary kiln incineration system capable of handling a variety of wastes. The pharmaceutical industry is concerned with public image and public perception and, therefore, this industry is paying close attention to demonstrating satisfactory waste management practices.

Wastes generated in the electronics industry usually include inorganic wastewaters and solutions and most always are cross-contaminated with organic solvents. A small amount of organic residues are also generated, typically in the form of spent solvents and cleaning solvents. Electronics encompasses very small firms and very large multi-national operations. Many of these wastes are discharged to municipal sewers by the smaller firms while on-site physical/chemical wastewater treatment is practiced by the large firms. Sludges are generated from both wastewater treatment and plating operations. Current and emerging trends for these waste materials include recovery of metals, stabilization/solidification, and secure land disposal.

Note: The overview of waste management practices in several major industrial groups is presented in the Appendix.

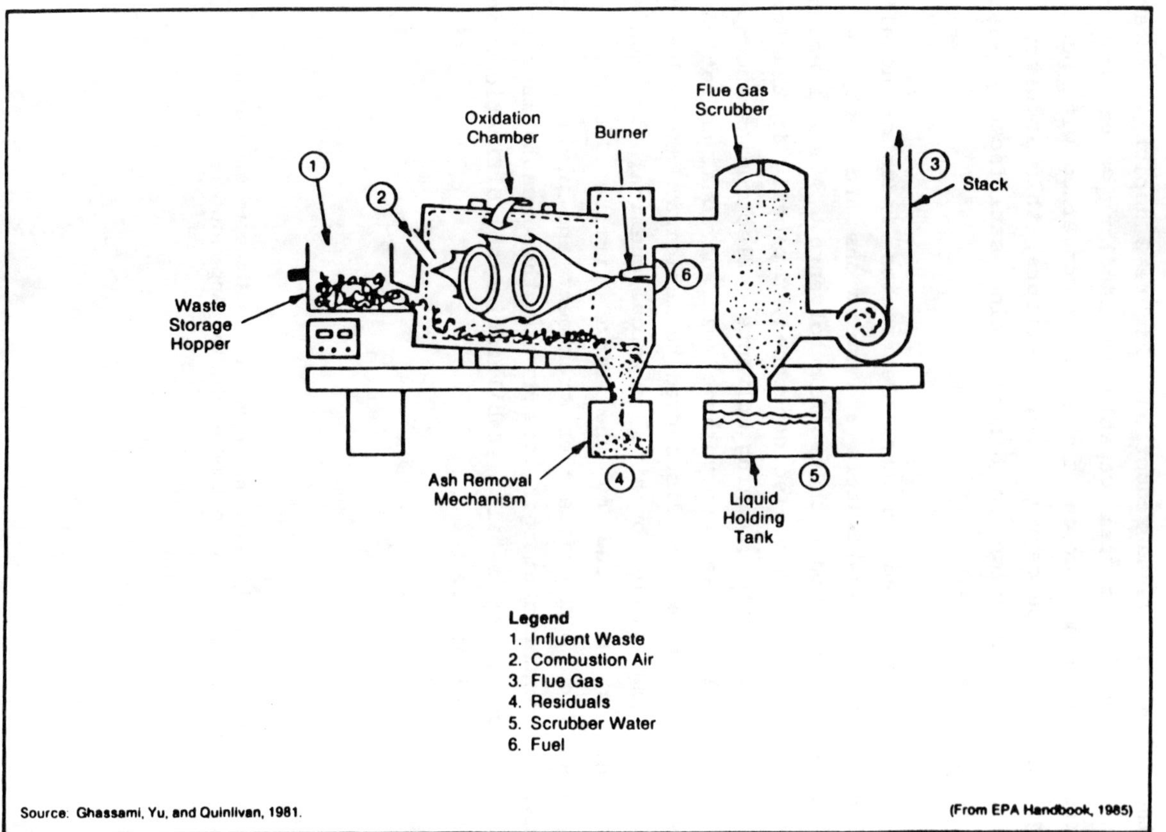

FIGURE 2   ROTARY KILN INCINERATOR SCHEMATIC

HAZARDOUS WASTE DISPOSAL AND
ENVIRONMENTAL MANAGEMENT ISSUES

Amir A. Metry, Ph.D., P.E.
Michael H. Corbin, P.E.
Roy F. Weston, Inc.
Weston Way
West Chester, PA 19380

ABSTRACT

Historically, industrial waste management practices have primarily utilized either on-site or off-site land disposal. Man of these land disposal sites were not designed or operated to handle these industrial residues and often were not run even according to sanitary landfill practices. This has resulted in contamination of private and public water supplies and general public health. This paper presents some innovative concepts related to alternatives to landfill disposal, means of remediation of inactive hazardous waste sites and rehabilitation of groundwater resources.

Since the start of the industrial revolution, industrial waste management practices have primarily involved the disposal and use of either on-site or off-site landfills. Industrial wastes were often co-disposed using typical so-called sanitary landfill practices. Many of these land disposal sites were not designed or operated to handle these industrial residues and often were not run even according to sanitary landfill practices. As a result, leachate resulting from these wastes was not properly contained or controlled and has migrated out of the landfill into underlying soils and contaminated groundwater or nearby surface water. This has resulted in contamination of private and public water supplies affecting drinking water and required use of alternate water supplies or treatment.

Several hazardous waste management alternatives are available other than land disposal and these include:

- Recovery and reuse.
- Waste minimization.
- Destruction via incineration.
- Biological degradation.
- Chemical detoxification.
- Thermal stabilization.
- Stabilization/soldification.
- Long-term storage.

Most of these technologies will generate a process residue (i.e., ash, sludge) which may ultimately require land disposal. However, this residue is typically non-hazardous, less leachable or less toxic, and is a smaller quantity than the original waste material. A properly designed secure landfill may be an acceptable option for handling these types of treatment residues.

The present design practice for secure landfill disposal facilities requires the use of sophisticated liner and leachate collection systems. These liner systems incorporate the use of clay or synthetic geomembranes to provide complete containment. Design features for these systems involve the use of two or more bottom liners, use of clay or geomembranes which assure compatibility with leachate, use of rigorous QA/QC protocol for liner installation, and use of composite materials for leachate conveyance and collection. Figure 1 depicts a schematic of a double-lined secure landfill cell.

In addition to sophisticated liner systems, present secure landfills also utilize engineered cover and cap systems to minimize infiltration during post-closure. The design objective is to utilize a high-efficiency cap which

Landfill
Double Membrane Liner System

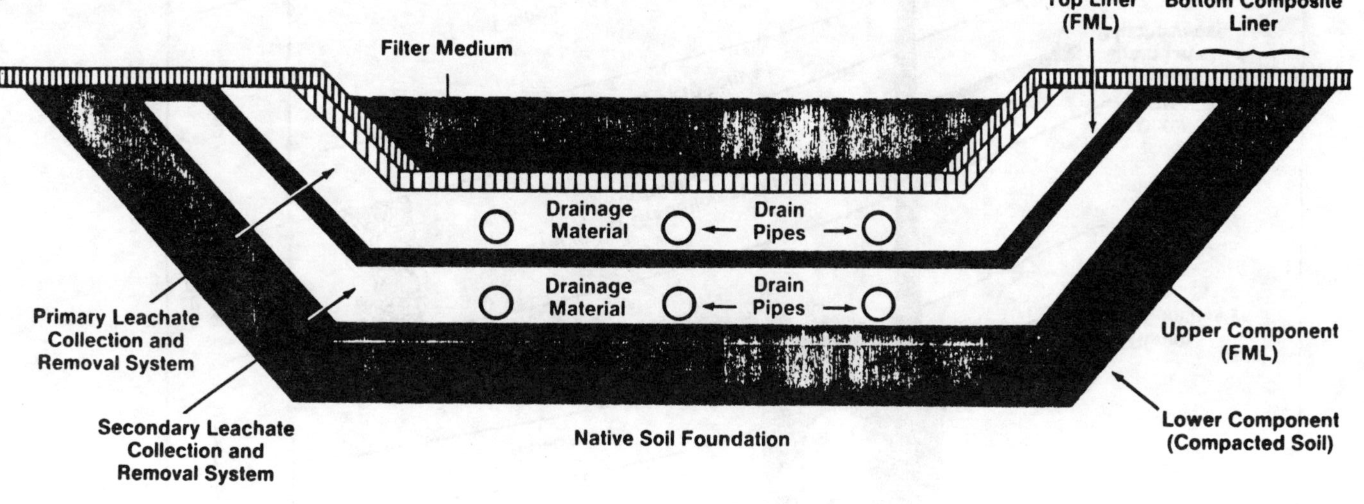

Figure 1

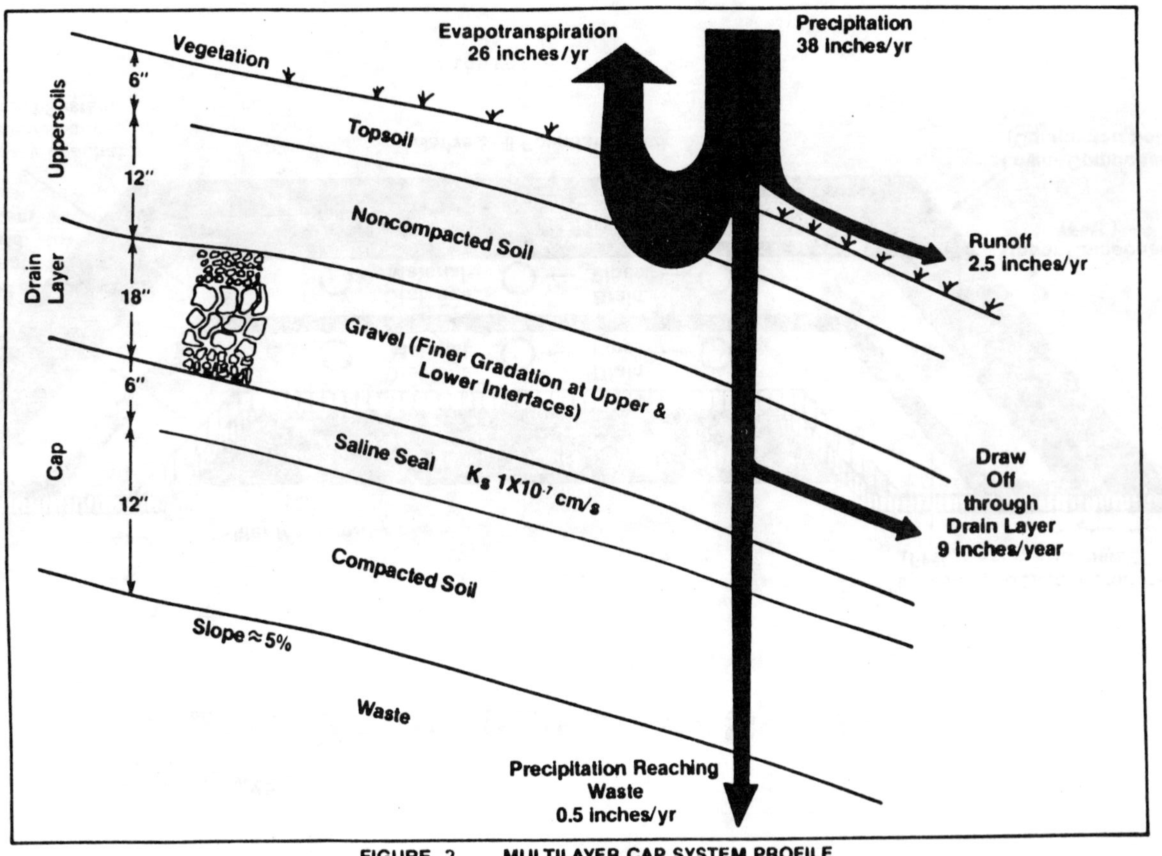

FIGURE 2   MULTILAYER CAP SYSTEM PROFILE

is designed to divert the maximum amount of precipitation. The thrust of this design objective is leachate minimization rather than collection and treatment. Figure 2 depicts the functional aspects of a high-efficiency multi-layer cap system.

To complement the advancement in landfill designs, emphasis is being placed on pretreatment and stabilization of waste including stabilization through the use of cements, pozzolanic or bituminous materials , thermal treatment, dewatering and moisture removal, and detoxification. Emphasis is on minimizing the amount of liquids contained in the waste material which is deposited in the landfill along with reducing overall toxicity of materials which are disposed.

Many new and effective technologies are being developed for addressing the environmental problems attributable to past landfilling practices. In general, they fall into two broad categories of:
- Source control, and
- Migration control.

The source control technologies are directed toward stabilizing the waste material or contaminant source to reduce leachate production. Several source control technologies include:
- Installation of a high-efficiency or impermeable cap.
- In-situ stabilization via injection grouting or biodegradation.
- Grading,diking, flood control berms.
- Retrofitting of liner systems.
- Excavation and removal.
- Retrofitting of leachate and gas collection systems.

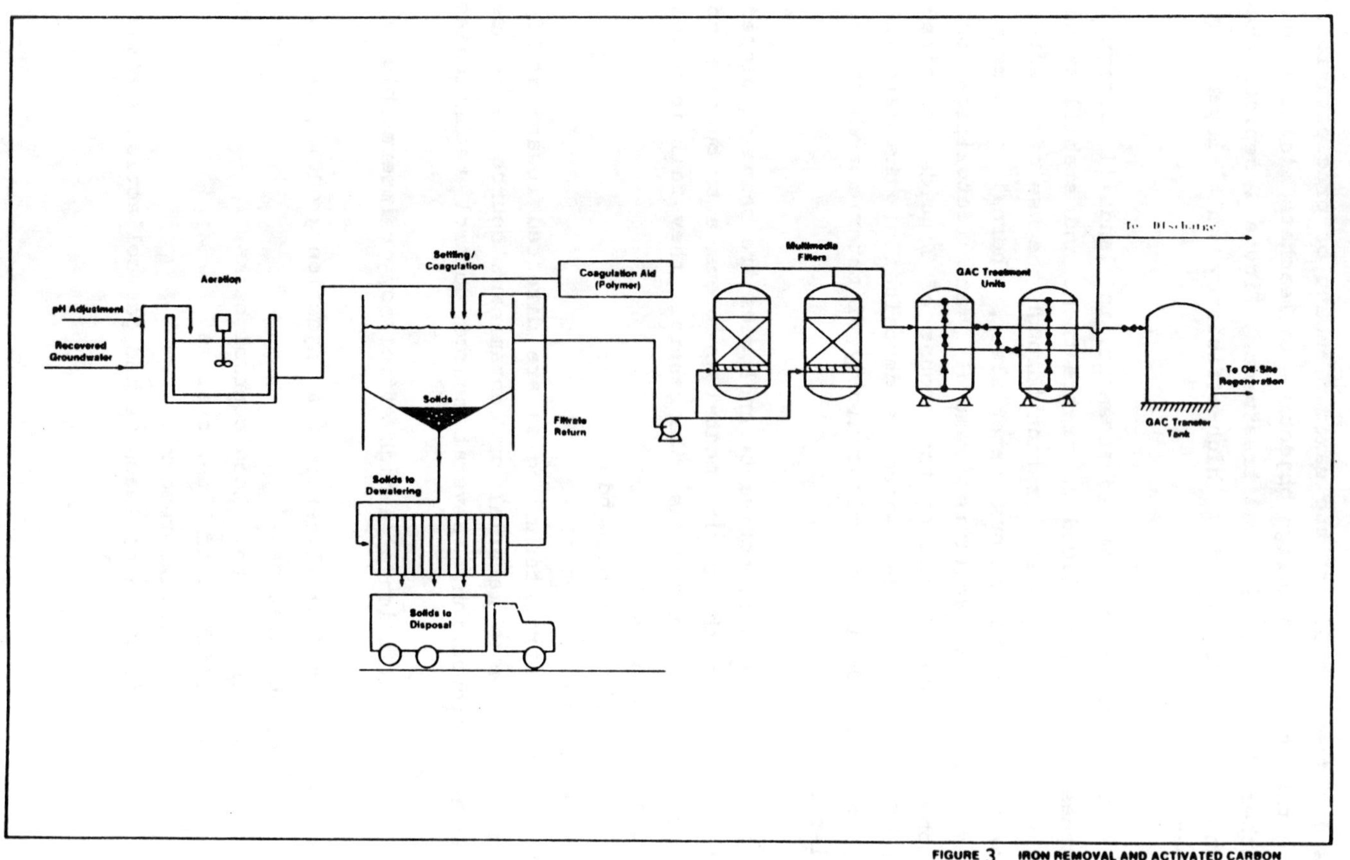

FIGURE 3   IRON REMOVAL AND ACTIVATED CARBON TREATMENT FOR RECOVERED GROUNDWATER

To address groundwater contamination problems caused by poor landfilling practices in the past, new techniques are being developed for groundwater recovery and treatment. These techniques involve use of one or a combination of the following concepts:

- Interception and recovery of leachate by pumping wells.
- Slurry walls and grout curtains.
- Groundwater interception trenches.
- Hydrologic manipulation to isolate the site.
- In-situ biotreatment and biodegradation.

The selection of a particular technique for groundwater remediation is highly site-specific. If contaminated groundwater is being recovered, it must be treated prior to discharge. Treatment techniques involve biological, chemical and physical treatment technologies similar to those employed in the wastewater treatment fields. Depending upon the composition of the groundwater, typical technologies may involve air stripping, carbon adsorption, metals precipitation, filtration, biological degradation, concentration, and ion exchange. The design of treatment systems must address the ability to handle varying flows and concentration in the groundwater which will occur as remediation progresses. Figure 3 depicts a typical groundwater treatment system for metals and volatile organics removal.

BIOLOGICAL TREATMENT OF INDUSTRIAL PROCESS WASTEWATERS
CONTAINING HAZARDOUS AND TOXIC CONTAMINANTS

Henryk Melcer
Environment Canada
Wastewater Technology Centre
Burlington, Ontario, Canada

ABSTRACT

The performance of conventional activated sludge systems was
reviewed with respect to their ability to treat industrial process
wastewaters. High removals of specific toxic trace contaminants can be
achieved. Improved performance may also be achieved through the use of
sludge retention time control and the addition of powdered activated
carbon. The development of fate-of-contaminant models and methods for
predicting loss of contaminants were reviewed. Volatilization,
adsorption and biological degradation were identified as the major
removal mechanisms for toxic trace contaminants in biological wastewater
treatment systems.

INTRODUCTION

The majority of data collected in North America to demonstrate the
ability of industrial wastewater treatment plants in removing toxic
contaminants is based on the 126 U.S. Environmental Protection Agency
(USEPA) priority pollutants. This practice stems from the 1976 Consent
Decree on priority pollutants when the emphasis in the U.S. Clean Water
Act for water pollution control shifted from regulation based on
conventional pollutants alone to regulation based upon both conventional
and non-conventional specific toxic trace pollutants. In 1983, the USEPA
published an updated "Treatability Manual" which collated data on the
priority pollutants in wastewaters from a wide range of industrial
sectors, the performance of treatment plants in these sectors, and the
treatment costs [1]. The Manual was prepared from numerous Development
Documents that were produced during the development of effluent
limitation guidelines and for industrial wastewater pretreatment
requirements in the U.S. Additional characterization data were generated
by the USEPA during a study to assess the impact of hazardous wastes
discharged to municipal wastewater treatment plants [2]. The occurrence
and concentration of contaminants in industrial wastewaters are typically
industry-specific.

Numerous studies have demonstrated the removal of toxic trace contaminants in biological treatment facilities. Richards and Shieh [3] presented the results of 21 studies on the fate of toxic trace contaminants in biological systems. They cautioned that the disappearance of contaminants in biological treatment systems should not always be attributed to biological degradation. Contaminants may be stripped from solution or they may adsorb on to biological solids. Alternatively, the persistence of a contaminant in a final effluent may not necessarily be due to recalcitrance. Rather, it may result from unsatisfactory environmental conditions in the treatment system. Grady [4] defined criteria that are essential to biodegradation of contaminants. These included the presence of a capable organism or microbial community and optimum environmental conditions. The concentration of the contaminant will influence biodegradation. Low concentrations may not induce the production of necessary enzymes and may provide too little energy to sustain growth. A high concentration may cause toxic inhibition unless acclimation can greatly increase the tolerance concentration of degradative organisms to the contaminant [5].

This paper addresses the level of success attained by biological wastewater treatment systems in removing specific contaminants from industrial wastewaters and reviews the development of fate-of-contaminant models and methods for predicting loss of contaminants as they pertain to improvement in biological degradation of toxic contaminants.

PERFORMANCE ASSESSMENT OF CONVENTIONAL INDUSTRIAL ACTIVATED SLUDGE TREATMENT PLANTS FOR REMOVAL OF TOXIC CONTAMINANTS

Biological wastewater treatment plants have long been used for industrial wastewaters. Traditionally, their role was regarded to be primarily the removal of organic matter and suspended solids. Plant performance was expressed as a removal efficiency in terms of gross organic parameters such as BOD_5, COD or TOC. Industry generally favoured the activated sludge system to address its needs. For example, phenol and thiocyanate removal has been achieved by activated sludge plants in the coking industry for 30 years, and ammonia and phenol have been removed in petroleum refinery activated sludge plants for an even longer period. The need to address a wide range of toxic trace contaminants has placed new demands on this technology.

Some pretreatment is desirable for most industrial wastewaters before biological treatment. Usually, this consists of segregation of wastewater streams, equalization and neutralization. The practice of treating complex chemical wastewaters combined with municipal wastewaters appears to be technically and economically attractive [6]. Typically, for well operated activated sludge plants receiving industrial discharges co-mixed with sewage, between 5 and 10 organic contaminants (primarily volatiles and phthalate esters) were found in the effluents at concentrations ranging from 1 to 20 µg/L for more than 50% of the time [7].

Metal contaminants in wastewaters are generally not biologically degraded. Rather, they are adsorbed by the biomass. Thompson and Prior [8] suggested that 80% of heavy metals in wastewater is ultimately attached to biomass and is concentrated by a factor of 3 000. This may present difficulties in cases where land disposal of waste sludge is

practiced. There is a paucity of data on heavy metal removal in
industrial wastewaters. The removal of metals in municipal activated
sludge plants may serve as a guide to the performance of industrial
plants. In a survey of treated effluents from 70 Ontario municipal
treatment plants, mean heavy metal concentrations ranged from <10 to 290
µg/L [9]. Between 3 and 6 metal contaminants were found more than 50% of
the time in treated effluents from over 100 municipal activated sludge
plants at concentrations ranging from 1.0 to 100 µg/L [7].

Probably the most comprehensive evaluation of toxic trace
contaminant removal from industrial wastewaters was conducted in 1981 by
the U.S. Chemical Manufacturers Association in conjunction with the
USEPA. Known as the CMA-EPA 5-Plant Study, it was conducted at 5
chemical plants that used activated sludge treatment plants [10]. Table
1 presents pooled treatment plant performance data for the removal of
volatile, acid and base/neutral extractable organic contaminant groups.
These ranged from 77% for acid extractables to 95% for the volatiles.
Individual organics in all groups were removed, with few exceptions, to
effluent concentrations very close to or below 10 µg/L. Of the 52
contaminants found in the treated effluents, 36 had median concentrations
less than or equal to 10 µg/L (Table 2). Only one contaminant,
pentachlorophenol, was present at a median concentration greater than 100
µg/L. This relatively good performance was achieved despite the high
influent contaminant concentrations. For example, 9 of the 52
contaminants found in the influent were present at concentrations greater
than 500 µg/L. Pooled data describing the performance of the treatment
plants in removing the organic contaminants that occurred at the highest

TABLE 1

CMA-EPA 5 Plant Study - Removal of major organic contaminant groups[10]

	Influent[1] (µg/L)	Effluent[1] (µg/L)	Removal (%)
Volatile	116	8	95
Acid	150	34	77
Base/Neutral	119	15	87

[1]Geometric means.

TABLE 2

CMA-EPA 5 Plant Study - Organic contaminant removal for all plants[10]

Average Concentration[1] in Treated Effluent (µg/L)	Number of Pollutants
≪ 10	36
> 10 - 50	10
> 50 - 100	5
≫ 100	1

[1]Based on geometric means.

concentrations in the influent are shown in Table 3. Those compounds present in the influent at concentrations in excess of 1 mg/L, acrylonitrile, bromomethane, toluene and nitrobenzene, were reduced to very low levels.

TABLE 3
CMA-EPA 5 Plant Study - Removal of organic contaminants[10]

Contaminant	Concentration (μg/L)[1]		Removal (%)
	Influent	Effluent	
Volatiles			
Acrylonitrile	10 300	65	99
Benzene	581	6	99
Bromomethane	1 250	ND[2]	100
Chloroform	348	13	96
1,2-Dichloroethane	524	9	98
Ethyl benzene	283	5	98
Toluene	4 500	7	100
Acid Extractables			
2,4-Dichlorophenol	347	24	93
2,4-Diethylphenol	270	5	98
2,4-Dinitrophenol	673	59	91
Pentachlorophenol	216	115	47
Base/Neutral Extractables			
1,2-Dichlorobenzene	331	28	92
Isophorone	650	ND	100
Naphthalene	802	6	99
Nitrobenzene	3 000	91	97
1,2,4-Trichlorobenzene	234	39	83

[1]Geometric means   [2]Not detected.

A detailed investigation of toxic trace contaminant removal from petroleum refinery wastewaters was conducted by the Petroleum Association for the Conservation of the Canadian Environment (PACE) during the period 1980-1985. Initially, six Canadian refinery treatment plants were screened for 58 chemical parameters that were found to be the most frequently occurring contaminants in an earlier American Petroleum Institute (API) study [11]. Five of the refineries had activated sludge systems. This work confirmed the API's findings that the majority of USEPA priority pollutants were not present in refinery wastewaters. The contaminants that were detected in treated effluents at concentrations greater than 1.0 µg/L for more than 50% of the time are shown in Table 4 [12]. The solvents, benzene and toluene, and the metals, chromium and zinc, were found at the highest concentrations. More than 90% of the phenolics were removed by the activated sludge systems. Except for naphthalene, polyaromatic hydrocarbons (PAHs) were present at very low concentrations. Further work by PACE examined the fate of toxic organic contaminants in a refinery wastewater treatment system [13]. Biological degradation was shown to be the major removal mechanism in the activated

TABLE 4
Contaminants in refinery effluents[12]

Contaminant	Detection Frequency (%)	Mean Concentration (μg/L)
Benzene	50	226
Chloroform	79	11
Hexachlorobenzene	64	0.03
Methylene Chloride	86	45
Naphthalene	57	7
Toluene	57	208
Arsenic	86	6
Chromium	86	190
Copper	100	13
Lead	71	7
Nickel	57	15
Zinc	100	85

sludge plant for the contaminants not removed by the API separator. A summary of major contaminants occurring in the final clarifier effluent is shown in Table 5. No chlorinated volatile organic species were present in this group of data. The most commonly found species were aromatic solvents, phthalate esters and phenolics. Similar results were also reported in further work by API and the USEPA at two refinery activated sludge plants [14].

TABLE 5
Organic contaminants in refinery final clarifier effluents[13]

Contaminant	Mean Concentration (μg/L)
Volatiles	
Acetone	800
Benzene	3
Cyclohexane	4
Ethyl benzene	3
Hexane	3
Methyl ethyl ketone	3760
Toluene	10
Xylenes	15
Base/neutral extractables	
Di-n-butyl phthalate	15
Di-n-octyl phthalate and	
Bis(2-ethyl hexyl) phthalate	25
Acid extractables	
Phenol	86
p-Cresol	160
2,5-Dimethyl phenol	10
2,3,5-Trimethyl phenol	22

CONVENTIONAL ACTIVATED SLUDGE SYSTEMS – MODIFIED OPERATION

The need to address these increasingly complex contaminants has promoted changes in the operation of activated sludge systems. Probably the most important of these has been the recognition that control should be based upon manipulation of solids retention time (SRT). SRT is inversely proportional to microbial growth rate and thereby provides a method of influencing microbial activity in the activated sludge system. Retention of microorganisms by SRT manipulation is important because the microorganisms that are responsible for degradation of complex organic contaminants may grow slowly and either be washed out of the system or be dominated by other microbial communities. A minimum SRT must be maintained to prevent washout. Bench scale activated sludge studies demonstrated the importance of SRT manipulation when treating complex process condensates for a slagging fixed bed gasifier [15]. Further work on condensates for the Hygas process [16, 17] clearly showed that, at SRTs of 15 to 40 days, phenolics, PAHs and heterocyclic nitrogenous compounds (HNCs) such as quinoline, pyridine and indole could be reduced from mg/L concentrations to detection limits (<10 µg/L). The removal of PAHs and HNCs from coking plant wastewaters has been reported when controlling SRT at 30 to 40 days in a predenitrification activated sludge system [18]. This approach has been used at full scale for wastewaters generated by the organic chemical industry [19], gas plants [20] and by-product coking plants [21]. Grutsch and Mallatt [22] cited the importance of controlling SRT in the treatment of petroleum refinery wastewaters. Effluent soluble COD was reduced on increasing SRT with negligible impact on BOD removal, indicating that the biomass was able to accommodate the refractory organic components of the wastewater.

Another major change that was introduced to the operation of activated sludge systems was the addition of powdered activated carbon (PAC) to the mixed liquor. The practice has been patented as the Powdered Activated Carbon Treatment process or PACT. It has attracted interest as a means for upgrading activated sludge systems with respect to BOD and COD removal. Other benefits which have been claimed for PACT over conventional activated sludge include improved sludge settling and improved stability to shock loads and toxic upsets [23]. It is speculated that these improvements are due to an apparent synergism that may arise through a PAC-mediated stimulation of biological activity and/or through biologically-stimulated regeneration of the adsorptive capacity of the carbon. Most published data on the performance of industrial PAC-augmented activated sludge systems are in terms of COD or TOC. For example, Grieves and Stenstrom [24] showed improved COD removal in PAC-augmented petroleum refinery activated sludge systems, as did Specchia and Gianetto [25] in PAC-augmented activated sludge treatment of dye works wastewater. The enhanced removal of organic contaminants from industrial wastewaters by PACT has been demonstrated at full scale only at the Dupont Chambers Works plant [26]. A wide range of organic chemical wastestreams are treated by this plant which produces low effluent contaminant concentrations (Table 6). Detailed bench scale activated sludge investigations [27] showed that the addition of less than 100 mg/L PAC to the influent did not enhance removal of the biodegradable compounds benzene, toluene, ethylbenzene, o-xylene, chlorobenzene and nitrobenzene. However, significantly improved removals of the poorly degradable and nonbiodegradable compounds 1,2-dichloro-benzene, 1,2,4-trichlorobenzene and lindane were observed.

TABLE 6
Organic contaminant removal in PACT system[26]

Contaminant	Concentration (µg/L)	
	Feed	Effluent
Volatiles		
Benzene	105	0.9
Carbon Tetrachloride	94	1.4
Chlorobenzene	1720	30
Chloroethane	280	12
Chloroform	201	21
Ethylbenzene	41	1.7
Methyl Chloride	1770	Nil
Tetrachloroethylene	24	1.7
Toluene	519	1.7
1,1,1-Trichloroethane	13	0.6
Trichloroethylene	41	1.9
Trichlorofluoromethane	155	3.0
Base-Neutral Extractables		
1,2-Dichlorobenzene	259	120
2,4-Dinitrotoluene	1900	243
2,6-Dinitrotoluene	1640	575
Nitrobenzene	454	2
1,2,4-Trichlorobenzene	523	169
Acid Extractables		
2-Chlorophenol	11	1.6
2,4-Dinitrophenol	161	5.0
4-Nitrophenol	1020	10
Phenol	489	38

FIXED FILM REACTORS

Fixed film reactors are gaining acceptance as an improved design alternative to activated sludge systems where longer SRTs are desirable. Fixed film processes provide long cell retention times and promote the presence of slow growing organisms. They also can provide cell concentrations an order of magnitude higher than those found in activated sludge systems, an effect which is thought to enhance the removal of adsorbable organic contaminants [28]. Physical attachment of micro-organisms improves cell retention during shock loading conditions which, in an activated sludge system, could lead to a poor settling sludge. Techniques for sampling and analyzing biofilm sludges have only recently become available. Consequently, few studies have been reported that address the application of biofilm reactors to industrial wastewaters, and even fewer that report the removal of toxic trace contaminants in biofilm reactors.

Phenol removal has been reported in RBC systems treating petrochemical wastes [29], petroleum refinery wastewater [30], coking plant wastewater [31] and coal gasification wastewater [32]. The fate of naphthalene in an RBC reactor was measured using labelled naphthalene [33]. Adsorption with subsequent biodegradation was the principal mechanism for removal of naphthalene.

A pilot-scale, two-stage coupled anoxic-aerobic fluidized bed system, using sand as a biofilm attachment medium, was operated in a predenitrification mode over a two year period to evaluate the feasibility of removing organic contaminants from coking plant wastewater [34]. Total nitrogen removal efficiencies in excess of 90% were achieved at a total system HRT of 0.67 days and SRTs in the range 20 to 40 days. Also, greater than 90% removal of total organic carbon, thiocyanate and phenolic compounds was measured. Table 7 lists the base/neutral extractable organic contaminants found in the raw wastewater and treated effluent. There was a high degree of variability in the data which reflected the problems associated with analysis of trace quantities of toxic contaminants in complex industrial wastewaters [7]. HNCs such as quinoline, isoquinoline, indole and carbazole occurred in the fluid bed feed at concentrations up to 10 mg/L. Only two contaminants were found in the effluent at or more than 10 µg/L. Minimal accumulation of the PAH and HNC contaminants was found in the sludge indicating that the majority of these contaminants were biologically degraded. The fluid bed system produced effluents with very low concentrations of toxic contaminants.

Anaerobic fixed film reactors have not been as readily endorsed as aerobic systems for the treatment of industrial wastewaters. Much of the reluctance apparently stems from the belief that methane fermentation systems are unable to tolerate shock loads of toxic substances that may be present in industrial wastewaters. Parkin and co-workers [35] suggested that engineers fail to recognize several important factors when considering anaerobic processes. By prematurely concluding that the methane bacteria are dead when gas production has ceased, the possibility

TABLE 7

Base/neutral organics removal from coking plant wastewaters[34]

Contaminant	Mean Concentration (mg/L± std.dev.)	
	Influent	Effluent
Polyaromatic hydrocarbons		
Anthracene and/or phenanthrene	0.08±0.09	T[1]
Fluoranthene	0.02±0.02	T
Pyrene	0.02±0.02	T
1,3 and/or 1,4 dichlorobenzene	0.06±0.10	ND[2]
Di-n-butyl phthalate	0.04±0.07	0.02±0.03
Di-ethyl phthalate	T	T
Heterocyclic nitrogenous compounds		
Carbazole	0.98±0.92	T
Indole	6.34±10.25	T
Isoquinoline	2.33±3.27	T
7A-methyl quinoline	0.54±0.82	0.01±0.01
2,6- and/or 2,7-dimethyl quinoline	0.04±0.03	T
2,4-dimethyl quinoline	0.02±0.03	T
3,4- and/or 5,6-benzoquinoline	0.09±0.07	T
7,8-benzoquinoline	0.03±0.04	T
Quinoline	7.22±5.89	T

[1]T - Trace amount (less than 0.01 mg/L) detected in more than one sample
[2]ND - Not detected in any sample

that the toxicity may be reversible is ignored. The significance of proper attention to SRT control is not appreciated, for example, adverse response to toxicity could be minimized or eliminated with SRT control. The methane bacteria have a considerable potential for acclimation to toxicant exposure. High SRT values promote acclimation. Reactor types such as the anaerobic filter, the anaerobic fluidized bed and the anaerobic rotating biological contactor provide very large values of SRT in small reactor systems which do not require high biomass recycle. Johnson and Young [36] pointed out that anaerobic fixed film systems operate at low HRTs, and therefore, the exposure time to a shock loading of toxic contaminants would be less than in a conventional suspended growth process.

The use of high rate anaerobic systems has been reported at full-scale for pulp and paper wastewater [37], and pharmaceutical wastewater [38]. Specific contaminant removal has been demonstrated in a small scale expanded bed anaerobic reactor using granular activated carbon as a biofilm attachment medium [39]. More than 99.9% phenol removal and between 75 to 99% removal of substituted phenol compounds from coal gasification wastewater was reported. Blum and co-workers [40] identified the most easily anaerobically degradable compounds of 24 coal conversion wastewater constituents that were tested using serum bottle and laboratory scale anaerobic filter procedures. Buisson and co-workers [41] demonstrated the anaerobic dechlorination of organochlorine pesticides in batch anaerobic degradation trials. The anaerobic degradation of 24 USEPA organic priority pollutants has been measured [36]. Only seven of these compounds produced up to 50% reduction in gas production at concentrations of 100 mg/L.

MECHANISMS OF CONTAMINANT REMOVAL

Research into the mechanisms of contaminant removal in wastewater treatment systems has shown that they are complex and difficult to model. Although work has focussed primarily on the activated sludge treatment of synthetic or municipal wastewaters, the results may be used to forecast similar trends for industrial wastewater treatment.

Metal Contaminants

Physical speciation of metals influences the mode of removal in activated sludge systems. Removal occurs by adsorption or precipitation of soluble species and by settling of precipitated and particulate forms. Some soluble species that are complexed tend to pass through the system. Many properties of metals and activated sludge systems have been found to influence metal removal but there is no single major criterion to which metal removal can be totally attributed. Brown and Lester [42] described three significant mechanisms of metal removal:

(i)   physical trapping of precipitated metals in the sludge floc matrix,
(ii) binding of soluble metal to bacterial cells by extracellular polymers, and
(iii) accumulation of soluble metal by the cell.

Data have been presented to substantiate the removal by precipitation [43,44,45]. Lester [46] compiled evidence to demonstrate that the toxicity of a metal contaminant is reduced by complexation with bacterial extracellular polymers and that the relationship between the biomass surface and the metal ions in solution could be described by Freundlich and Langmuir isotherms. Some activated sludge bacteria do not produce extracellular polymer. In these bacteria, metals may be accumulated in the cytoplasm or by adsorption to the cell wall. However, metal transformation by chelation with organic ligands will take place before accumulation within a cell [47]. Cheng and co-workers [44] suggested that hydrogen ions compete with metallic ions for binding sites on biomass. Therefore, an increase in pH will increase the number of free binding sites permitting greater metal binding to the sludge. Nelson and co-workers [48] concluded that pH was the most important factor governing metal adsorption by activated sludge, citing optimum pH values of 8 for copper and 10 for cadmium and zinc. They developed a model to predict the distribution of metals between species and surfaces in activated sludge systems. The model incorporated expressions which were of the form of a Langmuir isotherm, a conditional metal-ligand equilibrium and a conditional equilibrium between uncomplexed metals and a bacterial surface. Patterson and Kodulka [50], on the other hand, prepared an empirical model to predict the proportion of solids-bound metal to the total metal concentration. Expressions were generated by linear regression of data describing eight metals in activated sludge systems. The model was valid only over the range of experimental conditions from which it was derived and, therefore, is less general than that of Nelson and co-workers [48].

Lester [46] suggested that SRT might be a useful parameter for predicting metals removal from activated sludge systems. However, there is conflicting evidence for improved metal contaminant control through the manipulation of SRT. Some researchers [49,51] indicated that SRT manipulation had no beneficial effect whereas others [48] presented data to the contrary. However, a material balance can demonstrate that considerable flexibility exists in controlling effluent and sludge metal concentrations through SRT manipulation. This was elaborated by Melcer and Bridle [52] who assumed that biologically inert contaminants entering a complete mix activated sludge system were removed only by adsorption on to the biomass.

Organic Contaminants

Organic compounds present in wastewaters may be removed by non-biological mechanisms as well as by biodegradation. Consequently, both the air mass above a treatment process and the solids removed from a treatment process must be considered as potential sinks of influent organic compounds, in addition to the final aqueous effluent. Most models that have been developed to describe the distribution of organics to the three media during biological treatment incorporate adsorption, volatilization and biodegradation components. The extent of transfer to the solid phase is usually presented as a function of the compound's octanol/water partition coefficient and the relative stripping efficiency is described as a function of Henry's law constant. The most difficult component to model is that of biodegradation, due to the paucity of degradation rate data for specific organic compounds. It is likely that a limited number of microbial species may be responsible for biodegradation of a specific contaminant. However, isolation of the microorganisms or enzymes is almost impossible and no adequate mechanistic model has been developed. Consequently, semi-empirical models are used to describe biological degradation.

The contribution of adsorption, volatilization and biodegradation to the removal of specific organic contaminants from activated sludge systems has been estimated by several researchers [53]. The main observations drawn from these studies were that volatilization and biodegradation were the principal removal mechanisms for volatile compounds and chlorinated benzenes; phthalate esters were removed by biodegradation and adsorption; PAHs appeared to be removed by biodegradation and adsorption; biodegradation was primarily responsible for phenolics removal; and the removal of PCBs and pesticides was not well characterized. The relative importance of these mechanisms has been postulated, in a different study, to depend upon the degree of biomass acclimation [2].

Models developed for volatilization are based on assumed equilibrium conditions between the liquid phase and gas bubble in diffused air systems. Jones [54] established stripping rate constants based on mass balance-derived empirical models. A similar approach was taken by Blackburn and co-workers [55] who separated the equilibrium constant between liquids and gases from the stripping rate expression. They assumed stripping of volatiles to be controlled by a first order rate step. However, substances such as salts, oils, surfactants and industrial waste tended to decrease the stripping rate. A different approach was used by Roberts and co-workers [56] in which the transfer rate of an organic contaminant was estimated from the oxygen transfer rate using an experimentally determined proportionality constant. Among the predictions from the model by Roberts and co-workers [56], it was observed that surface aeration released more volatile organics to the atmosphere than did bubble aeration. Of these models, only Blackburn and co-workers [55] attempted to evaluate the stripping rate of organics in the presence of biomass which had been deactivated. They showed that biomass reduced stripping rates.

Organics may accumulate in sludges at concentrations several orders of magnitude greater than the influent concentrations. Compounds which have a special affinity for sorption include pesticides, phthalates and PAHs which have shown concentration factors of 13 to 13 000 [57]. In modelling the biosorption mechanism, a bioconcentration or solids partitioning constant is estimated from the octanol/water partition coefficient, and the concentration in the solids can be calculated from the equilibrium effluent contaminant concentration. The uncertainties of this technique lie mainly with the biomass used. Dobbs and co-workers [57] demonstrated that freeze dried solids did not exhibit the same sorption characteristics as viable biomass. They developed a relationship between the concentration of adsorbed organics on viable solids and corresponding octanol/water coefficients. Blackburn and co-workers [55] used freeze-dried sludge to test the adsorptive capacity of the sludge biomass. They indicated that this inactivated sludge had similar flocculating and settling properties as fresh viable sludge but, there was no assurance that the adsorptive properties were not altered. Use of viable biomass for adsorptive studies may be acceptable if the contaminants are mostly refractory (e.g. lindane). If compounds that are biodegradable are tested for adsorption with viable biomass, however, it is difficult to differentiate the removal due to biodegradation from that due to adsorption if only mass balance techniques are used.

The conceptual understanding of biodegradation processes can affect the modelling of the biodegradation mechanism. Biodegradation of an organic contaminant can mean the biological transformation to another form; no extent is implied [47]. Furthermore, biodegradation need not have a benign outcome. Instead, it could convert an innocuous chemical into a toxic one, change a readily metabolizable compound into a toxic one, or alter the toxicity of a chemical so that it acts against another organism. In this context, when a contaminant is said to be biodegradable it is generally understood that the organic carbon is broken down ultimately to carbon dioxide and a portion is utilized for cell synthesis. A reliable way of distinguishing between removal by biodegradation and removal by adsorption or volatilization is through the use of radioactive labelled organic compounds, although care must be taken to distinguish radioactive CO_2 from volatilized radio-labelled compound, or radioactivity incorporated into cell mass from adsorbed initial compound. Blackburn and co-workers [55] used radio-labelled organic compounds in the development of their model which discriminated between the proportion of the contaminant oxidized to carbon dioxide and the proportion used for cell synthesis in biodegradation. Jones [54], on the other hand, established an overall biodegradation rate, but, before this could be calculated in aerated systems, a volatilization rate constant had to be determined assuming negligible biosorption.

Some researchers have used a kinetic approach to describe biodegradation. Rozich and co-workers [58] used the Haldane equation, a variation of the Monod expression for cell growth incorporating an inhibition term, to model the removal of phenol from activated sludge systems. Kincannon and co-workers [59] used a kinetic analysis based on a material balance that described the rate of change of organic contaminants in petroleum refinery wastewaters. Generally, this type of empirical analysis provides a reasonably good estimate of the contaminant removal efficiency but the fate of the contaminant in terms of removal mechanisms is not considered.

Both Jones [54] and Blackburn and co-workers [55] integrated their models into an overall removal predictor. Very limited experimentally-determined data were available to verify these models. The National Council of the Paper Industry for Air and Stream Improvement [60] developed an overall model composed of various observed removal relationships including that of Blackburn and co-workers [55]. The model was intended for use as a screening method for estimating the fate of particular organic contaminants in pulp and paper industry biological treatment plants. Either activated sludge or aerated stabilization basin processes may be simulated with the model. For the compounds tested in activated sludge systems, phenol and chloroform were shown to be removed from the aqueous phase by biodegradation and stripping respectively, hexachlorobenzene via adsorption and stripping and, tetraguaiacol was not removed in significant quantities. Based on limited field sampling, the predicted overall chloroform removal of 96% for surface aeration compared well with measured average overall removals of 94% and 88% (Table 8). The impact of bubble saturation during stripping of volatile compounds

TABLE 8

Comparison of predicted versus measured removal of chloroform[59]

Aeration Type	Predicted Mean	PERCENT REMOVAL			
		Actual			
		Survey A		Survey B	
		N[1]	Mean/Range	N	Mean/Range
Surface	96	3	94/92-96	4	88/86-90
Subsurface	72	5	71/14-96	1	69

[1]Number of observations at each of 7 plants.

was demonstrated by the difference between the predicted stripping efficiency for surface and subsurface aeration.

The method of Melcer and Bridle [52] for predicting fate of contaminants was demonstrated for conservative organic contaminants. Using data from another study [61] they compared predicted and measured return sludge contaminant concentrations (Table 9). There are large differences between measured and predicted values for some organic contaminants and this is attributed to the greater difficulty in trace organic contaminant analysis. For the biologically resistant pesticides and the PCB there is relatively good agreement. The predicted values are within an order of magnitude of the measured values, indicating that the technique can be useful in predicting return sludge concentrations of non-volatile, biologically inert organic contaminants.

TABLE 9

Comparison of predicted and measured return sludge organic contaminant concentration[51]

	Sludge Concentration (mg/kg)		Predicted/Measured Ratio
	Predicted	Measured	
Pesticides/PCB			
Arochlor 1254	1249	844	1.48
Heptachlor	2950	822	3.60
Lindane	180	27	6.67
Toxaphene	708	259	2.73
Phthalate esters			
Bis-(2-EH)phthalate	462	153	3.02
Diethyl phthalate	649	31	20.93
Dioctyl phthalate	333	91	3.66
PAHs			
Benzo(a)anthracene	280	33	8.48
Naphthalene	1121	2.9	386
Pyrene	440	16	27.5

For all three removal mechanisms, the predictive equations cannot address dynamic behaviour because they are derived from steady-state conditions which are not representative of the typical operation of a treatment plant. Therefore, field testing of current models has not been undertaken. At the present time, the models are probably useful indicators of the fate of contaminants in activated sludge plants but should not be regarded as providing definitive results.

CONCLUSIONS

• Biological wastewater treatment systems have been demonstrated to treat industrial wastewaters successfully. High removals of specific toxic trace contaminants can be achieved.

• The concept of SRT control is considered to be essential to the stability of biological systems when treating wastewaters containing toxic and inhibitory contaminants.

• Fixed film reactors appear to have great potential for treating complex and inhibitory wastewaters in view of their capability of maintaining high SRTs.

• Volatilization, adsorption and biological degradation have been identified as the major removal mechanisms for toxic trace contaminants in biological wastewater treatment systems. These mechanisms are complex and interactive. To date they have not been adequately defined by mechanistic modelling.

REFERENCES

1.   U.S. Environmental Protection Agency, Treatability Manual. NTIS Report No. EPA 600/8-80-042, Springfield, VA, 1983.

2.   U.S. Environmental Protection Agency, Report to Congress on the discharge of hazardous wastes to publicly owned treatment works. NTIS Report No. EPA 530/SW-86-004. Springfield, VA, 1986.

3.   Richards, D.J. and Shieh,W.K., Biological fate of organic pollutants in the aquatic environment. Water Res., 20, 1077-90, 1986.

4.   Grady, C.P.L. (1985). Biodegradation: Its measurement and microbiological basis. Biotech. and Bioeng., XXVII, 660-674, 1985.

5.   Kim, C.J. and Maier, W.J., Acclimation and biodegradation of chlorinated organic compounds in the presence of alternate substrates. J. Wat. Pollut. Control Fed., 58, 157-164, 1986.

6.   Reynolds, L.F. and co-workers, The treatment and disposal of complex organic effluents. Water Pollut. Control, 84, 251-9, 1985.

7.   Melcer, H., Control of toxic trace contaminants in municipal wastewater treatment plants. Proc. of Ontario Ministry of Environment Tech. Transfer Conference, Toronto, ONT, 27-52, 1986.

8.   Thompson, K.C. and Prior, B.F., Some problems for a water authority caused by toxic metals in trade effluents. *Effluent and Water Treatment J.*, 24, 190-4, 1984.

9.   Atkins, E.D. and Hawley, J.R., Sources of metals and metal levels in municipal wastewaters. Canada-Ontario Agreement Report No. 80, Environment Canada, Ottawa, Ontario, 1975.

10. Engineering Science Inc., The CMA-EPA five plant study. Report prepared for Chemical Manufacturers Association, Austin, TX, 1982.

11. American Petroleum Institute, Analysis of Refinery Wastewaters for the EPA Priority Pollutants - Interim Report. API Pub. No. 4296, Washington, D.C., 1978.

12. Petroleum Association for Conservation of the Canadian Environment, Survey of Trace Substances in Canadian Petroleum Industry Effluents. PACE Report No. 81-4, Ottawa, ONT, 1981.

13. Petroleum Association for Conservation of the Canadian Environment, Fate and Transformation of Selected Trace Organics in Refinery Wastewater Treatment. PACE Report No. 85-7, Ottawa, ONT, 1985.

14. American Petroleum Institute, Refinery Wastewater Priority Pollutant Study - Sample Analysis and Evaluation of Data. API Pub. No. 4346, Washington, D.C., 1981.

15. Luthy, R.G. and co-workers, Biological treatment of synthetic fuel wastewater. *J. Env. Eng. Div. ASCE*, 106, No. EE3, 609-629, 1980.

16. Luthy, R.G. and Tallon, J.T., Biological treatment of a coal gasification process wastewater. *Water Res.*, 14, 1269-82, 1980.

17. Stamoudis, V.C. and Luthy, R.G., Determination of biological removal of organic constituents in quench waters from high BTU coal gasification pilot plants. *Water Res.*, 14, 1143-76, 1980.

18. Bridle, T.R. and co-workers, Biological treatment of coke plant wastewater for control of nitrogen and trace organics. Presented at 53rd Annual WPCF Conference, Las Vegas, NV, 1980.

19. Bridle, T.R. and co-workers, Operation of a full scale nitrification-denitrification industrial waste treatment plant. *J. Water Pollut. Control Fed.*, 51, 127-139, 1979.

20. Graham, N.A., The Waterton gas plant activated sludge installation. Presented at 30th Can. Chem. Eng. Conference, Edmonton, ALTA, 1980.

21. Roper, R.E. and co-workers, Biological treatment at Citizens Gas & Coke Utility. Presented at Indiana Water Pollut. Control Assoc. Conference, Indianapolis, IND, 1982.

22. Grutsch, J.F. and Mallat, R.C., Optimize the effluent system. *Hydrocarbon Processing*, 55, 105-112, 1976.

23. Sublette, K.L. and co-workers, A review of the mechanism of powdered activated carbon enhancement of activated sludge treatment. Water Res., 16, 1075-82, 1982.

24. Grieves, C.G. and Stenstrom, M.K., Activated carbon improvement of effluent quality in refinery activated sludge process. Industrial Wastes, July/August, 30-35, 1977.

25. Specchia, V. and Gianetto, A., Powdered activated carbon in an activated sludge treatment plant. Water Res., 18, 133-7, 1984.

26. Heath, H.W. Jr., Update on the PACT process. Proc. National Conference on Hazardous Wastes and Materials, Atlanta, GA, 163-170, 1986.

27. Weber, W.J. and Jones, B.E., Toxic substance removal in activated sludge and PAC treatment systems. NTIS Report No. PB86-182425, Springfield, VA, 1986.

28. Kobayashi, H. and Rittmann, B.E., Microbial removal of hazardous organic compounds. Environ. Sci. Technol., 16, 170A-183A, 1982.

29. Falkenback, P., Petrochemical industrial wastewater treatment - phenol removal by the rotating biological contactor process. Trib. CEBEDEAU, 35, 419-, 1982.

30. Congram, G.E., Biodisk improves effluent - water treating operation. Oil and Gas J., 74, 126-, 1976.

31. Blumenschein, C.D. and Helwick, R., Comparison of activated sludge to rotating biological contactors for treatment of coke plant wastes. Iron and Steel Engineer, July, 38-42, 1985.

32. Turner, C.D. and co-workers, Treatment of coal gasification wastewater using rotating biological contactors. Proc. Second International Conference on Fixed Film Biological Processes, Arlington, VA, 1257-76, 1984.

33. Glaze, W.H. and co-workers, Fate of naphthalene in a rotating disc biological contactor. J. Water Pollut. Control Fed., 58, 792-8, 1986.

34. Melcer, H. and Nutt, S.G., The application of predenitrification nitrification technology for trace contaminant control. Wat. Sci. Tech., 17, 399-408, 1985.

35. Parkin, G.F. and co-workers, Response of methane fermentation systems to industrial toxicants. J. Water Pollut. Control Fed., 55, 44-53, 1983.

36. Johnson, L.D. and Young, J.C., Inhibition of anaerobic digestion by organic priority pollutants. J. Water Pollut. Control Fed., 55, 1441-49, 1983.

37. Rantala, P. and Luonsi, A., Eds., Anaerobic Treatment of Forest Industry Wastewaters - Proc. of 1st IAWPRC Symposium on Forest

Industry Wastewaters. <u>Wat. Sci. & Tech.</u>, 17, 1985.

38. DeAngelis, D.F. and co-workers, Industrial waste pretreatment using a Celrobic anaerobic reactor. Presented at 41st Purdue Industrial Waste Conference, LaFayette, IN, 1986.

39. Suidan, M.T. and co-workers, Treatment of coal gasification wastewater with anaerobic filter technology. <u>J. Water Pollut. Control Fed.</u>, 55, 1263-70, 1983.

40. Blum, D.J.W. and co-workers, Anaerobic treatment of coal conversion wastewater constituents: biodegradability and toxicity. <u>J. Water Pollut. Control Fed.</u>, 58, 122-131, 1986.

41. Buisson, R.S.K. and co-workers, Behaviour of selected chlorinated organic micropollutants during batch anaerobic digestion. <u>Water Pollut. Control</u>, 85, 387-394, 1986.

42. Brown, M.J. and Lester, J.N., Metal removal in activated sludge: The role of bacterial extracellular polymers. <u>Water Res.</u>, 13, 817-837, 1979.

43. Sterrit, R.M. anc co-workers, Metal removal by adsorption and precipitation in the activated sludge process. <u>Environ. Poll. Series A</u>, 24, 313-323, 1981.

44. Cheng, M.H. and co-workers, Heavy metals uptake by activated sludge. <u>J. Water Pollut. Control Fed.</u>, 47, 362-376, 1975.

45. Brown, M.J. and Lester, J.N., Role of bacterial extracellular polymers in metal uptake in pure bacterial culture and activated sludge - I. <u>Water Res.</u>, 16, 1539-48, 1982a.

46. Lester, J.N., Significance and behaviour of heavy metals in wastewater treatment processes - I. Sewage treatment and effluent discharge. <u>Sci. of Tot. Env.</u>, 30, 1-44, 1983.

47. Sterrit, R.M. and Lester, J.N., Influence of bacterial growth on the forms of cadmium in defined culture media. <u>Bull. Environ. Contam. Toxicol.</u>, 24, 196-203, 1980.

48. Nelson, P.O. and co-workers, Factors affecting the fate of heavy metals in the activated sludge process. <u>J. Water Pollut. Control Fed.</u>, 53, 1323-33, 1981.

49. Elenbogen, G. and co-workers, Studies of the uptake of heavy metals by activated sludge. <u>Proc. 40th Purdue Industrial Waste Conference</u>, 493-505, 1985.

50. Patterson, J.W. and Kodulka, P.S., Metals distributions in activated sludge systems. <u>J. Water Pollut. Control Fed.</u>, 56, 432-441, 1984.

51. Brown, M.J. and Lester, J.N., Role of bacterial extracellular polymers in metal uptake in pure bacterial culture and activated sludge - II. <u>Water Res.</u>, 16, 1549-60, 1982b.

52. Melcer, H. and Bridle, T.R., Discussion of metals removals and partitioning in conventional wastewater treatment plants and fate of toxic organic compounds in wastewater treatment plants. _J. Water Pollut. Control Fed._, 57, 263-4, 1985.

53. Canviro Consultants Ltd., A novel application of a time series modelling approach to the management of dynamic fluctuations in trace contaminants in sewage treatment plants - Phase 1 Report. Prepared for Wastewater Technology Centre, Environment Canada, Burlington, ONT, 1986.

54. Jones, B.E., Fate of toxic organic compounds in activated sludge/carbon treatment systems (Volumes I & II), Ph.D. Thesis, University of Michigan, Ann Arbor, MI, 1984.

55. Blackburn, J.W. and co-workers, Organic chemical fate prediction in activated sludge treatment processes. NTIS Report No. PB85-247674, Springfield, VA., 1985.

56. Roberts, P.V. and co-workers, Modelling volatile organic solute removal by surface and bubble aeration. _J. Water Pollut. Control Fed._, 56, 157-163, 1984.

57. Dobbs, R.A. and co-workers, Partitioning of toxic organic compounds on municipal wastewater treatment plant solids. NTIS Report No. PB86-218427, Springfield, VA, 1986.

58. Rozich, A.F. and co-workers, Predictive model for treatment of phenolic wastes by activated sludge. _Water Res._, 17, 1453-66, 1983.

59. Kincannon, D.F. and co-workers, Removal mechanisms for toxic priority pollutants in activated sludge systems. _J. Water Pollut. Control Fed._, 55, 157-163, 1983.

60. National Council of the Paper Industry for Air and Stream Improvement, Inc., Simulation of organic compound removal in biological wastewater treatment processes. Technical Bulletin No.511, New York, NY, 1986.

61. Petrasek, A.C. and co-workers, Fate of toxic organic compounds in wastewater treatment plants. _J. Water Pollut. Control Fed._, 55, 1286-96, 1983.

RECENT ADVANCEMENTS IN THE THERMAL
TREATMENT OF HAZARDOUS WASTES

Harry Freeman
Hazardous Waste Engineering Research Laboratory
U.S. Environmental Protection Agency
Cincinnati, OH 45268

ABSTRACT

Thermal processes offer many advantages as hazardous waste treatment options. In this paper, the author discusses ten thermal processes that differ from conventional incineration. Included are
fluidized bed incinerators, waste as fuel in boilers, cement kilns,
pyrolysis processes, molten glass, Infrared incineration systems, supercritical water oxidation, plasma systems, mobile rotary kiln, and
electric reactors. The discussion includes process descriptions,
applications, and state of development for each of the systems.

Thermal Process Overview

Thermal processes can, in a matter of seconds or minutes, destroy
or significantly reduce the volume of wastes that might otherwise take
months or years to degrade in a landfill. Thermal processes can
greatly reduce environmental and health hazards associated with land
disposal, reduce the need for new landfill capacity, and eliminate the
possibility of problems literally resurfacing in the future. (1)

In its May 1980 Hazardous Waste and Consolidated Regulations, the
US EPA states:

"Incineration is a relatively well-developed and well-understood technology. Properly executed, it can accomplish safe destruction of
primarily organic hazardous waste, permanently reducing large volumes
of waste materials to non-toxic gaseous emissions and small amounts of
ash and other residues. Incineration can often provide an optimum,
permanent solution to hazardous waste management with minimal long-term
ecological burden." (2)

Thermal treatment processes are processes for reducing the volume
and/or toxicity of organic wastes by exposing them to high temperatures
in controlled environments designed to encourage material breakdown.
When the waste streams are subjected to high temperatures (usually
900°C or greater) they tend to break down into simpler and less toxic
forms. Thermal treatment can take many forms. The most common thermal
treatment systems for hazardous wastes in the U.S. are liquid injection
incinerators, fixed hearth incinerators, and rotary kilns. (3)

In addition to these processes, generally recognized as conventional incinerators, there are other somewhat less known thermal

treatment options that might be viewed as alternative thermal
processes. Some of these processes have been around for years, but
are becoming more known as interest in thermal processes increase,
and others are relativly new, having been developed in just the last
few years. A summary of a selection of these alternative processes
is shown below. Each of these processes differ from conventional
incinerators in some way. They may be attractive options for certain
categories of waste streams that do not lend themselves well to
conventional thermal destruction. They may be especially suitable
for being located at a generator's site. They may be useful to
reclaim some of the energy value of organic waste treatment, or may
produce a very stable residual.

ALTERNATIVE THERMAL TREATMENT PROCESSES

1. Fluidized bed incinerators
2. Waste as fuel in boilers.
3. Cement kilns.
4. Pyrolysis processes.
5. Molten glass.
6. Infrared incineration system.
7. Supercritical water oxidation.
8. Plasma systems.
9. Mobile rotary kiln.
10. Electric reactor.

It should be emphasized that this list is certainly not intended to
be inclusive of all of the good ideas in thermal treatment. However,
the processes chosen are some of those of which the author is aware,
and which can be discussed in the time allotted for this paper.

Fluidized-Bed Incineration

A fluidized-bed incinerator consists of a refractory-lined
cylindrical vessel containing a bed of inert granular material on a
perforated metal plate. Prior to waste incineration, the bed of
granular material is heated to a desired operating temperature and
air is passed through the perforated plate to "fluidize" the bed.
Fluidizing produces a very turbulent, boiling type of activity in the
bed. Liquid or granular wastes are injected into the bed for combustion. The temperature within the bed typically ranges from 500° to
900°C. Since the mass of the heated, turbulent bed is much greater
than the mass of the waste, heat is rapidly transferred to the waste
material and destruction of the waste occurs in a few seconds. In
addition to promoting the actual combustion of the waste, the hot
particles in the bed can act as a scrubbing device for selected off
gases and particulates. As the waste combusts, energy is returned to
the bed. Two types of fluidized systems are the bubbling system in
which air is passed through a stationary bed, and the circulating
system in which the whole bed circulates between two vessels. (5)
Applications:
Fluidized beds are designed for liquids, sludges, and solids
including contaminated soil. They are considered especially suitable
for sludges. Advantages are the compactness and low maintenance
aspects of the system, the ability to handle waste streams of varying
heating value, and low NO_x emissions relative to high temperature
incinerators.

State of Development:
Fluidized beds are used extensively in fields other than hazardous waste incineration, such as biomass and coal conversion. Over 160 units are used throughout the world to destroy sludges from municipal waste treatment facilities. (6) Over twenty systems are in operation world wide, destroying hazardous & toxic wastes. (5) A major development area is the use of active beds rather than inert beds. A two stage bed has been used with catalytic and dry off-gas clean-up applicability. (7) The EPA is currently testing ways to improve metal retention in bed particles. (8) A combustion process which captures phophorus compounds in bed media has also been developed. (9) Mobile fluidized bed systems have been developed and are in commercial operation. (5)

Waste as Fuel in Boilers

As land disposal options for hazardous waste disposal have become more restricted, the practice of cofiring hazardous wastes as supplemental fuel in boilers designed to burn oil, gas, or coal as a primary fuel have become more attractive. The practice of cofiring wastes predates the hazardous waste regulations since utilization of the material can have clear economic advantages that have nothing to do with disposal cost assistance. In 1983 some 700,000 metric tons of hazardous waste were fired in boilers in the United States. (10)

In most older boilers the cofiring capability was added after the unit was operational. Newer boilers have the capability designed into the unit. Most liquid fuel fired boilers use a system in which the waste material is introduced separately into the boiler from the primary fuel. While percentage energy contribution of waste streams vary according to the boiler and waste streams involved, they typically are less than 25%. A recent paper estimated that 65 percent of waste fired in boilers contributed less than 25 percent of heat input to the boiler. (11)

Applications:
Cofiring of waste in boilers is most applicable for liquid waste streams whose energy value exceed 4640 kJ/hg (5,000 BTU/lb). Examples of such waste streams are spent solvents and waste oils. Over 50% of the waste used as supplemental fuel in the chemcial industry exceeds 23,200 kJ/kg (10,000 BTU/lb). (11) Halogenated waste streams do cause corossion. Composite waste-fuel chlorine levels in excess of 1.5 to 2.5 percent can cause water tube corossion. Consequently, generators usually do not cofire waste streams with halogen content exceeding 5 percent. (11)

Can boilers achieve the same destruction efficiencies as hazardous waste incineration? It depends on the size and configuration of the boiler and the characteristics of the waste being fired. Tests seen by the EPA over a 5 year period on 42 boilers of different sizes and designs found that most industrial boilers, especially the large boilers did effectively destroy the waste, even some chlorinated

waste, a levels comparable to incinerators. (10) A later series of
tests to determine the effect of upset conditions on destruction
efficiencies found that for the most part the boilers still acted as
effective destruction devices under upset conditions. (10)
State of Development:
     While research is still being conducted by the EPA and others to
evaluate formation and fate mechanisms for a variety of flame configurations and waste types, the practice of cofiring hazardous waste in
boilers is widely recognized as an acceptable alternative thermal
technology. With new boilers increasingly being designed to accommodate a waste stream as fuel, it is estimated that the cofiring of
waste as fuel will increase in the future.

Cement Kilns

     An integral part of the process for manufacturing cement is
exposing limestone and several additives to temperatures above 1425°C
(2600°F) in a large rotary kiln fueled with a fossil fuel. The end
product of this process is a solid material called clinker. Ground-up
clinker is the major constituent of cement. Since the combustion
conditions and residence times in cement production are much more
severe than those present in many waste incinerators, cement kilns
are considered to be an acceptable disposal option for many organic
wastes. (1)

     When waste is used in a cement kiln, it is introduced directly
into the processing part of the kiln through a nozzle separate from
the one introducing the fossil fuel. Given the length of the kiln,
which can exceed 300 meters, combustion gases from the waste materials
can experience residence times in excess of 10 seconds, much more than
residence times in even the largest hazardous waste incinerators.
Applications:
     Cement kilns are considered appropriate for many high energy
liquid waste streams such as still bottoms, solvents, tars, chemical
by-products and oils. The cement industry, based on successful
experiences burning solid wastes such as tires and municipal refuse,
claims that many solid waste materials are also suitable candi-
dates. (12) Cement kilns are considered especially applicable for
chlorinated waste streams since the hydrochloric acid produced serves
to neutralize the clinker production process which is normally alkaline. (1) Over 250 organic compounds have been approved as fuel for
cement kilns. (12)

     The advantages of using cement kilns are that, in addition to
the wastes being destroyed, the energy value of the waste is reclaimed,
the capacity of the cement industry to consume chemical wastes is
quite large, and cement plants are already located near many waste-
generating sources. The disadvantages are that burning chlorinated
wastes in cement kilns appears to increase the production of particulates, requiring more extensive air pollution control devices, utilizing wastes as fuel in plants not used to handling wastes will require
an upgrading of the facilities.

The EPA has estimated that there exists enough cement-making
capacity within the largest waste-producing states to incinerate all
of the chlorinated wastes produced in those states. The California
Air Resources Board recommended the use of cement kilns as an acceptable means of destroying PCB wastes.
State of Development:
    Using the hazardous waste as fuel for cement kilns is practiced
at a dozen or so plants in the United States. (12) There is an
existing support industry that blends fuel utilizing the waste streams
for the industry. In the area in which these industries exist, such
as Southern California, the cement industry occupies a significant
niche in the alternative treatment capacity for the area.

Pyrolysis Processes

    Pyrolysis processes differ from incineration processes in that
in a pyrolysis process, the waste material is heated indirectly in an
oxygen starved environment to volatilize the organics to be recovered
or burned in an afterburner. In an incinerator an exces of oxygen is
provided to actually oxidize the organics directly. There are several
variations in the pyrolysis process that incorporate energy recovery,
valuable compound recovery, or post pyrolysis combustion. (13)

    In a pyrolysis process the waste streams are subjected to temper-
atures of 425 to 760°C (800 to 1400°F) in a quiescent environment.
Hazardous organic compounds can be volatized at such temperatures
leaving a clean residue. The volatiles are usually burned in a fume
incinerator at conditions to insure an acceptable destruction efficiency. Pyrolysis systems are normally designed for capacities of
450 to 4500 kg/hr (.5 to 5 tons/hr). (13)

    Advantages of the process include its ability to be very closely
controlled, its extremely low particulate emissions, its low temperatures avoiding metal & salt volatilization, and the potential recovery
of selected gases & materials. (14)

Applications:
    Pyrolysis can be applied to solids, sludges and liquid wastes.
Wastes with the following characteristics are especially amenable to
pyrolysis:
    °   Sludge material which is either too viscous, too abrasive,
        or varies too much in consistency to be atomized in a liquid
        incinerator.
    °   Wastes which undergo partial or complete phase changes
        during thermal processing, such as plastics.
    °   High residue materials, such as high ash liquids and
        sludges, with light, easily entrained solids that would
        generally require substantial stack gas cleanup.
    °   Materials that contain salts or metals which melt and
        volatilize at normal incineration temperatures. Materi-
        also like NaCl, zinc, and lead, when incinerated,
        may cause refractory spalling, heat exchanger surface
        fouling, and submicron aerosol emissions. (13)

State of Development:

Pyrolysis systems for the treatment of hazardous wastes are available in commercial scale units. The Midland Ross Company of Toledo, Ohio, U.S.A., markets units designed to treat either loose sludges or containerized materials. In a trial burn carried out in St. Louis in 1984, the company reported DRE's in excess of 5 nines for five organic contstituents with 50-70 percent chlorinated hydrocarbons. The unit tested is one of the dozen or so incinerators permitted under the Resource Conservation and Recovery Act in the United States. As the on site treatment of hazardous waste continues to be encouraged, it is envisioned that pyrolysis systems will come to occupy a more significant role as a thermal option.

Molten Glass

The integral part of this process is an electric furnace approximately 7 m long by 1.2 m wide (22 ft. by 3 ft.) that has a pool of molten glass covering the bottom. This type of furnace is used extensively in the glass manufacturing industry throughout the world to produce glass. When used as waste incinerator the unit utilizes the extremely high temperatures in the combustion chamber to destroy organic waste streams.

Waste materials, both combustible and non-combustible, are charged directly into the combustion chamber above the pool of molten glass. The waste can either be contained in fiberboard boxes, or uncontained in loose form. The temperatures of the pool of molten glass are maintained above 1260°C (2300°F) by use of electrodes immersed in the pool. Combustible wastes are oxidized above the pool, and inorganics and ash fall onto the pool where they are melted into the glass.

The furnace is designed to provide a residence time of 2 seconds for gases and up to hours for solids at the elevated temperature. Combustion gases are exhausted through a ceiling port and passed through a series of filters and scrubbers.

Shredded tires have been charged into the unit to provide supplemental energy value to the process. (15)
Applicability:
The proponents of the molten glass process claims it can treat virtually any waste stream, liquid or solid. Advantages cited for the process are: 1) the ability to accept a wide variety of solid and liquid waste streams; 2) it is based upon well understood glassmaking technology; and 3) the residual product is a fully stable glass. (14)
State of Development:
The developer of this system, Penberthy Electromelt, has obtained interim status for a 4 ton per day system in Seattle, Washington, U.S.A., to destroy selected hazardous waste streams, and for a 10 ton per day mobile system now under construction. The mobile system, which has been tentatively priced at $800,000, is to be ready in February.

Larger furnaces are possible using this technology. Units of 24 tons and 120 tons have been proposed by the developer, and an application for a 400 ton per day system is under review.

Finally, based upon a test melt of a 50 lb. sample of radium contaminated soil that resulted in a zero detectable release of radon from the glass, it appears that the molten glass technology may be appropriate for treating such contaminated soil. (16)

Supercritical Water

In the supercritical water process an aqueous waste stream is subjected to temperatures and pressures above the critical point of water, i.e. that point at which the densities of the liquid and vapor phase are identical. (For water the critical point is 379°C and 218 atmospheres). In this supercritical region water exhibits unusual properties that enhance its capability as a waste destruction medium. Because oxygen is completely miscible with supercritical water, the oxidation rate for organics is greatly enhanced. Also, inorganics are practically insoluble in supercritical water. This factor allows the inorganics to be easily removed from the waste streams. The result is that organics are oxidized extremely rapidly and the resultant stream is virtually free of inorganics. (16)

A patented process has been developed that incorporates the properties of superciritical fluids to oxidize organic contaminants in aqueous streams. The following is a brief summary of the process:

a. Waste is slurried with make-up water to provide a mixture of 5 percent organics. The mixture is heated using previously processed supercritical water and then pressurized.
b. Air or oxygen is pressurized and mixed with the feed. Organics are oxidized in a rapid reaction. (Reaction time is less than 1 minute). For a feed rate of 5 percent by weight of organics, the hat of combustion is sufficient to raise the oxidizer effluent to 500°C.
c. The effluent from the oxidizer is fed to a salt separator where inorganics are removed by precipitation.
d. Waste heat from the process can be reclaimed to provide sufficient energy for power generation and high pressure steam.

An advantage of the supercritical water process is its ability to detoxify waste streams quickly in a completely contained reaction. It is also very compact which contributes to its attractiveness as an on-site treatment option for waste generators. Cost figures for this process are not available. However, the developers claim that such costs will be competitive with other available options. (16)
Applicability:
Supercritical water processes are designed for aqueous waste streams with high levels of inorganics and toxic organics. The system's capability for treating aqueous waste streams with high percentages of halogenated material has not yet been demonstrated.

(14) However, bench scale experiments with the supercritical water
process have yielded organic carbon destruction efficiencies of from
99.970 to 99.996 percent for a range of chlorinated organic compounds.
For PCBs in transformer oil and a methyl ethyl ketone matrix, average
organic carbon destruction efficiency was 99.991 percent. (19)
State of Development:
     In the fall of 1985, the developer of the system, MODAR, Houston,
TX, demonstrated its process at the Niagara Falls plant of CECOS
International in a unit capable of treating up to 1000 gallons per
day of dilute organic wastes. The developers reported that all waste
constituents in a transformer oil contaminated with 1600 ppm of PCB
and a aqueous waste containing an organic solvent and 3 organics were
destroyed beyond limits of detectability. The developer expects a
full-scale plant to be available by mid 1987. (16)

Electric Reactor

     The Advanced Electric Reactor employs a new technology to
rapidly heat materials to temperatures between 2200-3300°C using
intense thermal radiation in the near infrared region. The reactants,
which can be in gaseous, liquid, or solid form, are isolated from the
walls of a vertical reactor by means of a gaseous blanket formed by
flowing nitrogen radially inward through porous graphite core walls.
Carbon electrodes imbedded in the reactor shell are heated and in
turn heat the reactor core to incadenscence so that heat transfer is
accomplished by thermal radiative coupling from the core to the feed
materials. Destruction is accomplished by pyrolysis rather than
odixation. (20)

     For solid waste treatment, the solid feed stream is introduced
at the top' of the reactor by means of a metered screw feeder connect-
ing the airtight feed hopper to the reactor. Nitrogen is introduced
primarily at two points in the reactor annulus formed by the external
containing vessel and the porous graphite core which creates the fluid
barrier. The solid feed passes through the reactor where pyrolysis
occurs at approximately 2200°C. After leaving the reactor, the
product gas and waste solids pass through two post-reactor treatment zones (PRTZ). The first PRTZ is an insulated vessel to provide
additional high temperature in excess of 1100°C, (2000°F) and
residence time (5 to 10 seconds). The second PRTZ is water-cooled.
It also provides additional residence time (approximately 10 seconds)
but primarily cools the gas to less than 500°C prior to downstream
particulate cleanup. (20)

     The advantages of the electric reactor are its transportability,
its very rapid start-up and shutdown times, its low gaseous emissions,
and its capability on producing a decontaminated stable residue.
Rough costs estimated by Huber Corp. for the treatment of 20,000
tons of contaminated soil at a typical remedial remedial site are
$736/ton. (16)
State of Development:
     The developer of the process, the Huber Corp., Borger, TX, has

reported several relevant tests that have been carried out over the
past two years. A test in the spring of 1984 to determine the destruction efficiency (DE) and DRE for the process with carbon tetrachloride
over a wide range of operating conditions demonstrated greater than six
nines (99.9999%) DRE. A September 1983 test on PCB contaminated solids
demonstrated better than seven nines DRE and resulted in Huber's
facility at Borger, Texas being certified to destroy PCB contaminated
solids. (21)

Infrared Systems

This system treats waste by metering the waste streams onto a
woven metal alloy conveyer belt and passing the waste under electrically powered heating elements to volatilize contaminants. Combustible
off-gases are then burned in a secondary chamber. The primary chamber
is rectangular and is lined with multiple layers of ceramic fiber
blanket insulation. Infrared energy is provided by heating elements
equally spaced along the length of the furnace. The electrically
powered heating elements can provide capabilities of 500-1850°F of
process temperature with material residence times variable between 10
and 180 minutes. Oxidizing, reducing, or neutral atmospheres can be
provided. A secondary chamber is equipped with a propane burner and
is sized to provide gas residence time of up to 2.2 seconds. The
process temperature capability is 2300°F. Scrubbers are provided for
off-gas cleanup.

Advantages of the system include its transportability, low particulate emissions, and rapid start-up and shutdown times. The
developers of the system say processing costs using the system depend
on the waste involved, but is in the range of $100 to $200 per ton.
(16)
Applicability:
The system is suited for sludges and contaminated soils. A wide
variety of such materials ranging in heat content from dioxin contaminated soils to heavily organic laden cresote sludges have been
treated successfully. (17, 18) Since much of this type of material
is common to abandoned chemical dump sites, the system is considered
especially suitable for the on-site treatment of these wastes.
State of Development:
The developer of this system, Shirco Infrared Systems, Dallas,
Texas, has completed construction of a mobile system capable of
treating 100 tons per day of contaminated soils. This unit is currently undergoing shakedown testing. Quoted processing costs are very
dependent upon the waste streams in question but range from $100 to
$200 per ton of material. A prototype of the system is currently
being operated on dioxin contaminated soil & chlorinated wastes for
the DeKonta Company near Hamburg, West Germany.

Plasma systems

Plasma systems are systems that use the extremely high temperatures of material in a plasma state to reduce waste materials to
their elemental form.

In a plasma arc device, a co-linear electrode arrangement creates
an electric arc which is stabilized by field coil medium through which
an electric current is passed. As air passes the arc, the electrical
energy is converted to thermal energy by adsorption by the air molecules
which are activated into ionized atomic states, losing electrons in
the process. Ultraviolet radiation is emitted when the molecules or
atoms relax from their highly activated states to lower energy levels.

The resulting gas is in the plasma state consisting of charged
and neutral particles with an overall charge near zero and with
temperatures up to 28,000°C (50,000°F). (2)

The plasma can best be understood by thinking of it as an energy
conversion and energy transfer device. As the activated components
of the plasma decay, their energy is transferred to waste materials
exposed to the plasma. The wastes are then atomized, ionized, pyrolyzed
and finally destroyed as they interact with the decaying plasma
species. Theoretically speaking, the destruction of wastes should
result in simple molecules or atoms such as hydrogen, carbon monoxide,
carbon and hydrochloric acid. The off-gases from the plasma system
are scrubbed to remove hydrochloric acid and are then flared. (2)

Advantages of a plasma system are its portability, its capability
of very quick start-up and shutdown, and the degree of material disasso-
ciation using such extremely high temperatures.
Applicability:
The plasma system is designed for liquid waste streams. It is
considered especially applicable for highly toxic, very refractory
wastes.
State of Development:
The operation of a plasma system constructed by Pyrolysis Systems,
Inc. has been tested in a series of short term runs at Kingston,
Ontario using MEK/ MeOH, carbon tetrachloride, and PCB's as feed
materials. DRE's of greater then six nines were obtained for the
chlorinated hydrocarbons, and the emissions from the system met
environmental discharge criteria established by Canadian regulatory
authorities. This unit has been moved to the Love Canal in New York
to begin the next phase of development. (25)

Westinghouse, Pittsburgh, PA, U.S.A., has completed the construction of the first commercial scale, 3 gpm, mobile unit and is currently
conducting performance tests using nonhazardous surrogate wastes. (16)

EPA Mobile Incineration System

EPA's Office of Research and Development has developed a mobile
rotary kiln incinerator capable of incinerating hazardous waste at
the point of generation. The unit, called the Mobile Incineration
System, consists of three major subsystems: 1) combustion and air
polution control equipment mounted on three trailers, 2) continuous
flue gas monitoring equipment located in a fourth trailer, and 3)
support equipment. (22)

The incineration system consists of a primary combustion chamber (rotary kiln) where solid or liquid wastes are introduced, volatilized, and partly oxidized and a clean fuel oil-fired secondary combustion chamber (SCC) for completing the combustion process initiated in the kiln. The combustion flue gases pass through a gas cleaning unit designed to collect and remove particulate matter and acid gases prior to discharge through the stack. The initial treatment of the combustion gases is accomplished by water quenching in the quench elbow, which is then followed by particulate removal in a cleanable high efficiency air filter (CHEAF). (The primary function of the elbow is to lower the gas temperature from 1200°C (2200°F) to adiabatic saturation, at approximately 85°C (185°F), and thus reduce the gas volume; some particulate and acid gas removal does occur in the quench elbow.) The final step in the gas treatment process is the removal of acid gas in a cross-flow packed-bed mass-transfer scrubber. The pH of each process water stream is controlled by the addition of sodium carbonate to enhance acid gas removal and to minimize corrosion of the gas cleaning equipment. The prime gas mover for the incineration system is a diesel-powered induced-draft fan that maintains a negative pressure in the entire system so as to prevent the leakage of toxic vapors. The process gas is discharged through a 12-m (40 ft) stack. (23)

In 1983, trial burns were conducted in Edison, New Jersey on RCRA-listed surrogates, including dichlorobenzene, trichlorobenzene, tetrachlorobenzene, Aroclor 1260, and tetrachloromethane (CCl_4). After a solids feed system was installed and tested in December 1984, additional laboratory tests were conducted. Currently, the mobile incinerator is installed at the Denny Farm site near McDowell, Missouri, where "cold" and "hot" tests were conducted using clean soil and soil contaminated with surrogates similar to those employed in the earlier liquid wastes tests (CCl_4 and hexachloroethane). Tests using dioxin-contaminated liquid wastes and soil verified the DRE and the effectiveness of the control devices. Interim delisting guidelines were established and analyses were conducted on ash, treated soils, filter materials, and process/ quench water to ascertain if the guidelines were attainable. (24)

The dioxin trial burns were successful, with DREs exceeding 99.9999 percent for the POHCs burned. During these tests 3.84 pounds of 2,3,7,8-TCDD, contained in 1750 gallons of liquids and over 40 tons of soil, were destroyed. Particulate emission permit limitations (<180 mg/Nm^3 @ 7% O_2) were achieved in three of four test runs. During the fourth run, particulate emissions exceeded the prescribed limit slightly, possibly due to the accumulation of submicron-sized particles in the air pollution control system. The observed CO emission values (1.3-7.7ppm) are equivalent to those from the best available incineration technologies. (24)

Conclusion

The ten thermal processes discussed herein clearly illustrate that there are many thermal options available for treating hazardous wastes. Such processes as the ones described should be investigated by generators confronting increasingly stringent environmental regulations.

REFERENCES

1. S. K. Stoddard, et. al. Alternatives to the Land Disposal of Hazardous Wastes: An Assessment for California. State of California Governor's Office of Appropriate Technology. Sacramento, Calif. 1981.

2. Federal Register, May 19, 1980. EPA Hazardous Waste and Consolidated Permit Regulations. Part VII.

3. A Profile of Existing Hazardous Waste Incineration Facilities and Manufacturers in The United States "PB-84-157072; EPA Washington, DC 1984.

4. W. S. Rickman, et al. "Circulating Bed Incineration of Hazardous Wastes." AIChE Summer National Meeting, Philadelphia, PA, August 1984.

5. George P. Rasmussen, Robert W. Benedict & Craig Young. "Fluidized Bed Thermal Oxidation." Chapter 8-3 in Standard Handbook for Hazardous Waste Treatment & Disposal. McGraw Hill, New York to be published 1988.

6. Mullen, John F. Fluidized Bed Combustion An Economic, Proven Method of Industrial Waste Disposal, LICA General Seminar 1985, Dearborn, Michigan September 4-6, 1985.

7. Meile, L. J., Meyer, F. G., Johnson, A. J., Ziegler, D. L., Rocky Flats Plant Fluidized Bed Incinerator, Rockwell International, Energy Systems Group, Golden, CO, Printed March 8, 1982.

8. Baston, N. F., Fluidized Bed Incineration of Waste, U. S. Patent 4,359,005, 1982.

9. Litt, Robert D., Tewksbury, Ted L., Trace Metal Retention When Firing Hazardous Waste in a Fluidized-Bed Incinerator, EPA Research & Development, EPA-600/S2-84-198, January, 1985.

10. Howard Mason & Robert Olexsey. "Hazardous Waste as Fuel in Boilers. "Chapter 8-4 in The Standard Handbook for Hazardous Waste Treatment & Disposal. McGraw Hill, New York, to be published 1987.

11. Moore, W. P. "Chemical Industry's Position on Regulatory Control of Burning Hazardous Waste in Boilers" Paper 85-78.4 presented at the 78th Annual Meeting of the Air Pollution Control association, Detroit, MI, June, 1985.

12. John F. Chadbourne. "Cement Kilns" Chapter 8-5 Standard Handbook of Hazardous Waste Treatment & Disposal, McGraw Hill, New York, to be published 1987.

13. J. K. Shah, T. S. Schultz, and V. R. Daiga "Pyrolysis Processes" Chapter 8-7 Standard Handbook of Hazardous Waste Treatment & Disposal. McGraw Hill to be published in 1987.

14. Harry Freeman. Innovative Thermal Hazardous Organic Waste Treatment Processes. Noyes Publication, Park Ridge, NJ 1985.

15. L. Penberthy. "Penberthy Pyro-Converter Detoxifies Hazardous Materials," Hazardous Materials and Waste Management, Vol. 4, No. 1, January/February 1986.

16. Harry Freeman. "Update on New Alternatives Hazardous Wastes Treatment Processes". Paper at Performance and Cost of Alternatives to Land Disposal of Hazardous Waste. Air Pollution Control Association, December 8-12, 1986. New Orleans, LA, U.S.A.

17. Mike Hill and Frank Rutkey. "Development of an Industrial Waste Incinerator System: Pilot Testing Through Full Scale Operations. ASME Waste Processing Conference Proceedings, 1984.

18. Phillip L. Daily. "Performance Assessment of Portable Infrared Incinerators." HMCRI Conference, Washington, DC, 1985.

19. R. A. Olexsey. "Supercritical Fluids Processing of Industrial Wastes," presented at the 54th Annual Conference, Water Pollution Control Federation, Detroit, MI, 1981.

20. Harry Freeman and Robert A. Olexsey. "A Review of Treatment Alternatives for Dioxin Wastes." APCA Journal. January, 1986. Volume 36, #1, APCA Pittsburgh, PA.

21. J. W. Boyd, et al. "Mobile Electric Reactor Detoxifies PCBs." Hazardous Materials and Waste Management Magazine. March-April 1986.

22. J. E. Brugger, et al., "The EPA-ORD Mobile Incineration Systems Present Status," 1982 Hazardous Material Spills Conference.

23. J. J. Yezzi, J. E. Brugger, I. Wilder, F. Freestone, R. A. Miller, C. Pfrommer, P. Lovell, "Results of the Initial Trial Burn of the EPA-ORD Mobile Incineration System," Proceedings of the 1984 National Waste Processing Conference Engineering: The Solution, ASME, New York.

24. "Dioxin Trial Burn Data Package: EPA Mobile Incineration System at the James Denney Farm Site, Mcdowell, Missouri," prepared by IT Corporation for EPA HWERL-Cincinnati, EPA Contract 68-03-3069, June 1985.

25. Nicholas Kolak. "Trial Burns - Plasmas Are Technology". Proceedings of 12th Annual Research Symposium. U.S. EPA/600/9-86/022 1986, Cincinnati, OH

CHEMICAL TREATMENT OF HAZARDOUS AND TOXIC WASTES

Lucio RIZZUTI and Mario SCHIAVELLO
Istituto di Ingegneria Chimica, Università di Palermo
Viale delle Scienze, 90128 PALERMO, Italy

ABSTRACT

A chemical treatment for the abatement of pollutants is a process in which chemical reactions and/or physico-chemical operations take place. The latter, usually called unit operations, are in fact used in the practice of the chemical industry as part of a chemical process or on their own. For this reason they will also be treated in this paper together with the pure chemical treatments.

A new and developing technique will be presented and examined in detail, that is the photoassisted treatment. It comprises both the photochemical and photocatalytic processes of wastes degradation. The former processes are well known and rather established due to their selectivity, although they can not be of widespread use.

The topics regarding photoassisted processes that will be discussed here are the basic theories of photocatalysis together with their thermodynamic requirements and kinetic constraints. These theories will be applied to processes of photocatalytic degradation of hazardous and toxic wastes.

A comparison of various waste treatment processes will be done by reporting a case study on the treatment of phenolic wastes. In some of these processes the Authors themselves have research experience.

In the final part suggestions of novel promising research areas will be given, both for the chemical industry researchers and the academic world, in order to increase the joint-efforts between applied and basic research.

INTRODUCTION

Wastes are produced in a large amount by all the industrial production activities.

For a long time the problems arising from this particular industrial side-product have been practically ignored and neglected. The air, the water and the land, were considered like immense receiving bodies, and pollution was not considered a real problem. There were not any particular precautions towards hazardous substances. The general attitude was that Mother Nature should have provided, by a sort of self-defence process, to overcome this danger caused by the development of the civilization. Of course this was a wrong attitude and shortly the management of wastes became a matter for the industry.

The dangers that may possibly arise from the presence of hazardous and toxic substances can be listed as follows:

A) The inherent danger of the waste itself;
B) The possibility that the material forms toxic compounds by a physical
   change or a reaction with other substances;
C) The production of toxic gases or dusts released in the atmosphere;
D) Contamination of surface or ground waters by the release of toxic
   materials.

In the following the influence of the type and nature of the waste on the choice of the decontamination process will be discussed.

The nature of the waste or contaminant and its physical and chemical properties influence also its toxicity and hazardousness. In fact two definitions of toxicity from the dictionary are:

"a substance that when introduced into or absorbed by a living
organism destroys life or injures health especially by rapid
action when taken in small quantities",

and

"a substance causing death or harm if absorbed by a living
thing",

while hazardous is a synonym of risky, i.e.

"full of possibility or chance of meeting danger".

From these definitions it is clear that while some substances can be considered toxic and/or hazardous for their chemical nature, in other cases their physical state is responsible for the possible dangers. For instance, sand is almost an innocuous material as is on beaches, and small amounts can be ingested without danger. But a material having the same chemical composition of sand and only a different dimension of particles, i.e. very fine ones, if breathed causes one of the worst industrial killer diseases, silicosis.

Moreover, in some cases the absorption of a small amount of substance is not harmful but the poisonous effect is cumulative and the lethal quantity is eventually reached after a period of time on several ingestions.

Thus the previous definitions can not be considered exhaustive, and must be changed to "any substance which is eventually harmful". It is obvious that it is not possible to completely list such substances. Just for reference the EEC list of toxic and dangerous wastes is reported in the following Table 1, which represents the Annex 15 of the EEC directive that defines toxic and dangerous waste as "any waste containing or contaminated by the substances or materials listed in the Annex 15 of this directive of such a nature, in such quantities or in such concentrations as to constitute a risk to health or the environment".

TABLE 1
The EEC list of toxic and dangerous wastes

1. Arsenic; arsenic compounds
2. Mercury; mercury compounds
3. Cadmium; cadmium compounds
4. Thallium; thallium compounds
5. Beryllium; beryllium compounds
6. Chrome VI compounds
7. Lead; lead compounds
8. Antimony; antimony compounds
9. Phenols; phenol compounds
10. Cyanides, organic and inorganic
11. Isocyanates
12. Organic-halogen compounds, excluding inert polymeric materials and other substances referred to in this list or covered by other Directives concerning the disposal of toxic or dangerous waste
13. Chlorinated solvents
14. Organic solvents
15. Biocides and phyto-pharmaceutical substances
16. Tarry materials from refining and tar residues from distilling
17. Pharmaceutical compounds
18. Peroxides, chlorates, perchlorates and azides
19. Ethers
20. Chemical laboratory materials, not identifiable and/or new, whose effects on the environment are not known
21. Asbestos (dust and fibres)
22. Selenium; selenium compounds
23. Tellurium; tellurium compounds
24. Aromatic polycyclic compounds (with carcinogenic effects)
25. Metal carbonyls
26. Soluble copper compounds.
27. Acids and/or basic substances used in the surface treatment and finishing of metals.

As the wasted material is produced in solid, liquid and gas phase, or in some cases in a combination of them, for the management of its disposal or treatment there is a variety of problems and a large number of possible solutions.

Waste treatment processes are, in general, classified in the following major types:

a) Physical and physico-chemical
b) Chemical, including the novel photocatalytic process
c) Biological
d) Thermal (pyrolysis and/or incineration)
e) Fixation and/or encapsulation.

This paper deals with chemical treatments and only the first two types will be examined, although type a) does not include chemical reaction. This is because the physico-chemical processes (unit operations) are normally used in the chemical industrial practice as part of chemical processes, for which the chemical reaction plays the major role. A particular effort will be given to the photoassisted processes, in particular to the novel photocatalytic one. On the other hand, even if types c) and d) comprise chemical reactions, they will be discussed elsewhere in this Congress. Finally type e), which includes a chemical reaction that does not play the primary role of transforming the hazardous and toxic material, is outside the subject of this paper.

In general, the following features are taken into account for the choice of a process for the abatement of pollutants and/or hazardous and toxic wastes:

a) the need to lower the concentration of the pollutant almost to zero or to a level compatible with the laws;
b) the need to transform the hazardous or toxic material to an innocuous substance;
c) the need, in most cases, to separate the antipollutant system from the recovered wastes;
d) economical evaluations;
e) the nature of the waste, i.e. its volume, homogeneity, hazardousness, toxicity, etc.

Taking into account these criteria, an overview of the most important and general processes will be presented, and some of the innovative ones will be discussed in a more detailed way. Moreover the parameters influencing the choice of a specific process for a particular situation and the chemical engineering implications on the chosen method will be discussed.

In theory, all the chemical and physico-chemical processes can be used for the treatment of wastes. Therefore a list of all the possible processes would be a very long one. In the current industrial practice this number is limited by the constraints imposed by the chemical or physical nature of the waste, particularly if the wastes are hazardous and/or toxic.

If the toxic matter is very stable, and also very difficult to extract from the waste as a whole, treatment is not convenient and the only way of handling the problem is the use of a controlled landfill. Luckily this case is not common and generally confined to problems of pollution. The well known case of Seveso represents a typical example of such an extreme situation.

However, cases of pollution will not be discussed here, as this paper deals with the treatments for avoiding pollution and not with the pollution itself.

PHYSICO-CHEMICAL TREATMENTS

In many cases the hazardous and/or toxic material is only a portion of the whole waste, that generally is voluminous and heterogeneous. Under these circumstances it seems convenient to use, if possible, a separation process in order to transfer the toxic material to a phase which is more suitable for the successive chemical treatments. In this way the dangerous material is concentrated, and both homogeneity and reduction of mass and volume are also obtained. Among the unit operations, adsorption over activated carbon, leaching, or steam distillation are the most suitable for achieving these goals. In some cases operations of crushing and/or grinding are useful before the treatment.

If the contaminant is in a gas phase, even if its partial pressure is very low, an absorption process can be used. In particular the use of a suitable solvent can also reach the objective of transforming, by means of a chemical reaction in the liquid phase, the toxic matter into an innocuous one, thus avoiding the problem of the disposal of the exhausted liquor. This is a combined physico-chemical and chemical treatment, frequently used for the control of emissions. A typical example is the removal of hydrogen sulfide from combustion effluent gases. Packed columns are used and both the technological assessment and design procedure are well established. The same type of reactor is used for the opposite case: the liquid contains the contaminant and the gas is the decontaminating agent. The chemical reaction into the liquid phase is the main process, although the whole process can be controlled by diffusion and mass transfer mechanisms. Oxidations, in particular with ozone, are typical examples of this type of treatment.

Liquid-solid separation processes, distillation, evaporation, solvent extraction, adsorption, etc., are widely used and their relative treatment cost is generally low.

Research efforts are needed in order to better develop some innovative physico-chemical processes as the use of supercritical fluids for solvent extraction, the membrane processes and the freezing processes. Some of these innovative treatments are subjects of reasearch papers presented in this Congress, thus confirming that the research in this area is active.

In particular it is worth noting that unit operations processes used for waste manegement allow, in some cases, the recovery of valuable chemicals. This feature is very important as a reduction of net costs is reached and in some cases a typical non-profit process, such as the waste treatment, can be transformed into a yielding one. For instance the recovery of metals, also as hydrates and hydroxides, by solvent extraction, or adsorption on ion exchanging resins, or by the use of membranes, or by suspension freezing, is a valuable process, in particular for rare and/or precious metals.

CHEMICAL TREATMENTS

It is obvious that, in this class of waste treatments, the major role is played by a reaction between the hazardous and/or toxic substance and a treatment chemical. This reaction is performed for the waste to lose both hazardousness and toxicity. As for the case of physico-chemical treatments, there is a large number of possible processes, as almost any type of reaction can be performed, depending on the type of transformation needed to destroy the hazardous and toxic substances contained in the waste or to transform them into substances more appropriate for successive treatments and/or disposal.

In general these chemical processes are used when the treatment regards only one substance, or a few substances chemically similar, as chemical reactions involve specific reactants at specific conditions. Interferences with the intended goal can arise if the chemical treatment is applied to a waste having a mixed composition: side reactions may consume the treatment chemical, impurities can interfere with the intended reaction, catalyst destruction and/or poisoning can occur, new hazards can be added by unexpected products, etc. These features, together with the costs of both reactor and treatment chemical, the kinetic and thermodynamic constraints and mass transfer limitations, if any, must be taken into account for the choice of the chemical treatment to be performed.

As for the case of the unit operation a complete list will not be given, and only the very innovative photoassisted processes will be discussed in more detail, as the technology and basic knowledge of the other processes is very well established.

Neutralization is a very common waste treatment process and is also used for hazardous and toxic wastes. It does not require any particular technology, but that of mixing two or more liquids. The control is the only complexity of the whole process, and a problem can be caused by the heat release in the case of concentrated solutions. The cost is low and commercial applications are frequent.

Precipitation is also a common process, with low cost. Its primary use is the removal of heavy metals. For the technological assessment the same considerations of the previous process can be done, while limitations can arise from solubility laws or substances interfering with the precipitation of the insoluble solid. A successive physico-chemical treatment (filtration, centrifugation, or others) is needed to separate the solid. As the pH frequently interferes with the precipitation mechanism, this process is used, in many cases, in conjunction with neutralization.

Electrochemical processes are also used for the removal of metals from process streams. Their cost is higher than that of the previous processes, due to the supply of D.C. power. They are particularly used for the copper removal.

Oxidation and reduction processes are widely used for the treatment of hazardous and toxic wastes, as the reactions involved can transform practically any type of chemical. The use of specific catalysts can overcome problems arising from the kinetics of the reactions involved.

Both liquid-phase and gas-phase processes can be performed and the only limitation is the cost of the treatment chemicals, that confines their utilization usually to dilute solutions. For the type of reactor and the process conditions to be used, well kown books [1-4] of Chemical Reaction Engineering provide both design procedures and technological assessment. A recent review on novel reactors [5] can also be useful for the choice of the reactor. Several types of specific reactors can be used, such as mixing tanks, fixed bed, fluidized bed, packed tower, etc., depending on the reaction to perform, and on the process conditions. Chlorine, chlorine dioxide, hydrogen peroxide, ozone and permanganate are some of the most common oxidising agents that can be used for the hazardous and toxic waste oxidation treatment. The choice of the reduction process depends much more on the substance to be destroyed or transformed and no general suggestion can be given. Besides the cost, oxidation and reduction processes have the two following additional disadvantages. Firstly the control of the process must be very severe in order to run the reaction under precise and specific conditions because the reaction tends to be incomplete, particularly in case of thermodynamic constraints. This goal is not difficult to reach, but causes an additional cost. Secondly, care must be taken on choosing the reaction to perform as the oxidized or reduced end-products must not be hazardous and toxic.

Fundamentals of the Photocatalytic Processes

Photocatalysis has much advanced in the seventies, during the world energy crisis, as a possible tool for the photochemical conversion and storage of solar energy.

Reactions as the photosplitting of water, dinitrogen and carbon dioxide photoreduction, as well as others have been studied in detail [6,7]. Although the main goal has not been reached yet, especially from the point of view of economical convenience, the basic theories progressed very much. In addition, a development of promising practical applications for the photocatalytic processes is fast growing in the field of pollutants photodegradation.

A photocatalytic process is based on the following main steps:
- photogeneration of electron-hole pairs, produced by photoexciting a suitable semiconductor;
- separation of electrons and holes by trapping them by means of species present upon the surface of the semiconductor;
- redox reactions by the separated electrons and holes, according to the thermodynamic constraints;
- displacement of the products and reconstruction of the surface.

It is easy to see that a pollutant adsorbed on the surface of a semiconductor can undergo an oxidation reaction, which may lead eventually, for the organic compound, to carbon dioxide.

In addition to this mechanism, which can be called "redox mechanism", the photodegradation of organic pollutants may occur in another way. It is well known that many organic pollutants possess chromophore groups such as to allow the absorption of light. Very often the light absorption occurs in the far ultraviolet range of the radiation spectrum, this being the cause of their photostability. If the pollutant is adsorbed, some of the adsorbed species may shift the absorption range to near UV or visible range. As a consequence, the pollutant molecules become prone to photoexcitation and thus to photodegradation by the solar radiation. This phenomenon, i.e. the shift of the absorption to longer wavelength, is well known to spectroscopists and is designated as "red shift" [8]. Of course the reverse may occur and the effect is called "blue shift", which is not useful for using the solar spectrum. For instance, ethanoic acid in a polar medium absorbs at 210 nm and gas phase ethanoic acid absorbs at 180 nm, while a species of adsorbed ethanoic acid absorbs at 300 nm, showing a red shift of about 90-100 nm.

In the following, this mechanism of photodegradation will be called "photolysis of adsorbed phases". In other words in the redox mechanism the light excites the solid, which must be a semiconductor. In the photolysis of adsorbed phases the light excites a species adsorbed onto the surface of a solid, which can be a semiconductor or an insulator. Of course, if the photocatalyst is a semiconductor, both mechanisms occur, in principle.

Now let us examine the main conditions under which the photocatalytic processes can be used for the photodegradation of pollutants.

The photocatalyst must be stable in acidic and basic media and under radiation. Among the semiconductors tested such as CdS, Fe_2O_3, WO_3, ZnO and many others, TiO_2 appeared to be the more suitable from the point of view of chemical and photochemical stability. Several papers do not report which TiO_2 modification, anatase or rutile, was used. From studies in which this point was checked, it appears that the anatase phase is the photoactive [9]. In one paper it is moreover explained why the anatase is active while rutile is not, for a specific reaction, i.e. for the phenol photodegradation [10]. The energy of the band-gap for the anatase is 3.2 eV, while for rutile is 3.0 eV, the difference being due to a difference in the valence band energies. Due to these band gap values, TiO_2 absorbs radiation in the range 385-400 nm, i.e. in the near UV range of the solar spectrum, utilizing therefore only a small fraction of the solar energy. Studies are in progress for trying to extend the absorption towards the visible range. Despite this drawback, we will see that TiO_2 is efficient in photodegradating most of the known pollutants.

Another point to fulfill relates to the thermodynamic constraints, when a semiconductor is used as photocatalyst. The photogenerated and separated electrons have, of course, a reducing potential dictated by the conduction band energy while the oxidant potential of the holes is determined by the valence band energy, for a n-type semiconductor (the reverse is true for a p-type semiconductor). Therefore the species which will be reduced and oxidized must have compatible redox potentials.

The laboratory experiments are carried out using aqueous oxygenated dispersions containing the pollutant at various pH values.

Several routes were suggested for the reduction and oxidation processes, according to the various conditions of the experiments. One likely hypothesis is that the electrons are trapped according to the following reaction:

$$O_{2(ads)} + e^- \rightleftharpoons O_{2(ads)}^-$$

The superoxide ion, which is unstable in aqueous solution, may evolve in several ways producing oxygenated species involved in the oxidation of the pollutant [9-11].

The photoholes are of course involved in the steps of the oxidation of the pollutants either directly or through intermediates [9-11].

A further point to consider is the interaction between the pollutant and the surface of the photocatalyst. In this context the existing experience suggests to be aware of the acid-base properties of both, photocatalyst and pollutant. For instance it has been reported that phenol in a solution at pH 5.9, the natural pH of phenol, is adsorbed on the surface of TiO_2 in a given amount. If the solution is at pH 13, phenate ions are present and they are adsorbed on the surface of TiO_2, acidic semiconductor, in a greater amount: moreover the adsorbed phenate ions absorb visible radiation [10].

A final point to outline is the separation between the photocatalyst and the decontaminated solution. Although specific studies on this point did not appear up-to-date, it is easy to see that the liquid-solid separation poses less problems than the liquid-liquid separation as it is the case for chemical and biological degradation processes.

As conclusion of this section we report two tables in which most of the results so far obtained are shown.

In Table 2 the half-lives for the total degradation of several pollutants are reported using TiO_2 slurries and under exposure to simulated sunlight [12]. The process leads to the total mineralization of the aromatic compounds. The natural half-lives of such compounds vary from days to weeks.

TABLE 2

Half-lives for the total photodegradation of contaminants
assisted by TiO_2 on exposure to simulated sunlight;
concentration of catalyst, 2.0 g/l; aerated
aqueous solutions; wavelength > 330 nm.
(By permission of the Authors)

Compound	Concentration (ppm)	pH	$t_{1/2}$ (min)
4-chlorophenol (*)	6	3.0	14
3,4-dichlorophenol	18	3.0	45
2,4,5-trichlorophenol	20	3.0	55
pentachlorophenol (*)	12	3.0	20
sodium pentachlorophenate	12	10.5	15
chlorobenzene	45	2.5	90
1,2,4-trichlorobenzene	10	3.0	24
2,4,5-trichlorophenoxy- acetic acid	32	3.0	40
4,4'-dichlorodiphenyl- trichloroethane (**)	1	3.0	46
3,3'-dichlorobiphenyl (**)	1	3.0	10
2,7-dichlorodibenzo- paradioxine (**)	0.2	3.0	46

(*) Wavelength > 310 nm.
(**) Adsorbed on TiO_2.

In Table 3 the rate of photodegradation of several pollutants are
reported as volume and as micromoles of CO_2 formed [11].

TABLE 3

Rate of CO_2 formation from illuminated aerated suspensions
of 0.2 g TiO_2 in 400 ml of 10^{-3}M solutions; 100W Hg lamp;
pH adjusted with dilute H_2SO_4 unless otherwise indicated.
(By permission of the Author)

		$R(CO_2)$	
Solute	pH	$(cm^3 h^{-1})$	$(\mu M\ min^{-1})(*)$
Benzene	2.8	1.91±0.07	3.25±0.11
Benzene	4.0(**)	4.75±0.07	8.10±0.12
Phenol	2.8	2.25±0.07	3.83±0.11
Phenol	4.0(**)	4.05±0.02	6.90±0.03
Monochlorobenzene	2.8	3.11±0.12	5.31±0.20
Monochlorobenzene	4.0(**)	5.24±0.11	8.92±0.18
Nitrobenzene (****)	2.8	2.06±0.08	3.51±0.13
Aniline	3.1	3.67±0.20	6.26±0.34
Benzoic acid	3.1	4.56±0.12	7.78±0.20
Catechol	2.9	3.21±0.04	5.47±0.06
Resorcinol	2.9	3.81±0.05	6.49±0.08
Hydroquinone	2.9	4.25±0.03	7.24±0.04
1,2-dichlorobenzene	3.0	4.48±0.10	7.64±0.16
2-chlorophenol	2.8	2.80±0.03	4.76±0.06
3-chlorophenol	2.8	3.17±0.03	5.41±0.04
4-chlorophenol	2.8	3.32±0.04	5.66±0.07
4-chlorophenol (***)	3.0	1.15±0.03	1.95±0.05
2,4-dichlorophenol	2.9	5.05±0.05	8.60±0.09
2,4,6-trichlorophenol	2.9	4.72±0.03	8.05±0.05
Methyl viologen (****)	2.9	1.67±0.08	2.84±0.13
2-naphthol	2.8	3.90±0.07	6.65±0.11
Chloroform	2.9	3.70±0.11	6.31±0.18
Trichloroethylene	2.9	3.42±0.09	5.83±0.15
Ethylene diamine	2.9	1.38±0.01	2.35±0.02
Dichloroethane	2.9	1.66±0.03	2.83±0.05
Dichloroethane	4.4(**)	3.16±0.03	5.38±0.04

(*)   The number of $\mu M\ min^{-1}$ assuming that all the CO_2 is dissolved.

(**)   No H_2SO_4 added.

(***)   La Porte Tiona grade TiO_2, 9 $m^2 g^{-1}$ surface area by BET method.

(****) Non-linear response curving upwards.

It is worth mentioning that, in addition to the laboratory experiments, several others were carried out by using solar radiation. The photodegradation was found to be comparable to that for the laboratory experiments or even higher.

CASE STUDY OF PHENOL ABATEMENT

An example of the possibilities which can be handled for the treatment of a specific pollutant will be reported.

Phenol is one of the most common organics contained in waste streams originating in the chemical process industries. It is well known that even low phenol levels, in the part per billion range, impart to water besides a "medicinal" taste and odor also a strong toxic effect. Therefore it is necessary to completely eliminate the phenol from the stream, before discharging. It is also important to note that the laws on this matter are very severe in all countries.

As far as it is known to-date, the interested industries use mainly biological and chemical (oxidation by ozone) treatments.

We will report a survey of the available methods, outlining with more detail the novel photocatalytic process.

Adsorption Process

The adsorption of aqueous phenol solutions over activated carbon has been recently studied [13]. The adsorption isotherms have been obtained in a phenol concentration range of 10^{-3}-2 mmol/l and on four different activated carbons at 20°C, and on one carbon at 55°C. An isotherm is proposed, in which the Redlich-Peterson equation is combined with elements of the Dubinin-Radushkevich analysis. All the parameters of the proposed isotherm are reported and the slight temperature dependency is outlined. So the isotherm at 20°C seems the most appropriate in case this adsorption process is utilized for waste treatment. The equipment design is easy as the thermodynamic equilibria are known.

Since this adsorption process on activated carbon efficiently removes traces of phenol from aqueous solutions, it can be used successfully as a final treatment to bring the phenol concentration as close as possible to zero.

Oxidation Process by Oxygen in Liquid Phase

A process of phenol oxidation in aqueous solution has been studied in a gas-liquid-solid CSTR reactor [14]. The solid catalyst used was supported copper oxide, and oxygen was continuously supplied at various pressures between 1 and 17 atm (absolute). The process temperature varied from 96°C to 146°C. An induction period has been observed, followed by a much higher steady-state activity regime. The reaction has been found to be first order with respect to phenol in both regimes, while the oxygen dependency is first and one-half order in the induction period and steady-state regime respectively. All the kinetic constants and activation energies are reported in the paper. Admittedly, oxidation rates at atmospheric pressure are too slow for a practical application, but moderate increases in temperature and pressure can bring the process in the range of practical importance.

Oxidation Process by Air in Vapour Phase

The vapour-phase oxidation of phenol over copper oxide supported on alumina was studied in a fixed-bed reactor, in the presence of 8.4 to 13.4 mol% water vapour [15]. The phenol concentration varied from 875 to 26 ppm (vol) and the temperature between 150°C and 270°C. At moderate temperatures (~ 250°C) and low conversions ($\sim 30\%$) about one-half of the phenol oxidized is converted to CO_2 and H_2O. The kinetic constants and the activation energy are reported. This vapour-phase catalytic oxidation can be useful to remove phenol from gas streams.

Oxidation Process by Ozone

This process has been extensively studied [16–18 and references therein], by using a gas bubbling apparatus or a wetted wall column. In all these studies it has been found that the oxidation rate of phenol by ozone was increased by increasing the pH of the solution. This result has been explained by proposing that the species attacked by the ozone is the phenate ion.

In particular the dependence of the kinetics from the pH has been investigated in the range 1.75–12, and both the phenol concentration and ozone partial pressure have been varied from 7 mg/l to 800 mg/l and from 800 ppm to 4600 ppm (by volume) respectively [18]. A complete kinetic equation has been proposed, with a very good agreement to the large number of experimental data obtained. This equation is based on the assumption that the ozone attacks the phenate ion instead of phenol. Under this hypothesis the reactions taking place in the liquid phase are:

$$C_6H_5OH \rightleftharpoons C_6H_5O^- + H^+$$

$$C_6H_5OH + OH^- \rightleftharpoons C_6H_5O^- + H_2O$$

$$C_6H_5O^- + O_3 \longrightarrow \text{Products}$$

The corresponding kinetic equation is:

$$r_3 = K_2k_3[O_3][OH^-][C_6H_5OH] \cfrac{1}{1 + \cfrac{K_2k_3[O_3][OH^-]}{k_1 + k_2[OH^-]}}$$

where k_1, k_2, and k_3 are the kinetic constants of the previous equations, whose values are reported in the paper, and K_2 is the equilibrium constant of the second reaction, whose value is $1.2 \cdot 10^4$ M^{-1} [19].

The basic assumption that the ozone reacts, in the liquid phase, with the phenate ion is verified by the results. Two different reaction mechanisms have been identified, that are substantiated also by observing the previous equation. The first prevails in acid solutions, giving rise to a first order kinetics in ozone, phenol, and hydroxyl ion. The second one prevails in basic solutions and gives a kinetics which is independent from the ozone concentration and first order both in phenol and hydroxyl ion. In the transition region (pH 6-8) the complete kinetic equation must be used as both mechanisms occur.

This study furnishes to the designer all the information needed for the development of the process and for its complete detailed design. It is obvious that the choice of the type of reactor and of the process conditions depends strongly on the waste to be treated and on the objectives to be reached. In any case, although the cost of this type of treatment is not low, its efficiency and the feature that there are no undesired side-products makes the process itself comparable with the other ones. The design of the reactor, either a stirred tank or a packed absorbing column, does not represent a serious problem in the chemical engineering design practice.

Alkylation Process

The catalytic alkylation of phenol to ether has been investigated [20]. By passing a gas phase mixture of phenol and alkyl halide through a fixed bed of K_2CO_3 (base) on Carbowax 6000 (catalyst), maintained at 170°C, the formation of the corresponding ether was observed. The phenol concentration before vaporization was in the range 0.55-0.11 M, with a liquid flow rate of 40 ml/h. Kinetic studies have not been performed, and only tables giving conversion % and yield % are reported. Besides this limitation of the investigation, a practical application for this process is not seen, both for the high temperature required and for the cost of the alkylation agent.

Biological Process

The biological treatment of phenolic wastes has been carried out by using activated sludge [21], pure cultures of Nocardia corallina and Pseudomonas fluorescens [22,23], and the commercial mixed culture "Phenobac" [24]. Both batch [21,23] and CSTR [22,24] reactors have been used. The temperature ranges were 17-37°C for the sludge and 22-42°C for the Nocardia corallina, while the Pseudomonas fluorescens and Phenobac experiments have been performed at 37°C. The initial phenol concentration ranged from 230 mg/l to 1010 mg/l for the sludge, 130-430 mg/l for the Pseudomonas and 40-410 mg/l for the Nocardia, while the inlet phenol concentration to the continuous Phenobac process was 170 mg/l.

For all the cultures zero order kinetics in phenol and first order in biomass has been found. All the kinetic constants, their temperature dependence, and the yield coefficients are reported in the referred papers.

With regard to the activated sludge, it is worth noting that, in batchwise conditions, the phenol concentration dropped to almost zero from a value of 350 mg/l in six hours, and from a value of 170 mg/l in two hours.

For the continuous experiment with <u>Nocardia</u> a residence time of 3.3 h was used. The inlet phenol concentration was varied in the range 80-410 mg/l, and the biomass concentration into the reactor varied accordingly, while the outlet phenol concentration was 3 mg/l, a value very close to that requested by the law. It is interesting to compare this value with that obtained for the Phenobac at the same temperature and inlet concentration. For a residence time of 3.3 h the outlet concentration is 4 mg/l, while a residence time of 3.9 h is required to reach 3 mg/l. Two considerations arise from these figures. Firstly that pure cultures are more active than mixed ones. Unfortunately the use of the former ones is confined to laboratory experiments, as the maintenance of a pure culture in a waste treatment plant would be very difficult and costly. Moreover the concept of "pure culture" contrasts with that of "waste". Secondly that, in order to reach the value of zero for the outlet concentration, an infinite residence time would be necessary. This means that biological treatments need a successive one, such as the previously described adsorption process, to fulfill the very severe law requirements.

Photocatalytic Process

This process was studied by using mainly aqueous TiO_2 dispersions and Pyrex batch photoreactors with immersed lamp [9,10,25,26]. The reasons for the choice of TiO_2 have been explained above.

All the studies revealed both the complete photooxidation of phenol to CO_2 with rates varying in accordance to the conditions, and the catalytic nature of the process. Several factors affecting the reaction rate have been investigated: TiO_2 and phenol concentration, pH of the solution, presence or absence of oxygen, presence of foreign ions as Cl^-, SO_4^{2-}, etc., presence of intermediate compounds, the TiO_2 nature. Several tests were performed with solar radiation: the results were satisfactory, showing thus the practical potential of the process. The results showed that the level of photodegradation was increased by the presence of oxygen and by the alkaline medium (pH $\sim$ 13, i.e. presence of phenate ions), while Cl^- ions in high amount (3-4% by weight in the dispersion) adversely affected the photodegradation. Intermediate compounds, as quinone and catechol, were detected, but they were unstable and photodegraded.

It has been found that TiO_2 under the anatase phase is active, while under the form of rutile is completely inactive. Reasons for this behaviour were given [10]. Attention to the kinetics and reaction mechanism was paid. It was considered that the rate determining step of the process was the reaction between $OH^\cdot$ radicals and phenol species adsorbed over the catalyst surface. Two distinct sets of sites were hypothesized to exist on the catalyst surface: one suitable for phenol adsorption and one for oxygen adsorption. Therefore the reaction rate was set as:

$$r = k'' \, \theta_{O_2} \, \theta_{Phen}$$

where k'' is the surface second order rate constant, θ_{O_2} and θ_{Phen} are the fractional sites coverages for oxygen and phenol. By developing the above equation, and using Langmuir-type adsorption isotherms, a final relationship was obtained for the rate equation:

$$r = k'' \, \frac{K_{O_2} \, p_{O_2}}{1 + K_{O_2} \, p_{O_2}} \, \frac{K_{Phen}[PhOH]}{1 + K_{Phen}[PhOH]_0}$$

which is a first order kinetic equation for phenol concentration and where K_{O_2}, K_{Phen} are the equilibrium adsorption constants, and $[PhOH]_0$ is the initial phenol concentration.

The following values for the constants have been obtained: $k'' = 2.04 \cdot 10^{-3} \, M^{-1} h^{-1}$, $K_{O_2} = 1.82 \cdot 10^{-2} \, KPa^{-1}$ and $K_{Phen} = 4.07 \cdot 10^3 \, M^{-1}$.

As for the reaction mechanism, the essential redox steps were as follows. Adsorbed oxygen molecules were considered traps for the photogenerated electrons:

$$O_{2(ads)} + e \rightleftharpoons O_{2(ads)}^-$$

The superoxide ions, being unstable in solution, evolve according to several routes:

$$O_{2(ads)}^- + H_2O \longrightarrow OH^- + HO_2^{\cdot}$$

and/or

$$O_{2(ads)}^- + OH^{\cdot} \longrightarrow O_{(ads)}^- + HO_2^{\cdot}$$

forming hydrogen peroxide radicals, known to be powerful oxidant agents for organic molecules. The photoholes are trapped by OH^- groups:

$$OH^- + h^+ \longrightarrow OH^{\cdot}$$

forming radicals which are reactive species responsible for the reaction, whose initial step is:

$$C_6H_5O^- + OH^{\cdot} \xrightarrow{\; H_2O \;} C_6H_4(OH)_2 + OH^- \longrightarrow \text{other intermediates} \longrightarrow CO_2 + H_2O$$

As seen, the method is promising for practical application. The complete mineralization of phenol also under solar radiation, the catalytic nature, the beneficial effect of oxygen, the relative facility of separating a liquid from a solid are all merits of this photocatalytic process.

Studies are still needed for proposing convenient solutions for the problem of the discontinuity of the solar radiation and for testing the photocatalysts by using real industrial wastes, which contain several contaminants together.

CONCLUDING REMARKS

There is no doubt that the concern for the environmental problems is fast growing at all levels: governments, industries, researchers, ordinary people.

While the discussion on planning and rebuilding a "clean" industrial apparatus is outside the scope of this Congress, it seems pertinent to outline few guidelines for the improvement of the management of hazardous and toxic wastes.

As seen during the Congress, there is a variety of methods and processes which can be handled: some of them are well established, while some are still in the state of research and development.

Without having the intention of making a classification of merits or of indicating precise lines of research for future work, we may draw your attention to the processes that make use of supercritical fluids, to those based on gamma-irradiation, to membrane processes, to the photocatalytic ones. All these methods, as well as others, still need to be developed and eventually tested in real problems, as the results obtained so far indicate some merits.

Apart from the field of research in which we all are involved, we must realise that, as for the nuclear energy problems, the maintenance of the ecological equilibrium is not a problem of a single country, as recently it has been clearly demonstrated.

Therefore a common effort of research, supported and coordinated by International Bodies, is highly needed. This is, according to our view, the message and the suggestion of this Congress.

Of course, apart from the suggestion, the scientists have the responsability of improving their efforts for the research, of exchanging ideas within groups in various countries, of stimulating the public opinion by proposing feasible and easy solutions to the problem.

REFERENCES

1. Hill, C.G., *An introduction to Chemical Engineering Kinetics and Reactor Design*, J. Wiley, New York, 1977.

2. Shah, Y.T., *Gas-Liquid-Solid Reactor Design*, McGraw-Hill, New York, 1979.

3. Froment, G.F., and Bischoff, K.B., Chemical Reactor Analysis and Design, J. Wiley, New York, 1979.

4. Astarita, G., Savage, D.W., and Bisio, A., Gas treating with chemical solvents, J. Wiley, New York, 1983.

5. Chaudhari, R.V., Shah, Y.T., and Foster, N.R., Novel Gas-Liquid-Solid Reactors, Catal. Rev.-Sci. Eng., 1981, 28, 431-518.

6. Graetzel, M., (Ed.), Energy Resources through Photochemistry and Catalysis, Academic Press, New York, 1983.

7. Schiavello, M., (Ed.), Photoelectrochemistry, Photocatalysis and Photoreactors, Fundamentals and Developments, Reidel Publ. Co., Dordrecht, 1985.

8. Nicholls, C.H., and Leermakers, P.A., Photochemical and Spectroscopic properties of organic molecules in adsorbed or other perturbing polar environments, Adv. Photochem., 1971, 8, 315-336.

9. Okamoto, K., Yamamoto, Y., Tanaka, H., Tanaka, M., and Itaya, A., Heterogeneous photocatalytic decomposition of phenol over TiO_2 powder, Bull. Che. Soc. Jpn, 1985, 58, 2015-2022.

10. Augugliaro, V., Palmisano, L., Sclafani, A., Minero, C., and Pelizzetti, E., Photocatalytic degradation of phenol in aqueous titanium dioxide dispersions, to be published.

11. Matthews, R.W., Photooxidation of organic material in aqueous suspensions of titanium dioxide, Wat. Res., 1986, 20, 569-578.

12. Borgarello, E., Serpone, N., Barbeni, M., Minero, C., Pelizzetti, E., and Pramauro, E., Putting photocatalysis to work, in Homogeneous and Heterogeneous Photocatalysis, eds. Pelizzetti, E., and Serpone, Reidel Publ. Co., Dordrecht, 1986, pp.673-689.

13. Seidel, A., Tzscheutschler, E., Radeke, K.-H., and Gelbin, D., Adsorption equilibria of aqueous phenol and indol solutions on activated carbons, Chem. Engng Sci., 1985, 40, 215-222.

14. Sadana, A., and Katzer, J.R., Catalytic oxidation of phenol in aqueous solution over copper oxide, Ind. Eng. Chem. Fundam., 1974, 13, 127-134.

15. Walsh, M.A., and Katzer, J.R., Catalytic oxidation of phenol in dilute concentration in air, Ind. Eng. Chem. Process Des. Develop., 1973, 12, 477-481.

16. Rizzuti, L., Augugliaro, V., and Marrucci, G., Ozone absorption in aqueous phenol solutions, <u>Chem. Engng J.</u>, 1977, <u>13</u>, 219-224.

17. Augugliaro, V., and Rizzuti, L., Temperature dependence of the ozone absorption kinetics in aqueous phenol solutions, <u>Chem. Engng Comm.</u>, 1978, <u>2</u>, 219-221.

18. Augugliaro, V., and Rizzuti, L., The pH dependence of the ozone absorption kinetics in aqueous phenol solutions, <u>Chem. Engng Sci.</u>, 1978, <u>33</u>, 1441-1447.

19. Weast, R.C., (Ed.), <u>Handbook of Chemistry and Physics</u>, 56th Edn., CRC Press, Cleveland, 1975.

20. Angeletti, E., Tundo, P., and Venturello, P., Gas-liquid phase-transfer synthesis of phenyl ethers and sulphides with carbonate as base and Carbowax as catalyst, <u>J. Chem. Soc. Perkin I</u>, 1982, 1137-1141.

21. Augugliaro, V., Gagliardi, L., Nicosia, S., and Rizzuti, L., Cinetica della ossidazione biologica del fenolo, <u>Ingegneria Sanitaria</u>, 1979, 11-18.

22. Rizzuti, L., Augugliaro, V., Dardanoni, L., and Torregrossa, M.V., Kinetic parameter determination in monoculture and monosubstrate biological reactors, <u>Can. J. Chem. Engng</u>, 1982, <u>60</u>, 608-612.

23. Rizzuti, L., Augugliaro, V., Torregrossa, V., and Savarino, A., Kinetics of phenol removal by Nocardia species, <u>European J. Appl. Microbiol. Biotechnol.</u>, 1979, <u>8</u>, 113-118.

24. Augugliaro, V., Finocchiaro, M., and Rizzuti, L., Comportamento transitorio di un reattore biologico in risposta a variazione a gradino della portata, Proc. of G.R.I.C.U. Meeting "Sviluppo della ricerca di Ingegneria Chimica", Politecnico di Milano, 1983.

25. Kawaguchi, H., Photocatalytic decomposition of phenol in the presence of titanium dioxide, <u>Environ. Technol. Lett.</u>, 1984, <u>5</u>, 471-474.

26. Okamoto, K., Yamamoto, Y., Tanaka, H., and Itaya, A., Kinetics of heterogeneous photocatalytic decomposition of phenol over anatase TiO_2 powder, <u>Bull. Chem. Soc. Jpn</u>, 1985, <u>58</u>, 2023-2028.

SUPERCRITICAL FLUID TECHNOLOGY IN HAZARDOUS WASTE TREATMENT

Michael Modell
Modell Development Corporation
Cambridge, MA 02138
USA

ABSTRACT

This paper describes theory and applications of supercritical fluid technology in the physical and chemical treatment of hazardous and toxic wastes. The background section provides a brief description on the supercritical fluid properties of interest. Two quite different types of applications are discussed: (i) extraction of organics with supercritical carbon dioxide from aqueous wastes, spent activated carbon and contaminated soils and (ii) oxidation of organics in supercritical water.

Supercritical fluids (SCFs) have been receiving much attention as potential extracting agents and media for chemical reactions. This interest stems from an unusual combination of thermodynamic and mass transport properties: solubilities of nonvolatile solutes in SCFs can be varied from moderately high to very low by relatively small changes in temperature and pressure while diffusivities are significantly higher than those of the same solutes in liquid solvents. The variability of solubility provides the basis for extraction methodology. The media to be treated (e.g., liquid or solid waste) is extracted with the SCF under conditions of high solubility; the temperature and/or pressure of the extract is then altered to reduce the solubility, thereby creating a second phase which is solute-rich; the phases are separated; the SCF solvent phase is recycled and brought back to the conditions of high solubility while the solute is recovered as concentrate.

As applied to hazardous waste treatment, SCF extraction may be considered where toxic contaminants are present in low to moderate concentrations in a nontoxic, nonhazardous medium. Removal of toxic organics from aqueous wastes is one application which has recently undergone successful piloting and the first commercial unit is under construction. A second application currently in research is extraction of toxic organics from soils. Preliminary economic estimates presented herein suggest that soil decontamination by SCF might represent a breakthrough. Data, flowsheets and economics will be described, to the extent that they have been made publicly available.

As a medium for chemical reactions, supercritical water (SCW) is being used for the oxidation of organics. Above the critical temperature ($374\,^{\circ}C$) and pressure (22 MPa), water is an excellent solvent for both organic matter and gases (e.g., oxygen, nitrogen, carbon dioxide). Dissolving waste and air or oxygen in SCW leads to a homogenous phase in

which oxidation is rapid. At 400 to 450°C, the kinetics are such that
99 to 99.9% conversion (to carbon dioxide and carbon monoxide) can be
achieved within 5 min; at 600 to 650°C, 99.9999% can be achieved in less
than 1 min.
 There are two applications of SCW oxidation under development: (i) a
surface treatment process for oxidation of organic wastes or moderately
concentrated aqueous wastes; and (ii) a downhole process for the oxidation
of organics in dilute aqueous wastes. Data, flowsheets and economics of
these two applications will be presented, to the extent that they have
been made publicly available.

INTRODUCTION

Over the past ten years, supercritical fluids have received
considerable attention as potential extracting agents and media for
chemical reactions (see, e.g., McHugh and Krukonis [1] and Paulaitas, _et
al_. [2,3]. Most of the activity has been research and exploratory
development; very few processes have reached commercial exploitation.
Some notable exceptions are propane deasphalting, tertiary oil recovery
with carbon dioxide, decaffeination of coffee with carbon dioxide and
hydrothermal manufacture of quartz.

Two types of applications are of interest in hazardous waste
treatment: extraction of contaminants from water and solids, and oxidation
of organic contaminants from aqueous solutions and concentrates. A list
of candidate materials is given in Table 1, along with critical
temperatures, pressures and densities. Aliphatic and olefinic
hydrocarbons are potentially attractive SCF solvents because they are
relatively inexpensive and their critical pressures are relatively low.
However, from the point of view of potential toxicity of residual solvent,
their use in environmental applications is questionable. If we focus on
environmental acceptability as well as low material cost, then we are left
with only two fluids of interest: carbon dioxide and water. Note that
both have relatively high critical pressures, which is a negative factor
when we consider capital cost of equipment.

The applications of SCF technology discussed herein are listed in
Table 2. The first three entries involve supercritical carbon dioxide for
removal of organics by extraction and/or desorption. The underlying
phenomena involve solubility and mass transfer characteristics of solutes
in SCF solvents. The last two entries concern the oxidation of organics
in supercritical water (SCW) and are based upon the kinetics of oxidation
in addition to solubility relationships.

TABLE 1

CRITICAL CONSTANTS OF SOME COMMON FLUIDS

	Critical Constants		
	Temp. (C)	Pres. (MPa)	Dens. (g/cm^3)
Ethylene	9.9	5.12	0.23
Chlorotrifluoromethane	28.8	3.87	0.58
Carbon dioxide	31.0	7.38	0.47
Ethane	32.2	4.88	0.20
Propylene	91.9	4.60	0.23
Propane	96.7	4.25	0.22
Dichlorotrifluoromethane	111.7	3.99	0.56
Ammonia	132.3	11.27	0.24
n-Butane	152.0	3.80	0.23
n-Pentane	196.6	3.37	0.23
Isopropyl alcohol	235.3	4.76	0.27
Methanol	240.3	7.99	0.27
Ethanol	243.4	6.38	0.28
Benzene	288.9	4.89	0.30
Water	374.2	22.12	0.32

The scientific bases of the underlying phenomena are treated in two parts. The section on properties of supercritical carbon dioxide is applicable to SCF extraction processes and supercritical fluids in general. Water is a notable exception to those generalities and, therefore, the properties of supercritical water are treated in a separate subsection.

Following the description of underlying phenomena, each of the applications listed in Table 2 is discussed. Flowsheets are presented and projected economics of commercial operation are given. Where costs have been published by the organizations developing the processes, they are provided herein. Where costs were unavailable in the open literature, the author has taken the liberty of providing his own estimates. The purpose of including such estimates is to provide the reader with some idea of the magnitude of costs associated with each of the processes. In spite of the fact that the author has been associated with the development of some of these processes, the estimates and extrapolations contained herein are not necessarily made with knowledge of the current state of technology development and do not necessarily represent the views of the organizations that are offering or will offer such processes commercially. For more accurate and current commercial and technical information, the reader should contact the organizations listed as sponsors in Table 2.

TABLE 2

APPLICATIONS OF SUPERCRITICAL FLUIDS IN HAZARDOUS WASTE TREATMENT

Methodology	Application	Fluid	Conditions	Sponsor
Desorption	Regeneration of adsorbents	Carbon dioxide	80-100°C, 8-15 MPa	Illinois Water Treatment Co. Rockford, IL
Extraction and desorption	Decontamination of soils	Carbon dioxide with and without cosolvents	40-80°C, 7-10 MPa	Louisiana State University Baton Rouge, LA
Extraction	Removal of organics from aqueous wastes	Carbon dioxide with and without cosolvents	20-30°C, 6-8 MPa	Critical Fluid Systems, Inc. Cambridge, MA
Oxidation	Destruction of organics and removal of salts	Water	450-650°C, 25 MPa	Modar, Inc. Houston, TX
Oxidation	Downhole destruction of organics in dilute aqueous waste	Water	300-450°C, 25-30 MPa	Oxidyne, Inc. Dallas, TX

SUPERCRITICAL FLUID EXTRACTION

1. Properties of Supercritical Fluid Solvents

The supercritical fluid region can best be visualized with the aid of
a reduced pressure-density diagram, shown in Fig. 1. The dashed curve is
the locus of liquid-vapor equilibria, which terminates at the critical
point (C.P.). Note that the critical density is about 40% of that of the
normal liquid.

The range of conditions of interest is reduced temperatures of 0.95 to
1.4 and reduced pressures of 1 to 6. As shown in Fig. 1, the region above
$T_r=1$ is a supercritical fluid; the region below $T_r=1$ is referred
to as a near critical liquid (NCL). Within these regions, there are
conditions which are optimal for extraction: densities high enough to
provide liquid-like solubilities, yet low enough so that diffusivities are
appreciably higher than those of normal liquids [4].

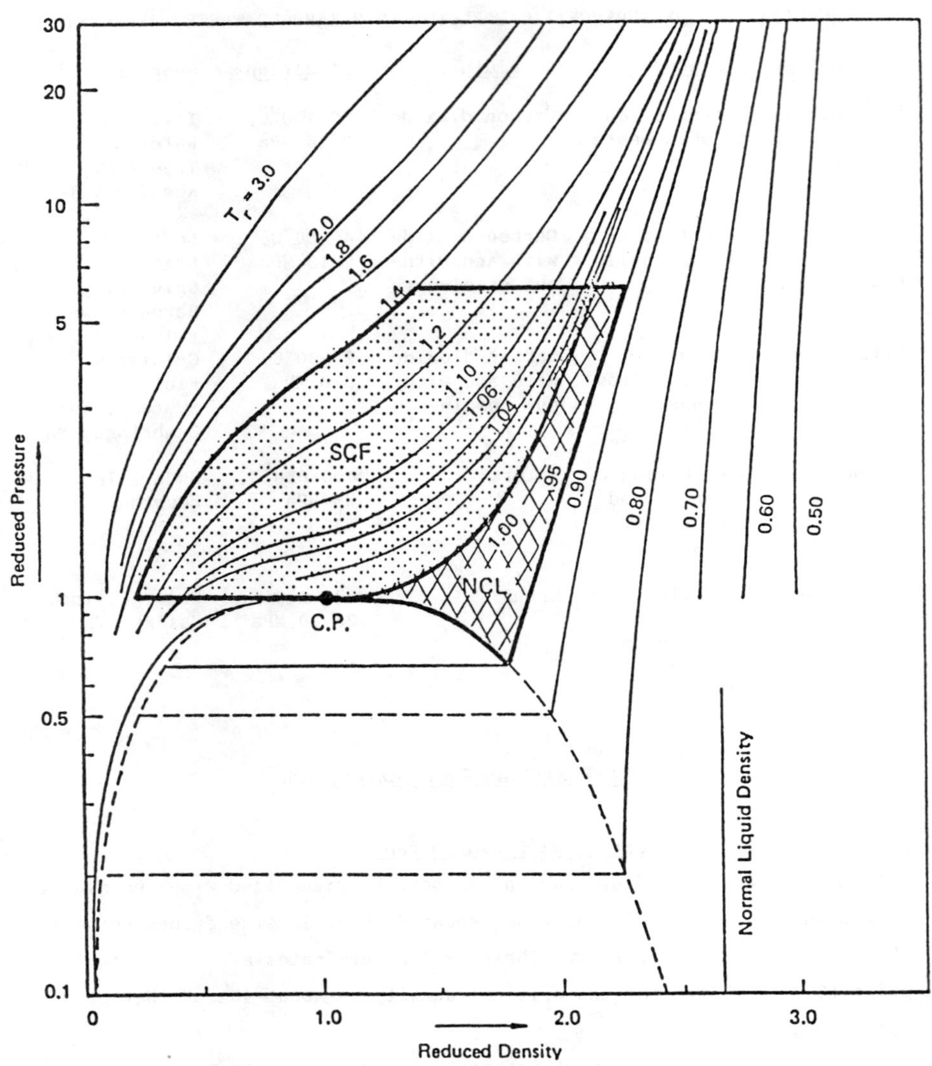

FIGURE 1.

REDUCED PRESSURE-DENSITY DIAGRAM. Supercritical fluid (scf)
and near-critical liquid (ncl) regions, as indicated

The extraction cycle can be illustrated by a solubility "map," as shown in Fig. 2 for naphthalene in CO_2 [4,5]. The dashed line is the solubility of solid naphthalene in coexisting liquid and vapor CO_2. A vertical tie line is shown for 25°C. The critical point indicated in Fig. 2 is actually the critical end point for the naphthalene-carbon dioxide mixture. It occurs near the critical point of CO_2 because the concentration of naphthalene is small.

Outside of the liquid-vapor envelope, the solubilities shown in Fig. 2 are isobars for solid naphthalene in equilibrium with a supercritical fluid mixture. The solubilities are directly related to fluid density. At temperatures slightly above the critical (32 to 50°C), the solubility drops dramatically with decreasing pressure because density decreases significantly with small decrease in pressure (see Fig. 1).

As pressure increases, it requires a higher temperature to obtain the sharp drop in density. At 15 to 30 MPa, the solubility increases with increasing temperature in the conventional manner (i.e., solute activity increases with higher temperature) because the fluid density does not decrease appreciably until the temperature is higher than 50°C.

An extraction cycle consists of a condition of high solubility, where extraction takes place, and a condition of low solubility, where solute is removed from the solvent. After separation, the solvent is returned to the first condition to complete the cycle.

Two extraction cycles are indicated on Fig. 2. Both start with extraction under conditions of high solubility at high pressure (e.g., 50°C, 30 MPa). In one cycle, pressure would be decreased to 8 MPa and the temperature adjusted to 45°C to provide for a condition of low solubility. In the second cycle, the mixture would be brought to subcritical conditions where a single stage or multistage distillation would be used to separate solute and solvent.

Fig. 2 illustrates general variations of solubility with temperature and pressure. However, the absolute values of solubilities can vary widely from one substance to another and the economics of extraction depend upon these absolute values. For CO_2, there is a large body of data for solubilities in liquid CO_2 that can be used as a guide. Table 3 is a partial compilation of solubilities of organics in liquid CO_2 at 25°C and 6.6 MPa [6]. Each of these solubilities represents a single data point on a solubility map. For example, from Table 3, the solubility of naphthalene in liquid CO_2 at 25°C is 2

FIGURE 2.

SOLUBILITY MAP OF NAPHTHALENE IN SCF AND NCL CO_2

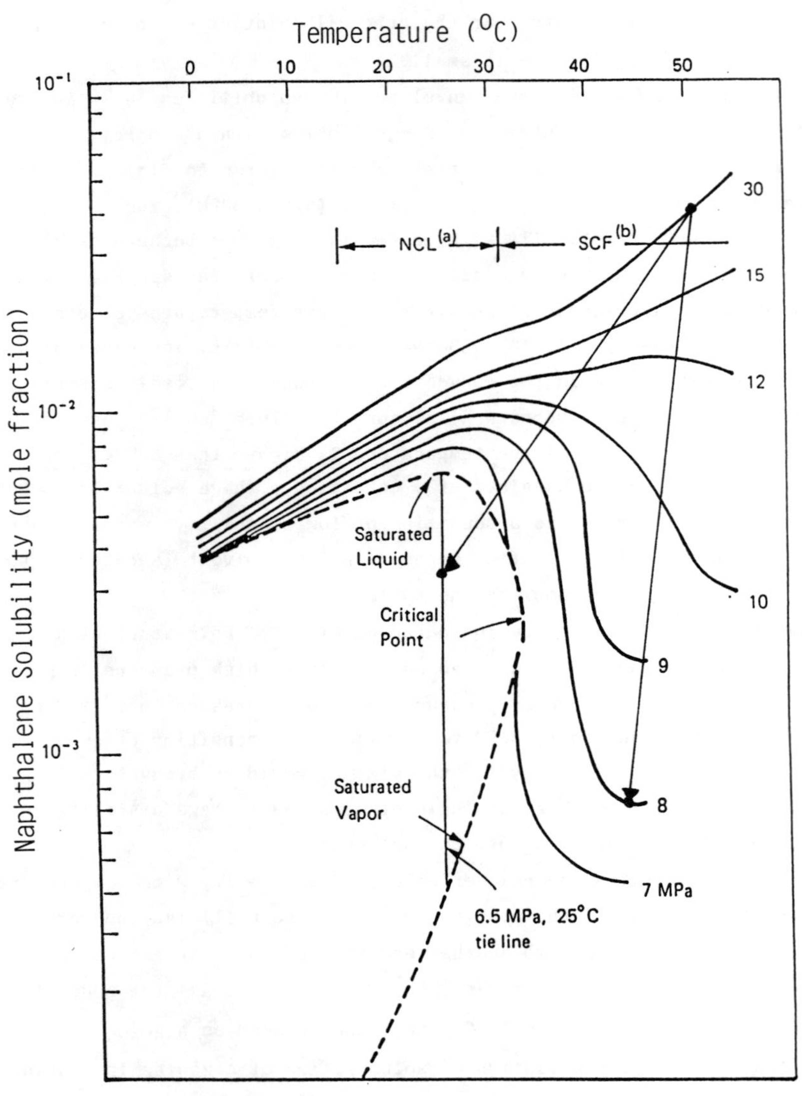

(a) NCL = Near Critical Liquid
(b) SCF = Super Critical Fluid

TABLE 3. SOLUBILITIES OF ORGANIC COMPOUNDS IN LIQUID CO_2 AT 25°C AND 6.6 MPa [6]
(Solubilities in weight-percent; M = completely miscible)

Paraffins

Cyclohexane	M
n-Dodecane	M
Ethane	M
n-Heptane	M
n-Hexadecane	8
Methylcyclohexane	M
n-Octadecane (mp 28 C)	3
Paraffin wax (mp 52 C)	1
Propane	M
n-Tetradecane	16

Olefins

1-Decene	M
1-Octadecene	10
Propylene	M

Aromatics

Benzene	M
Biphenyl (mp 71 C)	2
Chlorbenzene	M
a-Chloronaphthalene	1
p-Dichlorobenzene (mp 53 C)	M
p-Dimethoxybenzene (mp 53 C)	M
2-Methoxybiphenyl (mp 29 C)	5
B-Methylnaphthalene (mp 35 C)	9
Naphthalene (mp 52 C)	2
Nitrobenzene	M
o-Nitrocholorobenzene (mp 32)	21
a-Nitronaphthalene (mp 58 C)	1
o-Nitrotoluene	M
p-Nitrotoluene (mp 51 C)	20
Tetrahydronaphthalene	12
Toluene	M

Alcohols

Benzyl alcohol	8
B-Chloroethanol	10
Cinnamyl alcohol (mp 30 C)	5
Cyclohexanol	4
1-Decyl alcohol	1
2-Ethylhexanol	17
Ethyl alcohol	M
Furfuryl alcohol	4
Heptyl alcohol	6
Hexyl alcohol	M
Methyl alcohol	M

Aldehydes

Acetaldehyde	M
Benzaldehyde	M
Cinnamaldehyde	4
1-Heptaldehyde	M
Salicylaldehyde	M
Valeraldehyde	M

Ketones

Acetone	M
Acetophenone	M
Benzalacetone (mp 42 C)	5
Benzophenone (mp 48 C)	4
Chloroacetone	M
Cyclohexanone	M
2-Octanone	M

Ethers

n-Butyl ether	M
B,B'-Dicholroethyl ether	M
Phenyl ether (mp 28 C)	8

Phenols

o-Chlorophenol	M
p-Chlorophenol (mp 43 C)	8
o-Cresol (mp 30 C)	2
m-Cresol	4
2,4-Dichlorophenol (mp 45 C)	14
p-Ethylphenol (mp 46 C)	1
o-Nitrophenol (mp 45 C)	M
Phenol (mp 41 C)	3

Esters

Benzyl benzoate	10
Butyl oxalate	M
Butyl phthalate	8
Butyl stearate	3
B-Chloroethyl acetate	M
Ethyl acetate	M
Ethyl benzoate	M
Ethylene diformate	M
Ethyl lactate	M
Ethyl phthalate	10
Ethyl salicylate	M
Ethyl succinate	M
B-Hydroxyethyl acetate	17
Methyl phthalate	6
Phenyl phthalate (mp 70 C)	1
Phenyl salicylate (mp 43 C)	9

Amides

Acetamide (mp 82 C)	1
N,N-Diethylacetamide	M
N,N-Diethylformamide	M
N,N-Dimethylformamide	M
Formamide	0.5

Carboxylic Acids

Acetic acid	M
Caproic acid	M
Caprylic acid	M
Chloroacetic acid (mp 61 C)	10
a-Chloroproprionic acid	26
Isocaproic acid	M
Lactic acid	0.5
Lauric acid	1
Oleic acid	2

Amines and Nitrogen Heterocyclics

Aniline	3
o-Chloroaniline	5
N,N-Diethylaniline	17
N,N-Dimethylaniline	M
Diphenylamine (mp 53 C)	1
N-Ethylaniline	13
N-Ethyl-N-benzylaniline	4
N-Methylaniline	20
a-Naphthylamine (mp 52 C)	1
Phenylethanolamine	1
2,5-Dimethylpyrrole	5
Pyridine	M
o-Toluidine	7
m-Toluidine	15

Nitriles

Acetonitrile	M
Acrylonitrile	M
Benzonitrile	M
B-Hydroxyproprionitrile	1
Phenylacetonitrile	13
Succinonitrile (mp 54.5 C)	2

wt%. In Fig. 2, the corresponding point on the saturated liquid curve at 25°C is 0.62 mole-% or about 2 wt%. Thus, the data of Table 3 can be used as a guide for estimating the magnitude of solubility in the critical region. It should be noted that nearly half of the entries in Table 3 are completely miscible with NCL CO_2. The naphthalene solubility behavior discussed above is representative of the less soluble organics in CO_2.

Several generalities can be gleaned from Table 3: (i) for a given homologous series, solubility decreases with increasing molecular weight; (ii) increasing polarity and/or hydrophilicity tends to decrease solubility; and (iii) liquids tend to be more soluble than solutes that are solids at the temperature in question.

2. Regeneration of Activated Carbon

Regeneration of spent activated carbon from wastewater treatment is the first reported environmental application involving supercritical fluids [4,7-9]. Although the process has not yet been used on a commercial scale, it provides a basis for evaluating other applications involving desorption of solutes from solids. In particular, the economics of carbon regeneration have been used to estimate the costs for decontamination of soil with SCF CO_2.

A flowsheet for the process is shown in Fig. 3. The SCF CO_2 circulates through a recycle loop: first through a desorption vessel, where it picks up solute, then through an expander or valve to reduce pressure, and on to a heat exchanger to bring the temperature to the conditions required for solute precipitation. The solute is recovered in a separator and the solvent is brought back to the conditions required for desorption by heat exchange and recompression.

In the semi-batch mode of operation shown in Fig. 3, adsorption and desorption are carried out in different vessels. (Adsorbers are not shown in Fig. 3.) Three high-pressure vessels are used for desorption. At any one time, two vessels are offline for loading and unloading of activated carbon and carbon dioxide, while the other vessel is online undergoing desorption. In this manner, the SCF recycle loop is operated continuously.

Economic estimates for SCF CO_2 regeneration of activated carbon are given in Table 4. Two columns of annual cost estimates are given: the first, for a throughput of 1,650 ton/yr, were published by deFilippi, _et_

FIGURE 3. SUPERCRITICAL FLUID EXTRACTION OF SOLUTES FROM SOLIDS

(Source: Illinois Water Treatment Co.)

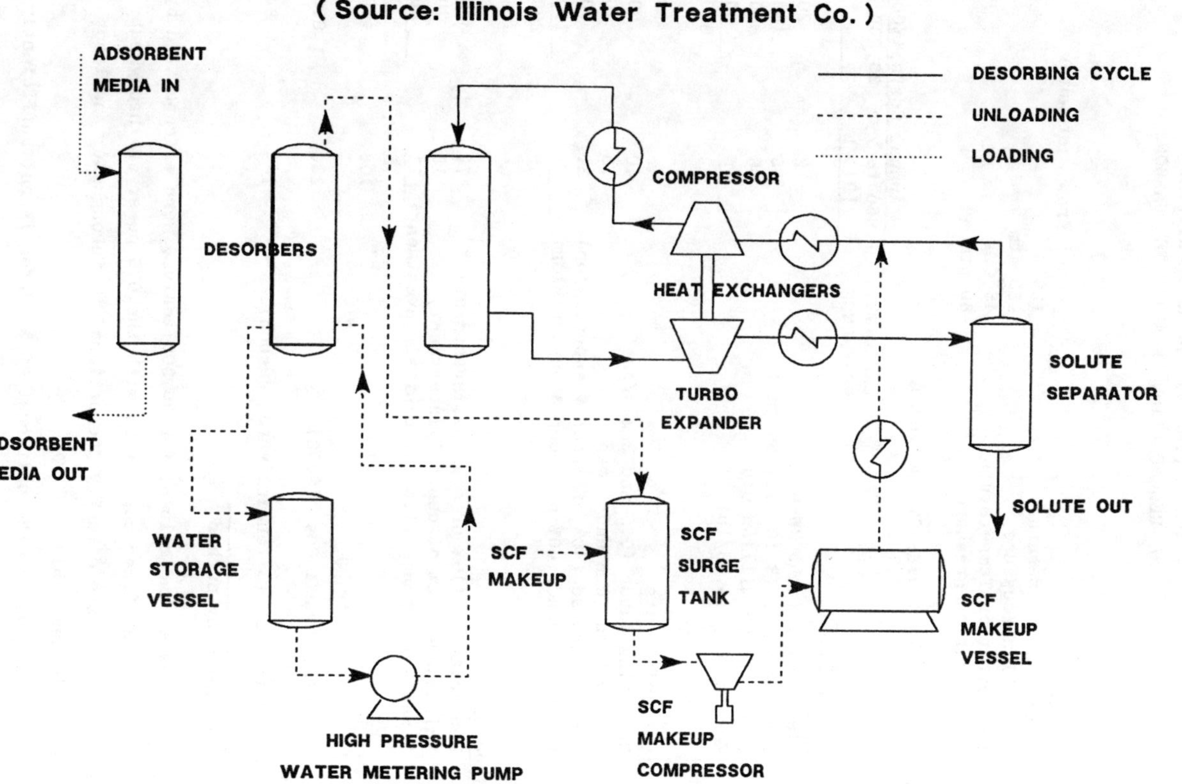

TABLE 4

COST ESTIMATES FOR SUPERCRITICAL
CO_2 REGENERATION OF ACTIVATED CARBON

BASIS [9]

Solute	phenol from wastewater
Desorption Temperature	$120^{\circ}C$
Desorption Pressure	150 atm
Precipitation Temperature . . .	$45^{\circ}C$
Precipitation Pressure	80 atm

ESTIMATED COSTS (1978–1979 dollars)

	Annual Costs ($000)	
CAPACITY:	1,650 ton/yr	20,000 ton/yr
REFERENCE:	[9]	*
Variable Costs		
Electricity ($.03/kW–hr)	2	20
Cooling water ($.10/million gal)	15	180
Steam ($3.50/million Btu)	50	600
Make-up CO_2 ($.03/lb)	5	60
	72	860
Semivariable Costs		
Operating labor (0.5 and 2 man/shift)	40	160
Supervision (0.5 and 1 man)	13	20
Labor overhead (40% of labor & supervision)	21	70
Plant overhead (60% of labor & supervision)	31	110
	65	200
Fixed Costs		
Depreciation (10% of capital investment)	77	340
Maintenance (2% of capital investment)	15	70
Taxes & insurance (1.5% of capital investment)	11	50
	103	460
Total Annual Cost ($000)	280	1,690
Unit Cost ($/ton activated carbon)	169	84

* Estimates by author

al. [9]; the second column, for 20,000 ton/yr, were extrapolated by the author assuming fixed costs vary with the 0.6 power of throughput, variable costs are directly proportional to throughput, and labor varies as shown in the table.

At 1,650 ton/yr, the unit cost of $169/ton is not sufficiently attractive to displace thermal regeneration of carbon. However, Table 4 indicates that there are large economies of scale to be had at higher throughputs.

Extensive bench scale studies [9] showed that strongly bound adsorbates could not be removed by SCF regeneration and that, depending on the solute and carbon, from 10 to 30% of the adsorbate might fall in that category. Physically adsorbed species were removed very rapidly and completely; so rapidly, that the column approached local equilibrium behavior [10].

As long as the regenerated carbon is used to adsorb the same species in subsequent cycles, the presence of the residual adsorbate does not alter the quality of the treated effluent (although it does reduce the capacity compared to virgin carbon). On the other hand, this phenomena restricts the potential applications of the process. For example, if the solute varies from day to day, it is conceivable that a more strongly adsorbable species could displace the residual solute that is not removed during regeneration. Thus, the process is not appropriate for use in centralized regeneration facilities, which are the ones that tend to operate at high levels of throughput where the economies of scale would be most beneficial.

3. Decontamination of Soils

The semibatch process illustrated in Fig. 3 could be applied quite generally for removal of contaminants from solids. In particular, there is interest in removing organic contaminants from soils. Knopf and co-workers at Louisiana State University have studied extraction of polychlorinated biphenyls (PCBs), DDT and toxaphene from contaminated topsoils and subsoils [11,12]. CO_2 extraction of wet subsoil (20% water) contaminated with 1,000 ppm PCBs was effective at 40°C and 10 MPa: over 90% of the solute was extracted in the order of 10 min at a space velocity of 0.07 g CO_2/g soil/s [11].

Topsoil contaminated with DDT (1,000 ppm) and toxaphene (400 ppm) could be decontaminated to about 70% of the adsorbates within 10 min, but the other 30% did not respond to longer extraction time. The authors suggested that the residual might be strongly adsorbed. In a recent publication, the same authors subsequently showed that over 95% of adsorbed DDT could be removed within 5 min at the same conditions by using a supercritical solvent mixture of 95 wt% CO_2 and 5 wt% methanol [12].

The cost of soil decontamination can be estimated, to a first approximation, by analogy to activated carbon regeneration. For the

20,000 ton/yr case given in Table 4, there are three factors which have
major effect on cost: (i) the operating pressure impacts the fixed costs
through the weight of vessels and the like; (ii) the supercritical fluid
recirculation rate (wt of solvent/wt of solid, which is proportional to
space velocity x desorption cycle time) influences the capital cost of
compressor and heat exchangers as well as the operating costs, which are
directly proportional to SCF recirculation rate; and (iii) the temperature
of desorption, impacts the heat exchanger capital cost and the heating and
cooling costs.

The temperature and pressure used for the activated carbon
regeneration process (see Table 4) are both more severe than the
conditions used by Knopf for soil decontamination. The space velocities
and desorption cycle times are the same order of magniture for the two
applications. Thus, the costs shown in Table 4 can be viewed as
conservative estimates of the cost of decontaminating soil by
supercritical fluid treatment. Since 20,000 ton/yr is a throughput
equivalent to a small remedial action site in the U.S. (typically 50,000
to 100,000 total tons of contaminated soil), the estimated cost of $84/ton
may be viewed as an upper limit. Soil decontamination costs below
$100/ton are viewed as extremely attractive; the cost of burial of
contaminated soils in Class I landfills exceed $200/ton, excluding
transportation, in the U.S. Thus, SCF decontamination appears to be
extremely attractive and potentially represents a breakthrough if these
preliminary economics can be confirmed by subsequent evaluation and
development.

4. Removal of Organics from Aqueous Wastes

All organics are not highly soluble in supercritical or near critical
CO_2. As a general rule, hydrophobic compounds are more soluble than
hydrophilic materials (e.g., compare alcohols with paraffins or anilines
with aromatics in Table 3). However, toxic compounds tend to be
hydrophobic. Thus, the processing concept is to extract toxic organic
contaminants from aqueous wastes and, if necessary, subsequently treat the
dilute aqueous effluent by biological oxidation or activated carbon to
remove the remaining organics.

Most of the research and development of this process has been done by
Critical Fluid Systems (CFS), Inc. [13-16]. Several pilot plants have

been build and the first commercial unit is under construction. Although the design details have not been made public, it is presumed that the commercial units will be of a design similar to that used for pilot tests, a schematic for which is given in Fig. 4 [15]. The aqueous waste is fed to an extraction column, which is under pressure and fed countercurrently with pressurized CO_2. The CO_2 is a near critical liquid if the extractor is operated at room temperature, as is the case for the process illustrated in Fig. 4. The treated aqueous effluent exits the bottom of the extractor and may require post treatment, as discussed above.

The CO_2 extract, containing organic and some water, is depressurized (to the pressure chosen for distillation) and fed to a decanter where an aqueous liquid phase may form and, if so, is removed. The CO_2 fluid phase is then passed to a distillation column where purified CO_2 vapor is removed overhead while a condensed organic phase (and possibly some water) is recovered from the reboiler. The purified CO_2 is recompressed to the extractor pressure, condensed in the reboiler and then recycled to the extraction column. The extracted solutes are depressurized to remove residual CO_2 and this concentrate is then detoxified (by, e.g., incineration).

The cost breakdown for treating aqueous wastes by CO_2 extraction has not been made public. However, CFS has published costs for extracting acetic acid from an aqueous phase [14] and those costs have been used herein as a guide to estimate costs for a hazardous waste application.

The costs are given in Table 5. The acetic acid extraction was evaluated as a separation step in acetic acid manufacture; the 20 wt% feed and 1 wt% effluent concentrations are out of the range of consideration in waste applications. On the other hand, a 45-tray extractor is not unreasonable for aqueous waste treatment.

With the exception of supervision and labor overhead, the costs for the acetic acid case were published by CFS [14]. The total annual cost of $1.9 million and unit cost of $0.04/gal were computed for an aqueous feed rate of 200,000 ton/yr which is equivalent to about 100 gal/min.

The acetic acid costs were extrapolated to the extraction of an aqueous waste from acrylonitrile manufacture. The conditions of extraction and separation as well as the variable costs were published by CFS following laboratory investigation [15]. Note that an 86-tray column is required for 99.9% removal of organics (2900 ppm influent to 4 ppm

FIGURE 4. EXTRACTION OF SOLUTES FROM AQUEOUS STREAMS WITH SCF CO$_2$
(SOURCE: Critical Fluid Systems, Inc.)

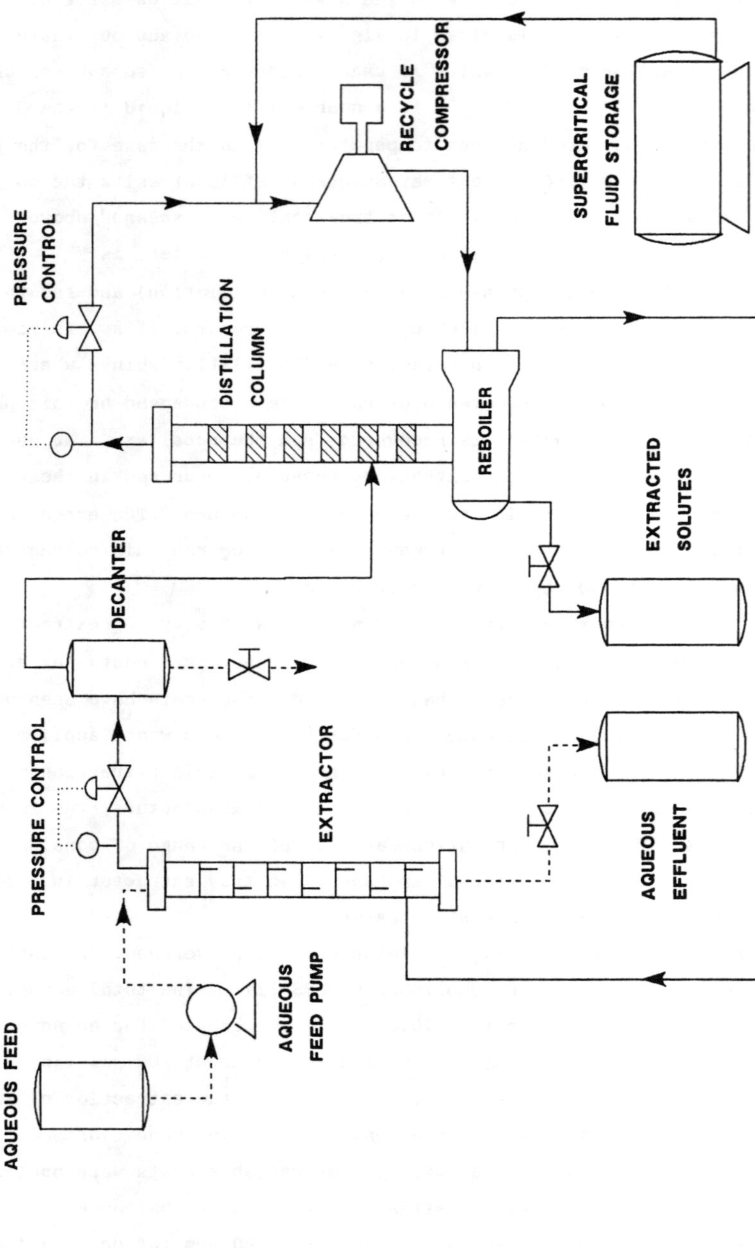

TABLE 5

COST ESTIMATES FOR CARBON DIOXIDE
EXTRACTION OF ORGANICS FROM AQUEOUS WASTES

WASTE COMPONENTS	Acetic acid	Acrylonitrile Acetonitrile
REFERENCE	[14]	[15]
BASIS		
Operating hours	8,000 hr/yr	8,000 hr/yr
Feed rate	200,000 ton/yr	40,000 ton/yr
	100 gpm	20 gpm
Feed concentration	20 wt%	2,900 ppm
Effluent concentration . . .	1 wt%	4 ppm
Solvent:feed ratio	2.0	1.5
Solvent flow rate	100,000 lb/hr	15,000 lb/hr
Cosolvent	Caprylic acid	none
Cosolvent loading	25 wt%	
Extractor temperature		$27^{\circ}C$
Extractor pressure		75 atm
No. of extractor trays . . .	45	86
Distillation temperature . .		$16^{\circ}C$
Distillation pressure		51 atm

ESTIMATED ANNUAL COSTS ($000)

Variable Costs		
Electricity	200	60
Steam	270	10
Water	20	20
Solvent make-up	40	10
	530	100
Semivariable Costs		
Labor	120	100
Supervision	20 *	20 *
Overhead	140 *	120 *
	280	240 *
Fixed Costs		
Depreciation	680	340 *
Maintenance	270	140 *
Taxes & insurance	100	50 *
	1,050	530 *
Total Annual Cost ($000)	1,860	870 *
Unit Cost ($/gal waste)	0.039	0.092 *

---

* Estimates by author

effluent). The semivariable and fixed costs were estimated by the author. The total annual cost of $900,000 and the unit cost of $0.09/gal are quite attractive, even for the relatively small feed rate of 20 gpm. (A typical world-class acrylonitrile plant will generate from 100 to 400 gpm.) Note that the fixed cost is 60% of the total; thus, the unit cost should decrease appreciably with increasing throughput.

The applicability of CO_2 extraction depends upon the type of organic present in the waste. The pertinent design parameter is the distribution coefficient of organic between CO_2 and water phases. Table 6 is a compilation from data published by CFS for organics in CO_2 with and without cosolvent [16]. Note that the distribution coefficient increases with increasing hydrophobicity. For the acetic acid separation described in Table 5 (25 wt% capyrlic acid cosolvent), the distribution coefficient was probably close to 0.5; for the acrylonitrile/acetonitrile case, the limiting solute was acetonitrile, which has the lower distribution coefficient of 0.75. Even for these relatively hydrophilic organics, CO_2 extraction is economically viable.

The last column of Table 6 is the calculated removal efficiency for a 25-tray extraction column. The effluent concentration given in Table 6 corresponds to an influent of 1 wt%. It can be seen from these results, taken together with the estimated costs given in Table 5, that CO_2 extraction of aqueous wastes may have broad applicability.

SUPERCRITICAL WATER OXIDATION

1. Properties of Supercritical Water

Above the critical temperature and pressure, the properties of water are quite different from that of the normal liquid or atmospheric steam. The critical point (C.P.), which lies on the vapor-liquid saturation dome, occurs at $374^{o}C$, 22.1 MPa and 0.3 g/cm^3.

Properties as a function of temperature at a supercritical pressure of 25 MPa are shown in Fig. 5. The density decreases slowly from room temperature to $200^{o}C$ and somewhat faster up to the supercritical fluid region. Above $380^{o}C$, the density decreases very rapidly with temperature. Beyond $410^{o}C$, the rate of change decreases again. The solubilities of benzene and heptane in water are shown in Fig. 5 [17,18].

FIGURE 5.
PROPERTIES OF WATER AT 25 MPa

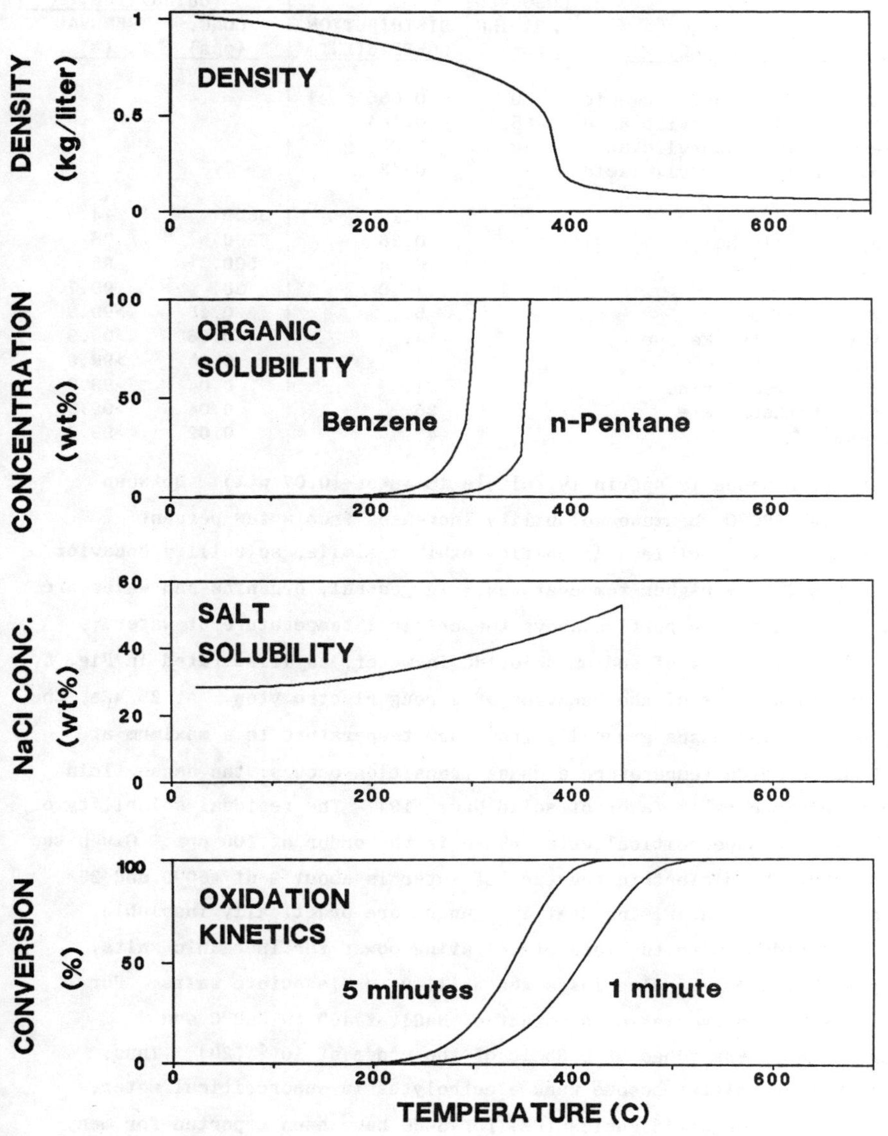

TABLE 6

DISTRIBUTION COEFFICIENTS AND EFFICIENCIES OF
CARBON DIOXIDE EXTRACTION OF ORGANICS FROM WATER [15,16]

SOLUTE	COSOLVENT	COSOLVENT LOADING (wt%)	DISTRIBUTION COEFFICIENT	EFFLUENT FROM A 25 TRAY EXTRACTOR (calculated)	
				CONC. (ppm)	REMOVAL (%)
Acetic acid	caprylic acid	0	0.066		
Acetic acid	caprylic acid	5	0.104		
Acetic acid	caprylic acid	10	0.35		
Acetic acid	caprylic acid	20	0.43		
Methanol			0.10	8600.	14
Isopropyl alcohol			0.26	6200.	38
Acetonitrile			0.75	500.	95
Acetone			1.10	30.	99.7
Acrylonitrile			5.	0.16	>99.9
Methyl isobutyl ketone			14.	0.06	>99.9
Trichloroethylene			17.	0.05	>99.9
Carbon Tetrachloride			21.	0.04	>99.9
Methyl Methacrylate			25.	0.04	>99.9
Xylene			45.	0.02	>99.9

At 25°C, benzene is sparingly soluble in water (0.07 wt%). Between
260°C and 300°C, benzene solubility increases from a few percent
to completely miscible. Aliphatics exhibit similar solubility behavior,
but at slightly higher temperatures. In general, organics and water are
miscible in all properties above the critical temperature of water.

The solubility of sodium chloride in water, as illustrated in Fig. 5,
is representative of the behavior of strong electrolytes. At 25 MPa, the
solubility increases gradually from room temperature to a maximum at
450°C, at which temperature a phase transition occurs: the dense fluid
phase disappears in favor of solid NaCl [19]. The residual solubility of
NaCl in the supercritical water phase is the order of 100 ppm. Given the
fact that the dielectric constant of water is about 2 at 450°C and 25
MPa, it is not surprising that inorganics are practically insoluble.

Coincident with the loss of solvating power for inorganic salts,
supercritical water also loses the ability to dissociate salts. For
example, the dissociation constant of NaCl at 400 to 500°C and
densities in the range of 0.35 is of the order of 10^{-4} [20]. Thus,
strong electrolytes become weak electrolytes in supercritical water.

Destruction efficiencies (DE) for SCWO have been reported for many
organics, a compilation of which is given in Table 7. These data were

TABLE 7

DESTRUCTION EFFICIENCIES OF ORGANICS
BY SUPERCRITICAL WATER OXIDATION

CLASS/ COMPOUND	TEMPER- ATURE (C)	RESIDENCE TIME (min)	DESTRUCTION EFFICIENCY (%)	REF.
Aliphatic Hydrocarbons				
Cyclohexane	445	7.	99.97	[21]
Aromatic Hydrocarbons				
Biphenyl	450	7.	99.97	[21]
o-Xylene	495	3.6	99.93	[22]
Halogenated Aliphatics				
1,1,1-Trichloroethane	495	3.6	99.99	[22]
1,2-Ethylene dichloride	495	3.6	99.99	[22]
1,1,2,2-tetrachloroethylene	495	3.6	99.99	[22]
Halogenated Aromatics				
o-chlorotoluene	495	3.6	99.99	[22]
Hexachlorocyclopentadiene	488	3.5	99.99	[22]
1,2,4-Trichlorobenzene	495	3.6	99.99	[22]
4,4-Dichlorobiphenyl	500	4.4	99.993	[22]
DDT	505	3.7	99.997	[22]
PCB 1234	510	3.7	99.99	[22]
PCB 1254	510	3.7	99.99	[22]
Oxygenated Compounds				
Methyl ethyl ketone	460	3.2	99.96	[21]
Methyl ethyl ketone	505	3.7	99.993	[22]
Dextrose	440	7.	99.6	[21]
Organic Nitrogen Compounds				
2,4-Dinitrotoluene	457	0.5	99.7	[23]
2,4-Dinitrotoluene	513	0.5	99.992	[23]
2,4-Dinitrotoluene	574	0.5	99.9998	[23]

fitted with first order kinetics to generate the oxidation conversion curves shown in Fig. 5.

In general, DE's of the order of 99 to 99.9% can be obtained at 400 to 500°C with 1 to 5 min residence time; 99.99% at 500 to 550°C with about 1 min; and 99.999% at 550 to 600°C with less than 1 min. It has also been stated that 99.9999% is achievable at 600 to 650°C with a residence time of seconds [24], but such data have not been published.

2. The Modar Process

This process has been under development since 1980. As practiced by Modar, it is intended for use with moderately concentrated aqueous wastes. A schematic for a process for treating an aqueous waste containing 10 wt% organic is given in Fig. 6 [23]. This process consists of the following steps:

1. The waste, as either an aqueous solution or a slurry, is pressurized and delivered to the oxidizer inlet. It is heated to supercritical conditions by direct mixing with recycled reactor effluent.

2. Oxygen is supplied in the form of compressed air, which is used as the motive fluid in an eductor to provide recycle of a portion of the reactor effluent. This inlet mixture is then a homogeneous phase of air, organics and supercritical water.

3. The organics are oxidized in a controlled but rapid reaction. The heat released by combustion of readily oxidized components is sufficient to raise the fluid phase to temperatures at which all organics are oxidized rapidly.

4. The effluent from the oxidizer is fed to a cyclone. Inorganic salts that are originally present in the feed or which form in the combustion reactions precipitate out of the fluid phase in the oxidizer and are separated here.

5. The fluid effluent of the solid separator is a mixture of H_2O, N_2, and CO_2. A portion of this is recycled through the eductor to provide supercritical conditions at the oxidizer inlet.

6. The remainder of the effluent is available as a high temperature, high pressure fluid for energy recovery. This stream is cooled to a subcritical temperature in a heat exchanger which serves to generate low pressure or high pressure steam.

7. Now at a subcritical temperature, the mixture has formed two phases and enters a high pressure liquid-vapor separator. Practically all of the N_2 and most of the CO_2 leaves with the gas stream. The liquid consists of water with an appreciable amount of dissolved CO_2.

8. The gas stream can then be expanded through a turbine to extract the available energy as power. A portion of the power is used for compression of the inlet air.

9. The liquid from the high pressure separator is depressurized and fed to a low pressure separator. The vapor stream is primarily

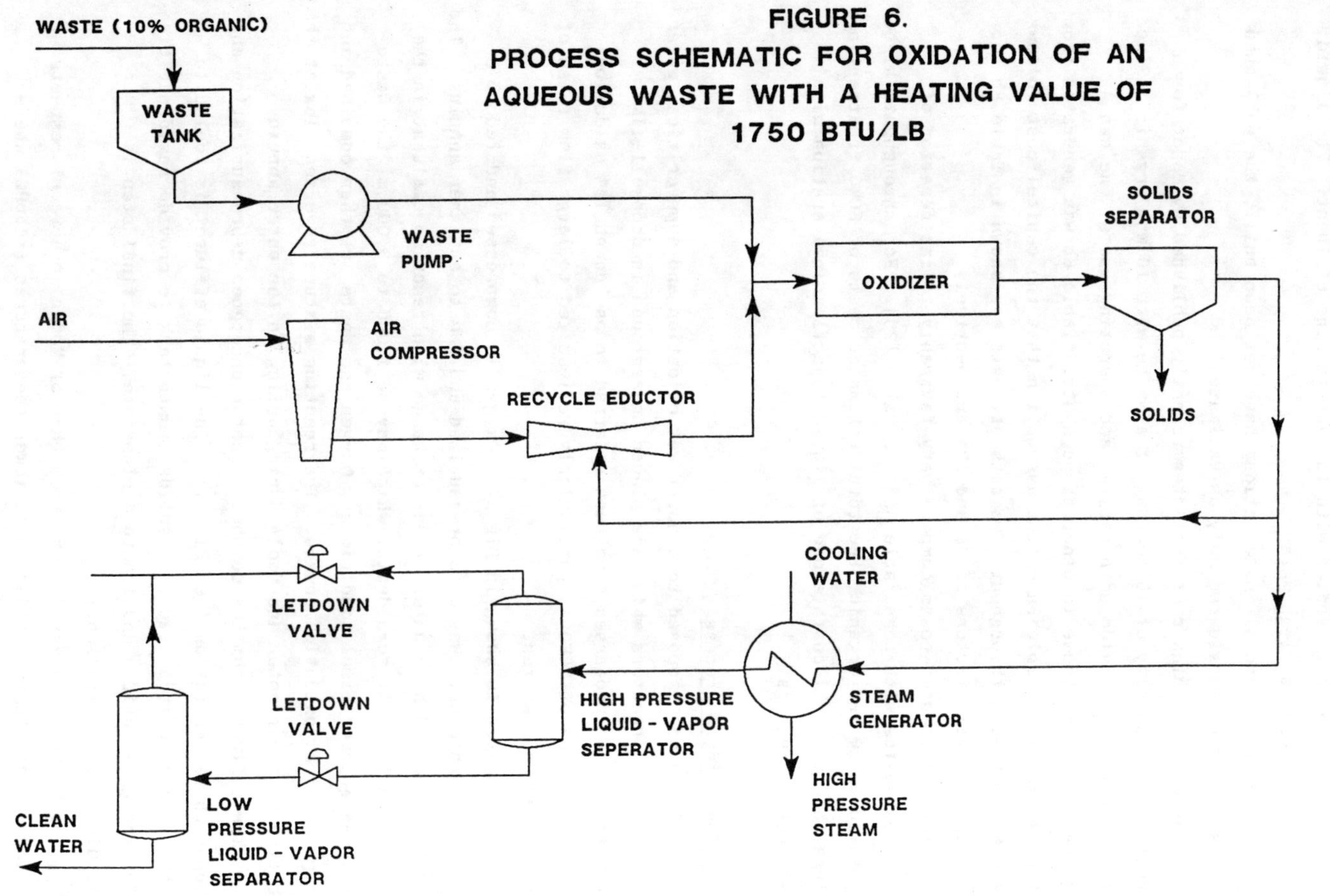

FIGURE 6.

PROCESS SCHEMATIC FOR OXIDATION OF AN AQUEOUS WASTE WITH A HEATING VALUE OF 1750 BTU/LB

CO_2 which is vented with the gas turbine effluent. The liquid stream is clean water.

Since no commercial installations have yet been built, the published costs should be considered only as estimates.

Table 8a contains cost estimates recently published by Modar for a unit with a capacity of 10 ton/day of a waste with 10 wt% organic. Since these estimates provide both capital and operating costs, one can extrapolate from these to other throughputs. Table 8b was generated from the estimates of Table 8a on the assumption that the capital cost follows a 0.6 power with throughput. Two sets of costs are shown in Table 8b for capital recovery factors of 25 and 12%, respectively.

These estimated costs compare very favorably to incineration of aqueous wastes with comparable heating value [25]. For throughputs higher than 10 to 20 TPD organic, the total estimated costs of SCWO treatment are less than the fuel costs alone of liquid injection incineration with effluent scrubbers.

3. The Oxidyne Process

Oxidyne has proposed to conduct wet oxidation and supercritical water oxidation in reactors which are placed underground in deepwell-like cavities. The processes have been referred to as "downhole" oxidation. This variation of SCWO is particularly well-suited to large flow rates of dilute aqueous wastes.

A schematic is shown in Fig. 7. Oxygen is compressed and fed to a central downcomer. Waste is pressurized and fed to an inner annulus. The waste is heated by countercurrent exchange with reaction effluent in the outer annulus. At some depth, which may be 1,000 to 3,000 m, the waste reaches supercritical conditions. Oxygen and waste are then combined and supercritical oxidation ensues. The reacting mixture reverses flow at the bottom of the reactor and cools during upflow in the outer annulus.

The effluent from the downhole reactor undergoes depressurization and vapor phase separation (see Fig. 7). The liquid effluent is cooled to ambient and recirculated in a solids growth tank to provide residence time for inorganic oxides and ash to agglomerate. The final step is liquid-solid separation.

The Oxidyne process differs from that of Modar in several respects. Oxidyne conducts the oxidation at lower temperatures ($<450^oC$) where

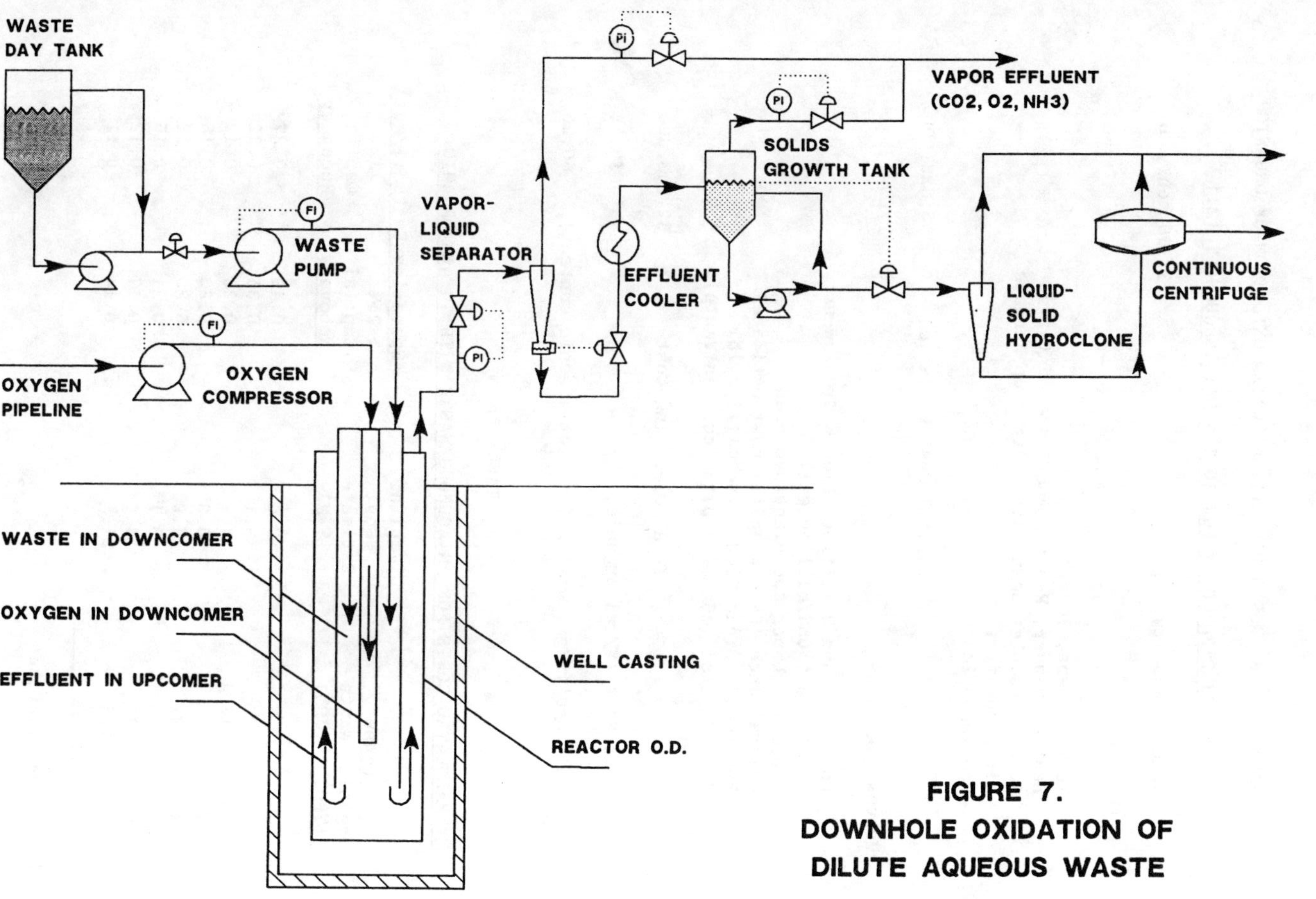

FIGURE 7.
DOWNHOLE OXIDATION OF
DILUTE AQUEOUS WASTE

TABLE 8a

COST ESTIMATES FOR SUPERCRITICAL WATER OXIDATION OF ORGANICS

PROJECTED COSTS FOR 10 TON/DAY THROUGHPUT [21]

Annual capacity . . . .   3,300 ton/year

Capital Costs

    Tanks, Pumps, Eductor
    Heat Exchanger, Reactor and Instrumentation
    Separators, Let-Down Units, Start-up Furnace
    Air Compressor
    Contingencies

                Capital Cost Estimate $5,320,000

Processing Costs

    Chemicals and Utilities (Elec @ $.04/kW-hr;
      Water @ $.055/million gal)
    Operating Labor and Plant Overhead
    Maintainance (5% of capital cost estimate)
    By-Product Steam Credit ($6/million lb)
    Cost Capital (25% of capital cost estimate)

                Total Operating Cost $2,143,000

Unit Cost ($/gal organic) $2.38

Unit Cost($/gal wastewater; 10% organic content) $0.26

TABLE 8b

EXTRAPOLATED COSTS FOR AN AQUEOUS WASTE WITH 10 wt% ORGANIC[*]

| Throughput | | Capital Cost Estimate ($ M) | Operating Costs ($/gal) | |
Organic (TPD)	Waste (GPM)		25% Capital Recovery	12% Capital Recovery
2	3	2.0	0.43	0.27
4	7	3.1	0.35	0.22
8	14	4.7	0.28	0.19
10	17	5.3	0.26	0.18
12	21	5.9	0.25	0.17
20	35	8.1	0.21	0.15
50	87	14.	0.17	0.12
100	174	21.	0.14	0.11

* Estimates by author

inorganic salts do not precipitate out in the reactor. As a consequence of the lower temperature, a longer residence time is required. The downhole process might be expected to require 1 to 5 min at 400 to 450°C to reach 99 to 99.9% destruction efficiency.

The downhole reactor is, in reality, a combined heat exchanger and reactor. The heat exchanger area provided by 1,000 to 3,000 m of pipe, combined with the excellent insulation provided by the earth, results in extremely efficient energy utilization. Thus, Oxidyne projects that the downhole process can be applied to aqueous wastes with as little as 1 wt% organic [26].

Processing cost projections for the downhole process are shown in Fig. 8 [27]. Note that the throughputs for the downhole process are considerably higher than those for the surface process, given in Table 8. For 100 gpm, where the two costs projections overlap, both are in the range of $.10 to .15 per gallon. At throughputs above 200 gpm, Oxidyne projects costs below $.10/gal, and possibly as low as $.07/gal. Such costs are not much higher than the present cost of deepwell injection. Thus, downhole SCWO presents the promise of an oxidative destruction at costs that are comparable to that of current disposal practice.

FIGURE 8.

PROCESSING COSTS FOR SUPERCRITICAL WATER OXIDATION OF AQUEOUS WASTES (Source: Oxidyne, Inc.)

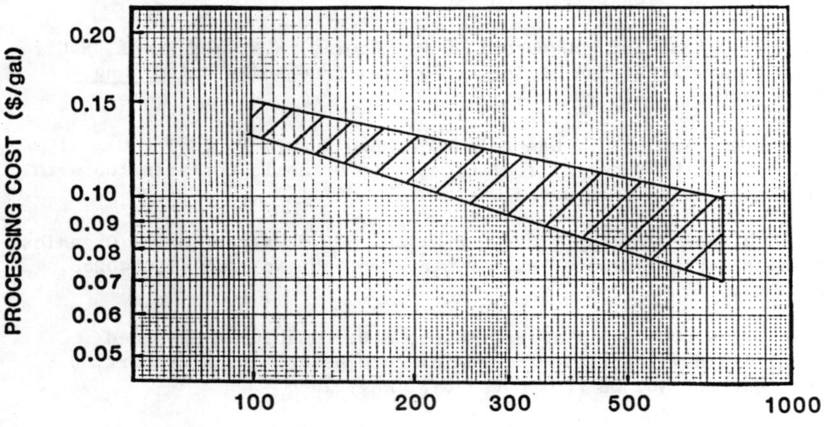

REFERENCES

1. McHugh, M.A. and Krukonis, V.J., _Supercritical Fluid Extraction; Principles and Practice_, Butterworths, London, 1986.

2. Paulaitis, M.E., Krukonis, V.J., Kurnik, R.T. and Reid, R.C., Supercritical Fluid Extraction. _Rev. Chem. Eng._, 1983, $\underline{1}$, 179.

3. Paulaitis, M.E., Penninger, J.M.L., Gray, R.D. and Davidson, P., editors, _Chemical Engineering at Supercritical Conditions_, Ann Arbor Science, Ann Arbor, 1983.

4. Modell, M., Robey, R.J., Krukonis, V.J., deFilippi, R.P. and Oestreich, D., Supercritical Fluid Regeneration of Activated Carbon. At AIChE National Meeting, Boston, Aug. 21, 1979.

5. Modell, M., Wastewater Treatment with Desorbing of an Adsorbate from an Adsorbent with a Solvent in the Near Critical State. U.S. Patent No. 4,147,624, issued Apr. 3, 1979.

6. Francis, A.W., Ternary Systems of Liquid Carbon Dioxide. _J. Phys. Chem._, 1954, $\underline{58}$, 1099.

7. Modell, M., deFilippi, R.P. and Krukonis, V.J., Regeneration of Activated Carbon with Supercritical Carbon Dioxide. In _Activated Carbon Adsorption of Organics from the Aqueous Phase_, 1980, $\underline{1}$, pp. 447-461.

8. Modell, M., Process for Regenerating Adsorbents with Supercritical Fluids. U.S. Patent No. 4,124,528, issued Nov. 7, 1978.

9. deFilippi, R.P., Krukonis, V.J., Robey, R.J. and Modell, M., Supercritical Fluid Regeneration of Activated Carbon for Adsorption of Pesticides. Final Report, U.S. E.P.A. Grant No. R804554, March, 1980.

10. Sherwood, T.K., Pigford, R.L. and Wilke, C.R., _Mass Transfer_, McGraw-Hill, New York, 1975, Ch. 10.

11. Brady, B.O., Kao, C.C., Gambrell, R.P., Dooley, K.M. and Knopf, F.C., Supercritical Extraction of Toxic Organics from Soils. _Ind. Eng. Chem. Research_, 1987, $\underline{26}$, 261.

12. Dooley, K.M., Kao, C.C., Gambrell, R.P. and Knopf, F.C., The Use of Entrainers in the Supercritical Extraction of Soils Contaminated with Hazardous Organics. In press.

13. Moses, J.M. and deFilippi, R.P., Critical-Fluid Extraction of Organics from Water. Final Report, U.S. D.O.E. Contract No. AC01-79CS40258, June, 1984.

14. Kingsley, G.S., Chung, M.E. and Moses, J.M., Cosolvent-Enhanced Critical-Fluid Extraction of Organics from Water. Final Report, U.S. D.O.E. Contract No. DE-FC01-84CE40673, May, 1986.

15. Rice, P.N., McGovern, W.E. and Kingsley, G.S., Innovative Methods for the Treatment of Hazardous Aqueous Waste Streams. Final Report, U.S. E.P.A. Contract No. 68-03-1956, May, 1986.

16. McGowen, W.E. and Moses, J.M., The Treatment of Solvent Contaminated Waste Using Liquefied-Gas Extraction. At Third Annual Hazardous Waste Source Reduction Conference, Worcester, MA, Oct. 23, 1986.

17. Connelly, J.F., Solubility of Hydrocarbons in Water Near the Critical Solution Temperature. J. Chem. Eng. Data, 1966, 11, 13.

18. Rebert, C.J. and W.B. Kay, The Phase Behavior and Solubility Relations of the Benzene-Water System. AIChE Journal, 1959, 5, 285.

19. Bowers, T.S. and Helgeson, H.C., Calculation of the Thermodynamic and Geochemical Consequences of Nonideal Mixing in the System H_2O-CO_2-NaCl on Phase Relations in Geologic Systems: Equation of State for H_2O-CO_2-NaCl Fluids at High Pressures and Temperatures. Geochimica et Cosmochimia Acta, 1983, 47, pp. 1247-75.

20. Marshall, W.L., Predicting Conductance and Equilibrium Behavior of Aqueous Electrolytes at High Temperatures and Pressures. In High Temperature, High Pressure Electrochemistry in Aqueous Solutions (NACE-4), 1987, p. 117.

21. MODAR, Inc., Detoxification and Disposal of Hazardous Organic Chemicals by Processing in Supercritical Water. Final Report, U.S.Army Contract No. DAMD 17-80-C-0078, 1987.

22. Modell, M., Processing Methods for the Oxidation of Organics in Supercritical Water. U.S. Patent No. 4,543,190, issued Sept. 24, 1985.

23. Thomason, T.B. and Modell, M., Supercritical Water Destruction of Aqueous Wastes. Hazardous Waste, 1984, 1, 453.

24. Modell, M. Supercritical Water Treatment of Organic Wastes. At HAZMACON 85 Conference, Oakland, Apr. 25, 1985.

25. Modell, M., Incineration of Low Heat Content Waste. At AIChE National Meeting, Boston, Aug. 27, 1986.

26. Smith, J.M. and Raptis, T.J., Supercritical Deep Well Oxidation of Liquid Organic Wastes. At International Symposium on Subsurface Injection of Liquid Wastes, New Orleans, Mar. 3-5, 1986.

27. Smith, J.M., Hartmann, G.L. and Raptis, T.J., Supercritical Deep Well Oxidation A Low Cost Final Solution. At APCA, Specialty Conference, New Orleans, Dec. 8-12, 1986.

MANAGEMENT OF HAZARDOUS WASTES
IN THE OIL REFINERY INDUSTRY

S. Haverhoek
Shell Internationale Petroleum Maatschappij,
Health, Safety and Environment Division.
The Netherlands

ABSTRACT

Modern oil refineries produce relatively small quantities of hazardous
waste. Adequate effluent treatment facilities and subsequent
destruction of sludges by either incineration or controlled landfarming
are key elements in proper waste management of refineries.
Additionally process integrated abatement technologies are expected to
be increasingly applied aiming at further waste reduction. Any waste
still generated will be rendered innocuous before final controlled
landfill disposal in order to avoid possible future liability.
Soil and groundwater management and control will be an integrated part
of waste management in refineries.

1. INTRODUCTION

Oil has from the outset been seen by industry as a valuable natural
product not to be squandered. Virtually all fractions can be used in a
wide range of applications. Therefore the quantity of waste generated
through petroleum refinery operations has always been limited in
relation to crude oil throughput.
Nonetheless some waste is generated in the form of gaseous, aqueous
liquid and solid materials, some of which are hazardous. Before World
War II these wastes were discarded into the environment without much
attention, due to the general lack of environmental awareness.
However, after the war the demand for oil reached levels at which even
the relatively minor waste quantities started to cause unacceptable
pollution. This initiated the development of additional and improved
treatment and control facilities, recycling and the use of modified
operational procedures to reduce waste generation.

2. SCOPE AND DEFINITION

This paper deals with the development of waste management inside
refineries from World War II, covering the current state of the art in
control, treatment and disposal of refinery hazardous waste. This will
be exemplified on the basis of experience mainly with European
refineries. It should be noted that waste management can vary from
location to location and from country to country due to differences in
local requirements and operative legislation. Also the type (simple,
complex, luboil, petrochemical) and level of modernization of the
refinery determine to a large extent the types and volumes of waste.

In this paper, waste is defined as everything that leaves the plant
which is not a useful product. This definition includes in principle
gaseous, aqueous and solid wastes. It is necessary to consider these
wastes in relation to each other, because treatment of flue gases and
effluents will generally lead to the generation of liquid and/or solid
waste.

Although consensus on the definition of hazardous waste is lacking, in this paper it is defined as everything that leaves the plant which is not a useful product and which has the potential to present a hazard to human health or the environment.

3. REFINERY WASTE SOURCES AND TYPES

Before World War II, oil production and its subsequent processing took place on a much smaller scale compared with today. The only hazardous solid waste which was produced in substantial quantities was acid tar from lubricating oil manufacture. This was usually deposited in lagoons on or adjacent to the refinery site.
Also, smoke and odour as well as discharge of some oil onto surface water were considered inevitable consequences of refining activities. Since 1950 a steep annual growth in oil production and refining has occurred and the relatively small oil discharges in the aquatic environment and the emissions to air started to cause concern.
The acid tar disposal problem and water and air pollution were tackled by the oil industry itself through the development and application of new technologies. This took place without much outside pressure.

The installation of effluent treatment facilities resulted in the production of substantial quantities of waste in the form of sludges, which had to be disposed of. In this case the environmental problem was partly shifted from the water compartment to the soil compartment, as improper disposal methods were often used.
During the 1960's sophisticated refinery installations appeared on the scene such as catcrackers and alkylation units. These plants generated new types of waste, such as spent catalysts and alkylation sludges.
The increased volumes of oil to be processed also led to increased amounts of tank bottom sludges including leaded sludges from gasoline tanks and spent chemical wastes such as acids, caustics, cleaning solvents, scales and sludges. All these wastes had to be disposed of.
Stricter product specifications led to the introduction of filter clays, which also had to be disposed of when saturated.
Recently another waste category has emerged, namely contaminated soil. This occurred as a result of accidental spills and leaks. With advances in scientific understanding and increased public concern about groundwater quality and possible impact of contaminated water on human health, legislation has been developed in North America and some European countries requiring industries to clean-up sites where wastes were buried even though they may have been considered acceptable at the time.
An indicative list of refinery wastes is given in Table 1.

As far as waste management within refineries is concerned, until the 1970's waste was regarded as a _transportation_ problem.
After the introduction of legislation concerning the storage, transport, treatment and disposal of different categories of waste it became an _administrative_ and an organizational problem.
Today, with the broad public and industrial awareness of the environmental consequences of industrial activities, waste is becoming more and more a _technological_ problem in the sense that cleaner technologies are required to minimize waste production or even to avoid the generation of hazardous waste altogether.

TABLE 1

Quantity and composition of refinery sludges

TYPE OF SLUDGES	TYPICAL QUANTITY (t/a)	COMPOSITION (%w)		
		Oil	Water	Solids
tank bottoms	500-3000	40-90	60-10	2-10
desalter bottoms	70	30-40	30-40	20-40
API bottoms	2500-20000	10-40	60-90	5-20
FFU sludges	2500-20000	1-10	80-98	1-10
final treatment sludge				
cake from filtration	3000	10	70	20
centrifuged sludge	2000	8-10	80-90	10-12
biological sludge	3000	0.1-0.5	80-99	1-20
slops and oil spills				
liquid slops	50-100	30-70	70-30	-
grease, wax etc.	100-200	200	-	-
liquid spills	30-70	30-40	-	60-70
asphalt spills	50-300	100	-	-
operating residues				
spent caustic	10000	traces	90	10
lime	5000	-	-	100
cooling tower waste	30-50	1-2	95	3-5
clays	300-1000	30-60	1-5	70-40
TEL/TML sludge	10-250	ppm	1-10	90-lead
other wastes				
gravel, sand etc.	500-1000	0-2	0-10	88-100
oily solids	1-200	10	-	90
general wastes	1-200	0	0	100
cracking fines	100-200	sometimes	-	100

4. HAZARDOUS WASTE CONTROL TECHNOLOGIES

As has already been mentioned, the installation of off-gas and
effluent treatment facilities has resulted in the production of
additional quantities of refinery waste.
On the other hand, waste minimization measures, good housekeeping and
process-integrated abatement technologies have resulted in considerable
reductions in wastes generated by refineries today.

The European hazardous waste legislation usually takes a dual track
approach by including processes and substances.

Refinery processes producing hazardous waste are listed and exemption
is only given on a proven and agreed substance composition.
The following refinery wastes are listed in the US Resource
Conservation and Recovery Act (RCRA) as hazardous:

1. Leaded tank bottoms
2. Neutralized hydrofluoric alkylation sludge
3. Dissolved air flotation (DAF) sludge
4. Kerosine filter clays
5. Luboil filtration clays
6. Slop oil emulsion solids
7. Exchanger bundle cleaning solvent
8. API separator sludge

Dissolved air flotation sludge and API separator sludge are produced
during refinery effluent treatment, and they usually represent more
than 90% of the total amount of hazardous waste generated in the
refinery. Accordingly, waste management in a refinery is dominated by
water management.

4.1 AQUEOUS WASTE CONTROL

Before 1945 refineries were using processes that generated large
amounts of waterborne contaminants per barrel of crude processed. A
large proportion of the total amount of oil lost (oil/water emulsions,
oily sludges, leakage oil) was discharged via the sewer system and it
was therefore obvious that mitigation efforts should initially be
focussed on the effluent treatment and control.
A significant research and development effort devoted to this topic by
organizations such as the American Petroleum Institute (API) resulted
in basic engineering guidelines for waste disposal. The API separator
stems from that work. Fundamental principles such as: "Control
pollution at source" and "Good water management is a key to lower cost
and a cleaner environment" were developed and translated into control
and treatment technologies.
The historic development of refinery effluent treatment technology is
shown in the following table:

Table 2:

1950 API separators (oil/water/sludge separators)
     In-plant controls/segregation of cooling and process water
1960 Aerated lagoons/sour water stripping/surface condensers
     Flocculation/dissolved air flotation/water recycling
1970 Activated sludge process/equalization/further segregation
     Optimization of chemical, physical and
1980 biological treatment processes enforced by legislation
     Tertiary treatment/polishing/Improved monitoring

The annual survey carried out by CONCAWE (report No. 1/86) during 1985 on refinery effluent has confirmed the trend of recent years, showing that Western European refineries are using less water, have installed much-improved water treatment facilities and are discharging less oil in their aqueous effluents than before. The overall effect of these continuing improvements is that, in general, water pollution by oil refinery operation no longer presents an environmental problem.

For the effluent treatment sludges (API, DAF and biotreatment) two basic treatment and disposal options are available: incineration (usually a fluid bed incinerator is employed) and land treatment, which basically consists of a combination of photodegradation, evaporation and microbiological degradation. The choice depends on the type and size of the refinery, the availability of a suitable land treatment area and on operative legislation. Uncontrolled landfill of oily sludges would not be acceptable by today's standards.

Incineration yields a solid waste in the form of fly ash, which can be rendered non-leachable by solidification prior to landfill or can be used in roadbuilding. Moreover, fly-ash can offer some possibilities in the waste-to-waste management concept. For example, scrubbing sludges combined with fly-ash can yield a stabilized product suitable for landfill.

Land treatment has the potential disadvantage of hydrocarbon emissions and possible soil and groundwater contamination by heavy metals and/or migrating hydrocarbons. Hence adequate control and monitoring are required. An example of a design to minimize pollution is shown in Fig. 1.

4.2   SOLID/LIQUID WASTE CONTROL

Solid wastes generated during past and present refinery operations which require attention entail:
Acid tars, treating clays, HF alkylation sludges, spent catalysts, leaded sludges, leak-bitumen, contaminated soils and spent caustic.

4.2.1 <u>Acid tar</u>

Acid tar is a hazardous waste generated in connection with the manufacture of lubricating oils. Oleum (concentrated sulphuric acid) treatment has been widely applied in the past to remove naphthenes and (polycyclic) aromatics from the distillation residues by sulphonation. The acid tar was initially dumped at sea, but after World War I it became practice to dump it in surface impoundments. Many attempts were undertaken to incinerate this waste but these largely proved impracticable due to its extreme corrosive properties and the massive SO_2 emissions which resulted.
In the early 1960's, alternatives for the oleum treatment were developed and the acid tar production has now been virtually eliminated. However, the acid tar ponds stemming mainly from the 1950's and earlier as indicated might present an environmental problem and may require a clean-up effort in the years to come. In-situ stabilization would probably be an acceptable treatment and disposal method.

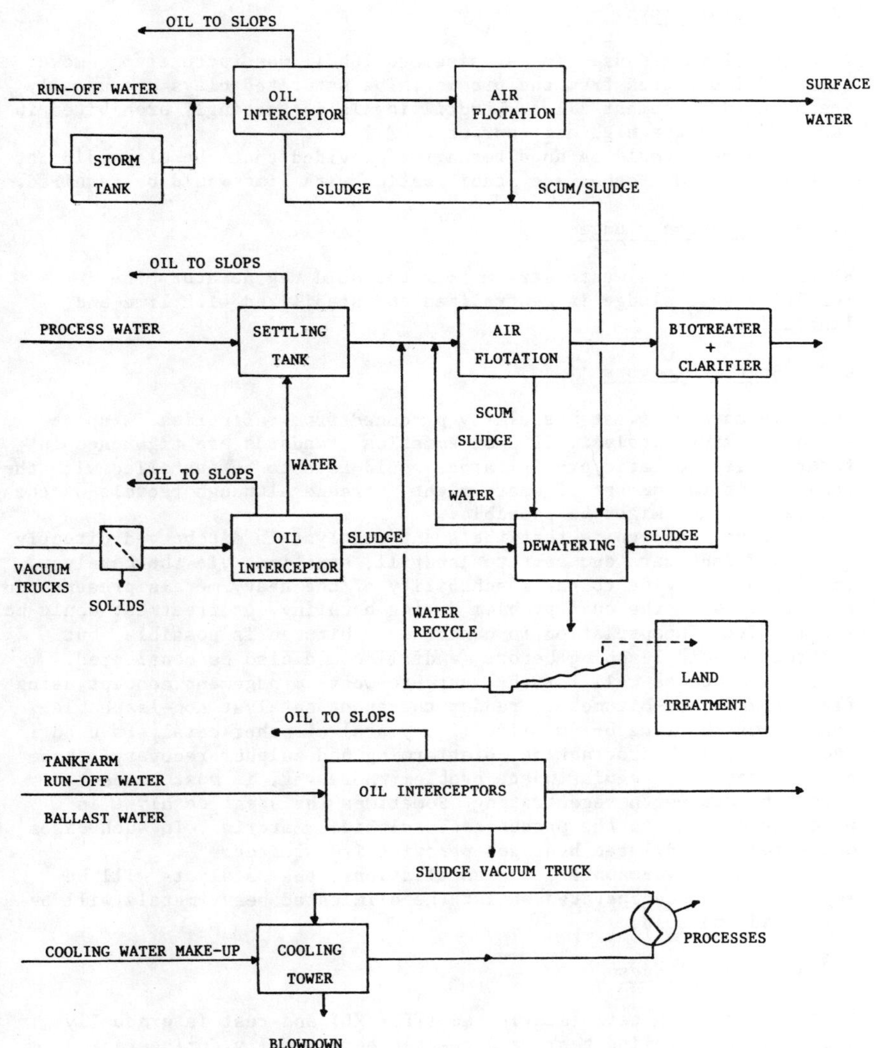

Fig. 1. SIMPLIFIED WATER TREATMENT SCHEME IN A REFINERY WITH LANDTREATMENT OF SLUDGES

4.2.2 <u>Treating clays</u>

Treating clays are used in kerosine and luboil manufacture to remove coloured impurities from the product. The saturated clays are usually incinerated in cement kilns. Landfilling is increasingly prohibited in Europe due to the high oil content (5%).
Land treatment would be an alternative provided that the clays did not contain acid tar, otherwise stabilization with lime would be required.

4.2.3 <u>Alkylation sludge</u>

HF units produce a waste stream from the acid regenerator. The resulting acid sludge is neutralized and stabilized with lime and landfilled.

4.2.4 <u>Spent catalysts</u>

The only catalyst waste regularly produced from refineries is spent fluid cracking catalyst. If dust emission standards are tightened in future, electrostatic precipitators would have to be installed with the result that the amount of waste might increase although recycle of the recovered fines might be possible.
The current practice is that the spent catalyst is discharged directly into road tank cars and sent to landfill, sometimes via the catalyst manufacturer. Owing to the leachability of the heavy metals present and also in view of the dust problem during handling, pretreatment would be recommended. Encapsulation in concrete or bitumen is possible, but controlled acid leaching before landfill could also be considered.
There is scope as well for the waste-to-waste management concept using fly-ash and leak-bitumen to render the spent catalyst non-leachable.
Because of the long useful life (2-5 years) of other catalysts used in hydrotreating, hydrocracking, platforming and sulphur recovery, these do not present a regular waste problem. Moreover, in most cases the catalysts are worth regenerating. Sometimes the spent catalyst is pyrophoric owing to the presence of sulphidic material. In such cases oxidation with diluted hydrogen peroxide is required.
In the latest hydroconversion installations, the catalysts will be rejuvenated or regenerated whilst the eliminated heavy metals will be recovered and sold.

4.2.5 <u>Leaded sludges</u>

Sludge containing tetra-alkyl-lead (TEL/TML) and rust is gradually deposited in gasoline tanks and removed periodically during tank cleaning for inspection and repair.
The handling of toxic sludge and scale requires special care, and is usually performed by a specialized contractor.

The Shell Argentina refinery in Buenos Aires has developed an interesting system, known as the "solar process chamber", in which the toxic TEL sludges are oxidized to the less toxic PbO_2. Subsequently the remaining sludge is stabilized by mixing with lime and cement and then landfilled.
Normal operation can entail collection of sludge in small drums and incineration of the total drum in a rotary kiln toxic waste incinerator equipped with a flue gas scrubbing system.
As lead in gasoline is to be phased out, this topic will disappear in the future.

4.2.6 Leak-bitumen

Some leak-bitumen/asphalt waste is produced due to leaking pumps, spills during sampling, etc. If this material is contaminated with sand and other debris, recycling is not always possible. At present disposal by incineration or landfill is practised.

However, bitumen is an excellent material for immobilizing other hazardous wastes and application in this field is being investigated, thus applying the waste-to-waste management concept.

4.2.7 Contaminated soils

Contaminated soil either from on-site spills or leakages or from pipeline and road tanker accidents is receiving increased attention, partly as a result of developing legislation.

Substantial quantities of contaminated soils can be produced as a result of building activities.

As transport and disposal costs are high, there is an increasing tendency to consider the possibility of controlled on-site land treatment even in regions with less favourable climatic conditions.

4.2.8 Spent chemicals

Spent caustic soda constitutes the main chemical requiring disposal from refinery operations, and there are a number of treatment processes and disposal routes to minimize the quantity involved:

(I) Whenever possible the spent caustic soda is treated and/or recycled for use within the refinery. There are also opportunities for the sale of certain types of spent caustics for (alkyl)phenol recovery.

(II) Injection to crude oil desalters or distillation units for pH and corrosion control is practised, but the quantities that can be disposed of by this route are limited by the necessity to maintain product specification.

(III) The spent caustic soda can be diluted or neutralized with spent acid to achieve an acceptable pH for subsequent effluent treatment.

4.3 OTHER SOURCE CONTROL METHODS

There are considerable environmental and economic incentives for the recovery of oil from refinery wastes for ultimate productive use and to reduce amounts of wastes to be disposed of. The techniques utilized in refineries include the following:

- crude oil storage tanks - sludge control by the use of side-entry mixers
- emulsion breaking by heat treatment/settling or distillation
- oil recovery from sludges by vacuum and/or pressure filtration, by centrifugal separation and by solvent extraction.

Examples of process integrated abatement technologies are:
- in-situ regeneration and recycling of spent catalysts
- replacement of the oleum treatment of base oils by solvent extraction methods
- the slops collection, treatment and recycle system.

5. FINAL DISPOSAL

If, despite attempts to reduce, reclaim and re-use it, there is still
waste to be disposed of there are only two appropriate methods:
1 - landfill after rendering the waste inert
2 - landtreatment, which only applies to oily and biological sludges.

Disposal at sea, mine and deepwell disposal as well as uncontrolled
landfill or dumping are considered generally inappropriate for the
European situation.

6. REFINERY WASTES - QUANTITIES GENERATED

There are only limited data available regarding the amount of refinery
wastes generated in Europe, and a specific study would be required to
obtain more definite information.

A modern refinery with advanced water treatment facilities including
incineration or land treatment typically generates 0.05% of the crude
throughput as solid waste to be finally disposed of in a controlled
landfill. This figure does not include household refuse and building
waste.

7. CONCLUSIONS

1. Owing to the installation of adequate effluent treatment and
   monitoring facilities in refineries, pollution of surface water with
   oil is currently under control.

2. Pollution control measures on air and water emissions have resulted
   in additional quantities of solid wastes to be disposed of.

3. Good water management, good housekeeping and waste minimization
   technologies have led to adequate control and reduction of hazardous
   wastes.

4. Incineration with flue-gas treatment is the most effective but also
   the most costly waste management technology for the elimination and
   destruction of (hazardous) refinery sludges.

5. Land treatment is currently developing towards a mature waste
   destruction technology in terms of optimization and control.

6. Uncontrolled landfill of oily sludges in licensed waste disposal
   sites is being gradually replaced by better waste management
   techniques as mentioned in 4 and 5.

7. The amount of solid waste for final disposal generated by a modern
   refinery might typically constitute 0.05% of the crude throughput
   where incineration and/or land treatment of refinery sludges are
   applied.

8. Although acid tar disposal in lagoons was considered to be an
   acceptable waste disposal practice at the time, there is evidence
   that some of these acid tar lagoons now require clean-up.

8. <u>FUTURE TRENDS</u>

1. There is a trend in refinery waste management towards
   - The gradual termination of inadequate disposal practices
     (landfill of oily sludges)
   - waste reduction by the application of process integrated
     pollution abatement technologies (e.g. solvent extraction
     processes instead of acid treatment of luboils) and recycling
     (e.g. spent catalysts, recovered oil)
   - the treatment of wastes on-site (incineration and landtreatment)
     as far as practicable
   - the rendering inert of any waste leaving the site by the
     application of fixation or encapsulation
   - adequate controls on waste disposal contractors and their
     facilities
   - the adoption of clear administrative procedures for waste
     collection, storage, transport and disposal.

2. Future developments in refinery operation will be aimed at more
   flexibility in terms of feedstock quality and product mix.
   Consequently, more and more sophisticated catalytic processes will
   be used in refineries and a moderate increase in spent catalyst
   quantities is anticipated.
   However, the current trend to regenerate and to rejuvenate spent
   catalysts will continue, thereby reducing the amount of spent
   catalyst to be disposed of.
   For the disposal of spent catalysts, metal leaching specifications
   are expected to be imposed by the authorities which will require
   treatment of the spent catalyst before landfill disposal either by
   fixation or by pre-leaching and recovery of the heavy metals.

3. In general, microbiology will play an increasingly important part
   in waste management within the refinery fence. For example:
   - Biotreatment of process waters
   - Landfarming of sludges and contaminated soils
   - Odour abatement and volatile hydrocarbon emission reduction with
     biofilters, e.g. in tank parks.

4. The in-house application of waste-to-waste technologies will be
   developed in the near future with a view to rendering any waste
   leaving the refinery for ultimate landfill disposal innocuous.
   Fixation/stabilization techniques using fly-ash and bituminous
   wastes might play a part in this.

5. With respect to integrated waste management in refineries
   (including air, water and soil), the cradle-to-grave concept will
   gradually be replaced by the conception-to-exhumation approach.
   Basically, this concept entails the integration of environmental
   aspects of new processes from the early stages of the process
   development onwards, but it also includes a properly designed
   treatment, disposal and monitoring system for any waste generated
   during the manufacturing process. This concept might also include
   waste products which can be recycled to the refinery for upgrading
   and waste treatment in order to reduce environmental impacts to a
   minimum. It also entails proper soil and groundwater management,
   leaving a clean site at the end of the useful life of the refinery.

THE BAVARIAN SYSTEM FOR SPECIAL WASTE MANAGEMENT: 15 YEARS EXPERIENCE IN COLLECTION, TREATMENT, DISPOSAL AND CONTROL - A CASE STUDY

Franz Defregger
Bavarian State Ministry for Regional Development
and Environmental Affairs, Munich
Federal Republic of Germany

ABSTRACT

Bavaria was the first State in the Federal Republic of Germany (FRG) to address central responsibility for the disposal of special wastes. At present 10 regional collecting points and 3 central special-waste disposal facilities are in operation, in which nearly 400 000 tons special waste are treated and disposed properly per year.

The central disposal facilities in Ebenhausen (near Ingolstadt) and Schwabach (near Nürnberg) are one of the most advanced of its physical treatment, incineration (rotary kilns) and secure landfill sites. The Landfill Gallenbach, in operation since 1975, will be also discussed in this paper in view of collection, analysis and treatment of leachate.

The first full-scale distillation unit in the world to purify leachates went in operation in 1986.

INTRODUCTION

The FRG was one of the first European countries to give serious attention to the problem of hazardous waste.
The Federal Waste Disposal Act was not enacted before January 1972.
Today this Waste Disposal Act has been actualizied by four amendments the last of which is just passing the Federal legislative procedure (Nov. 1986).

The Federal law being more or less only a kind of framework and guideline soon was completed by the corresponding laws on State's level.

Although here is not the place to give a more detailed survey about waste legislation in the Federal Republic one hint
may be allowed: The Federal Law together with the State's
legislation makes the towns and counties, i.e. the municipal
sector responsible for the disposal of waste.
There is however one important exception: Those types of
waste that cannot be codisposed with domestic refuse (because of their amount or characteristics) can be excluded from
the disposal obligation (hazardous or special waste).

If the municipalities make use of their exclusive right
– and that is with only few exceptions the regular case –
the waste generator himself automatically becomes responsible for the safe disposal of his waste.

As a consequence of this regulation a legislative "grey
area" remains concerning most of the waste material produced
by industry we are used to call chemical or hazardous waste.
This loophole was intended to be closed by the Waste Disposal Plans, each State is obliged to create for its area
according to article 6 of the Waste Disposal Act.

These Plans should be comprehensive in so far as they should
not only regulate the disposal of domestic but also of hazardous waste. But until today only few of the Federal States have published Waste Disposal Plans that come up to this
idea. Among these the State of Bavaria in South Germany was
the first to enact a Hazardous Waste Disposal Plan (in 1977)
strictly regulating statewide the safe disposal of hazardous
waste by offering industry a dense network of collection
stations and central treatment plants.

DISPOSAL ORGANIZATIONS

Within the system of waste treatment in the State of Bavaria
there has developed an elaborate charge system during its 20
years of operation.

Being the largest of the federal states (70.500 sq.km, nearly 11 million inhabitants), Bavaria has a relatively low
density of population and industry with respect to central
European conditions. Industry, however, does enjoy a comparatively high technological level combined with the near
absence of classic heavy industry. These factors lead to a
peculiar mix of advantages, problems and corresponding solutions.

With the foundation of a municipal cooperative (ZVSMM;
Zweckverband Sondermüllplätze Mittelfranken) working on a
strictly regional basis as early as 1966, and the foundation
of a state wide semipublic organization (GSB; Gesellschaft
zur Beseitigung von Sondermüll in Bayern; Association for
Disposing Special Waste) in 1970, Bavaria represents the
forerunner in hazardous waste treatment pracitices in Ger-

106

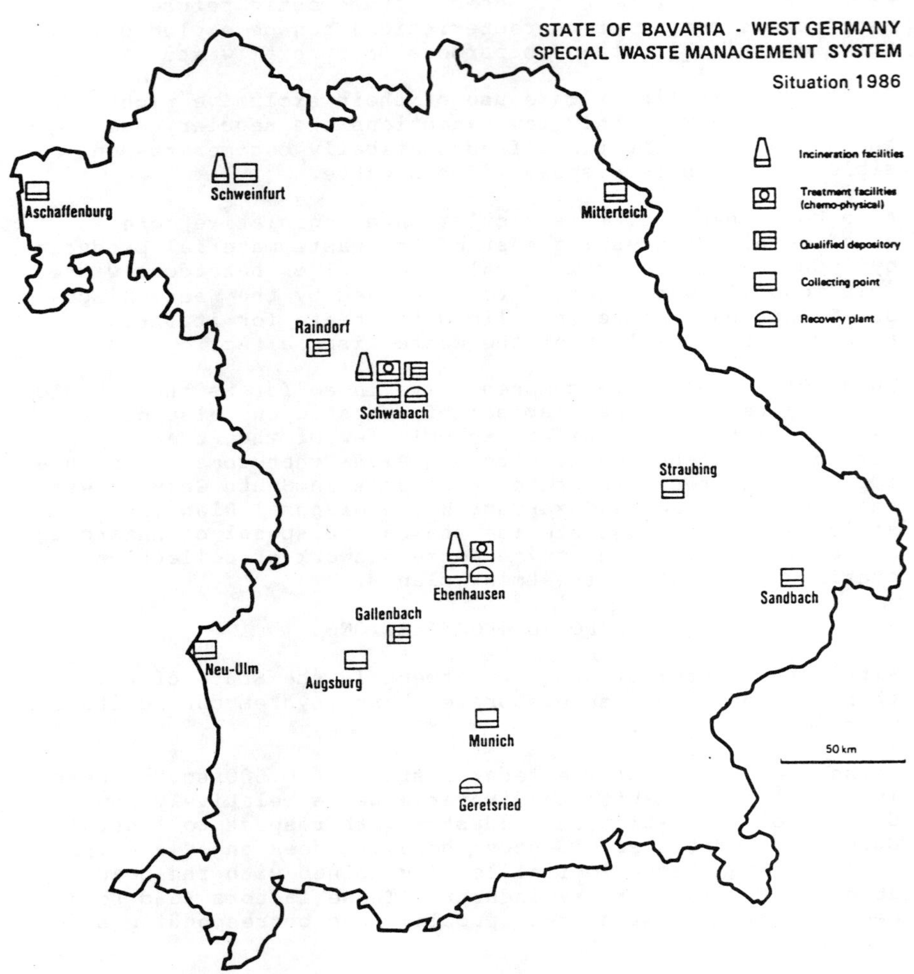

FIGURE 1
Special Waste Management System of Bavaria

many and beyond its frontiers, and was able, therefore, to shape federal legislation by presenting practicable solutions.

At present are in operation (see Figure 1):

- 10 regional collecting stations
- 3 central facilities (incineration, chemo-physical treatment, landfill)
- 1 recovery plant for contamindated solvents.

With the exception of the treatment plant of Schwabach (ZVSMM) all other facilities and collecting stations are run by GSB. Its stock fund had rose from 1 million in 1970 to 21 million DM today. Shareholders of the company are the Bavarian State (78 %), 3 municipal organizations (8 %) and 76 industrial firms (14 %). The activity of the company is conducted on a not profit-base but being a private company the GSB has to search permanently new ways to treat the hazardous waste in order to get a proper disposal or a recovery of raw materials from this waste.

As a consequence, waste treatment costs are relatively high thus creating potentially a substantial incentive for waste producers to ship their wastes to surrounding states or countries with lower treatment quality and costs. For this reason, waste exports from Bavaria are generally forbidden; the use of the above mentioned facilities is compulsory with only few exceptions granted for some already existing waste treatment plants or sites of large companies. In effect, GSB and ZSVMM have a monopoly in waste handling in Bavaria.

COLLECTION AND TRANSPORT

The collecting stations are fairly uniform in the service and in their equipment (see Figure 2). They provide an intermediate holding function and have equipment, which is fairly simple to run. They pretreat as much waste as possible to cut down the volume. They have a waste water treatment plant, in which the separation of the usually large volums of oilwater mixtures into water and solids takes place. The treated waste-water is discharged locally. The waste volume is cut back to 10 % of the original. Where necessary, facilities for neutralizing and sludge thickening are in place. All collecting points have an administration building and testlab, vehicle scales and an area for oily soil and containers to receive and provide intermediate storage for industrial sludges of all types.

FIGURE 2
Flow Chart: Collection Centre

Flow Chart: Collection Centre

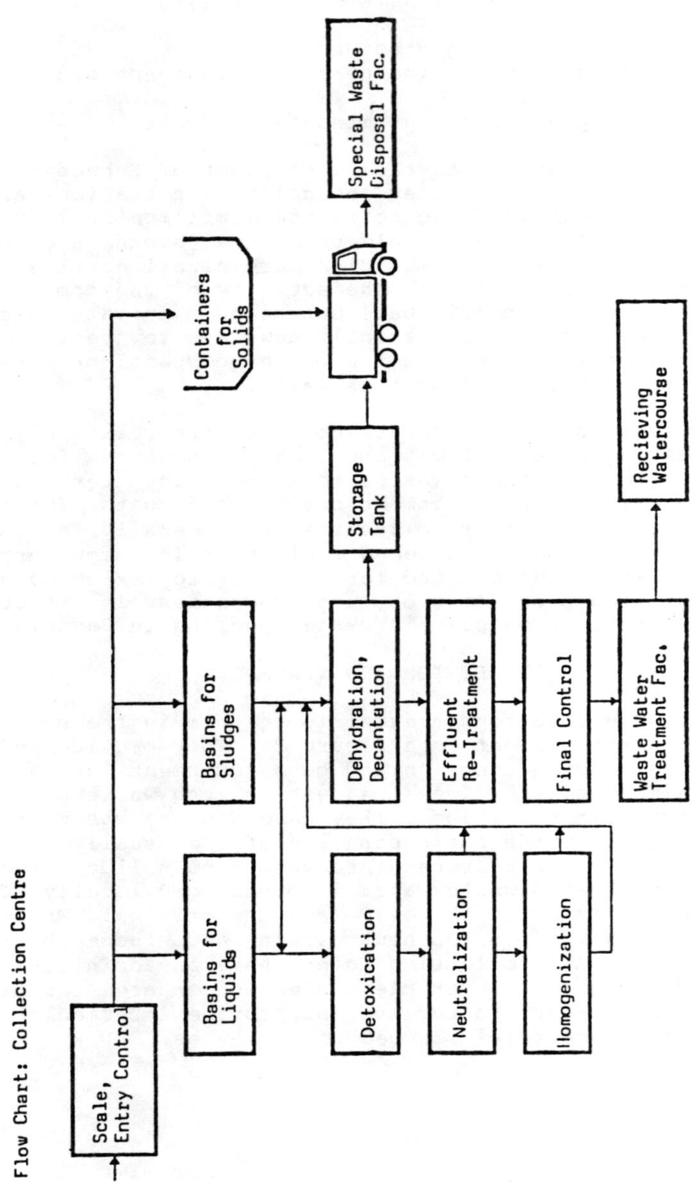

CENTRAL TREATMENT FACILITIES

Ebenhausen central treatment facility

The biggest and most modern of the 3 bavarian hazardous waste facilities is the Ebenhausen-Plant which started up 1976 on a 4 hectare site in Ebenhausen near Ingolstadt. The Ebenhausen plant comprises a laboratory, a chemo-physical treatment plant for organic and inorganic substances, a water purification plant, and Germany's largest waste incineration plant with a capacity of 100 times 10 to the sixth power btu/h or 80.000 metric tons (see Figure 3).

Incineration

The incineration plant has two parallel rotary kilns for solid an pasty wastes and a common burner-chamber with a set of six burners for liquid wastes. The heat of the off-gases from the after-burners is utilized in a steam boiler and in a steam turbine while the remainder is condensed in an air condenser; this electric energy not only the incineration plants entire power requirement but actually leaves an excess supply for the public grid. The steam from the turbine is utilized for heating the building and for process heat in the chemo-physical treatment plant.

Off-gases purification comprises an electrical precipitator for dust retention and a two-stage venturi-type scrubber which removes HCL and HF almost completely, while retention of SO_2 is 70 pc. is enhanced by flocculants. As regards airborne emissions and scrubbing water pollutants, this industrial incinerator has been reviewed several times. From simultaneously performed measurements of HCl, SO_2, HF and dust in raw gas and clean gas, the following fluctuation margins for routine operation ca be derived:

TABLE 1

Emissions of the Ebenhausen incinerator (1985)

Parameter	Raw gas mlg/m³	Clean gas mg/m³	limit mg/m³
HCl	1000-4000	22-90	100
SO_2	450-2100	40-300	400
HF	62-260	1-3	5
dust	1000-2000	10-20	50

A new legislative order for emission control has been introduced in March 1, 1986 sets new and still more stringent requirements as to the performance standards.

FIGURE 3
Ebenhausen Central Treatment Plant

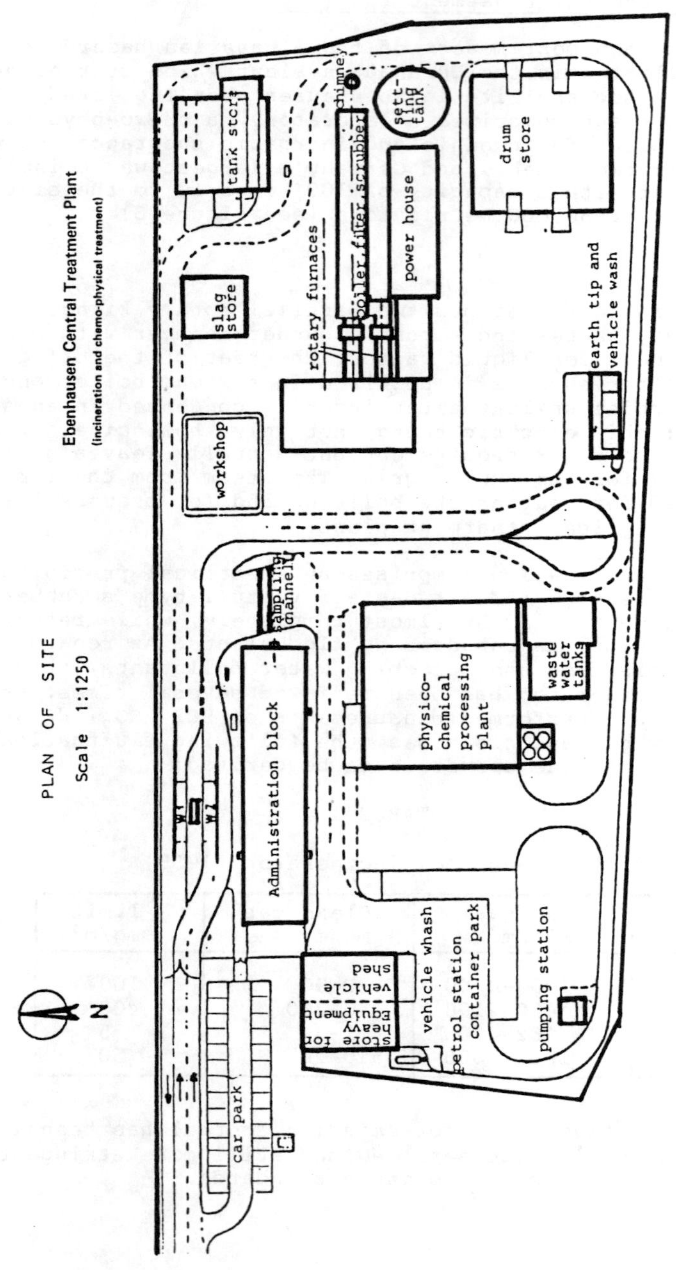

TABLE 2

Emission limits in the FRG (1986)
(values calculated for normal gas conditions, dry gas with
11 % O_2*):

Parameter	Limit		Remarks
CO*	100	mg/m³	* to be measured continously
$C_{org.}$*	20	mg/m³	** special inorganic elements in the par-
NO_x*	500	mg/m³	ticulate matter:
SO_2*	100	mg/m³	class I : Hg, Cd, Tl
HCl*	50	mg/m³	class II : As, Cr(VI), Co, Ni, Se, Te
HF*	2	mg/m³	class III: Sb, Pb, Cr, Cu, Mn, V
particulate			*** various polycyclic aromatic hydrocarbons
matter*	30	mg/m³	
class I**	0,2	mg/m³	
class II**	1	mg/m³	
class III**	5	mg/m³	
PAH***	0,1	mg/m³	

In order to meet these more stringent emission limits or
even to exceed them because of special orographic situations, the incineration plants Ebenhausen and Schwabach are
presently about to be equipped with additive air pollution
control technologies based on different processes. They
range from a quasi-dry scrubbing system combined with highly
efficient fabric filters over a wet, but waste water free
scrubber to an ionizing wet scrubber in the GSB central
facility supplementary to the existing, quite efficient
scrubbing system.

The waste incineration facilities in Ebenhausen are required
to install a computerized data gathering and processing unit
for those parameters to be measured continously indicated as
* in the list (including temperature in the afterburner
chamber). The resulting records are ultimately evaluated by
the Bavarian Environmental Protection Agency which is generally responsible for control and surveillance of hazardous
waste treatment and disposal facilities.

Chemo-physical Treatment

The inorganically polluted liquids and thin sludges especially from the electroplating industry is pretreated in the
chemo-physical plant.
These wastes are contaminated by cyanide, chromate, nitrite,
heavy metals, acid or basic solutions and normally need a
detoxification and a filtration to eliminate the toxic substances and the solids.
The detoxification is practized by oxidation (using sodium-
hypochlorite or hydrogendioxide) by reduction (using sodium-
hydrogensulfite or sodium dithionite) and metal precipitation as hydroxide (using lime) or sulphide (using sodium

sulphide). For separation of solids a filter press has been approved. The filter cake is landfilled and the filtrats after being analysed are discharged to the municipal sewage plant.
The annual capacity of the chemo-physical plant is about 50,000 tons per year.

The de-emulsification plant include the following treatment steps:

- Separating of solids first by sedimention in a four cham- bered discharge and sedimentation tank and second by a decanter.
- Chemical treatment by adding iron (III) chloride as de- emulsification reagent, neutralizing the solution by so- dium hydroxide and thus absorbing the oil particles at the surface of the precipitation of iron (III) hydroxide.
- Separation of sludge and water phase in vaccum-drum-fil- ters.

The residues of the treatment processes (dewatered, neutra- lized sludges, fly-ash, slag) and other solid wastes are de- posited at the two landfill sites in Schwabach and Gallen- bach by special security conditions. A third landfill site, located in Raindorf, went in operation 1985 (capacity: 800 000 m^3, basic-clay-cover: 2.0 m, owner: ZVSMM). While the landfill site of Schwabach is located together with the treatment facilities, the Gallenbach site is dislocated (over 40 km) from the Ebenhausen treatment facilities in view of hydrogeological conditions.

<u>Gallenbach Landfill</u>

The Gallenbach Landfill, in operation since 1975, comprises the following installations: (see Figure 4)

- Operational building with amenity rooms of various types for the personal.
- Laboratory for sampling of the substances dilivered.
- Vehicle weigh bridge.
- Vehicle and instrument shop.
- Control systems (drainage, retaining basins) for collec- ting surface water and leachate.

The site is operated by the 'area method' of landfilling on a 100 cm thick, pre-constructed clay pad with a reported permeability of 10^{-9} m/s. Watersoluble solid wastes con- taining heavy metals are desposited in drums and covered with concrete to reduce the contamination in the leachate.

Leachate is collected by a series of underdrains, located in the clay pad. It is then channeled to 2 closed-concrete-

FIGURE 4
Gallenbach Landfill Site

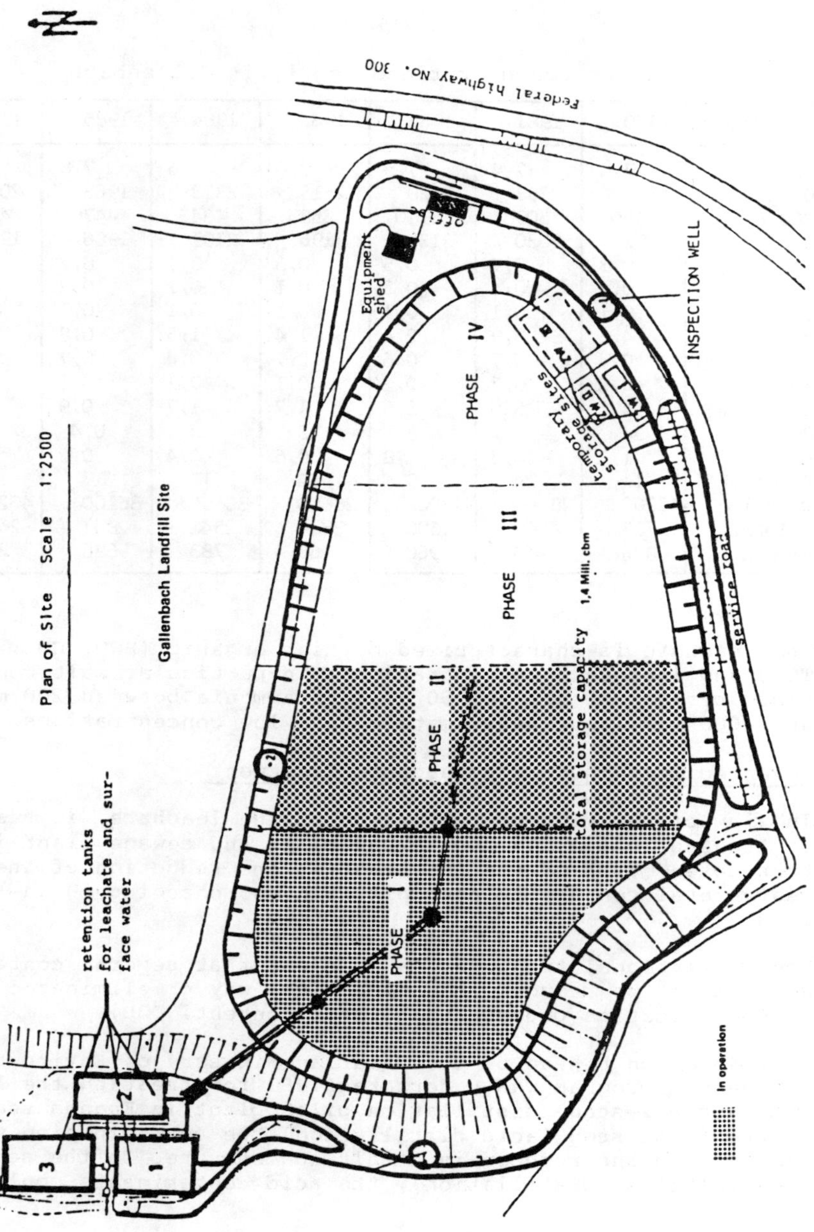

storages from where it is trucked to Ebenhausen-Plant for
treatment. Average leachate characteristics are identified
in Table 3.

TABLE 3

Average Leachate Concentrations (mg/l) in Gallenbach

Parameters	1980	1981	1982	1983	1984	1985	1986
pH	7,4	7,4	7,5	7,6	7,6	7,8	7,6
BOD_5	4526	4383	3520	2515	2853	1963	2075
COD	3850	3098	3900	3585	4343	4470	4640
TOC	3062	3530	3130	196	2109	1986	1940
Cr	0,3	0,7	0,5	0,3	0,1	0,2	0,1
Cu	0,4	0,4	0,3	0,1	0,1	0,2	0,2
Cn	0,1	0,1	0,1	0,1	0,1	0,1	0,1
Ni	4,6	3,9	2,4	1,4	1,5	0,8	0,6
Zn	0,4	0,7	0,6	0,7	0,4	0,7	0,6
Cd	0,2	0,7	0,2	0,1	0,1	0,1	0,1
Fe	4,5	3,0	3,5	1,7	1,7	0,9	1,0
Sn	34	3	3	3	3	0,4	1
Pb	1,2	4,8	1,8	0,6	0,4	0,2	0,1
Phenole	11	14	9	12	17	17	16
Chloride	47900	48900	4300	28600	33600	36100	43200
Sulphate	3717	4060	1600	3814	2543	2841	2600
Amonia	944	970	960	960	783	698	240

The leachate is characterized by high organic (BOD, COD,
TOC) and inorganic concentrations, in particular salt contents between 30 g/l and 50 g/l and ammonia between 200 mg/l
and 1000 Mg/l, but heavy metals have low concentrations.

Advanced Techniques for Leachate Treatment

In view of the high contamination of the leachate, it has to
be treated in the chemo-physical plant and sewage plant in
Ebenhausen. Task of the treatment is the reduction of the
heavy-metal-content by precipitation and changing pH-va-
lues.

The conventional treatment makes clear that several contaminants can't be reduced in a biological way or eliminated in
common oxidation process (high salt content, COD).

Therefore, on behave of the Bavarian Ministry of Environment
a research program was undertaken 1983 for treating the leachate by a 2-stage distillation pilot plant in Ebenhausen.
In the first step (acid distillation) the contamination were
cristalized and reduced in a salt concentrate. In the second
step (alkaline distillation), the acid contaminants could be

reduced again. The results made clear that TOC and COD could
be reduced over 90 %, particularly by using the discontinuos
distillation.

Based on this experiences, the first full-scale distillation
unit in the world went in operation 1986 in Schwabach to
purify leachates and comparable, aqueous hazardous wastes.
The treatment consists of purging the liquid (removal of
ammonia and highly volatile organics), a two-stage distillation process and separation of organic and salt residues.
The latter, together with $(NH_4)_2SO_4$ resulting in the
purging step, is disposed of in the landfill under special
previsions thus closing the loop between waste and leachate.
Volatile organics and organic residues are fed into the incinerator.

CHARGES SYSTEM AND INVESTMENTS

The disposal costs in Bavaria are reasonable, probably reflecting the facts that initial capital investments are not
being amortized and that later capital investments are sub-
sidized through interest-free or low-interests loans. The
pricing formulas are quite involved and are summarized as
follow. Waste treatment charges are established by process
for each of the processes used in the major waste categories. The charges are designed roughly to cover the transportation from the collecting centres and the cost of waste
treatment for each process.

Average treatment charges per ton (German Marks)

```
landfilling       DM   75,-  -   190,-*   (*deposition in
                                           concrete jacket)
chemo-physical    DM  120,-  -   300,-
incineration      DM  325,-  -  1500,-*   (*PCB)
```

The hazardous waste disposal facilities in Bavaria required
an investment of DM 150 millions in the last 20 years.
Financing is handled through subsidies, government loans and
favorable interest terms and the company's own recources obtained by economic activity.

CONTROL

The central task in each hazardous waste disposal system is,
and will continue to be to ensure that a hazardous waste is
directed along the right disposal route from waste generator
to waste transporter and to a disposal facility, where the
waste may be properly treated. The surveillance of this
waste transfer is perhaps the most difficult element to manage in the overall waste disposal control system, as far as
the manpower to do it efficiently and the great number of
individual checks which have to be performed are concerned.

FIGURE 5
Hazardous-Waste-Control-System
in the Federal Republic of Germany

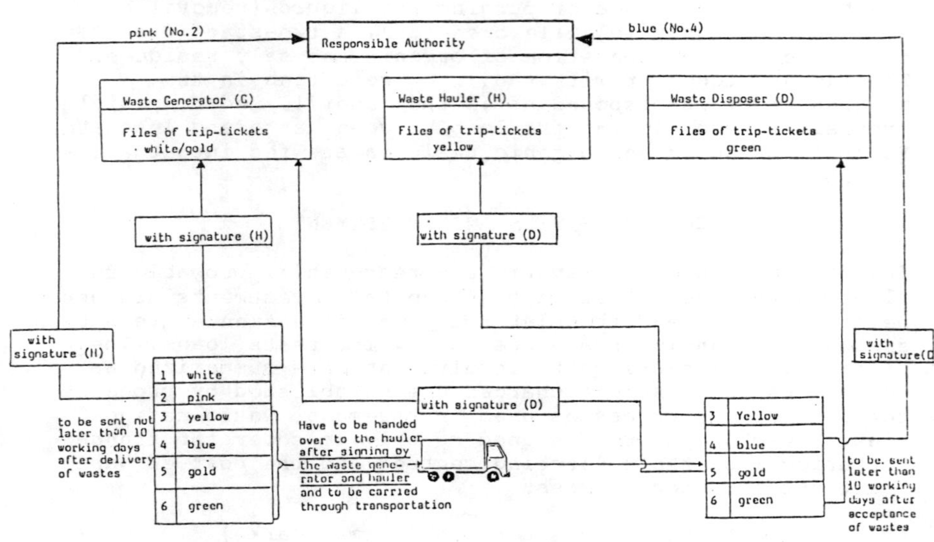

German Special Waste Manifest

The existing regulations in the FRG allow for thorough control of the wastes from their source to their disposal on the base of a 'trip-ticket-system'. Therefore in the FRG exists a control system in which hazardous waste generators can and usually are required to initiate a trip-ticket, cradle-to-grave recording system of special waste involving six separate copies. Waste haulers, waste disposers and responsible agencies receive and/or transmit appropriate copies. Thus there are numerous places for cross-checking the kinds and quantities of waste as it travels to its final destination (see Figure 5).

The basis for controling is the "waste catalogue", which has been in existence since 1975 containing all imaginable types of wastes (about 570 types of waste). In 1977 it was introduced a classification scheme from this general catalogue to separate out hazardous or 'special wastes'.

The list of 'special wastes' was designed to facilitate the practical application for both the waste generators and the controlling authority. 86 specific wastes considered hazardous have so far been identified and their individual characteristics and origins defined in the statutory regulation. Examples include: Tannery sludge, asbestous dust, spent acids, pickling bath, pesticides wastes, halogenated organics, halogenfree organics, paint-sludges, hardening salts etc.

This list had been extended 1986 to about 200 hazardous wastes, for which specific treatment and disposal methods are recommended.
In Bavaria, nearly 100 000 trip-tickets per year are controled by the Environmental Protection Agency with the help of a computerized system.

Chapter 2

WASTE MANAGEMENT METHODOLOGIES AND PRACTICES

INDUSTRIAL WASTE MINIMIZATION STRATEGY
IN THE ARTOIS-PICARDIE BASIN

Yvon RAAK
Director
Artois-Picardie Water Agency
764, boulevard Lahure - B.P. 818
59508 DOUAI - FRANCE

ABSTRACT

The Artois-Picardie Water Basin is one of the six French water basins
whose limits were defined by the 1964 Law on Water. The Water Basin
Agency is a public body created to fight against water pollution and to
manage water resources.

The situation at the end of the sixties being relatively catastrophic, an industrial waste minimization strategy has been defined : a
system of taxes on polluters has been set up, the funds collected this
way being redistributed to industries and local communities that construct sewage networks and treatment plants. In the field of industry,
the integration of pollution prevention at the very heart of industrial
production units has been favoured, in the frame of an "anti-pollution"
approach. Good results have been achieved : organic pollution discharged
has been reduced to a third with considerable success in several
sectors : discharged pollution reduced by 80 % in vegetable canning ;
by 70 % in paper plants ; by 85 % in sugar plants ; by 75 % in
wool-washing plants.

INTRODUCTION

For a long time and because of the abundant existing resources, the
French water policy originally amounted to realizing works which were
meant to meet quantitative needs. Later on, because the degradation
of the rivers, waterways and ground water had reached a critical point,
a sustained action for stream protection was needed to keep on supplying
water in economically sound conditions.

Based on this changing situation, a law was passed in 1964 which created 6 Financial Water Agencies and Water Basin Committees.In the Northern part of France, one of the main courses of action of the ArtoisPicardie Water Agency has been to fight against industrial pollution by favouring the integration of pollution prevention at the very heart of industrial production units.

THE ARTOIS-PICARDIE WATER BASIN AND THE WATER AGENCY

The Artois-Picardie Basin is the one that has the highest density of population ; more that 200 inhabitants per square kilometre which is twice the French average. Altogether there are around 4,5 million inhabitants. To the North of the Basin is situated the second most important industrial and urban concentration of the country after the Paris area. Among the best represented industrial sectors are : the mining industry ; the steel and metallurgical industries, the chemical industry, especially mineral chemistry ; the textile industry ; the food industry (sugar, vegetable canning, dairying and brewing).

The Artois-Picardie Water Basin Agency

It is a governmental body enjoying financial autonomy and situated in Douai, a town in the centre of the Artois-Picardie Basin. The Agency was created by the 1964 Law on Water to fight against water pollution and to protect and improve water resources. Its main missions are :

To fight against industrial and urban pollution.
To preserve water resources.
To monitor the quality of natural waters.
To maintain awareness of the aquatic environment.

The annual budget of the Artois-Picardie Agency is at the moment around 300 million francs. It employs more than 100 people, 25 of which are engineers or high level executives.

THE WASTE MINIMIZATION MANAGEMENT STRATEGY SINCE THE END OF THE SIXTIES

The situation at the end of the sixties was relatively catastrophic. Nowhere in France was the situation more serious for stream pollution than in the Artois-Picardie Water Basin. From 1954 to 1964, the lenght of streams completely deprived of fish had doubled ; the ground water tables had been affected by continuously rising nitrate contents, often by bacterial pollution and sometimes even by toxic pollution in certains places.

The cause of this pollution was the weakness of he hydrographic system composed of small receiver streams, but it was also the continuing idleness of local communities and industries with respect to the treatment of wastewater. Only one out of five towns had a sewage treatment station, which meant that only 15 % of urban effluents were treated ; the situation was scarcely any brighter for industries.
An immediate reaction was thus necessary.

For this purpose, a system of taxes on polluters has been set up.
Those who pollute have become those who pay. After a growing phase, the
taxes should have reached a level whereby it would have been more profitable for the polluter to invest in the elimination of the pollution
he was responsible for than to continue paying high taxes."Pollution"
taxes have not yet reached this level, but nevertheless today pollution
has its price. Roughly, for 1987, 1 kilogram of discharged oxidizable
matter costs for the polluter - 120 F ; 1 kilogram of suspended matter
- 60 F ; 1 kilogram of nitrogeneous matter - 40 F ; 1 kilogram of toxic
matter - 2,000 F.
With the sums collected in this way, the Agency operates like a mutual
insurance company and develops technical and financial courses of action.
Technical, as the Agency sends specialized engineers to advise the industrial firms and to work out with them the best treatment solutions, and
to study which techniques are the most suitable in each case. Financial,
because once the treatment projects are determined, the Agency can
provide money through subsidies and loans, up to 70 % of the investment
sum for industrial firms. Since its establishment, the Agency has thus
distributed 1 billion francs to industrial firms.

Meanwhile a waste management philosophy has been developed. Three
behaviours were possible :

The first one is to admit that pollution is the inevitable consequence of industrial activity. This leads to the addition of pollution
control devices to equipment already in place. These devices which treat
pollution at the output are generally easy to set up, available at any
time but expensive, the operating costs are often high since a large
pollution flux has to be treated, and moreover, the results aren't always
sufficient.

The second one consists in the use of polluted effluents as raw
materials : it can be an advantage not to destroy the produced pollution,
but to consider it to be a raw material that can be utilized, either
through its energy content, or through the by-products that can be made
with it. Anaerobic Methane Fermentation is the best example of this kind.

The third one is very different : more than pollution control, it
can be called "anti-pollution". In this strategy, pollution is eliminated
at the origin, not at the output, usually by three possible means :

. <u>Cleaning and maintenance of workshop</u>.
. <u>Modification of fabrication processes</u>.
. <u>New and less polluting processes</u>. The irreversibility thus
     achieved is the absolute guarantee of limiting pollution for ever.

Avoiding the creation of pollution seems to be the best way to fight
against it, technically as the anti-pollution system being integrated
with the manufacturing process absorbs automatically the short or long-
-term variations, with a constant reliability, and economically, especially if we look at the savings made on taxes.

Let's see some examples related to the second and third behaviours.

THE IMPLEMENTATION OF INDUSTRIAL WASTE MANAGEMENT TECHNOLOGIES IN THE ARTOIS-PICARDIE BASIN

Polluted Effluents = Raw Materials

The first example concerns the protein recovery by extraction in starch manufacturing. A plant located in the Artois-Picardie Basin uses grated potatoes : the starch is separated from the other elements by physical processes so as to obtain a dry product. The red waters obtained during the grating, and representing 95 % of the oxidizable effluents, were in fact eliminated by agricultural spreading or discharged to the river. For a processing capacity of 800 tons of potatoes per day for the above mentioned factory, 25 tons per day of oxidizable matter were produced. The idea was to try to recover the proteins contained in the rejected red waters. In order to do this, a recovery system was set up, mostly by centrifugation.

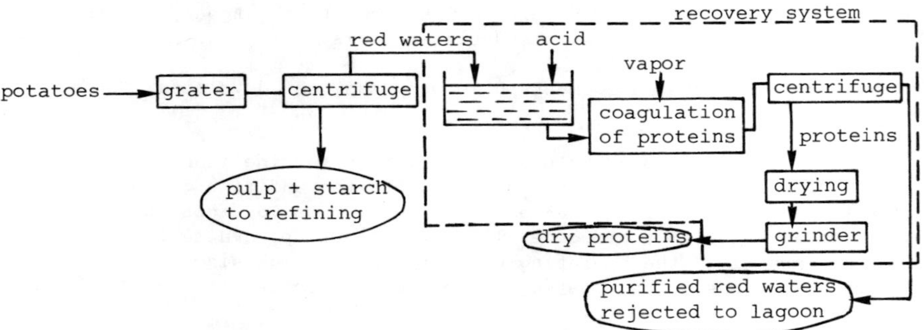

This operation enables the recovery of 50 % of proteins that were formerly lost. The red waters treatment produces a powder that contains 80 % proteins. This powder can be used for cattle feed, in the adhesive industry and in wood panel manufacturing. Improvments in the quality of this powder would lead to use for human feeding. This process modification also made it possible to reduce the pollution rejected.

TABLE 1
Pollution and Economic Balance

Rejects	Old	New		Old	New
Flow	3,5 m^3/t	3,5 m^3/t	Investment F74	5 500 000*	8 600 000
BOD	25 kg/t	15 kg/t	Annual Costs F80	1 000 000*	2 800 000
COD	40 kg/t	30 kg/t			
			Annual returns F80	0	1 800 000

basis : ton of potatoes.
* investment and annual costs necessary for the same pollution reduction with a standard treatment plant.

The second example concerns the recovery of demineralization effluents in sugar refineries :

In a sugar refinery which demineralizes the beet juice on an ion exchange, the regeneration of resins gives regeneration effluents loaded with proteins, potassium sulfate and ammonium sulfate :

For a plant in Ham (Somme) the regeneration effluents gave a pollution of 10 tons per day of oxidizable matter. The idea was to concentrate these effluents to produce on one hand potassium and ammonium salts, which can be used as fertilizers and on the other hand NSR (Non Sugar Reconstitute) for cattle feeding.

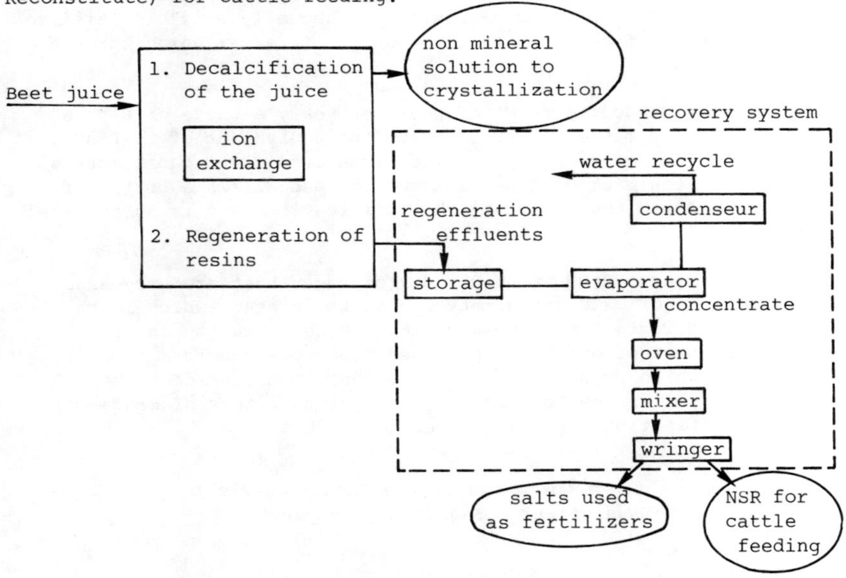

So the pollution has been quite totally suppressed. The investment cost is almost the same as that of a standard treatment but the annual returns are almost equal to the annual costs. Besides the recycle of the water makes it possible to reduce water consumption by 50 %.

TABLE 2
Pollution and Economic Balance

Rejects	Old	New		Old	New
Flow	0,75 m3/t	0	Investment	15 000 000*	8 200 000
BOD	4,5 kg/t	0	F74 Annual		1 950 000
COD	8,5 kg/t	0	Costs F80	1 900 000*	salts : 950 000
Salt	20 kg/t	0	Annual returns F80	0	NSR : 900 000

basis : ton of beet

The last examples deal with anaerobic methane fermentation. It is commonly known that methane fermentation is a process of degradation of organic matter by bacteria, in the absence of oxygen. As a result, biogas is produced containing, depending on the composition of the organic material, from 50 to 70 % of methane. So, simultaneously, we have production of energy and purification, since the degradation of organic matter leads to a more stable and less polluting product.

In a vegetable canning plant (100,000 tons/year of tinned food, 250,000 tons/year of frozen food) from 15 to 20 tons per day of oxidizable matter, or a waste equivalent to that of a town of 80,000 inhabitants, are treated, thanks to two digesters each with a capacity of 2,500 cubic metres, producing 500,000 cubic metres of gas annually which is utilized or burnt off. The purification rate is around 90 % after going through a final settling basin.

In a brewery producing 80 Ml of beer per year, a waste of 8 tons per day of oxidizable matter is treated, or the equivalent of that of a town of 80,000 inhabitants. Gas production is around 2,000 cubic metres per day. The gas contains 80 % methane and its use allows a profit of 0,5 MF per year. As to the effluent, the purification rate is an average of 80 %.

In a sugar plant, 5,000 tons per day of sugar beet are processed and 16 tons per day of oxidizable matter must be treated, which is the equivalent of the waste from a town of 160,000 inhabitants. The 120 cubic metres per hour of effluents produced are processed in a 1,200m3 digester which produces 3,500 m3 per day of biogas containing 80 % methane, which represents energy equivalent to 2 tons of crude oil per day. The purification rate is an average of 90 %.

Figure 1. Performances of the Thumeries Plant
(Sugar Plant - Beghin Say Company)

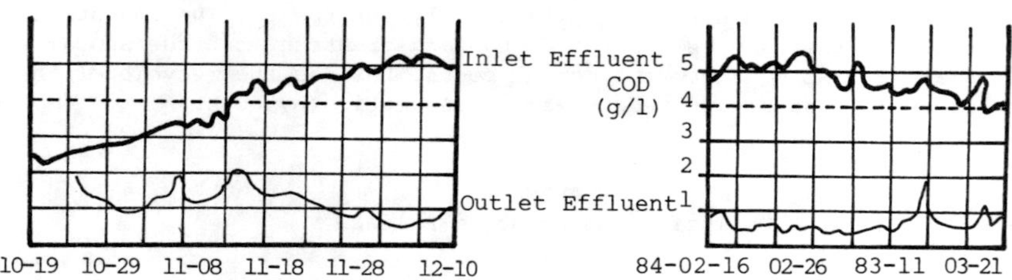

Of course, methane fermentation is not the answer to all problems, for example nitrogen is not eliminated and in many cases further treatment is necessary.

"Anti-Pollution" technologies

"Why make a strict rule of destroying pollution if it is possible not to create it ".

The first possibility is to use a recycle approach.

For example with the recovery and recycle of rinse baths in the cotton yarn mercerizing.
In the old procedure 600 l of water were necessary for a 7 to 8 kg cotton yarn cycle.

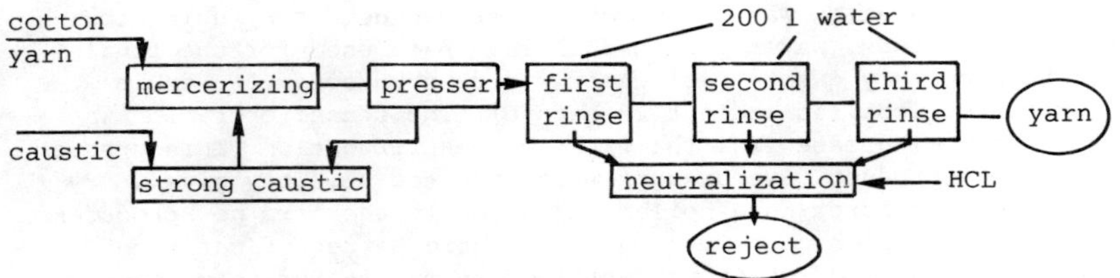

In the new procedure, only 200 l of water are necessary for a 15 kg cotton yarn cycle ; thanks to counter-current rinsing and recovery of caustic.

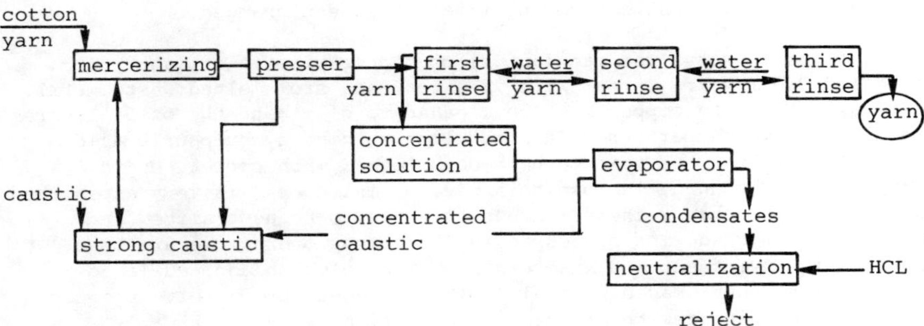

The new process consumes much less strong caustic since 60 % of the soda is recycled. The amount of soda to be neutralized is lower (100 kg/ton of yarn instead of 360 kg/ton) and the consumption of hydrochloric acid is reduced. The consequences in terms of pollution are very obvious.

TABLE 3
Pollution and Economic Balance

Rejects	Old	New		Old	New
Flow	90 m3/t	12 m3/t	Investment	11 000 000	1 450 000
Suspended materials	0,375 kg/t	0,070 kg/t	Annual Costs	1 880 000	1 250 000
DCO	22 kg/t	10 kg/t			
Neutralized Soda	360 kg/t	100 kg/t			

basis : ton of yarn

A second example of this kind deals with wool cleaning by one of the firms located in the Basin, Dewavrin, which processes around 30,000 tons of raw wool a year. Wool washing is a highly polluting operation that formerly necessitated vast quantities of water as the wool has to go through many washing tubs. In 1965, 13 cubic metres of water were necessary per ton of washed wool and the resulting pollution was equivalent to that made by a town of 200,000 inhabitants. What has been done? The first operation was to improve water management by using for the first washing tubs, water that had already been used for the final rinsing. The necessary quantity of water was thus reduced to 3 cubic metres per ton of wool (as against 13 before). The second operation was to recover the wool grease from the water by centrifugation. This grease was used as a by-product. Evaporators were then set up. They give a dry product that can be burnt in a boiler. As a result the firm now produces practically no pollution, consumes only 0.3 cubic metres of water per ton of raw wool and produces energy utilized in manufacturing processes. As to costs, the investment was around 15 MF and the annual costs around 1,5 MF per year. Twice this some would have been necessary for a standard treatment station.

Having described examples of water recycling, there now follows some cases where the fabrication process itself has been changed.

The first concerns the manufacture of copper wire in a plant producing 100,000 T/year. Copper wire if often made from melted waste metal. After laminating the copper must be cleaned to eliminate the oxide layers formed at a high temperature. This pickling is done by sulphuric acid. The problem is that the acid baths become loaded with copper. Up to 1 % of the produced tonnage is lost this way, whilst copper is very expensive and, on the other hand, these waste baths have to be neutralized and the sludge extracted. The corresponding treatment operations are thus expensive in investment and in operation. The solution adopted is to replace the sulphuric acid with alcohol which does not dissolve the copper oxides, but ensures their chemical reduction.

This technique has put an end to toxic waste and has avoided the setting up of a vast and expensive treatment system. The expense involved in the replacement of sulphuric acid with alcohol was almost non-existent. The annual alcohol consumption for a copper production of 100,000 T/year is 350 cubic metres.

TABLE 4
Pollution and Economic Balance

Rejects	Old	New		Old	New
Water Flow	0,75 m3/t	0	Investment	3 000 000*	–
Suspended materials	20 g/t	0	F79		
Oxidizable materials	15 g/t	0	Annual Costs F79	310 000*	760 000
Toxin	+ 0,3 equitox/t	0			
Sludge	2 kg/t	0			

basis : ton of wire.

The second experiment was in a brass foundry where taps are made. The oxidized parts were pickled with nitric acid, with the problem of the baths becoming rapidly highly charged in copper and zinc. A standard procedure of neutralizing and separating the metals could have been used but the investment and working costs were high. The solution adopted was to replace the acid pickling with friction by mechanical abrasion. For example, small plastic parts are mingled with the taps to be pickled, which are then all subjected to vibration. The friction between the objects eliminates the oxide layers on the surface of the taps. The inherent risks due to the use of nitric acid (a high concentration) are eliminated.

TABLE 5
Pollution and Economic Balance

Rejects	Old	New			Old	New
Flow	40 m3/t	2 m3/t	Investment	F79	516 000*	265 000
Cu	8 000 g/t	75 kg/t	Annual			
Zn	5 500 g/t	450 g/t	Costs	F79	180 000*	80 000

basis : ton of treated pieces

The last example deals with the dry recovery of Kieselguhr in breweries: Most breweries filter beer on kieselguhr on the basis of a filtration - cleaning up cycle.

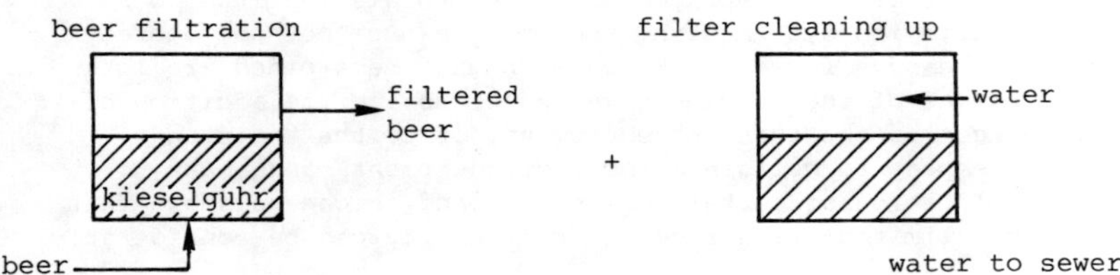

The new process suppresses all the problems related to cleaning up using water ·clogging, abrasion and pollution, thanks to cleaning up with compressed air, the thick paste produced being mixed later on with malt for cattle feeding.

TABLE 6
Pollution balance

Rejects	Old Procedure	New Procedure
Suspended materials	50 g/hl	
BOD	4 g/hl	
COD	8 g/hl	

basis : kl filtrate

THE PREVENTION OF ACCIDENTAL POLLUTION

After these examples of what has been done in the field of industrial
purification in the Artois-Picardie Water Basin. It's important to stress
the importance of the work that we have carried out within the Basin concerning the fight against accidental pollution and the reliability of the
operation of the treatment stations. To build a waste treatment station
is not enough, this station must also work correctly. In other words,
we must be sure that on the one hand it can treat the peaks of pollution,
and on the other hand it works all the year round without failing and
without breakdown.

The first stage is to recognize what sort of waste, what sort of
effluent is coming out of a plant. To do this long analysis campaigns are
run, thanks to automatic measuring stations that can be left working for
several days. Inside these measuring stations there is standard equipment
for measuring ph (acidity) and conductivity, nitrogen as NH_4 or total
nitrogen, conductivity and turbidity, and TOD (Total Oxygen Demand).
There is also a micro-computer that records information on tapes that
are then data-processed. This method allows an acute knowledge of
polluting discharges, for example, it can be seen problems. It is also
possible to see the cleaning up which takes place on a Saturday and the
lack of production on a Sunday.

The second stage in order to avoid pollution accidents, is to carry
out genuine reliability studies on the treatment stations, as is done in
the fields of nuclear industry or aeronautics. We especially applied
these reliability methods to a paper-mill, a dairy, a wool-washing plant
and a workshop where cars are sprayed, in order to discover the "weak
points" of the purifying system using the failure tree method, whereby
all the possible reasons for a breakdown wich must be avoided are looked
for. The conclusions of the studies have led either to the addition of
measurement devices to the existing equipment, or to the training of
operators with respect to certain aspects of equipment management, or
again to the duplication of certain equipment, and it can be shown that
with a relatively limited price rise the reliability can be considerably
increased.

CONCLUSION

In France, the creation of the Basin Agencies, together with the
regulatory action of the Government and together with the action of public opinion which is less and less tolerant of pollution has been very
profitable and in the last 15 years significant results have been
obtained in the struggle to protect our environment.
In the Artois-Picardie Basin, in the field of industry, where the
Agency has spent around one billion francs, organic pollution has been
reduced to a third thanks to a successful industrial waste minimization
strategy. Discharged pollution has been reduced by 80 % in vegetable
canning plants, by 70 % in paper-mills, by 85 % in sugar factories, by
75 % in wool-washing plants. As to toxic pollution, it has been reduced
by 70 %. So, even if there are still big industrial polluters, important
results have been obtained.
The consequence of all this has been not only to stop pollution
increases, but also to recover the initial natural charm of some of the
rivers which have thus become tourist and fishing attractions once more.

ADVANCED METHODS AND TECHNOLOGIES FOR THE TREATMENT AND DISPOSAL OF TOXIC OR HAZARDOUS WASTES

Claude CAMILLERI
Technical Department
and
Bernard VIGREUX
Waste Management Programme Department

SGN
78182 Saint-Quentin-en-Yvelines Cedex
France

ABSTRACT

Although toxic and hazardous wastes account for only about 4 % of total non-radioactive industrial wastes, they are a major and growing problem in terms of waste treatment and final disposal.

A proportion of these wastes can be processed to recover valuable or marketable substances, or destroyed by incineration using apparatus that must meet very stringent specifications concerning temperatures and leaktightness. However, the major part must be stored at final disposal sites in a manner that combines affordable costs with long-term guarantees of safety - requirements that impose maximum waste volume reduction plus a leaktight containment barrier between the wastes and the environment to preclude any risk of contamination.

These requirements are very similar to those encountered in the nuclear industry, which has developed satisfactory, proven solutions built around key processing operations such as volume reduction by incineration and waste encapsulation and packing to from an efficient containment barrier before disposal.

This paper presents advanced methods and technologies for the treatment and storage of undegradable and highly toxic (or hazardous) industrial wastes, based on long experience with nuclear wastes.

TOWARDS NEW METHODS IN WASTE HANDLING

Special working groups were recently established in France to examine issues in industrial waste handling and disposal. They estimate the total annual production of industrial wastes in France at 50 million tonnes. This includes 2 million tonnes of toxic or hazardous wastes, approximately one-quarter of which (500,000 tonnes) are generated by the chemical and related industries.

In France, as in other countries, toxic and hazardous wastes divide into three broad categories :

- organic wastes such as hydrocarbon wastes and solvents,
- liquid and semi-liquid mineral wastes such as surface treatment baths, acids and bases,
- solid mineral wastes such as sand from sand casting.

Problems have traditionally been encountered in safely transporting these wastes, especially when the production site is far from the storage or processing site. Their disposal can be an even bigger problem, involving one of three basic approaches :

- recycling to recover marketable or reusable products,
- destruction,
- storage.

Recycling would ideally be adopted for all toxic and hazardous substances. Unfortunately, separation and purification of marketable products are often not economically feasible. This is because many such products are present in low concentrations and complex mixtures, and also because waste production sites are highly dispersed geographically.

As a consequence, recycling is currently applied only for a few high-value pollutants, such as chromium.

Destruction of harmful substances can be achieved by chemical or biochemical processing to produce harmless residues (e.g. for cyanide) or by incineration (e.g. for PCB and dioxine). Special precautions must be taken during incineration, notably to avoid toxic or hazardous substances being released to the atmosphere in gases of combustion. The efficient solutions described further on and developed in the nuclear industry for incinerating radioactive wastes and cleaning off-gases before release are applicable to incineration of these toxic and hazardous wastes.

Waste storage is the only solution for substances that can be neither recycled nor destroyed. Requirements are again similar to those in the nuclear industry, which has developed advanced waste packaging technologies that have passed the test of time.

There are, however, two major differences between non-radioactive toxic (or hazardous) wastes and radwastes. The former do not emit radiation, obviating the need for shielding and radiological protection and thereby simplifying the design of process equipment and reducing capital costs. On the other hand, their harmfulness does not diminish with time, as is the case for radioactive substances. Waste packages must therefore be designed for infinite or, at least, extremely long storage with no significant release of dangerous products to the environment. As a minimum, engineers must fully understand the long-term migration behavior of harmful substances within and outside waste packages.

At present, toxic wastes that can be neither recycled nor destroyed are simply held at surface storage sites or sometimes in deep underground salt mines. The geographical location of these storage sites often imposes long transport and hence an additional risk of accidental dispersion in the environment. Moreover, the wastes are packaged very summarily, in containers wich rarely provide any guarantee against loss of confinement due to corrosion by waste products (even when neutralized) or external damage.

The processing and packaging technologies developed for radwastes do provide long-term guarantees concerning the non-release of harmful substances to the environment. They also assure near-absolute protection against damage by the external environment (corrosion, flooding, etc.), and safe transportation when the wastes are processed and packaged at (or very close to) their sites of production.

Nuclear methods and technologies are often dismissed because considered too expensive. However, much of their cost is due to the need for shielding and radiological protection. When applied to non-radioactive wastes, the same technologies are eminently simple. They deliver safety through in-depth knowledge of the compatibities or incompatibilities between wastes on the one hand and container or encapsulation materials on the other, rather than through complicated equipment. Built up through years of experience, this knowledge permits safe processing and packaging of diverse non-radioactive substances with reliable and hence guaranteeable prediction of long-term waste package behavior. Exact costs can only be precisely evaluated on a case by case basis, according to the nature and volume of wastes. Even if, as is probable, the costs are higher than with current approaches to non-radioactive waste handling, nuclear-developed technologies may be expected to dominate the market sooner or later. They guarantee safety and this is a key psychological argument in overcoming opposition from the local population around waste disposal sites.

NUCLEAR-DEVELOPED TECHNOLOGIES

No practical methods exist for eliminating the radiation emission from radwastes, apart from natural radioactive decay, and waste processing systems therefore divide mainly into two categories :

- waste volume reduction systems,
- systems for waste packaging prior to final storage and disposal.

This classification is used below in discussing the application of nuclear-developed technologies to non-radioactive toxic wastes.

Volume reduction

Volume reduction of aqueous radwastes usually involves, especially for saline solutions or fine dispersions :

- either removal from solution followed by separation of the solids from the purified liquid phase by conventional means. Use is made of techniques that are also fairly common in industrial waste processing, such as flocculation followed by decantation or filtration,

- or evaporation, a technique for which the nuclear industry has developed very advanced systems. Designed with special attention to cleaning vapors and obtaining high-quality, purified condensates, these systems are of great interest for processing non-radioactive toxic wastes.

Volume reduction of solid radwastes normally involves grinding, compacting, or incineration in the case of combustible materials. The greatest reduction in volume is obtained with incineration.

Concerning **grinding and compacting,** systems developed for radwaste processing could probably be applied beneficially to non-radioactive toxic wastes. Because such systems comply with stringent nuclear industry rules, they take a very sophisticated approach to hazard confinement, including controlled air removal from waste-containing areas and careful filtering.

Very high pressures, sometimes exceeding 500 bar, are employed in compacting solid radwastes for maximum volume reduction. The experience acquired with the corresponding presses is applicable to other types of waste.

Nuclear industry experience in radwaste **incineration** is equally of great value for processing non-radioactive toxic wastes. Incineration can not only reduce waste volume but, in many cases, also destroy the toxic substances.

SGN has developed, in collaboration with the French Atomic Energy Commission (CEA), an advanced incineration system built around an excess air combustion chamber followed by a postcombustion chamber. A string of safety devices assure totally safe and reliable operation of this system. Corrosion is avoided by careful selection of refractory materials and by holding temperature below the condensation point throughout the system up to the final high-efficiency filters. Waste feeding to the incinerator is assured under conditions of total confinement. The system operates at subatmospheric pressure for total confinement and controlled routing of all gases and vapors. Off-gas filtering is extremely effective, but postfilter scrubbing can be provided when appropriate.

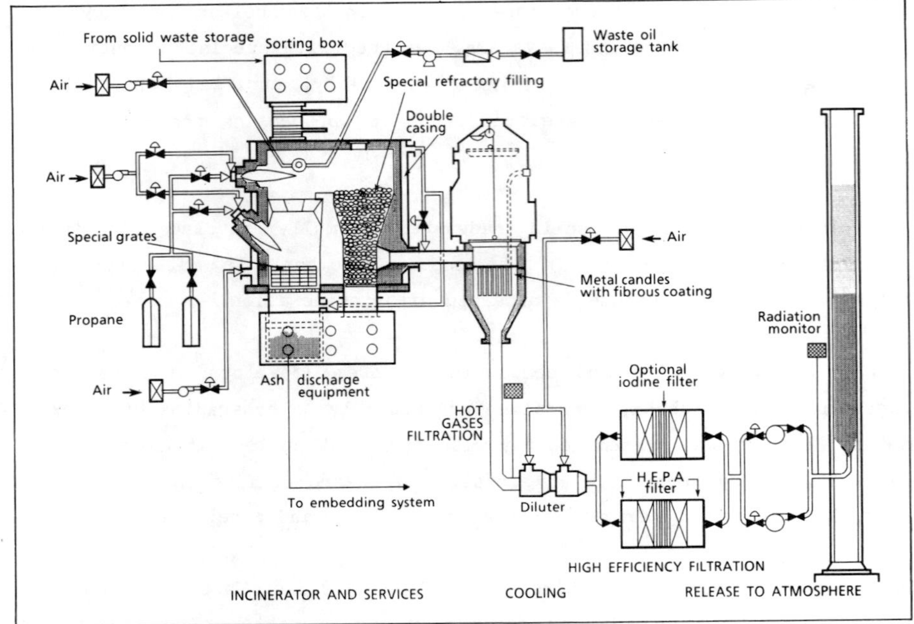

Fig. 1   Radwaste incineration system

The efficiency of off-gas filtering is partly due to special design of the postcombustion chamber filling and of the regenerable high-temperature ceramic filter immediately downstream.

The system was originally based on a design for incinerating municipal refuse, but the mentioned features are the result of long R & D and field experience to adapt this original design to radwate processing.

In the standard version, the incinerator operates at 1,000/1,200°C and the high-temperature filter at 800°C.

Again in the standard version, the system is available with waste capacities ranging from 20 kg to 100 kg per hour. This corresponds very closely to requirements for incinerating non-radioactive wastes in which the toxic substances can be destroyed by heat. The system can process waste oils and solvents, and the standard version includes a burner for this purpose. Other pumpable wastes can be handled subject to adaptation.

SGN has additionally developed 1 to 20 kg per hour incinerators for very special types of radwastes. These smaller-capacity systems use electric heating. They have the same safety features as described above and can be easily adapted for non-radioactive toxic wastes.

Waste packaging

Methods and systems developed for radwaste packaging are also applicable to indestructible toxic wastes that must be stored under totally safe conditions in final disposal facilities.

A first method uses **high-integrity containers** to form a hermetic barrier around the wastes. Studied in depth and quantified, the resistance of this barrier to out-migration or leaching of waste substances can be input to computational models for safety studies on very-long-term storage.

SGN has developed a container of the sort, qualified for storage of certain wastes from the Cogema spent fuel reprocessing plants at La Hague in France, and which can be employed without modification to store toxic wastes. It is pursuing related studies under a contract with the EEC.

A second, more common radwaste packaging method involves uniformly immobilizing the wastes in a matrix that assures the requisite resistance to leaching. Encapsulation can be performed in a polymer or glass matrix, but the following discussion is limited to less expensive bitumen and concrete solidification systems.

Bitumen encapsulation is a standard technique for radioactive waste packaging. The bituminization process and technology marketed by SGN under CEA licence employ simple, efficient solutions in which waste volume reduction and packaging are integrated in single unit. This unit has a falling-film vertical evaporator, heated with oil or steam. The waste, which must be pumpable, is introduced at the top of the evaporator simultaneously with hot bitumen. Within the single-stage evaporator, the water in the waste is evaporated and the remaining salts or suspended matter intimitely dispersed throughout the encapsulating bitumen, in a single continuous operation. The resulting product flows out of the bottom of the evaporator into drums, inside which it solidifies. The drums usually have a capacity of 200 liters and are positioned successively under the evaporator by a linear conveyor or by a turntable.

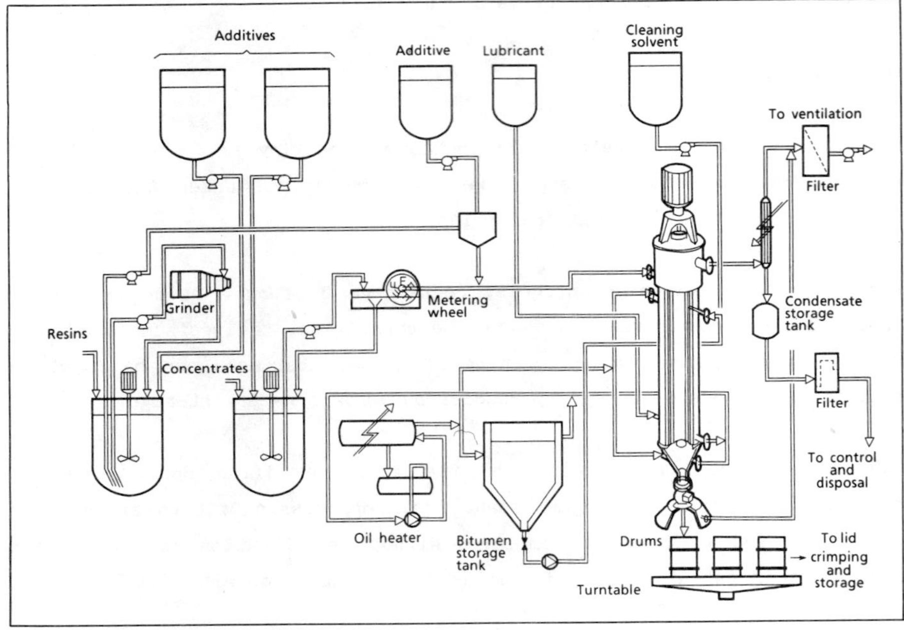

Fig. 2 Continuous bituminization process with vertical evaporator

The operating conditions, safety, and quality of the encapsulated product are perfectly controlled, benefiting from 20 years of development and substantial feedback from operating bituminization units handling a wide diversity of wastes. The bitumen matrix displays outstanding resistance to ion migration, ageing, and bacterial attack.

The pilot units operated by SGN for process adaptation-studies -using non-radioactive saline solutions or suspensions to simulate the radwastes generated by the nuclear industry -can be immediately applied in studying bituminization of non-radioactive toxic wastes.

Concrete encapsulation is the most widespread method of radwaste packaging, with the advantage of also immobilizing residual water in the wastes.

SGN has devoted large sums in recent years to developing and refining concrete encapsulation processes for radwastes, using the most appropriate technologies. The resulting systems are perfectly suited to packaging non-radioactive toxic wastes, both liquids and sludges, and also incineration ashes. The main potential difficulty in handling such wastes concerns the compatibility of the concrete matrix with the chemical waste composition. Extensive research on this subject has led to development of a methodology for adapting the process to each type of waste, and numerous tests have been performed on the interaction between the concrete and various anions (e.g. sulphates, chlorides and fluorides) or cations.

The quality of the encapsulated product depends very much on the technology chosen for mixing the wastes, cement and other substances, and this has also been the subject of considerable R & D by SGN.

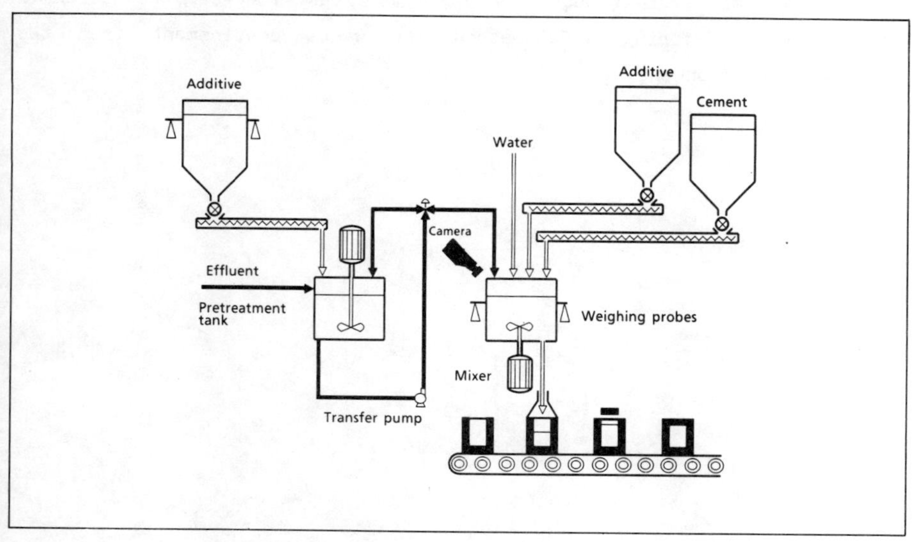

Fig. 3   Batch concreting process

SGN markets two concrete encapsulation systems :

- the first is a continuous system using a self-cleaning, twin-screw mixer to prepare up to 15 m3 per hour of concrete paste,

- the second is a batch system using a hermetic tank at the bottom of
which a propeller sets up vortex motion. Depending on the size of
this tank, the system prepares batches of 200 litres to 2 m3 of
concrete paste in about 15 minutes.

For each system, SGN operates a pilot unit for fast adaptation to new
types of waste. The long-term behavior and leaching resistance of the
waste-bearing concretes have been studied in depth by SGN and the CEA.

Final disposal

SGN, working with the CEA, has considerable experience with methods
of modeling and predicting the migration of substances from radwaste
packages in subsurface or deep geological storage, and also in the design
and engineering of such final storage sites. Long-lived radioactive wastes
are in this respect comparable to stable non-radioactive wastes. In both
cases, it is impossible to count on any significant long-term reduction in
toxicity when defining safe methods of waste confinement from the
surrounding environment.

TREATMENT OF WASTE IN INDIAN PROCESS INDUSTRIES

Y.I. Dave
Environmental Quality Department
Indian Petrochemicals Corporation Limited
Baroda
INDIA

and

K.T. Oza and S.A. Puranik
Department of Chemical Engineering,
Faculty of Engineering and Technology,
Baroda
INDIA

ABSTRACT

Due to the tremendous growth of Chemical Industries in India
during last decade, numerous types of toxic and hazardous wastes are
being generated in the form of solid, liquid and gas. Where as most
of them are being handled successfully, the disposal of some of them
still presents problems. Most commonly used techniques adopted are
extensive chemical treatment, biological conversion and incineration.
In this paper, an attempt is made to highlight briefly the application
of various techniques used to combate pollutants, latest technological
development in the field, their advantages and dis-advantages with
reference to Petrochemicals, tanning, textile, pulp and paper, pharmaceuticals, pesticides and distillery industries in general and toxic
waste in particular. It is expected that this paper will throw light
on the present picture of environmental management of wastes in the
Indian Process Industries.

INTRODUCTION

Industrial growth [1] from 1978 to 1986 in India (Table-1) has produced heavy pollution due to the discharge of toxic and hazardous wastes in the form of either solid, liquid or gas. Among such industries there are petrochemicals, pharmaceuticals, pesticides, tanneries, distilleries, dyes, refineries, fertilizers and other inorganic and metal industries. Main liquid and gaseous pollutants are cyanide, flourides, sulfides, pesticides, insecticides, heavy metals, carcinogenic hydrocarbons, SO_2, NO_x, CO, Cl_2, NH_3, particulate matters etc.

TABLE - 1.

Growth of Industries (Ton.)

Year	Sugar $x\ 10^5$	Paper $x\ 10^3$	Pulp $x10^3$	Pesticide (DDT) $x\ 10^3$	Alcohol $x\ 10^3$	Petro-chemicals $x\ 10^5$	Tann $x\ 10^3$	Ferti. $x\ 10^3$
1978-79	6.5	1250	240	12	400	39	6000	7300
1985-86	100	2100	500	20	600	70	7500	11000

The knowledge in India, as far as toxic and hazardous waste treatment is concerned is highly insufficient. The measures specified in existing acts for controlling discharges of wastes into the environment are inadequate. The easy way is 'Dumping'. It is done by pouring the wastes into a pond or a lagoon within the premises, storing wastes in solid, semi solid form or burying them in one's own premises, subjecting the wastes to partial treatment such as bio-degradation or evaporation and then dumping as above, conveying the waste to another site or water body for straight dumping, burning the wastes in open or in single cell burning chambers enormously called incinerators. Such practices are very harmful since there is no control on combution or products.

The current practice adopted for disposal of liquid wastes includes discharge into sewer, river, or sea. There was a blaze in the ganges due to huge effluent discharge from Barauni Refinery in 1967. The Bombay-Kalyan belt is polluted due to hundreds of sizeable industries. The Hoogly estuary is dirty due to wastes of jute mills, textile mills, tanneries, paper and pulp mills and distilleries. The Durgapur-Asansol region is spoiled due to steel coke-oven, fertilizer and distillery plants. Damodar and Gomti rivers have become two of the most polluted rivers in the country. The concentration [2] of toxic waste of flouride in Nalgada river has reached upto 12 ppm. The concentration of DDT and BHC in adipose tissues of general pollution in major cities has gone as high as 131 and 94 ppm respectively. The other problems caused are alkali soil degradation, micro nuetrient defficiency in soils, sedimentation in reservoirs and decrease of carbon to nitrogen ratio in soil due to increased use of fertilizers

in place of natural manure. Further, there are some additional factors
which are affecting these problems, e.g. lack of political commitment
and comprehensive environmental policy, poor environmental awareness,
functional fragmentation of the public administration, poor mass
media concern and prevalence of porerty are some of the major factors
responsible for increasing the difficulties of Environmental Management problems in India. Out of 2700 major industries in India about
1700 industries generates objectionable effluents, whereas only 460
industries have full fladge waste water treatment facilities and 251
under implimentation.

PROCESSING OF WASTES

Indian chemical industries discharge enormous amounts and
variety of wastes. Table-2 gives the physice-chemical characteristics
of wastes from selected industries in India.[3,4]

PETROCHEMICAL WASTE :

The Petrochemicals produced in India are numerous. They generate
effluents containing hydrocarbons, heavy metals and toxic chemicals
in large quantities. Normally before discharge, effluent is treated
in order to meet disposal standards specified by regulatory authorities. The gaseous emissions are monitored and the solid wastes are
suitably treated before using in landfill.

An integrated large Petrochemical Complex of India[5] producing
varieties of Polymers (LDPE, PP, PBR, PR, PVC), Fibers (AF), Fiber
intermidiates (ACN, ACR), and Petrochemicals (DMT, Benzene, Toluene,
Xylene, LAB, EG etc.) being a big complex in the country and known
for excellent environmental awareness is taken as case study.

Based on properties-general biodegradable waste, cyanide waste,
sulfide waste, acidic waste, flouride waste and sanitary waste are
separated at the source and treated with specialized processes.

Effluent stream from naphtha cracker, containing very high concentration of sulfide is separately treated by wet air oxidation, a
shell technology. This process converts sulfides into thiosulfates
and sulfates.

The effluent having low sulfide and high alkalinity is used for
neutralization of acidic waste, generated by DM water plant.

Petroleum Resin and Linear Alkyl Benzene plant generate effluent
containing high flourides. This effluent is treated with large quantities of lime in a separate unit to precipitate flourides as CaF_2.

This effluent with high alkalinity is then used to neutralize
acidic waste generated by DM Plant.

Effluent from Acrylonitrile Plant contains mainly hydrocyanic
acid, organic cyanides, acrylonitrile, acetonitrile and other organic
chemicals. The techniques used for such toxic waste are either incineration or extensive chemical treatment. Both these methods are
very costly. Such waste was handeled by incineration at high fuel

TABLE - 2.

Physico-chemical characteristic of Wastes from selected Indian industries in India :

Industry	Chemical Contents ppm	pH	BOD ppm	COD ppm	Total Solid ppm	Factory
Textile	$O_2=.8$, $N^-=17$, $S^{--}=10.5$, $PO_4^-=76$, $Cl^-=85$, $Cd^{++++}=26$	7.8-9.6	620	1225	1260	Modi textile/ Ambica Mill
Fertilizer	$O_2=.8$, $NH_3=720$ $N_3=65$, $N=12$ $SO_4^-=980$, $PO_3^-=4.2$, $CaCo_3^-=138$, $Cl^-=745$, $Ca^{++}=102$, $Mg^+=43.6$	8.5-8.9	280	535	2495	F.C.I. Nangal/ G.S.F.C.
Beveries Distilleries	$N^-=132$, $SO_4^{--}=560$, $S^{--}=77.4$, $PO_4^-=36.8$, $Ca^{++}=102$, $Mg^*=43.6$, $Cl^-=745$.	3.2-4.3	5400	7180	12980	Mohan Meaking
Pharmaceutics Antibiotics	$O_2=3.9$, $N^-=98$, $SO^{--}_4=18$, $PO_4^-=47.2$, $CaCo^+_3=185$, $Cl^-=140$	7.3-7.8	67	85	820	IDPL/HAL
Pesticides Sevin	$Cl^-=100$, $SO_4^{--}=100$, $Na^+=18000$, $N_2=500$	7-10	Nil	10000	40000	Sevin Pesticides
Paper & Pulp	Mercaptant Sulfide Lygnin	6.9-9.8	150-420	700-1000	600-2300	Oriental paper
Tanneries	Cromium tannin = 15-20	9.5	7000	-	3200	Kanpur Tanneries/ Vellore
Refinery	High concn. of sulfide oil and phenol	4-7	200	3000	300	Gujarat Refinery, HPCL
Petrochemical	Hydrocarbon oils, Grease metals S^{--}, F^-, CN^-	2.5-12	2000	4000	400-650	IPCL,NOCIL,

consumption i.e. one MT of fuel/5 m3 of liquid waste. After extensive research [6] and pilot plant study, an alternative method is developed to degrade toxic cyanide waste Biologically. Fig.(1).

In primary treatment, waste is digested with 2 to 4% hot alkali at 100-110°C for 1 to 3 hrs. and then chemical co-agulation. Further it is treated by 2 stage activated sludge extended aeration method. This biological process is not substantially different from that of standard activated sludge process. However seperation of strong CN decomposing bacterias, their propagation, cultivation and acctimatization is the art of this process.

Biodegradable chemicals such as benzene, toluene, xylene, methanol, LAB, paraffins, butanol, acrylates, oils, grease and a number of other chemicals, additives and sanitary wastes are primarily treated where in PH adjustment, removal of floatings, grits, oils, greases, hydrocarbons and suspended solids takes place. Secondary treatment is given by chemical floculation with the help of alum, ferrous sulfate, lime and polyelectrolytes. Further they are subjected to extended aeration biological treatment, secondary clarification and polishing before disposal. Characteristics of influent, effluent compared to disposal standard are shown in Table-3.

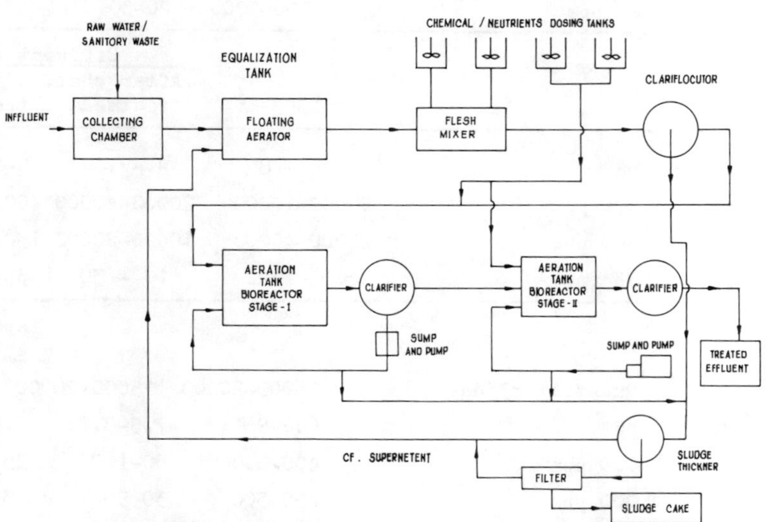

Fig.1 Biodegradation of Toxic Cyanide Waste

Waste from Paper and Tanneries :

The characteristics of waste from paper and tanneries are shown in Table-4.

Most of the Indian paper and pulp mills do not have adequate waste water treatment facility. The practice generally followed is to discharge these wastes into surface water or on-land. The capital and operating cost of every ton of paper for small unit is double that of

TABLE - 3.

Characteristics of treated and untreated waste

Effluent stream	Characteristics	Influent	Effluent	Disposal standard
Sulfide Waste	Quantity m3/day	24	24	
	Composition			
	(a) NaOH % wt.	4 - 4·5	4 - 5	
	(b) Na_2S ppm	500-4000	50-60 ppm	
	(c) Na_2CO_3% wt.	1 - 2.5	2 - 3	
	COD ppm	15000	2500	
Flouride Waste	Quantity m3/day	30	30	
	NaOH % wt.	3 - 4	3 - 4.5	
	COD ppm	250-300	150-250	
	F^{--} ppm	3000-5000	20-30	

			Effluent	
		Influent	After chem. treatment	After Bio treatment
Cyanide Waste	pH	6 - 8	8 - 10	9-10
	COD ppm	80000-100000	20000-40000	4000-6000
	BOD ppm	20000-40000	10000-20000	1500-2500
	CN ppm	80 - 120	10 - 30	3 - 5

		Influent	Effluent	Disposal standard
Common Waste	Quantity m3/day	15000-20000	15000-20000	-
	pH	6.6-9	7.5-8.8	5.5-9
	COD ppm	400-800	80-100	250 max.
	BOD ppm	250-500	30-50	30 max.
	Oil & Grease ppm	15-20	Traces	10 max.
	F^- ppm	2-3	1-1.5	2

large mills. Hence to install integrated full fledged water treatment
facility is not economical. Inplant survey of a mill of South India
[7,8], producing 100 tons of paper per day using bagasse, waste
cotton, rags etc. is carried out by NEERI as under.

The waste is given the following treatments. First the waste is
allowed to settle in a settling pond for 4 hours after removal of
screenings and grits. The reduction in BOD is 34%, COD is 40% and
Suspended Solid is 70%.

Then 6000 ppm of lime is mixed for 10 minutes and settled for
1 hour. The reduction in BOD is 53% & COD is 49.7%. Following lime
treatment, the waste is first treated in an anaerobic lagoon after
supplimentation of neutrients in the proportion of BOD : N : P : 100 :
5 : 5.1 The BOD removal is 30 to 70%. The detention time is 10 to 20
days based on BOD load. This is further treated in an aerated lagoon.
The detention time required for 90% BOD removal is 4 to 7 days. This
results in BOD of 30 ppm.

TABLE - 4.

Characteristics of Paper and Tannery Waste

Effluent Stream	pH	Alkalinity CaCO$_3$	COD	BOD	Total solid	Dissolved solid	Sus. solid	Total Voletile solid	Total non-voletile solid
			(All in ppm)						
Pulp & Paper waste	8.2-8.5	620-680	3400-4900	680-1250	3780-6000	2800-4900	980-1100	2520-3200	1260-2800

Effluent Stream	Waste	Alkalinity ppm	Total Solid ppm	S.S. ppm	BOD ppm	COD ppm	pH	Cl$^-$ ppm	Cr ppm
Tanneries.	Dyeing, vegetable, tanning, liming, seaking section	–	48340	–	6200	100	11.5	21250	1000
	Combined composite waste	630-1800	–	1300-5420	1698-3400	3640-3400	7.3-8.8	1740-6850	–
	Mean value	–	–	2762	2420	4730	–	3730	–

Tanning industry is one of the oldest industries in India. The
waste from it ranks among the most polluting of all industrial wastes.
Palar river has been considerably affected due to discharge of untreated waste water from tanneries. In India, considerable work has
been done on the treatment [9] of waste from tanneries, e.g. by

chemicals like alum, $FeCl_3$, lime, CO_2, anaerobic digestion, trickling filter, high rate biological filter, and oxidation ditch. These methods do not have wide applications on small tanneries due to the high cost involved. In such a situation, treatment in waste stabilization pond directly or after anaerobic lagoon is adopted. Tannery waste water and sewage are mixed in stabilization pond in the ratio of 1:1 to 1:6, sewage having BOD=120 ppm, COD=180 ppm. Acclimatized algal cultures are transferred to the ponds and stabilized for 18 days. The effect of detention time on BOD and COD is shown in Fig.2.

The BOD and COD values of the combined composite tannery waste ranged from 1698 to 3400 ppm and 3640 to 6740 ppm respectively. The effluent from soaking and liming contained Cl⁻=21250 ppm. The effluent from chrome tanning contain $\underline{Cr}$ =1000 ppm. The algal count during stabilization in different tannery waste sewage mixtures are varied from 38×10^4 to 176×10^4 ppm. A maximum count of 176×10^4 ppm are recorded on 12th day in 1:6 waste sewage mixture. The results clearly indicate that tannery waste is ameanable for treatment in stabilization pond in admixture with sewage. The greater the dilution, more efficient is the degree of treatment. Of the six dilutions studied, 1:1, dilution gave the best performance with reference to the BOD and COD removal compared to the other lower dilutions. Treatability studies by stabilization pond method indicates that the BOD and COD

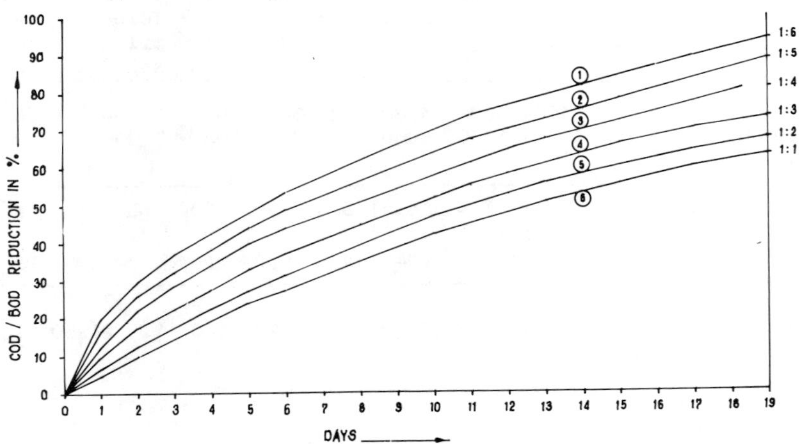

Fig.2 BOD/COD removal Vs. Detention time.

reduction of the waste sewage mixtures during treatment in experimental stabilization pond for 18 days ranged from 64 to 93 and 60 to 89 % respectively.

Textile Waste :

Cotton textile industry with 670 mills [10] is one of the biggest industries in the country. Effluents from textile mills are discharged into public sewers, sea, land, or municipal sewage farm. NEERI has carried out the survey on disposal problem in Textile Mills. It is evident from the above table, that the characteristics of the textile waste have wide variation. Therefore, the general treatment is as

under. Segregation of highly concentrated waste stream from the
dilute one, reduces its volume considerably for specific treatment.
About 15-20% water is reused. Study of one of the textile mills shows
that recovery of alkali, reduces PH, alkalinity, TDS and Na^- content.

TABLE - 5.

(A) Characteristics of Textile Wastes

(ppm)	Keir	Bleaching	Mercerizing	Dyeing	Printing	Combined waste
PH	10-11.8	8.4-10.9	8-9.8	9-11	6-8	6.7-8.1
TDS	12260-40000	2500-7900	2060-2500	3000-6000	1800-2500	2180-3600
BOD	2000-4000	100-525	100-150	500-1500	150-1500	218-300
COD	12832-20000	1350-1600	250-400	–	400-4000	400-800
Suspended	1000-2000	200-340	150-450 Cl-300-700 SO_4-100-350 Na-600-1200	300-1000	250-450	400-3100

(B) Effect of Caustic recovery

Characteristics	Before caustic recovery	After caustic recovery
PH	10.6	6.7 - 8.8
Alkalinity at PH = 8.3	2500	N.A.
Alkalinity at PH = 4.5	3200	710-880
TDS ppm	3475	1860-2660
SS ppm	1500	100 - 740
% Na^-	95	64 - 75

The substitution of chemicals such as carboxy methyl cellulose
and poly vinyl alcohol in place of starch in sizing cloth, use of
mineral acids instead of acetic acid in dyeing process and biodegradable detergent in place of soap have contributed towards reduction

of BOD from 10-15% in the combined waste.

The kier waste is segregated and contained in the cooling tank for 8-12 hours and released gradually at a uniform rate. Equalization of waste serves of self neutralize acidic and alkaline waste streams. Chemical coagulation, flocculation and sedimentation remove the colleidal impurities and colour. For this, alum, $FeSO_4$, $Fe_2(SO_4)_3$, $FeCl_3$, Chlorinated copper, and poly electrolytes are employed. Combined waste with sewage are treated in a single/two stage process by trickling filter and/or activated sludge. By this, BOD removal of 90-96% has been reported. Recently, this is followed by oxidation ditch and aerated lagoon.

OTHER MAJOR INDUSTRIES AT A GLANCE

There are about 104 distilleries in India[11]. They produce 550 million litres rectified spirit and 10-15 litres waste water per litre of rectified spirit. Their discharge causes deoxygenation, anaerobic situations, abnoxious smell of sulfides, indole, sketol and well water pollution. The distillery waste is treated by sludge aeration, sludge enzymatic digestion, formentation with .01 % urea + 5% gypsum, neutralization of spent with lime, spent wash evaporation for the by products recovery. Potassium salts, CH_4, S^{--}, yeast hydroleate, mixed feed, vitamins, dried yeast are the by-products which are recovered from distillery waste.

Chloro-alkali manufacturers are now switching over to diaphram cell process from mercury cell process to eliminate fusitive immissions of mercury vapours and generation of sludge containing chloro mercurous compound. Disposal of such sludge is still in question in India. New licences are not granted for mercury cell process.

Probably, India is the largest manufacturer of Pesticides[12] in the world, producing 44 types of pesticides in about 86 sizeable units. Harmful effects of pesticides on human life is known to every one and hence controlled use or ban on manufacturing activity is being imposed by developed country. CIBA(India) at its Goa unit, treats waste to disposal standard by extensive chemical treatment for the waste from organo phosphorous pesticides manufactured by alkaline hydrolysis, actual data on treatment method is not available. However $KMnO_4$ is used to remove heptachlor and two field DCP over a wide range of PH. Ozone dosage in excess of 4000 ppm are required for COD removal of around 80%. DDT can easily be removed by settling and coagulation followed by filtration.

Waste from refineries and oil field threatens the pollution of sulfides and hydrocarbons. Most of the refineries in the country are howing full fledged control to combat liquid or air pollution. However during start up, shutdown or during emergency, emission of pollutants through stack or flare cannot be avoided.

India being the country of farmers billions of tons of fertilizers are produced and used. Fertilizer waste contains major pollutants such as sulfides, florides, ammonia and cyanides. In most of the plants pollutants are being handled successfully.

Power plants, based on coal and cement industries are the largest polluting industries for suspended particulates matter. New plants are equipped with latest technique of electrostatic precipitators. However poor performance of ESD and collection and disposal of fly ash is the major constaint.

Effluents from various food industries such as sugar, dairies, slaughter house, yeast etc. have common[13] characteristics such as high BOD and easy digestion. An Anodex process, to produce methane gas by anaerobic digestion is becoming more popular, acceptable and economical method for handlihg such waste.

CONCLUSIONS

India as a developing country, is becoming self reliant in[14] its own needs in the field of chemicals, fertzilizers, petrochemicals etc. For handling concentrated toxic and hazardous waste probably incineration, fixation and encapsulation of such waste into insoluble matrix seems to be one of the promising and acceptable approaches. Number of organizations have given prime importance to these approaches, right from their inception stage. Solvent extraction to remove toxic organics from waste water followed by incineration of residue is another promising approach which is largely adopted and practiced by organic chemical manufactuerers in India. The zimpro process, i.e. addition of powdered and grannulated activated carbon to an activated sludge system which gives highly promising results for removal of toxic organic compounds is picking up very fast in Indian environment.

New techniques such as plasmid transformation, gene-cloning and transposon, mutagenesis to manupulate biodegradative activity of bacteria for toluene, xylene, n-octane, salicylate, chlorobenzene etc. are in conception stage.

ACKNOWLEDGEMENT

The authors acknowledge with thanks the help and co-operations rendered by Shri.R.Sethuraman, Director(Operations) and Shri.CS Patel, Deputy General Manager, Environmental Quality Department of IPCL. They also thank prof.MN Desai, Vice Chanceller of M.S. University for providing facility and permitting to present the paper.

REFERENCES

01. Waste Water Treatment, National Engineering Research Institute, Nagpur, India, 1982, 2, 6, 7.

02. Bowonder, B., workshop on Environmental Management Problems in India centre for Energy, Environment and Technology, Administrative Staff College of India, Hyderabad, India, 1984, 44, 46, 49, 55, 59.

03. Verma S.R., Tyagi A.K. and Dolela R.C., studies on characteristics and disposal problems of Industrial effluents with reference to ISI Standard Part-2. Indian J. Environmental Health, 1977, 19 3, 165-175.

04. Waste water treatment, National Engineering Research Institute,
Nagpur, India, 1982, 27.

05. Report on Environmental Management of IPCL submitted to FICCI
1985, 2 - 4.

06. Dave Y.I., Oza K.T. and Puranik S.A., Biodegradation of Toxic
Cyanide Waste, _Chemical Age of India_, 1985, <u>36</u>, 8, 775 - 777.

07. Sastry C.A., Kothandaraman and Aboo K.M., Treatment of Waste
water from small paper mill without soda recovery, _Indian J.
Environmental Health_, 1977, <u>19</u>, 4, 346 - 359.

08. Swamy V.M.K., Industry and Environment Pollution Monitoring
instrumentation, 10th workshop on Environmental Management,
Administrative college of India, Hyderabad, 1985, 48.

09. Govindan V.S., Treatment of Tannery Wastes, _Indian J. Environmental Health_, 1985, <u>27</u>, 1, 59 - 62.

10. Waste water treatment, National Engineering Research Institute,
Nagpur, India, 1982, 395-398, 401.

11. Waste water treatment, National Engineering Research Institute,
Nagpur, India, 1982, 423-425.

12. Saxena K.L., and Chakrabarti T., Organic pesticides and their
removal from aqeous system, _Indian J. Environmental Health_,
1978, <u>20</u>, 4, 334.

13. Lele, Anodek Process for the treatment of Food Industry,
Waste water, _Chemical Age of India_, 1985, <u>36</u>, 1, 81.

14. Course manual on Treatment and disposal of Petrochemical(ACI)
Industrial Waste, 1985, 21.

POLLUTION ABATEMENT AT GNFC -
AN APPROACH TO ZERO POLLUTION

VK Karia
Senior Manager
Gujarat Narmada Valley Fertilizers Company Limited
P.O. Narmadanagar-392 015
District Bharuch, Gujarat State, India

ABSTRACT

M/s Gujarat Narmada Valley Fertilizers Company Limited(GNFC) located in backward area of Bharuch District in Gujarat State of India is owning World's Largest Single Stream Ammonia and Urea plants.The plants were commissioned in December,1981, and commercial production started on 1st July,1982. Ever since its commissioning, GNFC are well satisfied with their record in the field of Environmental Management. Number of RESEARCH AND DEVELOPMENT oriented projects were undertaken as a result of which GNFC has been able to reduce the quantity of effluent by 71% (18205 M^3/Day). GNFC's effluent recycle/reuse programme has helped to reduce fresh water requirement by nearly 46%. GNFC's effluent is also utilised for the last five years for agricultural purpose in Company's demonstration farm and by farmers in nearby villages without any adverse effect. GNFC has also put to productive use its solid wastes and gaseous pollutants. GNFC has adopted a productivity oriented approach in managing the pollutants and hence, while ensuring proper quality of effluent, GNFC has been able to save nearly US$ 1 million so far.

INTRODUCTION

For a developing country like India, problem of pollution control has become a very complex issue due to conflicting need between improving standard of living on one side and maintaining ecological balance on the other side. On one side, there is a growing need to increase the production and profitability and on other side, a few factories have faced a closure due to violation of pollution control acts. Some have paid heavy compensation to villagers and farmers, either because the standing crop is damaged or because drinking water wells are polluted. Some animals are also reported to have died in different parts of the country due to drinking of polluted water.

In sharp contrast to this scenerio, Gujarat Narmada Valley Fertilizers Company Limited(GNFC), a joint sector Company, located in backward area of Bharuch District in Gujarat State owning World's Largest Single Stream Ammonia (1350 MT/Day) and Urea (1800 MT/Day) plants, has, through an innovative approach of "PRODUCTIVITY THROUGH POLLUTANTS", largely succeeded in putting most of its harmful and hazardous pollutants to productive use. As a result, there has been a direct saving of nearly 1 million US$ besides bumper increase in crop yield due to use of effluent water for irrigation purpose.

The text that follows gives a brief idea of GNFC's Approach To Pollution Control.

POLLUTANTS AT GNFC

From point of view of pollutants, GNFC has a worst combination of coal as boiler fuel and Furnace Oil/Low Sulphur Heavy Stock as feedstock for Ammonia production. As a result, many pollutants like particulate matters, sulphur dioxide, hydrogen sulphide, hydrocarbons, ammonia, cynide, vanadium,urea,oil,hydrochloric acid, sodium hydroxide, fly ash, clarifier sludge, lime sludge and carbon soot etc., are appearing in either gaseous, solid or liquid state. To minimise the pollution problems, built-in facilities like 100% recycle of carbon soot in Ammonia plant, setting up of a Sulphur Recovery Unit to prevent hydrogen sulfide pollution, use of Electrostatic Precipitators (ESP) to recover fly ash, cyclone separators followed by wet precipitators to minimise coal dust, a natural draft prilling tower in Urea plant in place of forced draft or induced draft tower, use of hydrolyzer stripper to minimise ammonia/urea in liquid effluent, an elaborate oil recovery system and tall chimney and flare stacks to adequately disperse off-gases etc. are provided in main plants.

TREATMENT OF EFFLUENTS

Inspite of built-in facilities mentioned above, some effluents bearing ammonical nitrogen, oil, suspended solids etc., continue to be generated. A full fledged Effluent Treatment plant has been provided to take care of all such effluents. A brief description of process involved are given hereunder:

Removal Of Ammonical Nitrogen

Ammonia bearing effluents at GNFC include approx. 8 to 10 M^3 per hr of grey water from Carbon Recovery Section, an equal quantity of condensate from CO Shift Conversion Section of Ammonia plant, effluents from hydrolyzer stripper and intermittent effluents generated by washings, leakages etc. in Urea plant :

TABLE-I

QUALITY OF EFFLUENT

Type	Ammonia plant		Urea plant	
	Grey Water	Condensate Blow-down	Hydrolyzer Stripper	Plant washing draining etc.
Flow M^3/hr	8-10	8-10	40	20
pH	8-9	6.5-7.5	8	7-8
Ammonia - ppm	1000	5000-10000	100	100-10000
Urea - ppm	-		200	100-10000
Carbon dioxide - ppm		20000-30000	-	50-60

These effluents are treated based on well established method of air stripping of free ammonia from waste water at high pH. In neutral solution ammonia exists as ammonium (NH_4^+) ion - but at higher pH it converts into dissolved ammonia (NH_3) gas which can be stripped by air to the extent of 90%. pH of the effluent is raised by addition of hydrated lime slurry or spent caustic soda effluent generated in DMW plant.

Removal Of Vanadium And Cynide

These pollutants do not appear in appreciable quantity.

Removal Of Oil :

Oil bearing effluent, generated in different oil handling area is pumped to gravity oil separator in Effluent Treatment Plant. Here due to approx. 2 hours resident time, oil separates out on surface and is collected with the help of pipe tappings. The effluent water, depleted in oil, is then made to pass through a bed of suitably sized hard coke where last traces of oil are removed by adsorption and the clear effluent is then sent to the treated effluent pond.

Effluent From Demineralized Water (DM) Plant

At DM plant approx. 1300 cubic meter per day of effluent is generated due to regeneration of cation and anion resins. Since acidic effluent neutralizes alkaline effluent, further treatment is not required.

Effluent From Raw Water And Filteration Plant

Approx. 1575 cubic meter per day of effluent having suspended solid level varying from 300 ppm to 30,000 ppm is generated partly in clarifloculator and partly during back wash of rapid gravity sand filters in Raw Water Treatment Plant. This effluent is pumped directly to ash pond where, due to large residence time, sludge gets settled and clear water overflows to treated effluent pond.

Effluent From Cooling Tower Blow-Down & Side Stream Filters

Approx. 7150 cubic meter per day of effluent is generated as cooling tower(CT) blow-down and back wash water from CT side stream filters in Ammonia and Urea plants.

Generator Quench Water

This is an intermittent effluent and is generated, for a short period, during start up and shut down of gasifiers in Ammonia plant. This effluent (about 1400 cubic meter per day containing 1 to 2% carbon) is stored in settling pond, where suspended carbon particles get settled down and clear effluent is pumped to ash pond. If, however, the level of pollutants is high, it is taken to Vanadium reaction tank for further treatment alongwith grey water before sending to ash pond.

Ash Slurry

Approx. 9000 cubic meter per day of ash slurry containing 2 to 5% suspended solids, is produced in Steam Generation plant. This slurry is stored in a large ash pond (admeasuring 600 M x 200 M x2M) where ash settles down and relatively clear effluent overflows to treated effluent pond.

Domestic Sewage

Approx. 3600 cubic meter per day of domestic sewage having BOD value (5 days 20°C) of 50 to 180 mg/l, is generated in factory and township. Township sewage is treated in Sewage Treatment Plant before discharging to effluent channel.

RESEARCH & DEVELOPMENT ORIENTED WORK UNDERTAKEN

When the plants were commissioned, actual quantity of liquid effluent generated was found to be 25,655 meter cube per day. Likewise 300 MT of fly ash, 400 MT of lime sludge, 20-50 MTPY of carbon soot, 200 MTPY of hard coke etc. were generated as solid waste. To manage these wastes, a conscious decision was taken not to view the problem of pollution from usual negative angle but to adopt a PRODUCTIVITY oriented approach. Three pronged attack was, therefore, launched :

- To reduce the quantity of effluent at source of generation wherever possible.

- To identify various applications for effluent and then to treat the effluent in such a manner that its quality becomes acceptable for the application identified.

- To find out market for waste so that not only the pollutants can be effectively disposed of but some revenue can also be earned.

Reduction In Quantity Of Effluent

Above efforts paid rich dividends. We could reduce quantity of effluent generated at source in plants like DM Plant, Steam Generation plant and Cooling Towers to the extent of 4030 cubic meter per day.A number of area could be identified where waste water could be beneficially used. In fact, we could recycle and reuse almost 14,175 cubic meter per day making a total reduction in efluent to 18205 M^3 per day or 71% of liquid effluent within main plant and demonstration farm (Effluent Balance - Table-II). As a result, Company's fresh water requirement could be brought down by 18205 cubic meter per day (Water Balance-III). Likewise, uses for solid waste and gaseous emissions were also established.

TABLE-II

EFFLUENT BALANCE

Source Of Effluent	Quantity in M^3 per day		
	Original	After R&D	Reduction
A) Reduction at source			
1. Cooling Tower Blow-down	7,150	4,150	3,000
2. DM Water Plant	1,305	600	705
3. Others	2,250	1,925	325
Sub total 'A'	10,705	6,675	4,030
B) Reduction by reusing/recycling			
1. Raw Water Treatment plant	1,575	775	800
2. Ash Slurry	9,000	-	9,000
3. Urea Hydrolyzer Effluent	1,000	-	1,000
4. Service Water	2,975	-	2,975
5. Lime Slurry	400	-	400
Sub total 'B'	14,950	775	14,175
Grand Total A+B	25,655	7,450	18,205

Notes: Reduction in total quantity of effluent by 71%
Balance quantity

TABLE-III

WATER BALANCE

Water required for (In M^3/Day)	Original	Revised		
		Total	Fresh water	Recycled effluent
1. Cooling Tower	17,000	14,000	13,000	1,000
2. DM Water plant	5,625	4,595	4,595	-
3. Lime Slurry preparation	400	400	-	400
4. Raw Water Treatment plant (Filter back washing)	1,575	1,575	775	800
5. Ash Slurry	9,000	9,000	-	9,000
6. Service Water	2,975	2,975	2,975	-
7. Demonstration farm for irrigation	2,975	2,975	-	2,975
Total ..	39,550	35,520	21,345	14,175

Notes:(i)   Consumption of fresh water could be reduced by 18,205 M^3/D(46%).

(ii)   In revised water balance, nearly 40% of company's fresh water requirement is met by recycled effluent.

A brief idea of various R&D projects is given hereunder:

Liquid Effluent

Effluents From Raw Water Plant A detailed analysis of effluent generated as filter back wash water revealed that after first 10 minutes, relatively clear water starts coming. Although the quality is not same as fresh water, specially with respect to suspended solids, we decided to recycle this water after storing in raw water reservoir where suspended solids were found to settle down due to large residence time (3-4 days). No adverse effect has been noticed since last 3 years, as observed from analysis report of raw water taken at different times during this period. On an average 800 M^3 of effluent could be reduced (or fresh water could be conserved) due to this every day (Table-IV).

TABLE-IV

EFFLUENT ANALYSIS AT DIFFERENT TIMES

Parameter		Effluent Water Time in Minutes			Av. quality of water in reservoir			
		05	10	15	1983	1984	1985	1986
Organic matter	ppm $KMnO_4$	25	12	5	4	5	4	4
Suspended particles	ppm	1100	650	400	20	15	12	12
Colloidal Silica	ppm	0.170	0.14	0.12	0.1	0.09	0.1	0.1
TDS	ppm	230	225	225	210	210	208	200

Alkaline Effluent From DM Water Plant   Analysis of alkaline effluent at different time interval revealed that first 20 M^3 of effluent is highly alkaline (pH more than 11) and, if recovered, can be used (Table-V).

TABLE-V

EFFLUENT ANALYSIS AT DIFFERENT TIMES

Time in minutes from start of regeneration	Before R&D programme			After R&D programme		
	% NaOH	pH	Qty (Cum)	% NaOH	pH	Qty.
10	0.9	12	3.0	0.9	12	3.0
20	2.00	12	5.7	Nil	6.8	5.7
30	2.60	12	8.5	0.89	11	8.5
40	4.0	12	11.0	0.80	11	11
50	0.48	12	13.5	0.79	11	13.5
60	0.60	12	16.0	0.37	11	11
70	40 ppm	11	20.0	Very low	9.0	20
80	6.6ppm	9.5	35.0	-do-	8.5	50
90	Very low	8.0	50.0	-do-	8.0	80

We are now using this effluent for :

- Regeneration of Weak Base Anion resins and
- Raising pH value of Ammonical Nitrogen bearing effluent in place of hydrated lime.

As a result, quantity of effluent has reduced by nearly 200 M^3/day. Consumption of Caustic Soda and Hydrated lime has reduced by 600 MTPY

160

(cost US$ 1,20,000) and 300 MTPY(cost US$ 20,000) respectively.

Acidic Effluent From DM Water Plant   Like alkaline effluents, we are separately collecting acidic effluents produced during regeneration of cation resins for initial 60 minutes and using the same for fixation of free Ammonia in final effluent whenever its pH value goes high due to any reasons.

Hydrolyzer Effluent From Urea Plant This effluent is being used as make-up water in Carbon Extraction Unit in Ammonia plant to replace boiler feed water. Earlier, it was used as cooling tower make-up water.

Close watch was kept on bacterial count, nitrifying bacteria and corrosion rate in cooling tower. No adverse effect was noticed. In Carbon Extraction Unit, ammonia is already present and hence no adverse effect is envisaged. Approximately 1000 M^3 of BFW(cost US$ 60,000) is thus conserved every day.

Water Required For Preparation Of Lime Slurry In ETP   Approx. 400 M^3 of fresh water was consumed in Effluent Treatment plant for preparation of lime slurry. Instead of fresh water, we are now using treated effluent reducing the quantity of effluent by 400 M^3 per day.

Cooling Tower Blow-Down   Upto middle of 1935, blow-down from cooling towers was relatively high due to high level of orthophosphate, high bacterial counts, etc. etc. On an average, total make-up water requirement was coming to nearly 17000 M^3 per day. Experiments carried out with various types of biocides and inhibitors, coupled with strict monitoring of all the parameters helped to improve the quality of circulating cooling water and thus reduce the blow-down by 3000 M^3 per day.

Use Of Treated Effluent For Ash Slurry Preparation   For preparation of ash slurry in Steam Generation Plant, approx. 9000 M^3/day of water is required. Initially, it was proposed to use a mix of cooling tower blow-down water and fresh water for this purpose. Since the CT blow-down was not available in sufficient quantity, due to very good quality of raw water only fresh water was used for this purpose.

A study revealed the possibility of using treated effluent for preparation of ash slurry. And hence, since last 4 years, only treated effluent is recycled and used for this purpose reducing the quantity of effluent by 9000 M^3/day (or saving 9000 M^3/day of fresh water valued at US$ 1,00,000).

Use Of Effluent For Irrigation Purpose   In order to use effluent for irrigation purpose, its quality should conform to Indian Standard 3307 (Table-VI)   with suitable adjustment depending upon local conditions.

TABLE-VI

TOLERANCE LIMITS AND ACTUAL VALUES

		Industrial Effluent	Effluent for agriculture use	Actual value
pH		5.5 to 9.5	5.5 to 9.5	7.5 to 9
Temperature	Max.	40°C	-	25-35°C
Total suspended solids	mg/l	100	-	30-150
Total solids	mg/l	-	2100(Inorganic)	1000-1100
BOD(20°C - 5 days)		30	500	Less than 30
Sodium	%	-	60-75	35-50
COD	mg/l	250	-	100-150
Oil & Grease	mg/l	10	-	Less than 10
Chloride as Cl	mg/l	-	600	300-500
Ammonical Nitrogen as N	mg/l	50	-	Less than 50
Free Ammonia	mg/l	5	-	Less than 5
Zinc as Zn	mg/l	5	-	Absent

As the quality is suitable for agricultural purpose, since last 5 years, approx. 6000 M^3 per day of effluent is being utilized both in Company's Demonstration Farm and by villagers without any noticeable adverse effect. Effect on soil is being monitored regularly by GNFC by carrying out soil testing.

<u>Solid Wastes</u>

<u>Carbon Sludge</u>   During gasification of oil, about 2% carbon is kept unburnt. Most of this unburnt carbon is recycled to gasifier. A small quantity which is bleed off is recovered and sold. On an average 20 to 50 MT of carbon is thus recovered and sold giving a revenue of approx. US$ 5,000 every year.

<u>Lime Sludge</u>   Approx. 4000 MT of lime sludge containing nearly 75% of carbonates and bicarbonates of calcium, is generated as waste product in Effluent Treatment plant. This sludge is being used for land filling with fly ash and as soil conditioner.

<u>Fly Ash</u>   Approx. 250-300 MT of fly ash is generated every day in Steam Generation plant. This fly ash is being used to reclaim low lying area in ravines by filling them with alternate layers of 230 mm of fly ash and 80 mm of yellow soil. So far nearly 72,000 M^2 of land in Company's township has been recovered using more than 2,00,000 M^3 of ash. It is planned to recover another 60,000 M^2 within next one-and-a half year in Bharuch city which has many ravines. Land thus reclaimed is being used for development of gardens, play-ground, forests and also for construction purpose. Filling of ravines has also helped in minimizing mosquito problems as water filled up in these low lying area was providing a breeding place for mosquito.

Fly ash is also being sold to cement industries for production of pozolona cement. Experiments were also conducted successfully to produce building bricks using a mixture of fly ash and soil.

Gaseious Pollutants

<u>Recovery Of Sulphur</u> Since GNFC's Ammonia plant is based on gasification of furnace oil containing 2.5 to 4% sulphur, large quantity of H_2S bearing gas is produced. A Sulphur Recovery Unit based on modified claus process is set up to recover 35 MT of 99.9% pure elemental sulphur every day from this gas. While eliminating gaseous pollution, valuable foreign exchange worth US$ 1 million is saved every year.

<u>Formic Acid Plant</u> Approx. 5000 Nm^3/hr of tail gas containing 40% CO, 13% Methane, 4.5% Hydrogen and 33% Nitrogen is produced in liquid nitrogen wash section of Ammonia plant. This gas is either burnt in power plant or flared. It is now decided to set up a 5000 MT/Year Formic Acid plant by using this waste gas as raw material. This plant is likely to be commissioned by middle of 1987.

<u>Use Of Impure CO_2 Gas In Nitrophosphate Plant</u> Company's Nitrophosphate plant will require 3,000 Nm^3/hr of Carbon Dioxide Gas. It is decided to use impure carbon dioxide, produced during hot regeneration of methanol in Rectisol Wash Section of Ammonia plant for this purpose.

<u>Recovery Of Ammonia Gas</u> Gaseous emission (200 Kg per hour) leaving Waste Water Treatment Section of Urea plant contained 77 percent Ammonia. GNFC decided to recover this ammonia by putting up an absorber using waste liquor as absorbent. Since November/December,1984, this scheme is in operation recovering 3.48 MT of ammonia every day saving US$ 4,00,000 every year.

ACKNOWLEDGEMENT

Permission granted   by Management of GNFC to publish this paper and the constant support and guidance by Mr.N Vittal, Managing Director and Mr. AN Aggarwal, Executive Director(Operation), is thankfully acknowledged.

Recycle and Biodestruction of Hazardous Nitrate Wastes

J. M. Napier
F. E. Kosinski

Development Division
Oak Ridge Y-12 Plant*
Martin Marietta Energy Systems, Inc.
Oak Ridge, Tennessee 37831

ABSTRACT

The U.S. Department of Energy (DOE) owns the Oak Ridge Y-12 Plant located in Oak Ridge, Tennessee. The plant is operated for DOE by Martin Marietta Energy Systems, Inc. One of the plant's functions involves the purification and recycling of uranium wastes. The uranium recycle operation uses nitric acid in a solvent extraction purification process, and a waste stream containing nitric acid and other impurities is generated.

Before 1976 the wastes were discarded into four unlined percolation ponds. In 1976, processes were developed and installed to recycle 50% of the wastes and to biologically decompose the rest of the nitrates. In 1983 process development studies began for in situ treatment of the four percolation ponds, and the ponds were treated and discharged by May 1986. The treatment processes involved neutralization and precipitation to remove metallic impurities, followed by anaerobic denitrification to reduce the 40,000 ug/g nitrate concentration to less than 50 ug/g. The final steps included flocculation and filtration. Approximately 10 million gallons of water in the ponds were treated and discharged.

---

* Operated for the U.S. Department of Energy by Martin Marietta Energy Systems, INc., under contract DE-AC05-84OR21400.

Introduction

The major responsibilities of the plant are: (1) producing weapon components and supporting DOE's design laboratories, (2) processing special materials, (3) supporting other DOE Oak Ridge Operations installations, and (4) supporting other U.S. agencies. The Y-12 Plant is within the city limits of the City of Oak Ridge and has several environmental challenges, all of which are being actively addressed. One of the challenges is the treatment and disposal of nitrate-rich liquids from a uranium recycle facility.

The plant uses a liquid-liquid solvent extraction process for purification and recycle of nonirradiated uranium wastes. One of the principal reagents is nitric acid, which is used to dissolve some of the uranium wastes. Since 1954, waste solutions from the solvent extraction process have been pumped into four unlined surface ponds. The ponds had no surface overflow; evaporation and percolation prevented the acidic wastes from overflowing the tops of the ponds.

In the past 20 years, many environmental laws have been enacted by the federal and state governments restricting the discharge of wastes to the air, water, and land. The continued use of open, unlined ponds for discharge of liquid wastes was essentially banned by the U.S. Congress by criteria established in the amended Clean Water Act of 1977. In anticipation of the future environmental restrictions, a management decision was made in 1971 by the Y-12 Plant and DOE to develop processes for recycling some nitrate wastes from the uranium purification process. It was also agreed that a process would be developed to biologically destroy some of the nitrate wastes. The remaining plant wastes from other operations were not recyclable and were to be discharged into the ponds until another disposal process could be developed.

Recycle Processes

The overall recycle process diagram is shown in Figure 1. Laboratory tests indicated that 30% of the nitrates could be recycled as nitric acid and 35% could be recycled as aluminum nitrate. The remaining nitrate (35%) could be biologically destroyed in a stirred-tank denitrification process.

The nitric acid waste comes from condensates produced in evaporation processes within the Y-12 Plant. The waste stream contains 3 to 10 wt % nitric acid, and the major impurities are approximately 100 mg/L of chlorides, 100 mg/L of fluorides, and 300 to 1000 mg/L of organics. To recycle the waste, the chlorides and fluorides have to be removed. The waste is passed through a vaporizer containing a mixture of 23.8 wt % aluminum nitrate nonahydrate, 53.9 wt % calcium nitrate tetrahydrate, and 23.3 wt % water (which removes the fluoride ions, as well as vaporizing the feed to the distillation column).

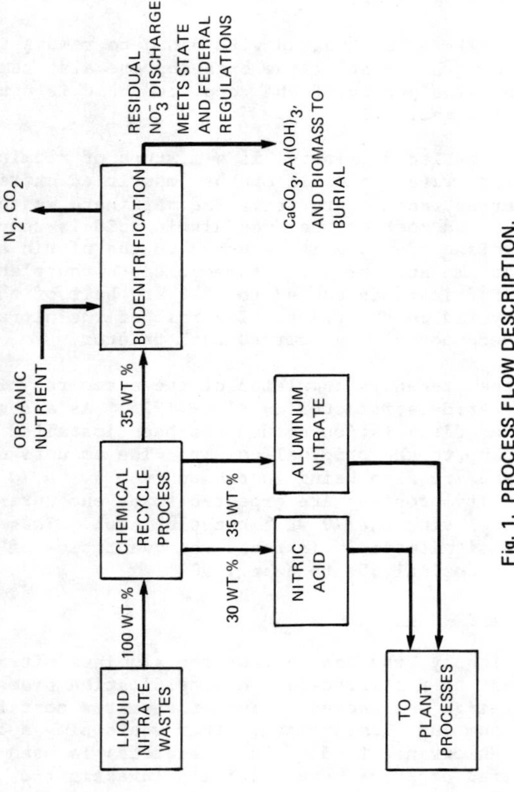

Fig. 1. PROCESS FLOW DESCRIPTION.

The vaporized salts are replaced when a 5:1 aluminum-to-fluoride weight ratio is obtained.

The dilute nitric acid vapors are then put into a glass distillation column where they are concentrated to 35 wt % nitric acid. The overhead distillate form the column is controlled at less than 50 mg/L of nitrates and is discharged to the environment at a pH of 6 to 9. The 35 wt % nitric acid product contains chloride and organic ions. The column is capable of producing 60 wt % acid, but because a reaction of the 60% acid with the organics can occur, a product of only 35 wt % acid was made.

The 35 wt % acid is batch treated with ozone to remove the chlorides to less than 10 mg/L. Most of the organics are also removed by the ozonation step. The purity of the recovered acid is usually better than commercially available acid.

A second waste, called raffinate, is a mixture of aluminum nitrate, free nitric acid, water, and all of the impurities extracted from the uranium solvent extraction process. The raffinate waste is vacuum heated at 60°C, and most of the free nitric acid is removed as condensate. This condensate is returned to the nitric acid distillation column and the acid is recycled to the plant. The vacuum-heated raffinate is cooled to 30°C and half of the aluminum nitrate is recycled to the plant. The rest of the nitrates are biologically decomposed in a stirred-tank reactor.

A review has been recently completed of these two recycle systems, which have operated satisfactorily since 1976. As a result of the review, a second distillation column has been installed that will serve as a backup to the original column. The amounts of recycled aluminum nitrate are also being increased from 50 to 70 wt %. No significant purity problems are expected to be encountered with the aluminum nitrate using the 70 wt % recycle level. These process changes should significantly increase the quantities of recycled nitrates from a nominal 65% to nearly 90%.

<u>Stirred-Tank Bioreactor</u>

As stated previously, the wastes from the aluminum nitrate recycle process are sent to a stirred-tank denitrification process. A stirred-tank design was chosen because the wastes contained high levels of calcium and aluminum salts that would plug a column denitrifier. An organic feed (calcium acetate) is used in the process. Several organics were initially investigated, but calcium acetate was selected. One of the reaction products is calcium carbonate. The nitrate waste feed contains about 3 wt % free nitric acid, which is not neutralized before injecting it into the bioreactor. The acid is neutralized by the calcium carbonate within the bioreactor.

The stirred-tank solution contains 50 mg/L of nitrates, 93 wt %
water, and 7 wt % solids. The dried solids typically contain 45%
calcium carbonate, 45% aluminum hydroxide, and 10% biomass, plus
trace quantities of metallic oxides and hydroxides.

The stirred tanks have performed satisfactorily since 1976, and
denitrification rates up to 5 g of nitrate per liter of reactor
solutions per day are common. This denitrification rate means that a
nominal 475 g (1045 lbs) of nitrate ions are decomposed each
operating day into a 95,000-L (25,000-gal) stirred reactor. To
sustain this rate, process conditions must be controlled. The
carbon-to-nitrogen (C/N) weight ratio is maintained at 1:1 to 1:2.
The solution pH in the tank must be above 6.8 and below 9.0. The
soluble phosphate ions must be 5 to 100 mg/L. The dissolved oxygen
has to be below 2 mg/L and the temperature above 10°C. The stirred
tank solution is relatively easy to control since the incoming feed
flow rate is small as compared to the volume of the tank (nominal 20-
to 30-day liquid retention time).

The liquid and solid wastes from the bioreactor are used to
neutralize other acid wastes and to seed the bioreactors, which is
discussed in the next section.

<u>Denitrification of Four Ponds</u>

As previously stated, some of the Y-12 Plant's wastes were not
recyclable and were discharged to the four open ponds. The
installation of the recycle and stirred-tank processes in 1976
significantly decreased the nitric acid flows to the ponds. In 1983,
a decision was made to stop discharging wastes into the ponds and to
develop a process to treat the acid waters in the ponds.

Laboratory tests were made and a denitrification reactor was
successfully operated in an open container. Batch tests using
volumes up to 208 L (55 gal) were denitried at rates up to 1000 mg
NO_3 per day per liter of reactor solution.

A pond chosen as a pilot plant had an initial nitrate level of 8300
mg/L and a pH of 2.7. The other ponds had nitrate levels up to
44,000 mg/L, and all were at pH 2.7 or below. The process steps used
for the in situ treatment of the pond are listed in Table 1. In
order to stir the pond, two large pumps (8000-L/min pumping capacity)
were installed to gently stir the ponds. The pond was neutralized by
adding calcium carbonate to a pH of 5 and then sodium hydroxide and
calcium hydroxide to obtain a pH of 7. The organic carbon was added
to the pond as acetic acid, which was also neutralized when the pond
was neutralized. Bacteria from the stirred-tank reactors were added
to seed the pond. The mixing pumps were operated to keep the
bacteria in suspension but not at a stirring rate fast enough to
inject large amounts of air (oxygen) into the water.

TABLE 1

Treatment Process for the Four Unlined Ponds

<u>Process Steps</u>

1. Neutralization

2. Organic carbon addition

3. Bacteria

4. Denitrification

5. Aeration (bio-oxidation of excess carbon)

6. Solids removal

    Settling of Solids

    Flocculation

    Filtration

    Storage

7. Release of water

After denitrification, the water was aerated by using floating aerators to biologically decompose the excess organic carbon that was added to ensure that all of the nitrates were biologically destroyed. The pumps were then stopped, and most of the solids settled. The water was pumped from the ponds to a ferric hydroxide flocculation process, filtered, and released. The water was chlorinated before flocculation to remove algae that had grown in the ponds after denitrification was complete.

The results of the pilot test are shown in Figure 2. As noted, all of the nitrates were destroyed in fewer than 38 days. About 14 of the 38 days were not useful because the pH of the pond was low and no denitrification occurred. During denitrification, the temperature and other process parameters were monitored. The other three ponds were then denitrified, and data on one of the other ponds are shown in Figure 3. Cold water interrupted the reaction for about two months. When the ambient temperature increased by about $10^{o}C$, the reaction restarted and continued until a low pH was achieved. Low pH conditions were common because acetic acid was injected at various times into the pond to adjust the C/N ratio. The denitrification reaction also forms carbon dioxide gas (carbonic acid), which lowers the pH. As noted in the data, a nominal denitrification rate of 300 mg of NO_3 per day per liter of pond solution was experienced. This rate is equal to decomposition of 2270 g of nitrates per day (5000 lbs/day) per pond.

The ponds were then aerated, using floating aerators, and the excess organic carbon was decomposed. Most of the solids were then allowed to settle in the ponds. However, algae formed in the water and had to be removed by chlorination before discharge to the environment. The water was then flocculated via ferric hydroxide and filtered to remove the total suspended solids to less than the discharge limit of 60 mg/L. The water was processed through the flocculation process and discharged at rates up to 200 L/min (60 g/min.) The ponds are now empty and the solids remain in the bottom of the empty ponds. The solids will be stored until a final decision has been reached concerning their disposition.

The water of the water being discharged to the environment was regularly monitored, and some of the data are shown in Tables 2 and 3. As noted, the quality was signficantly purer than required by the U.S. Environmental Protection Agency National Pollutant Discharge Elimination System (NPDES) permit.

Future Wastes

As previously mentioned, one of the program goals was to develop treatment processes for future wastes. A series of steel tanks (2,000,000-1 volume) was installed as treatment reactors. The acid wastes were neutralized with calcium hydroxide and put into one of

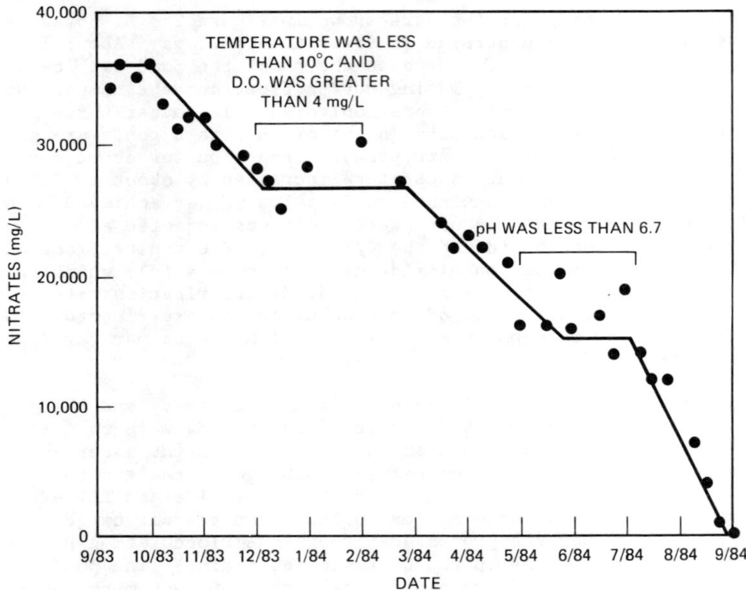

Fig. 2 . NITRATE CONCENTRATION, NORTHWEST POND.

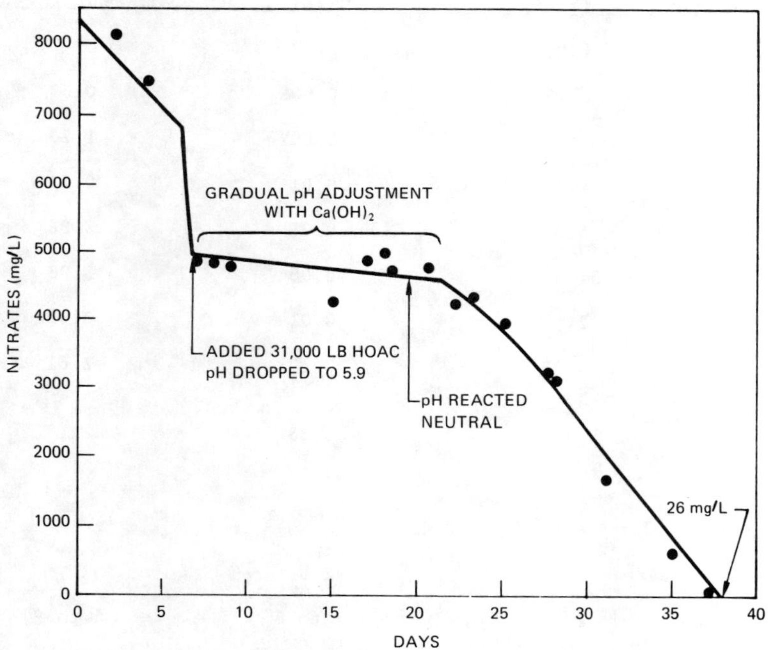

Fig. 3. NITRATE CONCENTRATION, SOUTHEAST POND.

TABLE 2

Contaminants	Original Pond (mg/L)	Treated Water (mg/L)	NPDES (a) Permit Limits Daily Max (mg/L)
Ag	0.006	0.02	0.43
Ca	0.23	0.002	0.69
Cn	-	0.002	1.20
Cr	4.8	0.01	2.77
Cu	5.6	0.01	3.38
Ni	31.3	0.08	3.98
Pb	2	0.01	0.69
Zn	4.25	0.13	2.61
Total Toxic Organics	-	0.33	2.13
Total Suspended Solids			
Oil and Grease	-	2.6	52
pH	2	7 to 8	6 to 9

(a) National Pollutant Discharge Elimination System.

TABLE 3

Other Pond Contaminant Concentration

Contaminants	Original Pond (mg/L)	Treated Water (mg/L)
As	0.14	.06
Al	583	0.15
Hg	-	0.002
Be	0.06	0.001
SO_4	-	3725
Ba	0.27	0.2
F	-	4.2
Ca	847	478
Cl	-	0.96
NO_3	8300 to 44,000	6

Note: Biological tests were required.

the stirred tanks. Acetic acid and biomass were added to the tank.
Very high denitrification rates (up to about 4000 mg of NO_3 per day
per liter are common. A tank is usually decomposed in less than 90
days at a beginning nitrate level of 45,000 mg NO_3 per liter. The
water and solids are then flocculated, and the solids are stored in a
steel tank until a decision is made on the final discharge process
for the solids. The treated water must meet the same chemical
specifications for release to the environment as required for the
ponds, plus two biological tests. The biological tests include
survival of four-day-old fathead minnows for seven days and a seven-
day test with ceriodaphnia. The ceriodaphnia must reproduce, and the
reproduction rate must be equal to the control test group.

<u>Conclusion</u>

The recycle processes are operating and some of the wastes are being
recycled. The waste ponds are now empty and will not be reused. The
new process tanks are operating satisfactorily, and all future wastes
will be batch treated and released. The water quality will meet the
chemical and biological specifications.

TOXIC WASTE AND EMISSION CONTROL MANAGEMENT AT RASHTRIYA CHEMICALS & FERTILIZERS LTD.

K. Mohan, Addl. Chief Engineer

and

S.S. Agarwal, Deputy General Manager
Rashtriya Chemicals & Fertilizers Ltd.,
Trombay Unit, Bombay - 400 074
INDIA

ABSTRACT

Rashtriya Chemicals and Fertilizers Ltd., (Commonly known as RCF in the short from) is the largest producer and distributor of fertilizers in India with a turn-over of around 500 million U.S. dollars per annum. Since some unavoidable toxic emissions do occur during the process of manufacture of fertilizers (which may vary depending upon the process or technology adopted), emission control management is considered as one of the top most functions of the management. This article deals with three specific case studies where reasonable success was achieved in the control of emissions.

INTRODUCTION

The initial operations of the company's plants started in 1965 and over the years the production capacities were increased with newer plants coming up with higher capacities and diversified product range. Presently there are two operating plants, one located at Bombay (known as Trombay Unit) and the other located at Thal (known as Thal Unit). Though management systems at both the units are alike, emission control management at Trombay Unit has got spacial significance not only because of the diversity of products manufactured at this site but also because of its geographical location near a crowded metropolitan city. Actually the decision to locate the plant at Trombay was taken may back in the year 1959 keeping in view the available infrastructural facilities, easy availability of feed stock naphtha (from nearby refineries) and also other imported raw materials such as rock phosphate, sulphur etc. At that time, i.e., in the year 1959, the area adjacent to the factory site was sparsely populated and the population within a 3 kilometer radius was hardly 50,000. However over the years the population of the area has increased to half a million people. Because of this single geographical factor, toxic emission control is considered as one of the top most functions of the management. To get an overall idea of the present day operations of the plants at Trombay Unit an account of the production capacities of the plants is given at Table-1 on the following page.

TABLE-1

			Tonnes/day	Tonnes/annum
A.		**Capacities of plants producing Fertilizers.**		
	i)	Urea (Old)	300	99000
	ii)	Urea (New)	1000	330000
	iii)	Suphala (NPK 15:15:15)	1000	300000
	iv)	ANP (NPK 20:20:0)	1200	361000
B.		**Capacities of plants producing intermediate products**		
	i)	Ammonia	320	105600
	i i)	Ammonia (New)	900	297000
	iii)	Nitric Acid (Old)	320	105600
	iv)	Nitric Acid (New)	750	247500
	v)	Sulphuric Acid	300	99000
	vi)	Phosphoric Acid	100	30000
C.		**Capacities of plants producing industrial products**		
	i)	Methanol	120	36000
	i i)	Ammonium Bicarbonate	12	4000
	iii)	Conc. Nitric Acid	60	20000
	iv)	Sod. Nitrate/Nitrite	12	4000
	v)	Methylamines	12	4000
D.		**Requirement of raw materials and feed stock**		
	i)	Rock Phosphate	925	277500
	ii)	Di-ammonium phosphate	250	75000
	iii)	Muriate of Potash	260	78000
	iv)	Sulphur	1000	33000
	v)	Naphtha	50	16500
	vi)	Associated gas	1800 to 2000	600000

EMISSIONS AND THEIR CONTROL

During the process of converting raw materials into finished products certain unavoidable residual gases are emitted out into the atmosphere. Which are mainly Sulphur-dioxide, Oxides of Nitorgen, Ammonia and Suspended Particulate Matter all of which are considered as pollutants. Ambient air quality standards exist for all these pollutants (excepting for ammonia). It may be noted here that the ultimate objective of all pollution control measures incorporated/adopted is to see that the ambient air quality in and around the factory is within the national standards. For this reason, toxic waste and emission control is considered to be one of the top most functions of the management and a separate department, headed by a senior officer and assisted by a small group of experienced engineers, has been made responsible for the efficient performance of various emission control measures already incorporated and also monitor the progress of those under implementation.

In addition there is a four tier system of Environment Management, Emission control and Safety as given in the organization chart below (Figure-1).

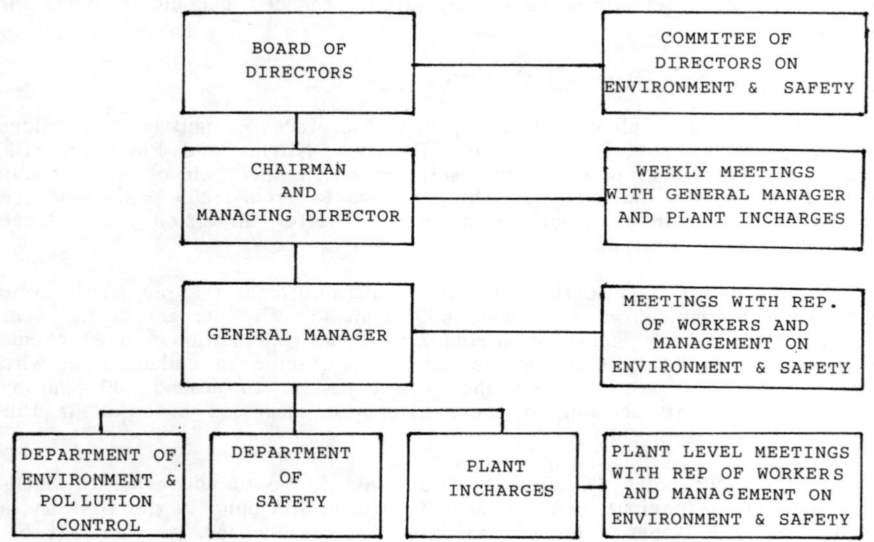

Figure-1. Organization chart for environmental management, pollution control and safety.

<u>Combating Sulphur Dioxide Emissions</u>

The Sulphuric Acid and the Steam Generation plants were the main sources of sulphur dioxide emissions in the the past. However in the year 1977 the old and conventional (single conversion and single absorption type) Sulphuric Acid plant, which was emitting more than 2000 ppm of sulphur dioxide through the stack gases, was modified into double conversion and double absorption process to limit the sulphur dioxide emissions to less than 600 ppm. Another measure that was adopted for reduction of sulphur dioxide emissions was replacement of fuel oil with Low Sulphur Heavy Stock (LSHS) in the first phase and later on in the second phase (ie., in the year 1980) by sulphur-free Bombay High gas in all the boilers of the Steam Generation plant. As a net result of all these measures, in quantitative terms, sulphur dioxide emissions from the factory have come down from about 7.5 tonnes per day in the year 1976 to about 4 tonnes per day in the year 1978 and finally to less than one tonne per day in the year 1980.

<u>Combating Particulate Emissions</u>

Particulate emissions occur mainly from the salt group of plants consisting of Urea, Suphala (NPK:15:15:15), Phosphoric Acid and ANP (20:20:0) plants. In the new Urea plant particulate emissions are kept within the acceptable limit of about 25 mg. per cubic meter by wet scrubber dedusting system.

In the ANP and Phosphoric Acid plants, dust emissions are kept to the minimum by the use of dust collectors containing fabric filter elements. To combat dust emissions from the conventional hot air rotary granulators of Suphala plant, venturi scrubbers were installed in the year 1977 thereby reducing dust emissions from around 2000 mg. to around 60 mg. per cubic metre. All these measures have substantially reduced particulate emissions into ambient air.

Combating Emissions Of Oxides Of Nitrogen (NOx)

The nitric acid plants are the principal sources of emissions of oxides of nitrogen from this RCF comple. The new Nitric Acid Plant at RCF employs the technology of extended absorption with chilled acid or alternatively alkali scrubbing for limiting the NOx emissions to around 300 ppm. and are let out at a height of about 65 meters for better dispersion and reduced ground level concentrations.

By a process modification the NOx emissions from the old Nitric Acid Plant were brought down to around 600 ppm. (on an average) in the year 1985 from the earlier values of around 2000 ppm. In addition a new scheme of Selective Catalytic Reduction was also implemented in collaboration with M/s. ESPENDISA, Spain to limit the NOx emissions to around 300 ppm by February 1986. An account of these efforts is discussed in detail in this paper as a case study.

In the ANP plant NOx levels which used to be of the order of 3000-8000 ppm. in the begining were reduced to around 700 ppm. in the first phase and to around 300 ppm in the second phase by urea addition. Similarly visible plume from the calcium nitrate conversion reactors of the same plant was almost eliminated by making use of an unutilized scrubber with a few minor modifications. Details pertaining to these success achievements are also discussed in this paper as case studies.

Prevention Of Episodal Discharges

Apart from emission control at source points, safety aspects are also given top most priority to see that there are no episodal discharges from the factory into the ambient air, which may be harmful to the residents nearby. Two such areas where episodal discharges can occur are given below. The gas involved in this case is ammonia.

1. Pressurized ammonia storage tank (Hortonsphere).
2. Atmospheric pressure ammonia storage tank.

Realizing the importance of prevention of episodal discharges lot of preventive check-ups are being done at regular intervals to ensure total safety of the system. The Hortonsphere is being inspected externally at intervals of two or three years by ultrasonic devices to ensure that there is no reduction in the shell wall thickness below the allowable safe level. In addition to the regular checking, the storage tank was also emptied twice during the last twenty years for thorough physical internal inspection. The last time when this job was taken up and completed was in March 1985. At the time of this last inspection all the weld joints were thoroughly checked up by die-penetration tests and a few of the joints were also checked radiographically. All these tests indicated no abnormalities worth the mention and the storage tank was found to be in perfect condition (as good as new).

The atmospheric pressure storage tank had also undergone rigorous physical internal inspection during the year 1985 after emptying it out completely and no abnormalities worth mentioning were noticed and the tank was found to be in perfect health. However the external thermocole insulation of the tank was found to have absorbed lot of moisture and consequently lost its insulation characteristics. Also it was falling off. So after the inspections the tank was reinsulated completely with poly-urethane material (cast in situ and backed by aluminium cladding) which is considered to be one of the best insulation materials available today in the world for this service.

Monitoring

Emission control efforts do not stop at providing the pollution control measures outlined earlier but go further to monitor these emissions round the clock at the source points. In addition ambient air quality in and around the factory is monitored with the help of a mobile monitoring van not only to keep all emission control installations under continuous surveillance but also to ensure that they give correct and optimum performance at all times. In addition to this RCF is also financing the operation and maintenance of two fixed ambient air quality monitoring stations near to the factory by the Bombay Municipal Corporation. Also action has been taken to install three automatic (computerized) fixed ambient air quality monitoring stations around the factory premises for early detection of any emissions from the factory so as to plan for preventive/corrective actions before the situation gets aggrievated. These are expected to be operational by middle of 1987.

CASE STUDIES

Reduction Of NOx Emissions From Nitric Acid Plant

Our old Nitric Acid plant is designed to produce 320 tonnes per day of nitric acid of 56% acid concentration in two identical and parallel streams. The plant is an old conventional design of Chemical Construction Co. of U.S.A. with catalytic oxidation of ammonia at about 8 ata. pressure followed by absorption in a bubble cap column to produce 56% acid (Please see Figure-2). A catalytic combustor has been provided in each of the two trains for converting oxides of nitrogen (NO_2 and NO) in the tail gas into NO and nitrogen (and make the stack colourless) with the help of residual gas coming from liquid nitrogen wash section of ammonia plant (which contains about 46% CO, 6% CH_4, 6% H_2 and the rest as inerts). During normal operation of the plant under design conditions, the NOx emissions from the plant (in either of the two trains) used to be of the order of 1700-2200 ppm and at around 2000 ppm on an average (mostly as NO) and the stack emissions were not visible. However local air pollution control authorities insisted on bringing down these emissions to half the earlier levels at the earliest.

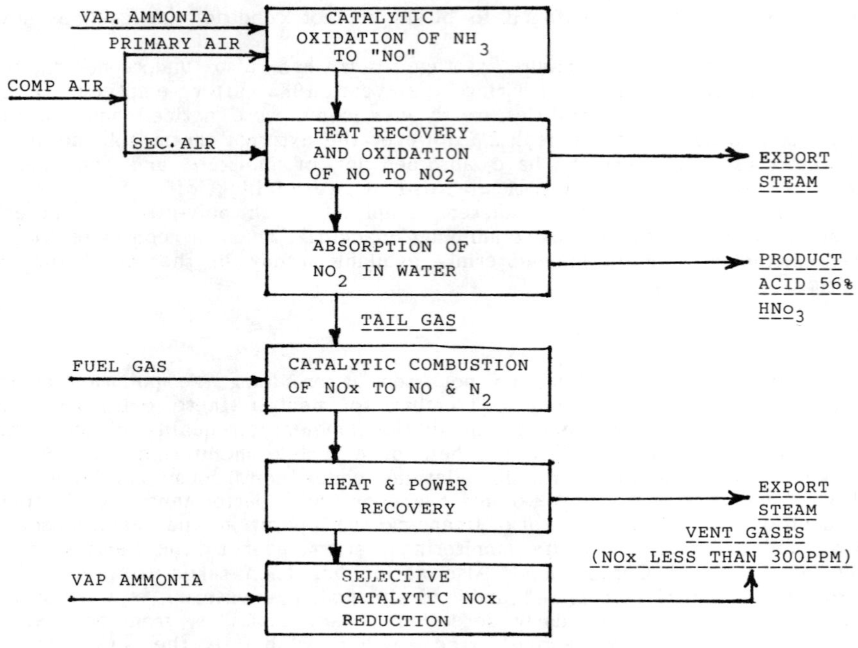

Figure-2. Process flow scheme of
Nitric Acid plant.

To achieve this objective RCF decided on a scheme with know how from M/s. ESPENDISA, Spain where‑in NOx can be reduced to nitrogen with ammonia on a platinum based catalyst. This new scheme envisaged minor Process modifications and was scheduled to be completed in a period of 18 months time (i.e. by Feb. 1986). But quicker results could be obtained due to innovative thinking and it was possible to bring down the NOx emissions to around 700 ppm by minor modifications to the process virtually without any investment. The methodology adopted in this case is explained below with reduction reactions taking place in the catalytic combustor reactor with CH_4.

1) Decolorization $CH_4 + 4NO_2 \;---\; CO_2 + 2H_2O + 4NO$
2) Combustion $CH_4 + 2O_2 \;\;\;---\; CO_2 + 2H_2O$
3) Abatement $CH_4 + 4NO \;\;---\; CO_2 + 2H_2O + 2N_2$

The above reactions in the catalytic combustor take place in a sequential manner and all are exothermic. The oxygen in the tail gas has to be completely consumed before reaction (3) can take place. Usually, the oxygen content in the tail gas was being maintained at about 3% as per design conditions and complete consumption of oxygen was resulting in a temperature rise of around 400°C i.e. for every 1% oxygen, the temperature rise is around 130°C. Since the catalytic combustor is designed to operate at an inlet temperature of about 475°C and a maximum of 600°C (for prevention of damage to the catalyst), an inlet temperature of around 475°C shall always mean a temperature of the order of 875-900°C in the combustor which is not acceptable. To limit and maintain the temperature of combustor below 600°C flow of residual gas to the combustor was being curtailed which was resulting in lesser reduction of NOx and consequent higher emissions. To overcome this problem of high temperature (without many constraints) and at the same time to increase the efficiency of NOx reduction, oxygen content in the tail gas was optimized to less than 3% in the first place and in the second place residual gas from Ammonia plant was replaced with methane rich associated gas.

Though it took considerable time in explaining the new methodology to the plant operators and supervisors, in the end it was a success and NOx levels could be reduced to around 600-700 ppm.

Notwithstanding the above, the original scheme of catalytic conversion of NOx with NH_3 over a platinum based catalyst with know - how from M/s. ESPENDISA was implemented and was made operational in Feb. 1986. Since then the NOx emissions have further come down to levels between 200-300 ppm and the air pollution control authorities are happy about the outcome of the efforts put in the right direction.

<u>Reduction Of NOx From ANP Plant</u>

The ANP plant was built with know-how from M/s. Stamicarbon (Netherlands) and M/s. Hoechst (West Germany) and detailed engineering was done by M/s. UHDE (West Germany). The plant is designed to produce 1200 tonnes per day of nitro-phosphate of grade 20:20:0 ($N:P_2O_5:K_2O$). The process involves digestion of rock phosphate with nitric acid followed by crystallization and separation of calcium nitrate etc., as per flow scheme given at Figure-3. During the process of digestion of rock phosphate some NOx is liberated, the quantity being proportional to the inpurities present in the rock phosphate used and also the operating conditions. The design of the plant provides

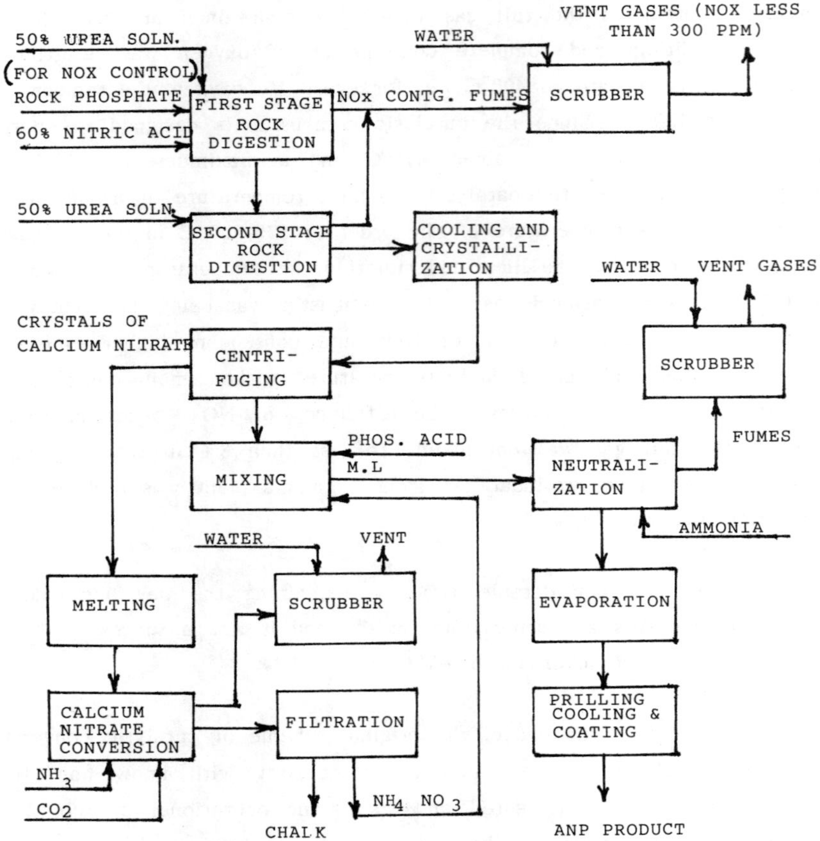

Figure-3. Process flow scheme of ANP plant.

for a scrubbing system with a spray tower with guaranted NOx emission levels of less than 1800 ppm. However during the actual operation of the plant the emissions were very much higher (of the order of 3000 to 8000 ppm) and were not acceptable to the air pollution control authorities. Since no immediate solution was coming-forth from the consultants, RCF went ahead on its own to solve the problem. At the outset it was noticed that the reaction temperature at the rock digestion stage was very much high (i.e. 80-90°C as against design limits of 65°C) and was resulting in higher Nox emissions. The immediate solution therefore was to cool the reactor contents and as there was no such provision, external cooling was resorted to (with cooling water). This reduced the NOx emission levels substantially but were still higher (2000-3000 ppm). So further efforts continued.

As literature survey indicated that it is possible to control NOx emissions with the help of urea, a quick and cautious approach to the problem was attempted and bench scale experiments were conducted in right earnest to establish the reduction in NOx levels based on urea addition. The results obtained with Jordan rock phosphate are shown at Figure-4. From these

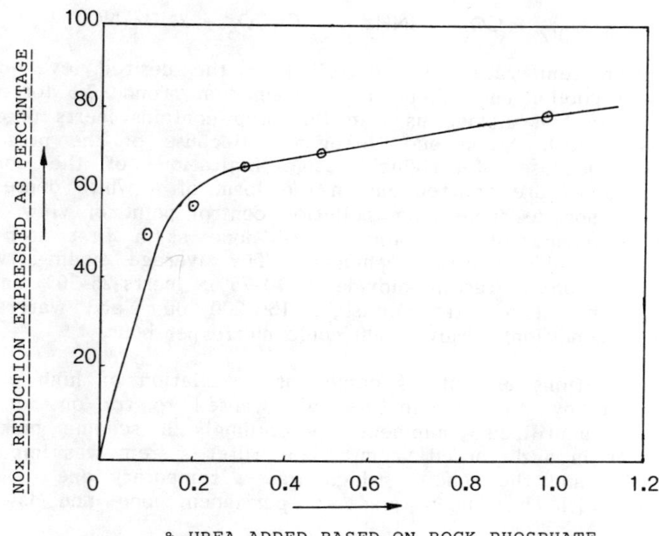

Figure-4.   Bench scale Experimental resutls
of NOx reduction with urea.

results one can easily see that the optimum quantity of urea required to reduce NOx levels is of the order of 0.3% by weight of rock phosphate used. Immediately afterwards plant scale tests were also conducted by adding urea by manual means at the rock digestion stage. Encouraged to these results which almost tallied with the ones obtained from bench scale tests, it was decided to add urea continuously and addition as a solution was found to be more convenient instead of addition as a solid. Accordingly arrangements were made to add urea as a concentrated solution by a permanent system of pumps, flow meters etc.

Though these steps resulted in appreciable reduction of NOx emissions, experience gained in the operating plant indicated that the same quantity of urea added in two steps gave much better results than added in a single step and this method is being followed currently.

Thus the problem of NOx emissions from the nitro-phosphate plant was overcome to the satisfaction of all concerned.

Reduction Of Visible Plume From Calcium Nitrate Conversion Reactors

The calcium nitrate conversion reactors of ANP plant convert calcium nitrate solution into chalk and ammonium nitrate by simultaneous ammoniation and carbonation and the resultant slurry is later filtered to separate ammonium nitrate (from chalk) for recycle in the plant's process. The plant has been provided with two conversion reactors one of which is a stand-by while the other is in line. The overall reaction at the conversion stage is given below.

$$Ca\ (NO_3)_2 + CO_2 + 2NH_3 \longrightarrow CaCO_3 + 2NH_4\ NO_3$$

The reaction temperature is controlled at the desired level of $80°C$ by indirect water cooling and the pH is maintained at around 7.5 for efficient conversion. The carbon dioxide used in this step contains inerts upto 4-6%. (mostly Hydrogen, CO, Argon and Nitrogen). Because of the presence of inerts and also because of original design limitations of the conversion reactors, toxic fumes are emitted out in the form of a white dense plume. As this plume is not desirable from pollution control point of view, analysis and flow measurements of the plume were done as a first step before contemplating a suitable control technique. The average findings were as follows:- Composition : carton dioxide : 70-75%, inerts-25-30%, ammonia : 200-2000 ppm, ammonium nitrate mist : 150-200 ppm and water vapour (as per saturated condition). Flow : 400 cubic metres per hour.

Though suggestions and offers came for installation of high efficiency scrubbers followed by mist eliminators, we wanted to try on our own a scheme of using unutilized equipment. Accordingly a scheme making use of spray scrubber in a discarded vacuum belt filter system was implemented with good results and the scheme which was a temporary one (on a trial basis) to start with has been made a permanent one and is working satisfactorily now.

CONCLUSIONS

Toxic waste and emission control requires lot of concerted efforts on the part of the management, supervisors, workers and technical personnel not only to maintain good and safe working atmosphere inside the manufacturing plants but also to ensure that the ambient air quality in and around the factory area is maintained within the allowable safe levels. If there are any problems in achieving these objectives proper solutions should be found within a reasonable time frame work by all concerned.

HAZARDOUS WASTE MANAGEMENT:
THE MUNICIPAL INTEREST - CADMIUM

Steven O. Salley and Ralph H. Kummler
Department of Chemical Engineering
Wayne State University
Detroit, Michigan 48202
USA

ABSTRACT

The Detroit Water and Sewerage Department (DWSD) provides wastewater collection and treatment services for 3,200,000 people and over 1500 industrial dischargers. The wastewater flow is generally toward the Detroit or Rouge Rivers where major interceptors divert dry weather flow to a single, large treatment plant near the confluence of the two rivers. Significant rainfall creates runoff in excess of the interceptor capacity which, by design, overflows to the Detroit and Rouge Rivers. Potential contamination by hazardous materials, such as cadmium, could be large.

The City of Detroit has an Industrial Waste Control Division to enforce the Federal, State, and Local laws and regulations.

A Data Management System (DBMS) has been designed to assist the Industrial Waste Control (IWC) division of the City of Detroit Water and Sewerage Department in implementing an Industrial Pretreatment Program (IPP). A system mass balance routine provides enforcement engineers with an ability to quantify flow and pollutant levels throughout the sewerage system. Its use for cadmium is described herein.

INTRODUCTION

The Detroit Water and Sewerage Department (DWSD) provides wastewater collection and treatment services over an area encompassing 1,683.5 km^2 (650 square miles) and including 3.0 million people and over 1500 industrial dischargers. About 62 percent of the service area is served by separate sewers; the rest is served by combined sewers.

Once the wastewater reaches the City, flow is generally towards the Detroit or Rouge Rivers where major interceptors divert dry weather flow to a single, large treatment plant near the confluence of the two rivers. Significant rainfall creates runoff in excess of the interceptor capacity which, by design, overflows to the Detroit or Rouge River at approximately 80 locations.

The water quality of the Detroit River is important because of its impact on Lake Erie, as it accounts for 93 percent of the total inflow to Lake Erie. The Detroit River is a major navigational route in the Great Lakes; serves as a major gathering area for waterfowl during the migration season; and is used for recreational purposes, sportfishing, and as a source of water for cities and industries in Southeastern Michigan.

In this work, we will examine environmental protection of the Detroit area from the municipal point of view. In particular, we will look at Detroit's Industrial Waste operation, using the hazardous compounds of cadmium as a specific example.

The City of Detroit has had an Industrial Waste Control (IWC) Division as a separate section of the Water and Sewerage Department since 1972. The primary purpose has been to enforce the Federal, State, and Local laws and regulations to protect the above resources and implement the Industrial Pretreatment Program (IPP) pertaining to the commercial and industrial dischargers.

Under this program, the Publicly Owned Treatment Works must develop and implement procedures to ensure compliance. We have previously discussed the database management system designed to assist the IWC in performing compliance functions (1,2).

BACKGROUND

Mass balancing is an essential feature in the utilization of the Industrial Waste Control data base. Using these modules, IWC personnel can identify the individual contributing elements in the sewerage system. Elements from the data base as well as other known information for other inputs to the wastewater system are aggregated to give a complete picture of the entire DWSD service area or a detailed picture of a portion of the system.

The function of the system mass balance is to give the IWC engineer the ability to quantify flow and pollutant levels for each system basin, for the entire system, and for the major steps in the treatment process. The user of the mass balance program will be able to select the geographical area and chemical species of interest and receive a report on the source and magnitude of each contributor of that species within that area.

The Basin Balance

The mass balance program is formulated around a system of drainage basins. The DWSD service area was divided into 39 drainage subdistricts based on the natural topographical patterns of the region as described in the DWSD Segmented Facilities Plan (3).

A map of these subdistricts is shown in Figure 1. These basins average 2200 acres in size and their boundaries are based on ground slope and small sewer grades. The basin boundaries as well as the locations of

industrial users in the basins have been digitized using the Michigan
Plane Coordinate (MPC) system. The MPC coordinates for the basins and
nearly 1000 industrial users are available from the Detroit Section 201
Final Facilities Plan data files and have been incorporated into the IWC
data base.

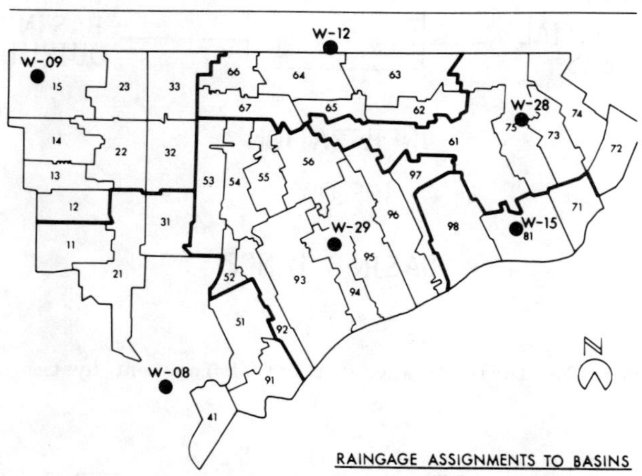

Figure 1. Basin Layout for Detroit Final Facilities Plan

For each basin, the mass balancing program computes the input from
each of the major contributing factors (Figure 2). These factors include
industrial users, residential users, infiltration and inflow sources, and
storm water runoff. Flow from some of the basins is lost via combined
sewer overflows (CSO's). The individual basins, when combined, comprise
the entire DWSD service area. Thus, the summation of the basin components
(4,5) will provide flow rates and concentrations for the wastewater
streams at the Oakwood Northwest Interceptor and the Detroit River Interceptor (Figure 3).

Available Data

A complete mass balance on the system requires input terms that are
outside the Industrial Pretreatment Program data base. Aside from the
industrial user inputs to the system (Figure 4), significant volumetric
and mass inputs are contributed by residential users. The other, non-user

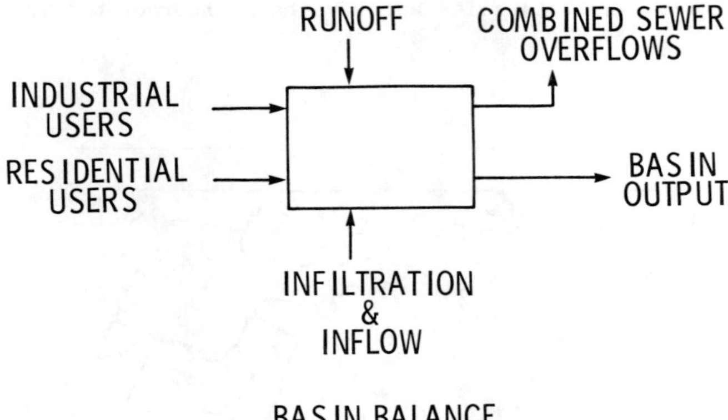

Figure 2.   Basin Balance for Data Management System

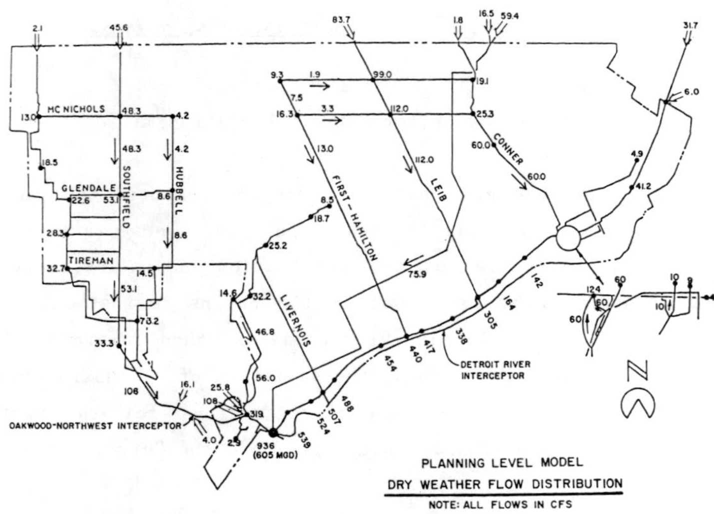

Figure 3.   Dry Weather Flows in Detroit Collection System

contributors to the system are runoff, infiltration, and inflow.   These
latter sources are often significant determinants of the quality of certain wastewater constituents and necessitate the addition of several new

files to the data base and also the incorporation of a large quantity of
information from previous efforts in performing mass balances on the DWSD
service area. The inputs to the mass balancing program that are available
from previous work of Giffels/Black and Veatch and Urban Science Applications have been described previously (1,2,5,6).

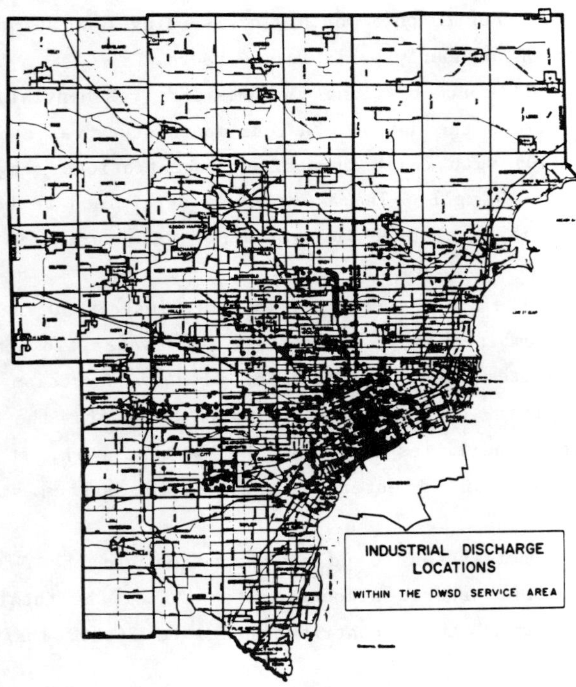

Figure 4. Dot Map Illustrating Industrial Locations in the Detroit Area

CADMIUM BALANCES

Cadmium, given the Chemical Abstract Service number, 7440-43-9, is a
"listed" pollutant, included in the consolidated list of OSHA's Right to
Know Chemicals (8). It is a type B species on the Annual Report on Carcinogens, National Toxicology Program, which means that cadmium compounds
are "reasonably anticipated to be carcinogens". It is on the International Agency for Research on Cancer's list as a "probable carcinogen"
based upon animal evidence. Cadmium dust is listed in 29 CRF 1910 subpart
Z from OSHA and there are threshold limit values for many cadmium
compounds.

In the Detroit Section 201 Final Facilities Plan (9) cadmium was chosen as the single hazardous chemical to be modelled. Hence there is an excellent data base of field data and simulated operations for Cd in the Detroit sewer system, treatment plant, Rouge and Detroit Rivers.

Wastewater from 144 IU's contributed to Cd concentrations during the 1980/81 study period. Analyses of Cadmium were conducted by the Detroit Water and Sewerage department using standard EPA methodology. Sampling at IU locations is conducted on a monthly basis using ISCO automatic samplers which composite on an hourly basis for 24 hours.

Values of Cd concentrations in domestic (residential) sewage have been examined across the nation. A four month study was conducted in the Detroit Area (10) with daily composite concentrations ranging from not detectable to 0.009 mg/l. The average was determined to be 0.004 mg/l, resulting in an input of 3900 lbs of Cd per year to the DWSD collection system.

In Detroit, the average rainfall is approximately 30 inches per year. In the Detroit combined sewer system, the stormwater flow amounts to 14% of the total system flow (9), carrying with it corrosion products and solid debris. Stormwater concentrations were taken from the Giffels/Black and Veatch Final Facilities Plan (5). From that work, it is estimated that 21,500 lbs/yr of Cd enters the collection system via stormwater runoff.

The major contributor to Cd concentrations in the DWSD collection system is industrial users, accounting for 64% of the total Cd loading. Rippel (11) estimated their contribution to be 24,200 lbs/year for the year 1980/81.

The inputs to the collection system are illustrated in Figure 5. The total estimate from all sources is 49,600 lbs/year. Based upon the SWMM model 17,400 lbs/year of cadmium are directly deposited into the Detroit and Rouge Rivers in combined sewer overflows in a typical year. 37,800 lbs/year can be estimated to enter the treatment plant influent. 14,700 lbs per year are estimated to leave the treatment plant in sludge residuals. 24,000 lbs/year leave the plant in the plant effluent. Thus 49,600 lbs/year of Cd are estimated to enter the collection system and 55,200 are estimated to leave the collection system, a difference of 10%. It is unlikely that these numbers are accurate to better than 10%.

Using SWMM model, Qual II and the Plume model, as described by Kummler (9), concentrations are predicted for the mouth of the Rouge River and the Detroit River below the confluence of the Rouge River and the

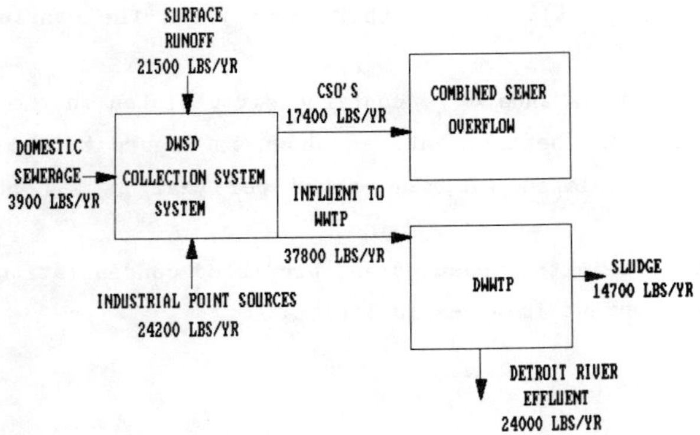

Figure 5. Detroit Collection System Cadmium Material Balance

effluent of the DWWTP. An average concentration of 2.2 mg/l is predicted at the mouth of the Rouge River and an average concentration of 0.6 mg/l is predicted in the Detroit River. The Michigan standard for Cd is 1.2 mg/l.

In the Detroit 201 Final Facilities Plan, the Rouge and Detroit Rivers were divided into five segments each along the borders of Detroit for analysis and modelling results. Then the water quality was assessed as the composite of concentrations within each segment. The segments were chosen on the basis of both input loading and river characteristics.

The resulting profiles for the two rivers are presented in Figures 6 and 7, with the concentration in µg/l displayed as a function of river segment. In the Rouge River, the minimum and average concentrations predicted for the 1979 model rain year are virtually identical because of the dominance of upstream cadmium sources unrelated to the Detroit collection system. The maximum concentrations occur during wet weather events in the upstream segments where the river flow is small. Near the mouth in segments 4 and 5, where Detroit's heavy industrial location lies, the river flow has been significantly augmented with relatively clean cooling

192

water. The Detroit River is comparatively clean, but exhibits a steady increase in the near shore (plume) Cd concentration from the headwater (Segment 1) to the confluence of the Rouge River (Segment 5). Significant northeast side electroplating inflates the Cd concentration during CSO events; the river quickly dilutes that input until the confluence of the Rouge is reached.

The Michigan Cd standard is continuously violated in the Rouge. The Detroit River is much better, but, as shown in Figure 8, the standard is predicted to be violated for many hours per year at all but the head water.

A summary of selected measured and predicted concentrations for Cd in the Detroit environment is given in Table 1.

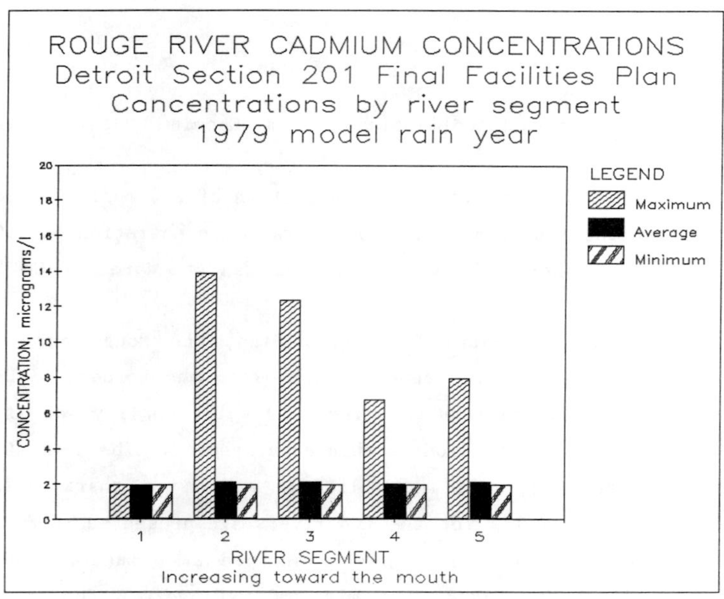

Figure 6. Concentrations of Cadmium by River Segment in the Rouge River

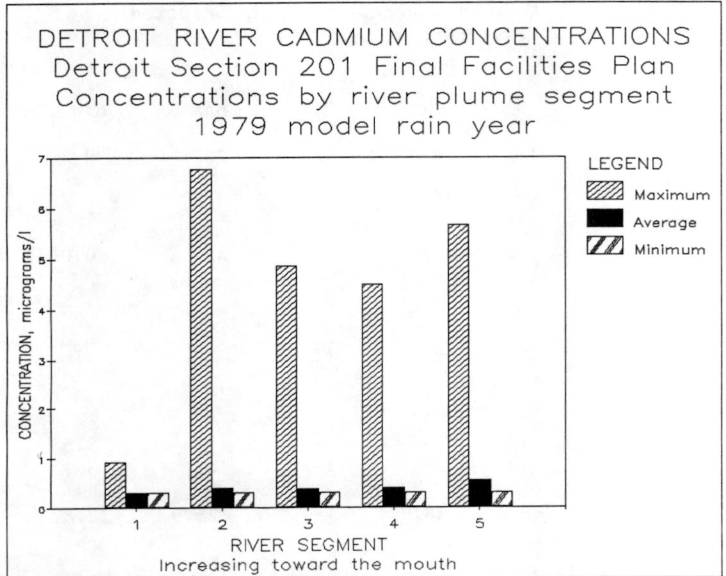

Figure 7.   Concentrations of Cadmium by River Segment in the Detroit River

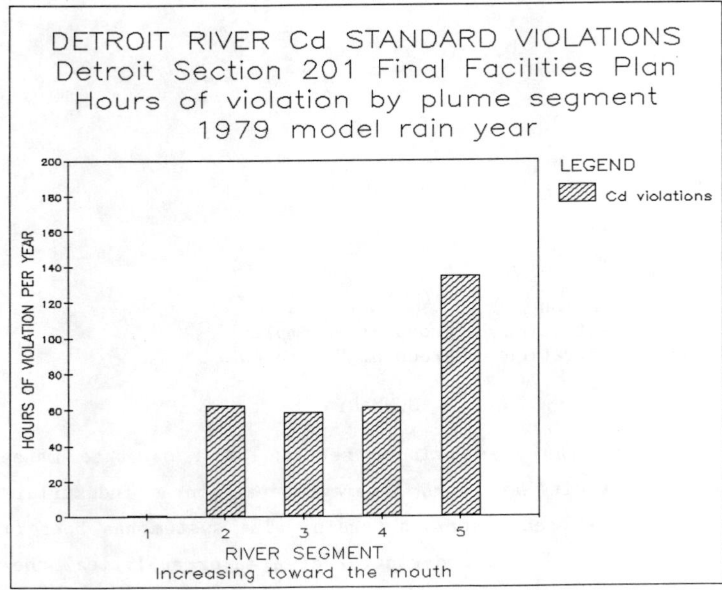

Figure 8.   Hours of Cd Standard (1.1 g/l) Violation Predicted for the
Detroit Detroit River in the Detroit Section 201 Final Facilities Plan Model.

Table 1. Cadmium Concentrations in Selected
Environments, μg/l

	N	$\bar{m}$	σ	Max	Min	Reference
Sweden Combined Sewer Overflows	115	44	220	2200	0	7,8
USA, Various sites combined sewer overflows	22	23	-	350	0.0087	9
Detroit-201 Final Facilities Plan CSO's	125	32	59	-	-	4
Detroit Interceptor Concentrations	12	18	-	34	3.5	10
	12	12	-	42	3.7	11
Predicted DWWTP Influent for Nominal Year	-	18.9	-	25.9	12.82	6
Predicted at the Mouth of the Rouge River	-	2.2	0.4	8.0	2.0	6
Predicted in the Detroit River below the confluence of the Rouge River	-	0.6	.5	5.7	0.3	6
Water Quality Standard		1.2				6

$\underline{N}$ = number of samples
m = means value
σ = standard deviation
Max = maximum concentration in group of N samples
Min = minimum concentration in group of N samples

SUMMARY

The design of the mass balance section of a database management
system for the Detroit Water and Sewerage Department's Industrial Waste
Control Division has been described herein. The system has been created
to assist in the scheduling and control of a major analytical chemistry
sampling and analysis program, to preserve alphanumeric records on each
of 7000 industrial users, and to perform enforcement, surcharge and permitting functions based upon the housekeeping and analytical data values
stored in the data base. It will also perform basin-by-basin or system-

wide mass balancing, potential spill analysis and removal credit analyses for the IWC.

Herein a description of the mass balancing program and its calibration using data obtained in the Detroit Section 201 Final Facilities Plan for cadmium is provided.

REFERENCES

1. Kummler, R.H., and Salley, S.O., SWMM Modelling for Industrial Waste Control Management. In Proceedings of the (SWMM) Stormwater Management Model User's Meeting, Ecole Polytechnic de Montreal, Montreal, 1986.

2. Salley, S.O. and Kummler, R.H., A Data Management System for the Detroit Industrial Pretreatment Program. Elsevier Science Publishers, Amsterdam, 1986.

3. Giffels, Black and Veatch, Detroit Waste and Sewerage Department Segmented Facilities Plan. Book VI, June 1977.

4. Giffels/Black and Veatch, Alternate Facilities Interim Report. Report on the Detroit Section 201 Study, June, 1981.

5. Anderson, J.A., Harlow, C.D., Baranec, J. and Anderson, H.M., Combined Sewer Overflow Modeling in the Detroit 201 Study Using SWMM. In _Proceedings of the USEPA (SWMM) Stormwater Management Model User's Meeting_, Austin, Texas, January 19-20, 1981.

6. Kummler, R.H., SWMM Modelling for the Detroit 201 Final Facilities Plan: Final Results. In _Proceedings of the USEPA (SWMM) Stormwater Management Model User's Meeting_, Ottawa, Ontario, October 1982.

7. Berndtsson, R., Hogland, W. and Larson, M., Mathematical Modelling of Combined Sewer Overflow Quality. In _Urban Drainage Modelling_, Pergamon Press, New York, 1986, 305-315.

8. Berndttsson, R., Hogland, W. and Larsen, M. Combined Sewer Overflow Discharge Quality - State of the Art in Sweden. In Proceedings of the Nordic Hydrological Conference, Rejuyavic, Iceland, 1986.

9. Hogland, W. Berndtsson, R., and Larson, M., Estimation of Quality and Pollution Load of Combined Sewer Overflow Discharge. Proceedings of the Third International Conference on Urban Storm Drainage, Gothenburg, 1984.

10. Funk, Stephen, Chromium and Cadmium Pollution to the Detroit River. A senior thesis submitted to R.H. Kummler, Wayne State University, Detroit, MI., 1979.

11. Rippel, G., A Study of Heavy Metals in the Detroit Wastewater System. Detroit Water and Sewerage Department Report, June 1982.

A WASTE MINIMISATION MANAGEMENT STRATEGY IN CRUDE OIL STORAGE TANKS BASED ON A SUBMERGED JET TECHNOLOGY

S T Kolaczkowski

School of Chemical Engineering
University of Bath
Claverton Down
Bath BA2 7AY
U K

and

A Kidd

Progressive Technical Services Ltd
Factory Road, Sandycroft
Deeside, Clwyd
U K

ABSTRACT

The accumulation of sludge on the bottom of crude oil storage tanks presents a waste disposal problem when the tanks require cleaning for mechanical inspection/repair work.   In the paper, existing sludge management strategies are reviewed and a new strategy is presented based on the application of a submerged jet sludge resuspension technique. This homogenises the sludge with the crude oil in the tank, enabling it to be processed on the refinery. In this way, the valuable hydrocarbon content in the sludge is recovered and does not pose a disposal problem. Available desalting treatment facilities on the refinery are utilised to remove the salts and sediment from the sludge. This process minimises the volume of waste for disposal and oil loss as hydrocarbons in the sludge.

The paper describes the submerged jet technology and its application as an effective technique for sludge control and tank cleaning.

NOTATION

D	nozzle diameter	m
h	distance from nozzle centre-line to the ground plane	m
u	velocity component in x direction	$m\ s^{-1}$

x	longitudinal distance from the orifice	m
y	vertical distance measured from the wall	m
z	transverse direction measured from the centre plane	m

Subscripts:

o	denotes fluid velocity at nozzle exit	
m	denotes maximum jet velocity	
½	denotes the condition when $u = u_m/2$	

INTRODUCTION

In petroleum refineries and storage terminals a deposit known as sludge accumulates on the bottom of crude oil storage tanks (see Figure 1). The crude sludge is a mixture of organic and inorganic products together with water forming a stable emulsion of oil, water and solids. A typical mixture consists of:

hydrocarbons	≈	90%
inorganic solids	≈	3%
water	≈	7%

and in some cases the hydrocarbon content may be as high as 98%. These hydrocarbons contain a complex range of constituents, as encountered in petroleum crude oils. Since the deposits have settled from crude oils, a high paraffinic wax and asphaltene content is expected in the sludge. In a recent study [1] of sludge components identified in stored crude oils, the asphaltene content (hexane insoluble, but benzene soluble fractions) of crudes tested was less than 10%, whereas the sludge contained more than 20%.

As the crude oil is transported from the oil wells and is stored in tankage, its bulk temperature generally decreases as it loses heat. As the temperature drops, paraffin waxes may start to be precipitated and settle as deposits on the bottom of storage tanks.

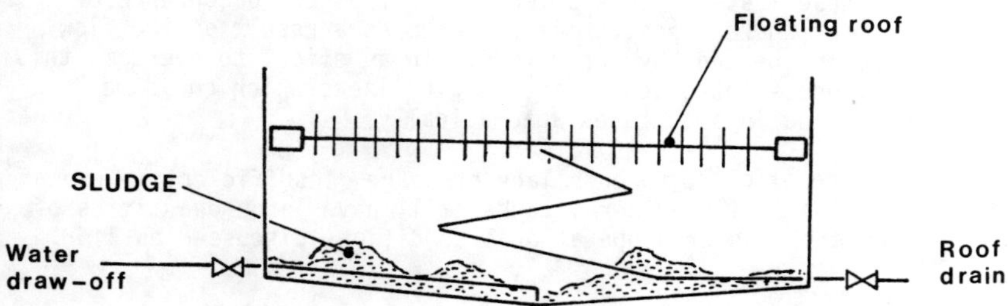

Figure 1   A schematic of a crude oil storage tank - floating roof design

Since the mechanism by which asphaltenes remain "in solution" or are dispersed in the oil medium is complex, and very dependent on the composition of other fractions [2], it is not surprising that asphaltenes are also precipitated and found in the deposits.

The quantity of sludge in crude tanks can build up over a number of years to heights of 4-5 metres, the quantity depending very much on the nature of the crude stored and the sludge management strategy employed. With a trend towards the processing of heavier crude oils with higher wax and asphaltene contents, the rate of sludge build-up is likely to increase.

The presence of sludge in tanks is undesirable for a large number of operational reasons. For example, these may include:

- reduction in storage capacity

- crude unit process upsets due to water/sludge slugs which may increase exchanger fouling and upset desalter operations

- trapped water in the sludge causing tank bottom corrosion

- inaccuracies in tank gauging

- potential roof damage (floating roof tanks)

- trapped water in sludge creating a potential for boil-over in the event of a tank fire

- an increase in tank maintenance and cleaning costs

- hazards associated with tank cleaning

- disposal problems of sludge after cleaning

SLUDGE MANAGEMENT

A variety of techniques have been used to control sludge build-up in crude oil storage tanks.

(a) Propeller Mixers

Traditionally, side entering mixers have been used to homogenise the contents of the tank and control sludge build-up. Earlier mixer installations were of the fixed angle type, and Figure 2 illustrates a typical installation. These installations often suffer from the occurrence of regions in the tank where deposits may build up as a result of the flow patterns induced in the tank by the mixers. In an effort to overcome this problem, mixer vendors introduced swivel angle mixers which could be adjusted manually and in some cases automatically.

Despite these efforts, through lack of usage, insufficient maintenance and undersizing of propeller mixers, tanks still have large quantities of sludge which cause the adverse operational conditions discussed earlier.

(b) Fixed Jet Nozzles

A small number of crude oil tanks have fixed jet nozzles to mix and resuspend the sludge. The nozzles may be fixed on a manifold ring and energised by a recirculating pump or during tank filling. In our experience, the number of these installations is small, and they are likely to suffer from sludge mountain regions as a result of inducing fixed flow patterns, and limited pumping energy.

(c) Tank Heating

Some of the crude oil tanks that were designed to store heavy crude oils that may need to be kept warm to remain pumpable have steam heating coils or manifolds. Where possible, operators will utilise heat in combination with mixers to dissolve and disperse sludge build-up.

Some locations have utilised externally positioned heat exchangers through which the contents of crude tanks have been circulated and heated to raise bulk temperatures. The combination of heat and mixing energy is an effective method of dispersing sludge, although the associated energy costs may be substantial. Refinery operators may not always appreciate the magnitude of these costs. The high energy costs mean that operators generally use this technique once sludge has built up to a significant level. Consequently, they would have experienced some of the operating problems outlined earlier.

TANK CLEANING

Having failed to manage the build-up of sludge, tank operators need to clean their tanks periodically (5-15 years) for maintenance or inspection. The following range of techniques is practised:

(a) When the floating roof cannot be landed safely

In some cases, sludge build-up prevents operators from being able to land the floating roof safely. In this instance, the following range of techniques has been used to reduce sludge levels:

- Circulating and heating the contents by means of internal heating coils or external heat exchangers. Where possible, mixers are energised and in some instances proprietary

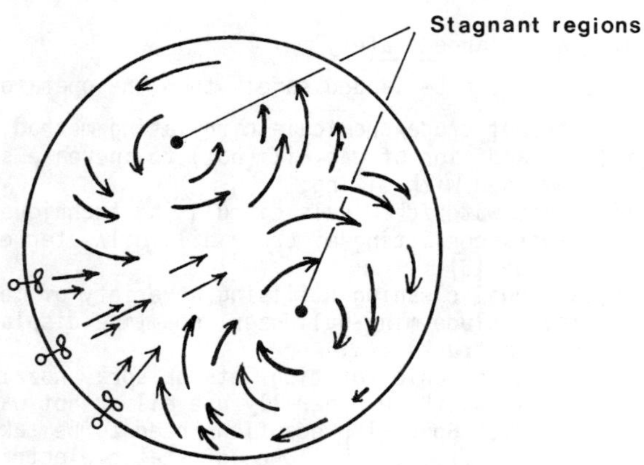

Figure 2   Schematic of a typical side entering mixer
installation

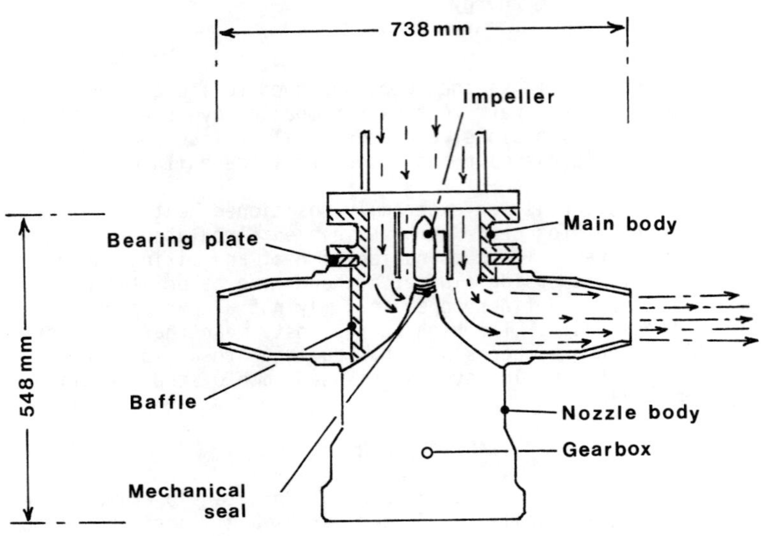

Figure 3   Sectional drawing of a shell mounted P-43
indicating fluid flow

       chemicals (solvents/dispersants) are added to assist with the
sludge dispersion process [3].

-    Circulating the tank contents and injecting hot crude/chemicals
through the roof-legs to reduce sludge levels in the tank at
identified problem areas.

(b)   <u>Roof can be landed safely</u>

If the roof can be landed safely then the operator may consider:

  (i)   the hot crude/chemicals circulating method;

 (ii)   the addition of gas oil (hot) to increase sludge solubility
combined with mixing;

(iii)   a hot water/chemicals circulating technique, resulting in four
layers consisting of crude oil, oil/water emulsion, water and
sludge [3];

 (iv)   a manual cleaning utilising a variety of mechanical aids; these
may include mini-bulldozer, positive displacement pumps,
vacuum trucks *etc*;

  (v)   using portable rotating jets or spray nozzle tank cleaning
machines; these generally use oil or hot water as a washing
medium. Special precautions need to be taken to minimise the
dangers from the build-up of static electricity since the jets
are discharge in a vapour space. In some cases chemicals are
added to assist in the dispersion process.

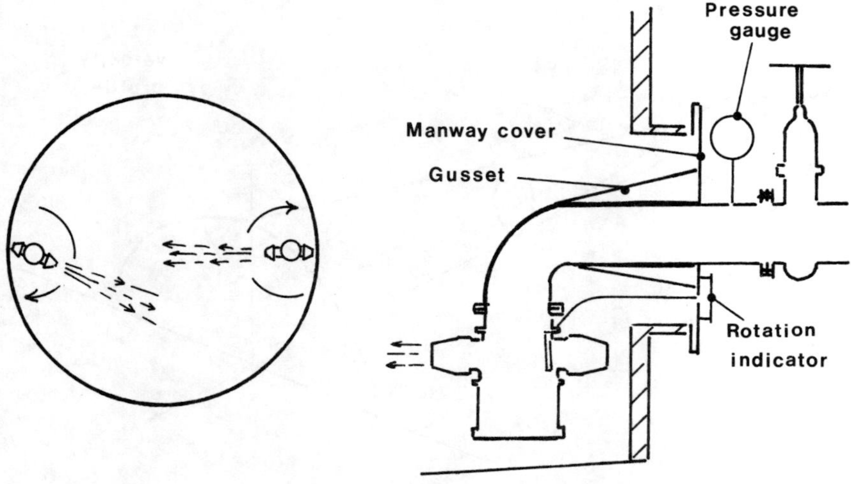

Figure 4a   A schematic of a side (tank shell)
mounted installation

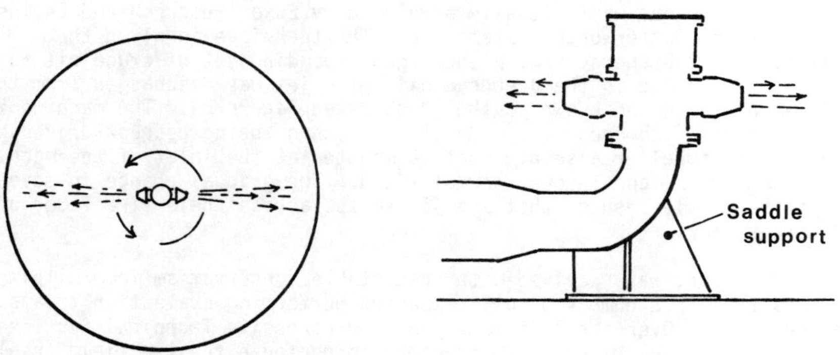

Figure 4b   A schematic of a centre mounted
installation

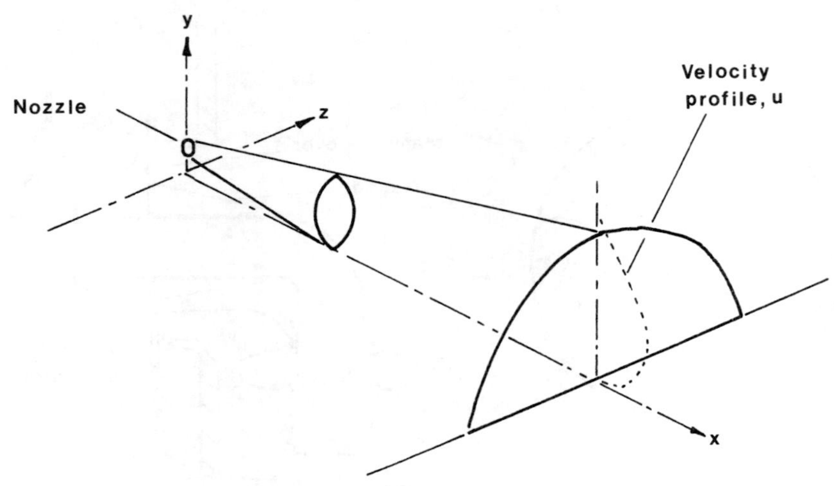

Figure 5   Transition of a free jet to a three-
                  dimensional wall jet

NEW SUBMERGED JET SLUDGE CONTROL TECHNIQUE

Background

   This technique was initially developed by Exxon Research and Engineering Company and Butterworth Systems Inc. The technique involved the
utilisation of the energy from a submerged expanding jet of crude oil to
resuspend the sludge in the overhead oil. The jet was discharged from the
nozzle(s) of a machine known as the 'P-43' (see Figure 3). The machine is
fluid dynamically-powered, the body of the machine being gearbox-driven by
a fixed-blade impeller assembly that is mounted at the inlet of the machine
and rotated by the continuous flow of fluid. The gradual change in movement of the nozzle ensures that the jet sweeps across the entire floor of
the storage tank.

   The product was received with predictable conservatism from within
the industry with a number of oil companies performing evaluation trials at
selected sites. Over the last four years, Progressive Technical Services
Ltd, a major UK tank cleaning contractor, conducted extensive field trials
with the P-43 developing a new approach to on-stream tank cleaning and
sludge control. Having previously had many years of experience with
conventional tank cleaning techniques, the P-43 offered the way forward
with a more efficient, less hazardous technique.

   In 1986 the company acquired all rights to the P-43 (patents, marketing, manufacturing) and now market the acquired expertise on tank cleaning
and sludge management techniques throughout the world.

Machine location

The P-43 may be positioned either at the side or in the centre of
a crude oil tank as shown in Figures 4a and 4b. In the case of the centre
mounted installation, both nozzles normally discharge throughout the entire
$360°$ rotation of the machine. One nozzle however can be blanked off to
accommodate pump flow limitations. In the case of the shell mounted installation (Figure 4a) only the nozzle facing away from the tank shell discharges as the machine gradually rotates in a clockwise direction. An
internal baffle in the machine prevents the nozzle facing the tank shell
from discharging. The flow'change-over' between the nozzles is smooth and
commences when the nozzles are approximately tangential to the tank shell.

Jet expansion

In both shell and centre mounted installations the jet issuing from the
nozzles will initially behave as a free unbounded jet, having an initial
zone where smooth core flow is surrounded by a progressively thickening
turbulent shear layer [4]. However, at a distance which will depend on
the clearance h/D, the jet and the ground plane will begin to interact,
and a new structure of flow will begin to develop approaching the characteristic shape of a three-dimensional wall jet. Figure 5 illustrates this
transition. As the maximum velocity (u_m) moves towards the plane, a steep
velocity gradient develops close to the surface of the plane. It appears
that although the shear on the ground plane is known to play a very small
part in the overall momentum balance, the interaction with the ground
plane has the effect of increasing jet width. However, the rate of decay
of the maximum velocity with distance from the nozzle is very close to
that of a free jet [5].

Experimental results [6] indicate that for a wall jet the maximum
velocity (u_m) as a function of jet length (x) may be represented by:

$$u_m = \frac{u_0}{0.125} \left(\frac{D}{x}\right)$$

(1)

for $\frac{x}{D} > 10$, where D is the nozzle diameter

and the growth of the half-velocity widths $y_{\frac{1}{2}}$ and $z_{\frac{1}{2}}$ in the y and z
directions may be approximately represented by:

$$y_{\frac{1}{2}} = 0.045x + 1.111D$$

(2)

and $$z_{\frac{1}{2}} = 0.19x + 0.987D$$

(3)

for $\frac{x}{D} > 15$

The application of these equations to estimate the characteristic
shape of a jet is illustrated in Table 1. In this example, three P-43
machines are equi-distantly positioned on the circumference of a 72 m
diameter tank. As estimate is made of the shape of one of these jets at
a distance of 36 m from the nozzle. This would correspond with the centre
of the tank.

Considering the shape, velocity and gradual radial movement of this
jet, it is no wonder that considerable improvements have been obtained in

TABLE 1

Characteristics of a jet at a distance of 36 m
from the nozzle for a specified nozzle exit condition

$x = 0$ (nozzle)	Flow rate $= 750$ m^3 h^{-1} $D = 0.1$ m $u_o = 26.5$ m s^{-1}
$x = 36$ m	$u_m = 0.58$ m s^{-1} $y_{\frac{1}{2}} = 2$ m $z_{\frac{1}{2}} = 7.5$ m

resuspending sludge and homogenising the contents of a crude oil tank. The
entire floor area of the tank is swept by direct jet action rather than
relying on a combination of direct action and the establishment of weaker
secondary circulation currents in the tank.

Crude/sludge blend processing requirements

Once the sludge has been homogenised in the crude oil it can then be
processed on the refinery. If the P-43 is operated on a routine basis
(instead of mixers) to control sludge and homogenise tank contents, then
normal processing requirements are followed. However, if the technique is
applied to the cleaning of a tank with a high sludge content, then the
resulting crude/sludge mixture needs to be further diluted with a fresh crude
mixture to reduce the sediment and water content (B S & W) to a level acceptable for processing.

In Progresive Technical Services Ltd's experience to date, of having
cleaned more than 30 tanks with this technique over the last $2\frac{1}{2}$ years, they
have not been advised of any adverse effects on fouling in crude oil pre-
heat exchangers or process upsets. Refinery reports have said that no
noticeable changes were observed in the performance of the heat exchangers
after or during the processing of the sludge/crude mixtures.

Recent case study - tank cleaning operation

In 1986 the submerged jet sludge resuspension technology was applied
in an 81 m diameter crude oil tank that had such a severe sludge problem
that the operators were afraid to land the floating roof, fearing roof
damage. This example has been selected as a case study to illustrate the
effectiveness of this technology in dealing with a severe problem. The
tank had mixers installed. However, it is not possible to comment as to
the manner in which they had previously been operated. Sludge levels were
known to be in the region of 2.3 m in places with an average of 1.2 m
across the tank (see Figure 6). Before the P-43s could be inserted, it was
necessary to reduce the high sludge regions in order to land the roof
safely. This was achieved by installing a hydraulically operated pump on
the surface of the roof, and through a proprietary designed nozzle (inserted
through roof leg supports) the submerged jet technology was applied to

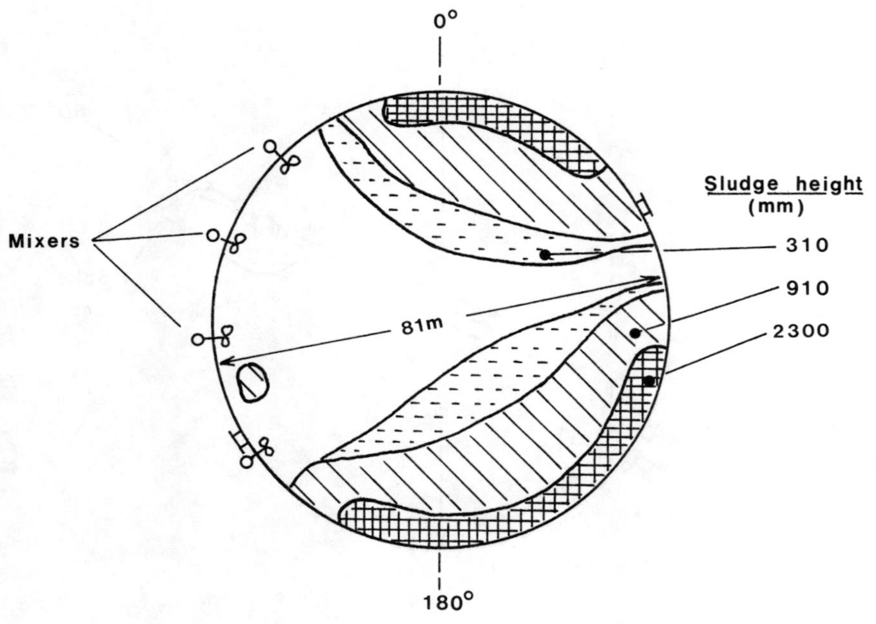

Figure 6   Case study - estimate of sludge levels in
                an 81 m diameter crude tank

resuspend and disperse the sludge in the peak areas. This technique was
extremely successful, and after 7 days the roof could be landed. There
were only two suitable positions where the P-43s could be installed, and
these were at 180° apart (see Figure 7). The P-43s were energised by mobile
pumps using flexible cargo hoses for suction and discharge connections.
During the first phase of the P-43 operation, only one of the machines was
energised, using two mobile pumps in parallel to provide the necessary
additional energy to reduce overall sludge levels at the maximum range of
the jet. The technique successfully reduced the sludge levels from an
estimated height of 2 m down to 250 mm in a 4 day period. The operating
characteristics for the P-43 operation are illustrated in Table 2.

The blend of sludge and crude was then processed as a 10% stream with
fresh crude on the crude distillation unit (CDU). Key operating parameters
were carefully monitored on the CDU and there were no detectable changes in
the performance of heat exchangers either during or after this processing
operation. The resuspended sludge was successfully processed.

CONCLUSION

The application of a submerged jet sludge resuspension technique
(*via* the P-43) where the jet sweeps across the entire floor of the tank
provides an effective method of overcoming the problems of stagnant regions

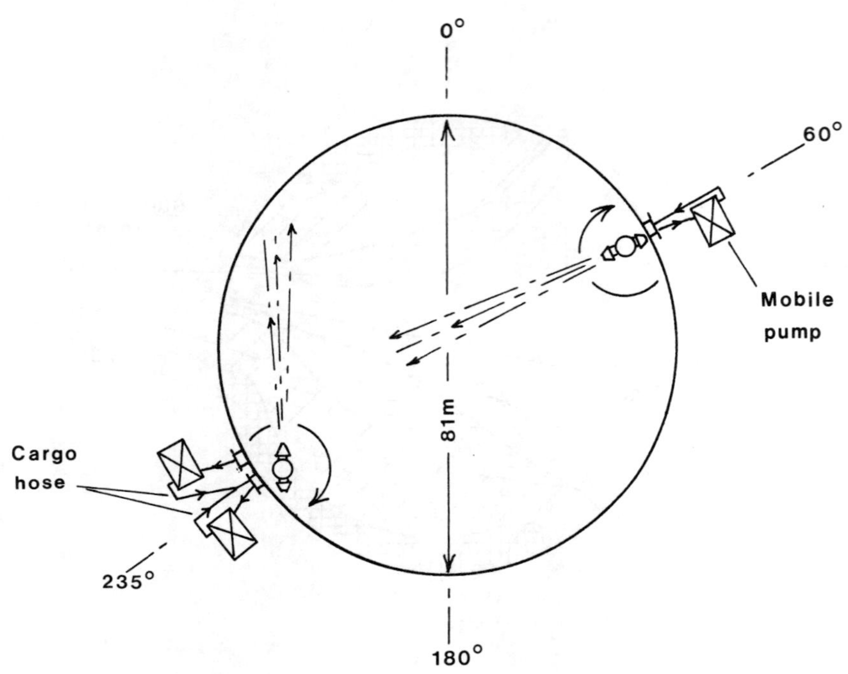

Figure 7   Case study - location of P-43s during the
tank cleaning operation

TABLE 2

Case study - operating characterictics for
the P-43 tank cleaning operation

P-43 location, degrees	235	60
Nozzle, mm	100	100
Flow, $m^3 h^{-1}$	1254*, 1090+	1090+
Time, h	24*, 24+	24+
Nozzle rotation, $h\ rev^{-1}$	4-5	4-5

* First phase of operation - one P-43 energised by two pumps for
a 24 hour period

\+ Second phase - 2 P-43s energised simultaneously

in the tank where deposits may build up.

The rapid way in which sludge was resuspended in a tank cleaning operation illustrates the effectiveness of this technique. Routine operations of the P-43 in a permanently mounted installation will provide the basis for an effective sludge control management strategy. The frequency of P-43 operation will be dictated by the severity of sludge deposition, but a 12-24 hour period of operation every month should be sufficient.

REFERENCES

1   Mochida, I, Sakanishi, K, Fujitsu, H, "Stored crude oil sludge components identified", *Oil and Gas Journal*, 1986, Nov 17, 58-63

2   Speight, J G, "The Chemistry and Technology of Petroleum", Marcel Dekker Inc, 1980, pp245-252

3   Barnett, J W, "Better ways to clean crude storage tanks and desalters" *Hydrocarbon Processing*, 1980, Jan, 82-86

4   Davies, P O A L, Fisher, M J, Barratt, M J, "The characteristics of the turbulence in the mixed region of a round jet", *J Fluid Mech*, 1963, 15, 337-367

5   Davis, M R, Winarto, H, "Jet diffusion from a circular nozzle above a solid plane", *J Fluid Mech*, 1980, 101, 201-221

6   Narain, J P, "Three-dimensional turbulent wall jets", *The Canadian Journal of Chemical Engineering*, 1975, 53, 245-251

BUILDING A PLANT FOR THE SAFE
REMOVAL AND DISPOSAL OF
CROCIDOLITE ASBESTOS INSULATION

David Chambers
Director
DCE Group Limited
Thurmaston, Leicester, LE4 8HP,
U.K.

ABSTRACT

The removal of blue crocidolite asbestos lagging and insulation is a
difficult and potentially dangerous operation. It is a known carcinogen
and great care is needed to keep airborne concentrations to an absolute
minimum. In 1983 British Rail needed to find a contractor to design and
build a plant for the safe stripping and hygienic disposal of this
material from old railway vehicles. The chosen contractor adopted an
unconventional approach to the project. Specialist suppliers were
selected at the earliest possible stage and the relevant health and
environmental authorities were encouraged to have an active and close
involvement throughout. Together with the contractor the result was a
very effective team to tackle this difficult project from start to
finish. The plant which is described has been in continuous operation
for three years, and has achieved all expectations. It is suggested that
this approach could be successfully applied to similar toxic waste
problems in the process industries.

INTRODUCTION

When a process has been identified as a source of unacceptable pollution,
the conventional approach is to call in the company or consultant
engineer responsible. He then tries to define the problem and develop a
specification against which competing specialist suppliers can bid.
There are several reasons why this method is unsatisfactory when toxic
waste is involved. It is the purpose of this paper to describe an
alternative approach which can produce a much better solution, and
demonstrate it using a case study.

In any pollution problem involving toxic materials, there are three main
interested parties ~ the plant management, the specialist suppliers and
the relevant health and environmental authorities. Although the latter
may actually have instigated the action they then have little involvement
until they inspect the finished installation to ensure that it conforms
to the required standards. For their part the plant management may
regard the authorities as rather like policemen and believe that
involving specialist suppliers to early risks premature commitment to
those suppliers, and therefore to their (perhaps high) prices.

The suppliers themselves may be reluctant to divulge too much in the
early stages in case all their work is subsequently used by a less
scrupulous customer as a basis for competitive quotations. In the
meantime the authorities, confined to a consultative role, think their
powers of enforcement may be compromised, and therefore are cautious and

legalistic in their advice.

Although these attitudes are understandable, surely when combatting a
toxic waste problem a team approach which draws the maximum contribution
from all parties concerned is what is really needed. For example greater
involvement of the authorities in the design stage and throughout the
project should in fact minimise the need for their enforcement powers, as
well as increase their own expertise. Furthermore, how often has a
customer bought, not the best solution to his problem, but simply the one
that seemed to come closest from several competing suppliers? Overcoming
that means selection of key suppliers must come much earlier, as soon as
the problem is identified, and not in the normal way just before
contracts are placed. Obviously this choice is vital and the following
is a suggested check-list:-

Is the supplier able to measure actual emissions or indicate the
likely emission from a new plant?

Can he get his priorities right? Is he really trying to solve the
problem or, simply hard sell his product?

Does he understand precisely what is required in terms of reduced
risk?

Is his equipment compatible with the pollutant involved? Has he
references on similar applications?

Is his equipment compatible with the process? (Continuous processes
require it to operate continuously).

If he knows about collection for example, does he also know about
disposal? (There is no point in solving one problem to create a
secondary problem elsewhere).

Is he open about his limitations? Few can be experts in all
aspects.

Is his equipment suitable for handling toxic materials? It is of
good quality mechanically reliable and fail-safe? (A visit to his
factory should quickly answer this one).

Is his equipment easy to maintain, to inspect and to test?

Is the supplier commercially sound and competitive?

Once all the main parties have been assembled at this early stage, they
can begin work as a team to tackle the problem. The author has
experience of alternative approach,and the following case history may be
of interest in proving its effectiveness.

CASE HISTORY

For some years an engineering company had held a contract with
British Rail for the dismantling and disposal of its old railway
vehicles. The contract included only those carriages that had been
certified asbestos free. However, British Rail also possessed hundreds

of carriages fitted with lagging and insulation made from blue
crocidolite asbestos.

Recognising the potential danger, British Rail looked for a contractor
who was prepared to make the necessary investment for its safe removal
and hygienic disposal. The sheer size of railway carriages demanded
large secure storage areas and close proximity to a rail head. The
company had a site with the necessary facilities, although it was near to
a city. Nevertheless, it was decided to enter into negotiations with
British Rail offering to accept vehicles containing asbestos, to strip
them and dispose of the waste. In so doing the company undertook to
design and install a completely new purpose built plant, and to give
guarantees to British Rail that there would be no risk to the work force
and public at large.

Once agreement was reached in principal, and before starting the design
of the actual installation the company identified all of the interested
parties. In particular it was clearly recognised that the control and
the filtration of asbestos laden air within the plant would be of major
importance, and one of the first steps was a survey of dust control
companies capable of bringing the appropriate expertise to this aspect of
the new plant. At the same time an engineering architect was employed to
design the overall installation. After a series of preliminary
discussions about the proposed plant, a nucleus of an engineering team
representing the several disciplines involved was formed. It then
commenced a series of meetings with the U.K. Health and Safety Executive,
the Planning and Environmental Health Departments of the local city
council, and other interested parties.

It soon became obvious that a risk of incompatibility existed between the
various designs and types of equipment being put forward. The company,
therefore, insisted that regular meetings be held between all the
suppliers to ensure frank and frequent communication. Everyone involved
quickly became enthused and responded most positively to the challenge.
Visits were made to a number of similar, but less sophisticated
installations, and the benefits of mutual cooperation were quickly
appreciated. The result was a working document that covered every aspect
of the installation from the design of the building itself right down to
the type of personal protective equipment the workforce would have to
use. Operational procedures were written up with the involvement of the
Health and Safety Executive and refined to an exceptional degree before
any major contracts were signed. A description of the plant and its
operation based on that working document indicates the scope of that
detail.

PLANT DESCRIPTION

The building is capable of completely accommodating a full length
passenger carriage. It is weather proof and air tight with smooth ledge-
free internal surfaces. To protect the internal rendering in the main
work zone, the walls are lined with heavy steel plates welded into
position and fully sealed. Figure 1. shows a plan view of the facility.

It is possible to vacuum clean and finally wash down all internal
surfaces. The solid floor is sealed and waterproofed and gullies and
drains lead to an internal underground sump. The work zone just after

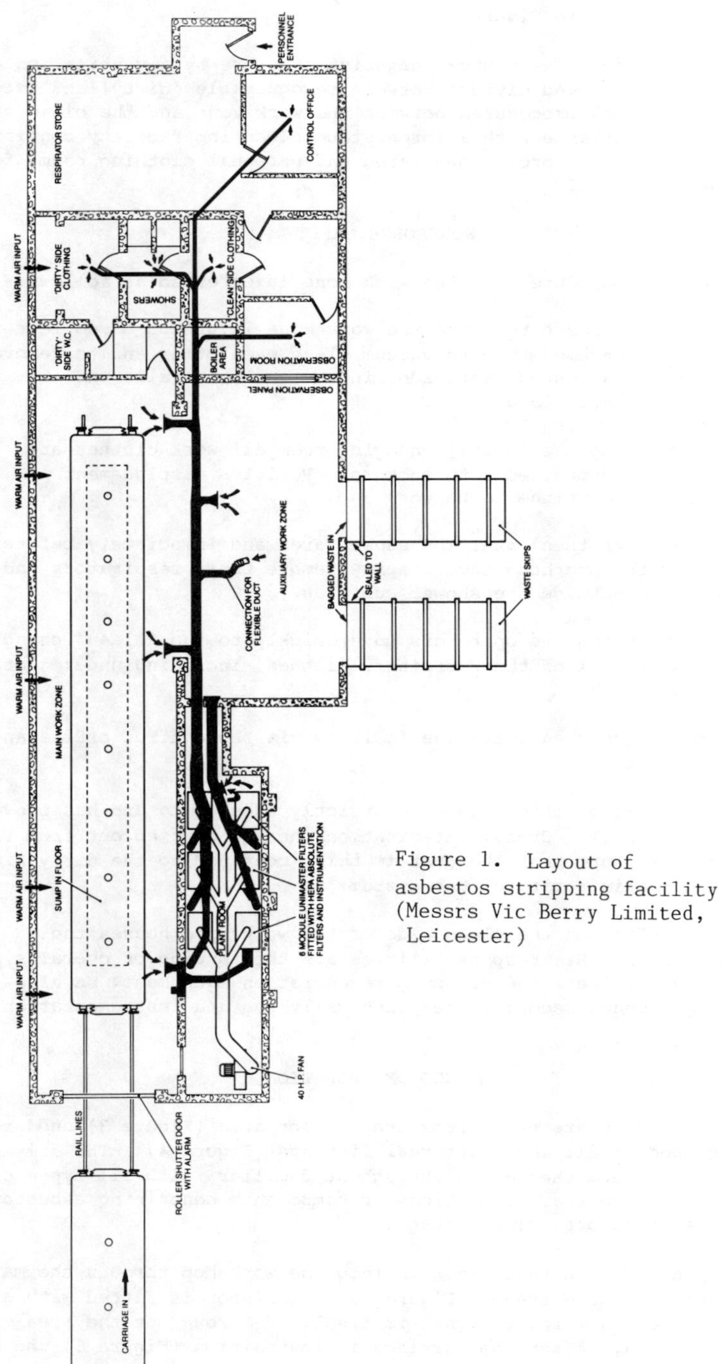

Figure 1. Layout of asbestos stripping facility (Messrs Vic Berry Limited, Leicester)

completion in shown in Figure 2.

The whole building is kept under negative pressure by a ventilation and dust control plant, and divided between recognisable 'dirty' and 'clean' areas. The control procedures between the work zone and the clean side are of prime importance with a three stage cleansing facility consisting of a dirty clothing store, shower area and personal clothing room, for use by the workforce.

WORKFORCE SAFETY

A strict exit proceedure from the work zone is enforced as follows:-

> The operatives are to leave the work zone in groups of no less than two. This enables them to vacuum clean each other and to remove most of the asbestos waste adhering to their overalls and the outside of their face masks.

> After entering the 'dirty' changing room all work clothes are removed and deposited into lockers. Positive displacement respirators continue to be worn.

> The operators then enter the shower area and immediately before entering the overhead shower spray remove their respirators and place them outside the shower cubicles.

> After showering the operators move quickly to the 'clean' changing rooms where all of their ordinary clothes, including under clothes, are kept.

> All operators then leave the facility via the control office and are 'logged out'.

This procedure in practice has been strictly adhered to and has proved to work extremely well. Cross contamination checks, carried out from time to time to make sure that the dust is being confined to the dirty side, have proved the integrity of the system.

The control office on the clean side of the work zone houses the plant supervisor. His responsibilities are the issuing of overalls, new face masks and filters and air pump regeneration equipment. He also maintains personnel records, the plant daily log and instrumentation data.

METHOD OF OPERATION

The rail carriages are taken from the storage area (Figure 3) and are first stripped of all loose internal fittings (Figure 4). The company's engineers have made themselves throughout familiar with all types of carriage to ensure that no sections or components containing asbestos are disturbed at this preliminary stage.

Each carriage is then taken in turn into the workshop through the main roller shutter door entrance (Figure 5). This door is fitted with a limit switch and a warning signal is displayed throughout the area when the door is open. After the carriage is in position (Figure 6) the main

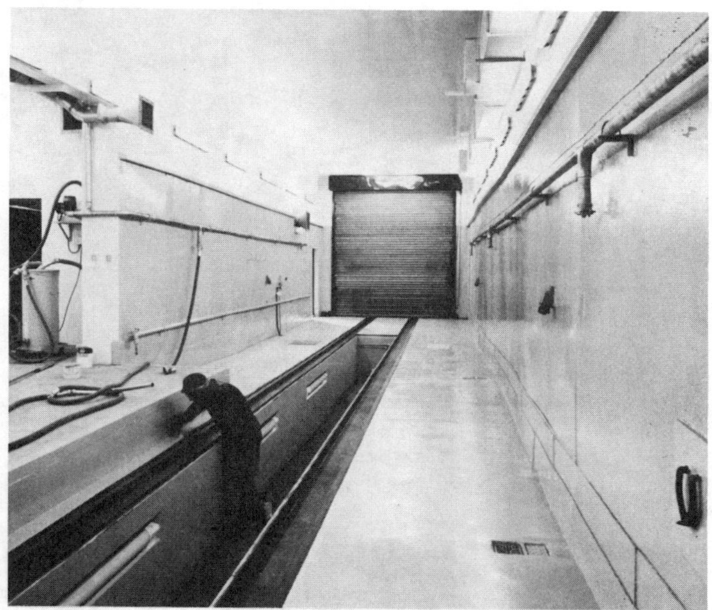

Figure 2. Main work zone just after completion.

Figure 3. Storage area for carriages

Figure 4. Removal of loose fittings.

Figure 5. Carriage ready to enter plant.

door closed, all other openings into the building are secured and seals checked for integrity. The operators then enter and the extraction and ventilation plant is switched on.

The task of removing the asbestos lagging is labour intensive. The outer skin of sheet metal is first removed to expose the lagging, which is then soaked with water and stripped out using both manual and electrically powered tools such as scrapers, wire brushes, chisels and similar implements. The waste removed is promptly bagged and the bags sealed (Figure 7). The bags are made of heavy gauge polythene and marked to indicate they contain toxic waste. These are finally wet wiped clean before loading into disposal skips (Figure 8). The skips, specially designed by the company's engineers, are located outside and seal on to two ports in the work zone wall (Figure 9).

After all the asbestos has been removed, the entire railway carriage and all of the internal surfaces of the building are vacuum cleaned. High pressure water sprays are then used to wash down completely all of the equipment within the building and the fabric of the building itself. The extraction and ventilation system is of course left running during this phase.

As the carriage is being finally dismantled, the pieces are reduced to handleable sizes, approximately 1 M^2 and any free asbestos which may still remain on the sheets is bonded to it by varnish. These sheets are also bagged or wrapped in polythene prior to being deposited in the waste skips.

It has been found in practice that the cleaning procedure ensures an asbestos fibre count in the atmosphere of zero before to the main door is opened for the next cycle to commence. Again the ventilation plant is still operating at this stage to ensure an in-draft through the door opening.

DUST CONTROL AND VENTILATION PLANT

The work zone, the dirty side changing facility, showers and clean side offices and plant room have an internal volumetric capacity of 900M^3. Following initial discussions with members of the Health and Safety Executive it was decided that a minimum air change rate of 15 per hour would be adequate to keep the entire plant under negative pressure and to ensure good ventilation. The extraction and filtration plant was therefore designed to handle an air volume of 15,000 M^3/hr.

As the intention is to keep the dirty side working zone under negative pressure, the filtration and extraction plant is installed outside of the main building and has a final discharge to atmosphere. A ducting complex is fitted around the work zone to provide high level extraction. Long lengths of flexible ducting may also be fitted to the system to provide localized extraction when the operatives are working in a long enclosed vehicle.

The route taken by a flexible duct can be both long and tortuous, so, some experiments were done as to ascertain and provide for the likely system resistance when these ducts were in use.

Figure 6.   Carriage in position.

Figure 7.   Stripping and bagging of asbestos.

Figure 8. Special disposal skips.

Figure 9. Special mechanism sealing skip against work zone wall

The main manifold duct system was carefully designed so that the air
input points are distributed around the building to avoid any 'dead
areas'. The extraction plant is connected to a six module dust filter
(Figure 10) fitted with woven polypropolene filter pads. This fabric was
chosen to ensure a satisfactory level of primary filtration and because
it is unaffected either by moisture or chemical attack from any of the
materials likely to be handled.

After each of the six primary filter modules, a HEPA filter is fitted
rated to give a penetration on sodium flame test (EUROVENT 4/4 procedure)
of no more than 0.01%. The type of HEPA filter chosen is suitable for
high humidity levels (up to 100%) because of the moist air that could be
induced into the system during the washing down procedure. The whole of
the filtration system is kept under negative pressure by a centrifugal
fan with discharge stack to atmosphere.   The discharge velocity is
17m/sec which gives a good dispersion rate and is a suitable velocity for
isokinetic sampling.

To ensure that the primary filters are cleaned regularly a cleaning
mechanism is automatically activated by a built in starter/controller
each time the fan is shut down. This device ensures that the onus for
cleaning the filters, and maintaining them at proper efficiency is
removed from the operator. The dust collected by the filters is
deposited into polythene bags located inside sheet metal collection bins.
This minimizes the risk of the bags being punctured. A pressure
balancing pipe between the bin and filter case ensures the bags remains
open and do not collapse under negative pressure. At regular intervals
the bags are first sealed and then removed from the bins and deposited
along with the other asbestos waste in the skips.

The efficiency of the filters is carefully monitored by various
instruments. The pressure drop of each primary and secondary filter is
measured and automatically recorded on charts (Figure 10). The pressure
drop of each HEPA filter alone is also continuously measured and is
connected to a visual 'go', 'no go', warning system, equipped with two
red and one green light. The 'upper' red light indicates an excessive
pressure drop, and the 'lower' red light an absence of pressure drop (if
for example the HEPA filter has been removed). The green light means
the filter is working within the defined parameters. The pressure drop
of the primary filter is also displayed by a differential magnahelic
gauge. As part of the discipline imposed by the management the plant
supervisor must visit the filtration plant room at least once per shift
to check that all of the recording devices are functioning correctly,
that all green lights are showing, and to note down the pressure drop of
the primary filter. This latter value is kept in a daily log that is
inspected regularly by management, the the company's safety engineers and
the enforcement agencies. From the beginning the company has insisted
that the fullest records on the plant performance are kept. This permits
computer aided forecasts about the remaining life and effectivness of
both the primary and the HEPA filters to be made.

As the building is air tight an air make-up system is installed which
includes a warm air boiler capable of supplying clean warmed air (when
appropriate) up to 90% of the minimum air extract volume. Some tolerance
was allowed on the air volumes of both the input and extract systems to
assist final balancing. In fact, it has been found that the original

Figure 10. Dust extraction plant room.

decision to rate the input at 90% of the output was correct.

RESULTS

Regular air sampling has been carried out both in the work room, the surrounding offices and outside the building. The results within the work zone have been extremely good. After quite long exposure periods background counts of .25 fibres/cc have been recorded whilst personnel samplers have indicated values of .87 fibres/cc. Of course all the operators in the work zone are fitted with positive displacement respirator suits. Samples from the adjacent offices have indicated traces of fibres to a value of 0.008f/cc, and outside the building to a level of .003f/cc, but in all probability these were not crocidolite.

CONCLUSION

This plant has now been in operation for nearly three years. During that time hundreds of railway carriages have been safely and efficiently stripped of crocidolite asbestos. Regular monitoring by both the company's engineers and the relevant enforcement agency has not revealed any cause for concern. The only items that have been renewed during its life are the HEPA filters, which were changed after 27 months operation because of some slight doubt about the integrity of one seal. Overall the plant has performed extremely well and the engineering team who were involved in its design are well pleased with the results.

It was refreshing for the author to be dealing with a client who adopted such a positive approach. For example the risk of public sensitivity to the installation of such a plant near to a city centre could well have deterred most companies. However, the cooperation, enthusiasm and above all teamwork of all those involved has succeeded in providing a unique and very safe toxic waste handling plant.

ACKNOWLEDGEMENTS

The author acknowledges the help and assistance of Mr. V. Berry of Vic Berry Ltd., Leicester, in the preparation of this paper, and for allowing the techniques to be described in such detail.

POSSIBILITIES AND LIMITATIONS FOR THE DETECTION OF UNKNOWN SUBSTANCES AT HAZARDOUS WASTE SITES USING DETECTOR TUBES

Wolfgang May
Drägerwerk AG, 2400 Lübeck
D

ABSTRACT

In risk assessment – especially at hazardous waste sites – one often has to deal with the problem of unknown and sometimes also hazardous components being present in the atmosphere at a spill scene environment. Among other portable instruments detector tubes can be an easily applicable tool to detect a broad variety of hazardous contaminants on the spot. They can be used even under full body protection including protective gloves. This may help to find out certain potentially harmful substances or at least a family of substances in order to prevent danger from people working at waste sites.

The investigations of hazardous waste sites by means of detector tubes are not only limited to the gaseous phase but may also be extended to qualitative determinations of some volatile contaminants (e. g. several aromatic or chlorinated hydrocarbons) in liquids and even soils.

For a series of common families of hazardous substances a sampling strategy has been developed and will be discussed in detail to aid as rapid approach for preliminary classification of certain chemicals at spill scenes or hazardous waste sites.

INTRODUCTION

Compared to measurements at workplaces, where a wide range of comprehensive knowledge of the compounds of interest is present, the detection of pollutants from hazardous waste sites represents a very special kind of measurement because of the often unknown nature of the chemical components. The problem of risk assessment at hazardous waste sites or during accidental releases is also based on a generally multified nature of airborne pollutants. In order to approach the task of sampling or monitoring airborne pollutants portable instrumentation is necessary. Several instruments are provided by manufacturers including portable total hydrocarbon vapor monitors with either photoionization or flame ionization detector, portable gas chromatographs, portable infrared analyzers, combustible gas measuring devices with catalytical sensors to measure in the range of the lower explosion level, gas detector tubes with direct-reading, substance-selective alarm devices and dust monitors. Besides the need for portable instruments at various chemical releases into the environment personal protection suites and other protective equipment should be highly recommended.

Detector-tubes can be regarded as one possibility among other measuring devices tackling the need to measure or to identify airborne pollutants on the spot. It is to be taken into consideration that there is no measuring instrument available that could be used as a simple unit for dealing with all demands of practical activities, especially with the background of unknown pollutants from accidental chemical releases or hazardous wastes. Sampling strategies should help as systematic guidelines for preliminary classification of a certain percentage of the most common chemicals.

DESCRIPTION OF THE DETECTOR TUBE METHOD

Detector tubes are hollow tubular shaped glass bodies consisting of one or more filling layers of chemically impregnated carrier material[1]. For this purpose silica gel or aluminum silicate are the most commonly used types. Both ends of the detector tube are normally fused. By means of an aspirating pump a certain air volume is drawn through the opened tube. According to international detector tube standards tube and pump together form the detector tube unit and are usually supplied by the same manufacturer.

If an airborne contaminant is present, the chemicals of the filling material undergo a chemical reaction during the intake procedure of every pump stroke producing a discoloration. After a certain number of pump strokes have been carried out, the concentration of a contaminant is determined by evaluating the stain length by calibrated scale which is printed on the surface of the detector tube (length-of-stain type detector tube). The calibration of the scale figures is conducted in the facilities of a tube manufacturer according to established rules of generating gas concentrations[2, 3]. It is of importance that the scale figures are adapted to the sampling volume and the air flow characteristics of a given aspirating pump so that detector tubes and detector tube pumps of different manufacturers are not interchangeable.

A second group of commercially available detector tubes is color comparison or color matching tubes. With these a certain number of strokes are to be carried out until the color intensity of the indicating layer matches the color of the comparison layer or any other color standard which is given in the operating instructions of the tube or as another part of the detector tube. The number of pump strokes then is used to calculate the point at which the concentration of the airborne contaminant matches a given color standard. Normally the operating instructions supplied by the manufacturer are containing a table with the information on the measured concentration based on the corresponding number of pump strokes. Color comparison tubes are calibrated by the manufacturer according to the same rules of generating known gas and vapor concentrations.

Detector tubes respond to the presence of airborne gases and vapors and to some extend also to particles, for example, aerosols of both liquid and solid particles. Aerosols that can be measured with tubes are e. g. chromic acid, arsenic trioxide, potassium and sodium cyanide, nickel dust, oil mist and sulfuric acid. The measuring range of detector tubes covers a broad range of concentration, beginning even far below 1 ppm at the lower end of the range and going up to some volume percent at the upper end. As there are more than 220 different types of detector tubes commercially available, a broad variety of at least more than 350 different hazardous compounds can be measured: inorganic gases as well as organic vapors.

As detector tubes are calibrated at the facilities of the tube manufacturer there is no need for calibration or recalibration of the tube in the field, i. e. the detector tube can be regarded as a preserved analytical method. If field calibrations should be necessary, known concentrations of test gases can be carried in the field, e. g. using reference gases in small pressure bottles. Usually a detector tube is designed for single use, although it is possible that some tubes may be used twice or more often during the same day.

APPLICATION OF DETECTOR TUBES

Detector tubes have originally been developed for measurements of airborne contaminants at workplaces so that most of the commercially available detector tubes are designed to measure in the range of the threshold limit values of airborne pollutants[4, 5]. Other applications are possible with some high-range detector tubes with measuring ranges up to several volume percent of compounds like carbon monoxide, sulfur dioxide, hydrogen sulfide etc. With tubes of this kind process control and emission control can be performed.

The application of the detector tube measuring method is
worldwide accepted for occupational industrial hygiene purposes. Established detector tube standards have been published in several countries[7, 8, 9]. A great number of
papers dealing with applications of detector tubes at workplaces have been reported elsewhere and should not be discussed in detail (see reference 4 to 6). If acute concentrations of pollutants are present in the environment (air,
water and soil), detector tubes can be used to identify the
presence of certain chemical compounds and can also be extremely useful in locating the origin of a suspected pollutant although a precise concentration measurement is often
not possible.

INVESTIGATIONS AT HAZARDOUS WASTE SITES

Airborne pollutants

For detection of airbore pollutants at hazardous waste
sites detector tubes can be used without any other additional
equipment besides the detector tube pump. The investigations
can be performed in the same manner like control measurements
at workplaces although there are several important differences compared to measurements at workplaces. In the latter
case there is a lot of comprehensive knowledge based on previous measurements and additional know-how with respect to
relevant process data, whereas investigations at hazardous
waste sites normally have to start at the very beginning.
Another influence on the measurement in ambient air can be
caused by high air currents, which lead to some statistical
uncertainty of the concentration of the airborne pollutants.
But at least the tube reading can be taken as a semi-quanti-
tative measuring result.

Detector tubes can further be regarded as a cost effective
method to find the location of a suspected hazardous chemical
on hazardous waste sites, they can also be very useful to get
information on the size of a gaseous cloud caused by release
of certain pollutants from a waste site or any other accidental release of chemicals. Detector tubes are never to be seen
as a competitive method against other portable instruments or
even high sophisticated analytical chemical methods in a
laboratory, they should rather be considered as one method
within the whole range of gas analytical procedures. There
are several advantages of the tube method over the other
instrumentation:

Due to the direct measuring result detector tubes can be
used as a cost effective and more important highly time saving on-site method if quick decisions are to be made in case
of an emergency. The analytical turnaround time from stationary or mobile laboratories usually is the time determining

factor for further operational activities especially in the
case of airborne pollutants. These can have a direct influence on both occupational and public health so that it is
very important to keep the analytical turnaround time as
short as possible. Detector tubes and more sophisticated analytical methods should be used as supplements, the detector
tubes lead to an immediate result after the sampling for airborne pollutants and further decisions, e. g. if people are
to be evacuated from their homes, can be made very soon.
Sampling and subsequent analysis in a chemical laboratory
then can lead to the precise concentration of an airborne
pollutant.
If, e. g. polyvinyl chloride products begin to burn on a
waste site, there will be high amounts of hydrogen chloride
released during this fire. The detector tube method will show
the presence of high hydrogen chloride concentrations in the
ambient air in a measuring time of less than 3 minutes. Additional sampling for HCI with impingers and pumps and subsequent titrimetric or photometric analysis will then give the
more precise measuring result a few hours or a few days later
without the sometimes dangerous time pressure.

Another advantage of the detector tube method can be seen
if the detector tube shows for sure that there is no or only
a minor concentration of an airborne pollutant present, there
is no need for subsequent and time and money consuming analytical procedures in a chemical laboratory. So the tube
method can be a useful and cost effective quick test not only
for detections of hazardous compounds on waste sites but also
for control measurements in hazardous waste management, e. g.
chemical warehouses, trucks, storage tanks etc.

It should be pointed out clearly that a detector tube can
of course not replace high sophisticated portable gas chromatographs or even gas chromatograph mass-spectrometer-combina-
tions in a stationary laboratoy. Although more than 350 different chemicals can be measured with commercially available
tubes, this does not mean that the tube method will always be
the method of the choice because there are other chemicals
that cannot be measured with detector tubes or there are
several airborne pollutants present at the same time and the
tube method does not lead to a definite measuring result on
the spot. This underlines again the need for more sophisticated portable, mobile or stationary analytical devices but
it also can be seen that the detector tube method has found
its place as an alternative analytical method with time and
cost effective applications for investigations of airborne
pollutants on hazardous waste sites.

Water and Soil

Detector tubes can also be used for investigations of contaminated liquids (water) and even contaminated soil. As the
tubes are designed for measurements of airborne pollutants
both liquid and soil cannot be investigated directly as there
would be a severe influence on the reagent layers. The impregnated chemicals of the filling material would be washed
out, if water was drawn through the detector tube. So for
these applications a sample of the liquid from a hazardous
waste site is placed in a gas wash bottle and the detector
tube unit is connected to the outlet tube of the wash bottle
by means of a short rubber hose. Squeezing the pump causes
ambient air to be drawn into the gaseous phase above the
liquid. This causes the evaporation of light volatile components of the contaminated liquid from a hazardous waste which
components are then swept into the detector tube. The concentration can then be evaluated as a stain length in a comparable manner. As the scale figures are usually calibrated
on a gas concentration basis a conversion factor is necessary
for determination of contaminants of liquids in mg/L. A
selection of contaminants that can be detected with the discussed method are given in table 1[10].
A special application of detector tubes for the determination
of sulfides in drilling fluids is known as the Garrett-gas-
train and is recommended by the American Petroleum Institute[11].

As seen in table 1 the investigation of contaminated
liquids is limited to light volatile chemicals, strong polar
compounds and salts (e. g. heavy metal salts) cannot be determined with this method. Also, it should be pointed out
that the lowest detectable concentrations given in table 1
have been determined by adding only one compound to a water
sample at room temperature. If several other components
should be present at the same time this may have an influence
on the lowest detectable concentration. Recently investigations have shown the influence of petrol hydrocarbons on the
lower detectable concentration of perchloroethylene determined in water by means of detector tubes[12]. It should
also be mentioned, if there are some time and temperature
dependant chemical reactions, e. g. hydrolysis, that those
reaction products will not be detectable by the described
method.

Investigations of contaminated soil can be done in a comparable way, if the soil is emulgated with distilled water in
order to form an air permeable matter. Alternatively the soil
can be extracted by a suitable solvent, mostly water is recommended, and the solvent is subsequently investigated for
contaminants by means of detector tubes.

TABLE 1

Contaminated Liquids

Contaminant	U.N. - I.D. number	lowest detectable concentration (mg/L)
Acetaldehyde	1089	10
Acetone	1090	1
Acrylonitrile	3022	2
Ammonia	2073	0.5
Benzene	1114	1
Carbon Disulfide	1131	0.5
Carbon Tetrachloride	1846	0.1
Chlorobenzene	1134	1
Chloroprene	1991	0.1
Cyclohexane	1145	5
Diethylether	1155	1
Dimethylsulfide	1164	1
Epichlorohydrine	2023	0.1
Ethanol	1986	500
Ethylacetate	3075	50
Ethylbenzene	1175	1
Hexane	1208	10
Methylene Chloride	1593	1
Methylmercaptane	3073	0.2
Octane	1262	5
Pentane	1265	10
Perchloroethylene	1897	0.1
Petrol Hydrocarbons	1115	10
Toluene	3108	0.5
Trichloroethane	2831	3
Trichloroethylene	1710	0.1
Triethylamine	1296	30

Some minor modifications of the described method may extend the method for further applications. If a contaminated liquid or soil contains sulfides or cyanides, a known amount of sulfuric acid may be added in order to form hydrogen sulfide or hydrogen cyanide respectively. Both can then be detected with detector tubes. It has been reported in the literature[13] that the detection of sulfides can be performed quantitatively but for most of the applications detector tubes should be used for qualitatively identifications of certain chemicals as contaminants of water and soil from hazardous waste sites. The lower detectable concentrations given in table 1 should be used as a guideline to get a feeling for the order of magnitude of concentrations found by means of detector tubes.

Also for investigations of contaminated water and soil the philosophy of the detector tube method for application at hazardous waste sites or hazardous waste management should be seen in the light of a quick analytical method, which is again time and cost effective to find the order of magnitude of a contamination. The tubes do not replace the analytical possibilities of stationary chemical laboratories but they can help for rapid identification of environmental pollution caused by hazardous waste sites, train derailments, leaks of underground storage tanks etc.

SAMPLING STRATEGY FOR DETECTION OF UNKNOWN SUBSTANCES

As mentioned above one of the most important differences between measurements at workplaces and measurements at hazardous waste sites or spill scenes is the usually unknown nature of the pollutants from these locations. People would be quite happy if there would be a portable or mobile measuring device that could be used under every possible circumstances that may arise from potentially harmfull areas. As this is not available now and will probably be not possible in the future based on today's knowledge of different measuring techniques, one has to live with some degree of uncertainty in the detection or identification of unknown substances. But nevertheless, an approach can made based on comprehensive statistical knowledge about which substances could be involved at hazardous waste sites or during transportation on rail, water, road and in the air for potentially accidental releases of chemicals.

It should be taken into consideration that in the industrial countries between 6000 and 8000 different chemicals are transported each day. Those are chemicals from high to low toxic level, chemicals that are transported in tons scale down to a few grams. A sampling strategy has earlier been reported[14, 15] based on the knowledge on the kind and amount of transportated chemicals, comprehensive knowledge of policemen and firebrigades drawn from their experience of activities on hazardous waste sites, spill scenes and accidental releases of chemicals. Between 10 and 20 different types of detector tubes are necessary for preliminary classification of the most commonly used and transported groups of chemicals and those reaction products from accidental fires on hazardous waste sites or other accidents.

As can be seen from table 2 detector tubes give positive indications with certain groups of chemicals or quite selective with only one chemical. Indications should be evaluated as the presence of a certain chemical family after using at least more than one tube with respect to cross-sensitivities.

TABLE 2

Detector Tubes for Different Pollutants

Detector tube	Positive indication of the following chemical families
Polytest	Easily oxydizable chemicals like esters, aromatic hydrocarbons, ketones, alcohols, aliphatic hydrocarbons, carbon monoxide, halogenated aliphatic hydrocarbons
Ethylacetate 200/a	Esters, aromatic hydrocarbons, ketones, alcohols
Benzene 0.05	Aromatic hydrocarbons
Acetone 100/b	Ketones
Alcohol 100/a	Alcohols
Hydrocarbons 0.1 %/b	aliphatic hydrocarbons
Carbon Monoxide 10/b	Carbon monoxide
Methylbromide 5/b	Halogenated aliphatic hydrocarbons
Amine test	Amines, ammonia
Acid test	Acid gases and vapors
Hydrogen Sulfide 2/a	Hydrogen sulfide
Hydrogen Cyanide 2/a	Hydrogen cyanide
Nitrous fumes 0.5/a	Nitrous fumes

All tubes of table 2 should be used equivalent starting with polytest as the most unselective tube to identify easily oxidizable pollutants. With the tube amine test and acid test the presence or absence of several basic or acid gases and vapors can be tested. With the need in mind for keeping the time in identifiying a chemical family at hazardous waste sites or during accidental releases as short as possible it is more important to identify a group of hazardous chemical rather than individual compounds which is nearly impossible in the field. If it can be seen that e. g. aromatic hydrocarbons are present it is of minor importance if benzene, toluene or xylene or a mixture of all is present. For safety reasons activities for the worst case should be started in order to prevent any damage from workers or public health and safety. Using all detector tubes from table 2 means a total sampling time on-site of less than 20 minutes, less time than is consumed for high sophisticated analytical methods.

For convenience of the users the manufacturers are providing so-called hazardous material kits with 20 different types of detector tubes including pumps and other measuring or sampling equipment like glass bottles for taking a sample for subsequent laboratory analysis. But also these kits are no fool-proof equipment, they should be used by well-trained personal only, although no calibration nor further maintenance is needed for detector tubes. If none of the detector tubes shows any reading that does not mean that no harmful substance is present. In these cases other analytical methods are highly recommended.

The biggest advantage of the detector tube method for tackling with unknown substances at hazardous waste site or spill scenes is that they can be used time and cost effective for preliminary classification of potentially harmful pollutants. The tube reading should always be taken as a qualitative indication, regardless of the scale figures. After the rapid test on the spot further activities with more sophisticated methods can be performed for identification of individual chemicals and their precise concentration measurement.

CONCLUSION

Detector tubes can be used for risk-assessment at hazardous waste sites or spill scenes as an alternative analytical approach for preliminary classification of the unknown chemical groups. They do not replace other portable instruments and do not compete with high sophisticated analytical procedures. Due to their time and cost effective usage they have found their place for investigations on the spot. Based on their immediate indication, further activities for safety reasons can started before any analytical turnaround from mobile or stationary laboraties is possible.

ACKNOWLEDGEMENT

The author would like to thank Mr. B. Mußmann and Dr. W. Bäther for their experimental research work with contaminated soil and water.

REFERENCES

1   K. Leichnitz, <u>Detector Tube Handbook</u>, Publication 4340e, Drägerwerk AG Lübeck, 1985

2   K. Leichnitz, Generating test gases for calibration of methods used in environmental analysis, <u>IUPAC Pure & Appl. Chem.</u>, <u>55</u>, 1239 (1983)

3   NIOSH Technical Report, Gas and Vapor Generating Systems for Laboratories, <u>U.S. Department of Health and Human Services</u>, NIOSH Publ. No. 84 – 113, Cincinnati Ohio 1984

4   K. Leichnitz, Vergleich von Kurzzeit- und Langzeitmethode
    zur Beurteilung der Schadstoffsituation an Arbeits-
    plätzen, Staub-Reinh. Luft, 40, 241 (1980)

5   K. Leichnitz, Luftuntersuchungen verbessern Arbeitsplatz-
    bedingungen, Humane Produktion, 3, 24 (1983)

6   L. Grupinski, Ölkonzentrationsmessungen in der Raumluft
    von Arbeitsräumen, Arbeitsschutz 1, 12 (1972)

7   Japanese Industrial Standard JIS K0804, Detector Tubes,
    Tokyo 1985

8   A. J. Collings, Performance Standard for Detector Tube
    Units used to monitor gas and vapors in working areas,
    IUPAC Pure & Appl. Chem. 54, 1763 (1982)

9   British Standard 5343 Part 1, Gas Detector Tubes,
    Specifications for short term gas detector tubes,
    BSI, London 1986

10 B. Mußmann, Analysen für den Umweltschutz,
    Drägerheft 325, 18 (1983)

11 American Petroleum Institute, Standard Procedures for
    Testing Drilling Fluids, 7th edition, Dallas, Texas 1978

12 W. Bäther, Bestimmung von halogenierten Kohlenwasser-
    stoffen im Abwasser mit Prüfröhrchen – Anwendung und
    Grenzen, Paper presented at Analytica, Munich 1986

13 R. W. Handy et al., A method for determining the
    reactivity of hazardous wastes that generate toxic gases,
    Hazardous and Industrial Solid Waste Testing:
    4th Symposium ASTM STP 886, Editors: J. K. Petros et al.,
    ASTM, Philadelphia 1986, pp. 106 – 120

14 D. Schneider, The Draeger Gas Detection Kit,
    Draeger Review 46, 5 (1980)

15 M. V. King, P. M. Eller and Richard J. Castello,
    A Qualitative Sampling Device for Use at Hazardous
    Waste Sites, Am. Ind. Hyg. Assoc. J. 44, 615 (1983)

DISPOSAL OF WASTE-WATER AND OIL-CONTAINING SLUDGES IN SOILS

Erzsébet Pap Kránitz
Department of Plant Production
Agricultural University
Mosonmagyaróvár, H-9200 Vár St. 2
H

and

János Pap
Department of Plant Production
Agricultural University
Mosonmagyaróvár, H-9200 Vár St. 2
H

ABSTRACT

Taking into consideration soil- and environmental protection, the disposal of waste-water sludges of various origin in soil has been examined in model, petri-dish, plot and
plant experiments for many years. The results of plot and
plant experiments are mainly described in this paper. With due
regard to it, the disposal of hazardous wastes into soil are
reasonable under controlled circumstances. But the disposal
on farmland is at the same time a special solution, then pe-
riodical and qualitative characteristic features are limiting.
During the yearly sludge loadings metal, heavy metal enrich-
ments in the 0-60 cm thick layers have not been experienced
at the receiving areas. The crop yield of the indicator plants
can be attributed principally to the years' affects, the cultivated plants respond to the various sludge types differently. The metal and heavy metal content of the yields were not
considerably changed by the effect of the sludge treatments.
Principally after sludge loading of various soils /depending
on the behaviour of metals and time/ the quantitative and
qualitative development of the absorbable forms can determine
the further disposal of hazardous wastes.

INTRODUCTION

Concentration of settlements, industrial and agricultural production has led to a considerable accumulation of by-products and waste-materials. The greatest anxiety is perhaps caused by the quantity of waste-water and waste-water sludges.

Every by-product, also these materials as well form an integral part of biomass, their return into the natural circulation would be desired and necessary, but this is prevented and influenced by more factors.

This work will inform you about our latest results in the disposal of waste-water and oil-containing sludges. Their reusage in soil is one of the possible ways of harmless disposal of concentrated polluting sources, then their deposit or destruction - because of environmental and financial reasons - is not possible.

At the same time the following factors must be taken into consideration in the case of reusage of sludges in soil:

- the opinion that waste-materials must be at any price disposed on farmland should not be predominant;

- we must be careful because of the detrimental components;

- waste-materials must be primarily used in those cultures without coming directly into the food-chain;

- the interests of the agricultural utilizer should be in every case predominant;

- disposal of waste-materials should occur only under controlled circumstances.

MATERIALS AND METHODS

The objective of this experiment was to study the effects of waste-water sludges of industrial origin but produced by different treatments,of leather factory sludges reduced with heavy metal salts, and of oil-containing sludges produced during oil-processing and in transport and cleaning vehicles on soil-properties, load bearing capacity of soil, and on plants cultivated in model-, petri-dish and plot experiments.

Normal and above normal quantities were used during the experimental treatments to estimate the possible detrimental effects.

Each kind of sludge was examined before disposal on farmland, and these results were taken into account in determining the quantity to be applied. Observations were taken during the growing season, results were estimated both professionally and mathematically. Whole soil and plant analyses have been made annually.

RESULTS AND CONCLUSIONS

Waste-Water Sludges of Industrial Origin

On examination of the treatment grades of waste-water sludges of industrial origin /concentrated, conditioned and burnt sludges/, conditioned sludge was found to be the type most suited for farmland disposal. This is justified by more facts, which are as follows:

- conditioned sludge has no negative influence on plant and soil;

- damaging microelement accumulation was not found though above rates of 50 tonnes per hectar the heavy metal content of plant and soil increased, staying under the allowed critical value;

- treatment of conditioned sludges is relatively simple, there is no need for a special machine stock, it can be spread out with an organic manure distributor;

- the management of water-supplies will be influenced

favourably.

Our own experiments have proved, that fertilizer addition is reasonable, in most cases waste-water sludges posses a higher nutritive material content, but their yield increasing effect is much less, than that of fertilizers. In our experiment giving fertilizer together with waste-water sludges gave a favourable effect and yield increase was significant compared with fertilizer application only. It must be noted as well, that additional fertilizer application does not induce an increased microelement-uptake.

This experiment suggests that it is necessary to add fertilizer along with waste-water sludge primarily to ensure the availability of absorbable nutrients. The application of sludges according to our assumption causes no problems in the short run, but a following build up of sludge quantities over time a microelement accumulation is revealed to a certain extent. The amount of conditioned sludge which can be applied without adverse effect, is 50-80 tonnes per hectar per year at a dry substance content of 25-27 %. The application rate of 50-80 tonnes per hectar could be realised in every 4th-5th year, as because as the conditioned sludge is transformed in the soil during this period, its detrimental effect discontinues. Further increase in amounts is not reasonable because of organizational limitations.

Leather Factory Sludges Reduced with Heavy Metal Salt

During the examination of the leather factory sludges reduced with heavy metal salt, treatment variations on a dry basis of 20-260 tonnes per hectar were applied with mustard and maize as indicator plants. A characteristic of the sludge is, that it contains mainly chrome, thus these experiments have been performed accordingly.

In the first experimental series it could be proved - similarly to the data of the literature - that the uptakable Cr turns into non absorbable form in the soil according to the Hungarian conditions and therefore toxicity is not to be feared. The uptakable Cr values after sludge treatment are presented in table 1.

TABLE 1

Uptakable chrome in soil /ppm/ at mixing and after harvest
of indicator plants

Treatments t/ha dry substance	Uptakable Cr /ppm/ at mixing	As indicator plant in soil after harvest	
		mustard	maize
1.	19,1	19,2	0,00	0,00
2.	42,9	43,2	3,00	0,00
3.	73,5	74,0	6,00	3,70
4.	114,3	115,2	8,30	5,30
5.	171,5	172,7	10,00	9,00
6.	257,3	259,1	14,00	12,30

There were substantial differences in the air-dry material in mustard and maize in relation to the control, but in the plant Cr could be measured only in mustard at the three highest doses. At the same time the chrome content of the root - both at maize and mustard - increased in proportion to the raise of doses.

According to us the physiological inhibition has developed because as a result of the high chrome uptake of the root, the other nutritional elements can be taken up by the root only in amounts. It could be important, as through negligence yield decreases are possible, but no toxicity exists. The unfavourable effect on air-dry material can be observed mainly in mustard. The growth habit of plants is not only favourable at the 3rd sprouting, but also the air-dry material has developed considerably, particularly in maize, where the differences are significant. It is important in the farmland utilization, to spread sludge in autumn and sow maize in spring, so that 5-6 months are disposable for chrome binding. On the basis of the experimental years it can be established, that there were no unfavourable changes in the quantity and quality of yields from the effect of sludge treatments. The characteristic measures of soil's value properties were not affected unfavourably by the treatments. Heavy metal enrichment has not been revealed. Changes in the microbiology of the soil has not been detected. The amount of 50 tonnes per hectar of sludge with a dry material content of 25 % - under controlled circumstances - can be disposed of without any unfavourable changes. But controlled circumstances constitute the precondition of the disposal, and it is chiefly important to pay attention to the development of heavy metals being brought into the soil.

Dewatered Oil-Containing Sludges

The farmland disposal of dewatered oil-containing sludges was examined to improve the soils of mainly unfavourable properties or non cultivated soils. On the basis of experiments lasting for more years it has been established, that the agricultural receiving conditions must be ensured fundamentaly by taking into consideration the properties of the cultivated area.

The results and experiences of the plot and plant experiments in the years of 1983-1986 are summarized in this work. Sludge treatments of 20-200 m^3 per hectar were applied in the plot experiments, in four repetitions. The objective of these experiments was to analyse the possible damaging effects. The indicator plant was maize. No anomalies have been experienced during the growing season. The crop yields resulting from the application rate of 100 m^3 per hectar were higher than the control, but not significantly. The development of the crop yield was not influenced by the further increase of the sludge doses during the examined years /table 2/.

As to the analyses of the soil samples, taken in spring and autumn the total metal and heavy metal content increased in soil during the years, but at the high doses not in direct proportion exept for the zinc content /table 3/.

TABLE 2

Crop yield and microelement content of maize cultivated on calciferous alluvial soil /means of four repetitions/

Treatments	Crop yield tonne/ha	Zn	Cu	Mn
		p p m		
1. Control	12,7	15,5	1,1	2,8
2. 20 m³/ha	13,7	14,7	1,3	2,7
3. 100 m³/ha	14,3	17,0	1,3	2,0
4. 200 m³/ha	10,7	17,2	1,0	2,0

TABLE 3

Data from calciferous alluvial soil tests /means of four repetitions/

Treatments	pH	of Spring	of Autumn in 20-25 cm layers	in 40-60cm layers
1. Control	7,00	0,007	0,009	0,006
2. 20 m³/ha	7,08	0,01	0,01	0,003
3. 100 m³/ha	7,07	0,01	0,2	0,005
4. 200 m³/ha	7,07	0,02	0,2	0,003
Values before treatment	7,04		0,003	

Total extract %

TABLE 4

Microelement content of calciferous alluvial soil, autumn soil samples from 0-25 cm layers /means of four repetitions/

Treatments	Pb	Cr	Cu	Cd	Zn	Mo	Ni
				p p m			
1. Control	37,5	12,5	33,3	2,6	41,7	8,3	50,0
2. 20m³/ha	38,0	16,0	33,5	2,4	83,7	7,6	52,5
3. 100 m³/ha	38,1	15,6	61,5	2,3	108,9	7,8	50,6
4. 200 m³/ha	42,2	17,2	51,7	2,6	133,6	8,6	56,0
Values before treatment	10,3		18,8	0,5	60,0	6,1	34,5

On soils treated with oil-containing sludges the decrease in total extract content can exceed in the soil 75 % annualy; the decomposition is nearly entire on soils of lighter mechanical composition.

The moisture content of the soils was always higher in the experimental soils than the control's value. This is particularly important for growths in dry years.

According to the soil-microbiological analyses the count of total bacterium and paraffin decomposers decreased at the sludge treatments of 100 and 200 m^3 per hectar.

TABLE 5
Data from soil-microbiological analyses

Treatments	In x^6/g absolute dry soil	
	total bacterium count	paraffin decomposer count
1. Control	12,039	18,650
2. 100 m^3/ha	9,096	14,790
3. 200 m^3/ha	10,170	15,510

With lower sludge amounts the numbers of bacteria are higher than in the samples of the control. This can be explained by the presence of large numbers of microorganisms in the micro-flora of soils, which not only tolerate the oil-pollution, but also utilize it in their metabolistic processes. The oil-digesting efficiency of soils is satisfactory. In evaluating the limits it must be taken into consideration, that the negative changes in the count of microorganisms on soils treated with oil-containing sludges were not mainly caused by the oil content, much the more by such additional polluting materials, as heavy metals, plant protection chemicals and also by various organic solvents and detergents.

The effect of the oil-containing sludges on soils depends particularly on the sludge releasing factory, soil properties, but also on the species of the indicator plant.

ACKNOWLEDGEMENT

The authors wish to express their appreciation to Mrs. Wendy Macaskill /New Zealand/ and their Hungarian Colleagues Mr. Rezső Schmidt and Mrs. Judit Koltai for their help in completing the translation of this manuscript.

REFERENCES

1. Papné-Kránitz, E. and Mucsy, Gy. and Urbányi Gy., Die Behandlung und unschadhafte Unterbringung von ölhaltigen Abschlämmen. Schriftenreihe der Technischen Universität, Wien, Altöl-Entsorgung und Verwertung, 1982.

2. Pap, J., A különböző kezelésü szennyviziszapok talajban történő felhasználása. Doktori értekezés, Mosonmagyaróvár, 1984.

3. Pap, J. and Papné-Kránitz, E., Nehézfémsó mentesitett és a nyers szennyviziszap felhasználás hatásának vizsgálata tenyészedényes és parcellás kisérletekben. Jelentés, Bőrgyár, Pécs, 1986.

4. Papné-Kránitz, E. and Pap, J., Olajtartalmu iszapok mezőgazdasági elhelyezés vizsgálata. Jelentés, ATEK Mezőgazdaságtudományi Kar, Mosonmagyaróvár, 1985-1986.

A REVIEW OF HAZARDOUS WASTE MINIMIZATION
IN THE UNITED STATES

Harry M. Freeman, Research Program Manager
Hazardous Waste Engineering Research Laboratory
U.S. Environmental Protection Agency
Cincinnati, OH 45268

Elaine Eby
Office of Solid Waste
U.S. Environmental Protection Agency
Washington, DC 20460

ABSTRACT

This paper summarizes the findings and recommendations of a recently prepared U.S. EPA report, Minimization of Hazardous Waste; Report to Congress. The authors explore the current waste minimization practices in the U.S. and discuss the potential for future waste reduction. Incentives for waste minimization in the United States are strong. Between now and 1990 it will become much clearer as to what role waste minimization will play as a major environmental improvement strategy in the U.S.

INTRODUCTION

There is a national policy in the United States to eliminate hazardous waste. The U.S. Congress stated in the Hazardous and Solid Waste Amendments of 1984 to the Resource Conservation & Recovery Act of 1976, a very broad piece of legislation that authorizes the Federal hazardous waste regulatory program, that

"The Congress hereby declares it to be the national policy of the United States that, wherever feasible, the generation of hazardous waste is to be reduced or eliminated as expeditiously as possible. Waste that is nevertheless generated should be treated, stored or disposed of so as to minimize the present and future threat to human health and the environment."

Reflecting the intent of this policy, there has been adopted by the EPA and other public agencies similar variations of the hierarchy shown below as a guide for hazardous waste management options.

1. Waste Reduction: Reduce the amount of waste at the source, through changes in industrial processes.

2. Waste Separation and Concentration: Isolate wastes from mixtures in which they occur.

3. Waste Exchange: Transfer wastes through clearinghouses so that they can be recycled in industrial processes.

4. Energy/Material Recovery: Reuse and recycle wastes for the original or some other purpose, such as for materials recovery or energy production.

5. Incineration/Treatment: Destroy, detoxify, and neutralize wastes into less harmful substances.

6. Secure Land Disposal: Deposit wastes on land using volume reduction, encapsulation, leachate containment, monitoring, and controlled air and surface/subsurface water releases. (2)

The term "waste minimization" has been defined differently by different organizations. The USEPA in its October 1986 Report to Congress on the minimization of hazardous waste defined waste minimization as:

The reduction, to the extent feasible, of hazardous waste that is generated or subsequently treated, stored, or disposed of. It includes any source reduction or recycling activity undertaken by a generator that results in either (1) the reduction of total volume or quantity of hazardous waste, or (2) the reduction of toxicity of hazardous waste, or both, so long as the reduction is consistent with the goal of minimizing present and future threats to human health and the environment. (2)

In this paper the authors summarize hazardous waste minimization in the U.S.   Much of the information in the paper has been extracted from the USEPA report, <u>Report to Congress: Minimization of Hazardous Waste,</u> October 1986.

<u>Profile of Hazardous Waste Generated in the U.S.</u>

In 1983 the Congressional Budget Office estimated that U.S. industry generated some 266 million metric tons of hazardous waste. 98% of this amount is estimated to be generated by large facilities that are generating in excess of 100 kg/month of hazardous waste. Table 1 shows the percentage contribution of various industrial sectors to the total amount of hazardous waste generated.

Existing information indicates that the predominant waste streams generated by U.S industries are corrosive wastes, spent acids, and alkalines used in the chemical, metal finishing, and petroleum refining industries. Many of these waste streams contain heavy metals, thus making them toxic. Solvent wastes are generated in large volumes both by manufacturing industries and a wide range of equipment mainte-nance industries that generate spent cleaning and degreasing solutions. Cyanide/reactive wastes come primarily from the chemical industry, and the metal finishing industries. (4)

The hazardous waste contribution of the ten largest generating sectors according to characteristic waste type i.e., solvents, corrosives, etc., is shown in Table 2.

Table 1. Industry Ranking by Hazardous Waste Generation

Rank	Major industry	Percent of total waste generated
1	Chemical & Allied Products	47.9
2	Primary Metals	18.0
3	Petroleum & Coal Products	11.8
4	Fabricated Metal Products	9.6
5	Rubber & Plastic Products	5.5
6	Miscellaneous Manufacturing	2.1
7	Nonelectrical Machinery	1.8
8	Transportation Equipment	1.1
9	Motor Freight Transportation	0.8
10	Electric & Electronic Machinery	0.7
11	Wood Preserving	0.7
12	Drum Reconditioning	< 0.1
	Total	100.0

Source: Congressional Budget Office (CBO 1985) (3)

Source Reduction Potential

In reviewing such data as is contained in Tables 1 and 2 one might be led to ask two significant questions: First, how much have the reported amounts already been reduced because of adoption of improved processes and operational procedures, and second how much more can we expect these amounts to be reduced in the future if the generators adopted all known and available source reduction options to reduce their waste generation. The U.S. EPA addressed these two questions in preparing the support document for its report to Congress.

To provide an approach to answering the questions the EPA developed two qualitative indices, the current reduction index (CRI) and the future reduction index (FRI). Both indices vary from 0 to 4. The CRI is a qualitative measure of how much source reduction has already taken place, i.e., a CRI of 3 would indicate that 75% of the waste that would otherwise be produced if the generator had not undertaken any source reduction efforts is not being produced because of the source reductions options already adopted. The FRI is an estimate of future reductions possible i.e., an FRI of 3 would indicate that 75% of the waste stream could still be reduced. The indices are independent of production rates and pertain to specific waste generation expressed in units of waste per unit of product. In gereral the indices are based upon the averages of the indices for various processes in the industrial sector that were evaluated by an engineering firm under contract to the Agency. A list of the various industrial indexes is shown in Table 3.

It appears that U.S. industry has reduced their waste considerably. However, there is room for more reductions. Another opinion on this

Table 2.

Profile of RCRA Characteristic Waste Generation by the Ten Highest Volume Wastes Generating Industries in 1981

Industry description	Volume of waste generated (M gal)			
	Ignitables	Corrosives	Toxics	Reactives
Chemical and Allied Products	140	8,200	1,200	15,000
Machinery, Except Electrical	67	2,300	71	< 0.1
Transportation Equipment	40	950	530	< 0.1
Motor Freight Transportation and Warehousing	< 0.1	< 0.1	0.7	NR[b]
Petroleum and Coal Products	8.3	51	1.8	< 0.1
Primary Metals	2.9	220	560	31
Construction, Special Trade Contractors	< 0.1	870	430	< 0.1
Fabricated Metal Products	5.4	310	320	240
Electrical Equipment Manufacture	15	170	40	40
Electric, Gas, and Sanitary Services (includes POTWs)	< 0.1	32	38	38

Source: U.S. EPA, 1986 (4)

Table 3

Table 3-3 National Hazardous Waste Generation and Reduction Profile

Industry	Percent of total waste generation	Waste reduction index Current	Future (probable)	Future (maximum)
Chemical and Allied Products	47.9	2.7	0.7	1.2
Primary Metals	18.0	2.7	0.7	1.2
Petroleum and Coal Products	11.8	2.0	0.6	1.2
Fabricated Metal Products	9.6	1.6	0.9	1.5
Rubber and Plastic Products	5.5	2.6	0.7	1.4
Miscellaneous Manufacturing	2.1	1.9	0.9	1.4
Machinery Except Electrical	1.8	1.9	0.9	1.4
Transportation Equipment	1.1	1.9	0.9	1.4
Motor Freight Transportation	0.8	1.6	1.3	1.7
Electric & Elec. Machinery	0.7	1.2	0.6	1.3
Wood Preserving	0.7	2.6	0.5	1.8
Drum Reconditioning	<0.1	2.6	0.7	1.4
OVERALL	100.0	2.4	0.7	1.3

NOTE: This data in some instances was extrapolated from small industrial groups in the sector. Caution should be exercised in utilizing
this information.

TABLE 4
Ten Highest Volume Waste Generating Industries- Generation and Recycling Volumes During 1981

Industry	Volume of Waste generated (M gals)	Total volume recycled (M gals)	Volume recycled on site (M gals)	Volume recycled off site (M gals)
Chemicals and Allied Products	28,000	340	300	32
Machinery - Except Electrical	4,200	26	18	7.9
Transportation Equip.	2,300	900	880	22
Motor Freight Transportation	1,700	NR*	NR*	NR*
Petroleum and Coal Prod.	1,300	36	32	4.2
Primary Metal Ind.	1,000	170	18	150
Construction - Special Trade Contractors	870	0.2	0.1	0.1
Fabricated Metal Prod.	820	24	14	9.6
Electric and Electronic Equipment	670	47	0.4	46
Electric, Gas, and Sanitary Services (includes POTWs)	470	3.3	0.1	3.2

* NR - No recycling of this type of waste reported.
Source: RIA Generator Survey data.

question was contained in the report, "Serious Reduction of Hazardous Waste" produced by the U.S. Congress, Office of Technology Assessment. The OTA stated "Thus substantially more waste reduction is feasible and more will become feasible. Getting a national voluntary waste reduction goal of perhaps 10 percent annually for 5 years could be useful." (5)

Hazardous Waste Recycling Profile

Hazardous waste recycling, especially on-site recycling, is included within the definition of waste minimization. Recycling of waste materials can be characterized by three major practices (1) direct use or reuse of the material in a process, (2) reclamation by recovering secondary materials for a separate end use (e.g., recovery of metal from sludge material), and (3) removing impurities from a waste to obtain a relatively pure, reusable substance (e.g., removal of impurities from a cyanide plating bath solution results in a bath that can be reused). The recycling of hazardous waste by U.S. industries is not a dominant waste management practice. Over 95 percent of hazardous waste generated in 1981 was disposed of by practices other than recycling and reuse. (4)

A rough breakdown of recycled waste according to waste categories (4) is:

    24% Solvent (halogenated & non halogenated)
    <.1% Halogenated waste (non-solvent)
    28% Metal bearing waste
    29% Corrosive Waste
    20% Cyanide reactive waste
    100%

Table 4 lists the volume of hazardous waste generated & recycled in 1981 by the 10 highest volume hazardous waste generator. The data suggest that the volume of waste recycled on-site by a generator increases with the total volume of waste recycled.

This pattern of recycling has both industry-specific and waste-stream specific components. Some major industries are more likely to recycle than others; that is, they recycle a substantially larger fraction of the waste they generate. Within an industry category, some wastes are more likely to be recycled than others (e.g., solvents more than pesticides), and the patterns o onsite and offsite recycling vary with the size of the industry and the waste stream generated. (4)

The pattern of recycling in the United States is predominatly an onsite waste management practice, accomplished either by using the waste directly without prior processing or by reclaiming the waste to recover constituent materials that then can be used directly. Eighty-one percent of the volume of hazardous waste recycled by U.S. industry in 1981 (the RIA Mail Survey study year) was performed onsite. However, the profile of recycling in the United States is changing to include offsite commercial recycling operations and direct transfers of waste from generator companies to others who can reuse the waste. (4)

Incentives and Disincentives for Waste Minimization

Strong incentives currently exist in the U.S. to promote waste minimization in the private sector, including (1) dramatic increases in the price of all forms of hazardous waste management, partially caused by Federal and State standards, (2) difficulties in siting hazardous waste management capacity, (3) permitting burdens and corrective action requirements, (4) financial liability of hazardous waste generators, (5) sharp increases in the cost of commercial liability insurance, coupled with a steep dropoff in its availability, and (6) public pressure on industry to reduce the production of waste. (2)

Probably the most significant regulatory incentive to encouraging hazardous waste minimization is the EPA's current program to evaluate all hazardous waste streams to determine if the continued land disposal of those materials is appropriate. This action is authorized by the most recent version of the RCRA, a set of amendments passed in 1984 that redirected the nation away from its dependence on land disposal for hazardous waste. The EPA is to make its determinations for the various wastes in accordance with the schedule shown in Table 5.

Table 5
Schedule for Land Disposal Prohibitions Decisions

Dioxin-containing wastes	November 8, 1986
Spent or discarded solvents	November 8, 1986
Metals and cyanides, corrosives, halogentated organics	July 8, 1987
At least one-third of all ranked hazardous wastes	August 8, 1988
At least two-thirds of all ranked hazardous wastes	June 8, 1989
All hazardous wastes	May 8, 1990

In essence, RCRA requires that EPA restrict RCRA wastes from land disposal unless the Agency determines that land disposal is more protective of human health and the environment than are available alternative treatment technologies.

There are also disincentives to waste minimization. Even though waste minimization processes can lead to cost savings, availability of capital for plant modernization is often an obstacle. Many firms are understandably hesitant to make any modification to their production process for fear of risking the quality of their product. And there are technical barriers related to process design. There are some products which simply cannot be manufactured, given current technology, without producing hazardous wastes. Finally there are barriers which result from a lack of engineering information on source reduction and recycling techniques. The EPA Report to Congress stated that this last barrier is most often the case with small companies. (2)

Table 6 Selected State Programs With Waste Reduction Activities

State: Program name and/or coordinating body	Program components	Annual budget	Waste reduction as percent of activities
California: Waste Reduction Unit (Alternative Technology & Policy Development Section of Dept. of Health Services)	Research grants Technical assistance	$1.5 million	<25
Connecticut: Office of Small Business Services (Dept. of Economic Development)	Technical assistance Loans	$50,000	<10
Georgia: Hazardous Waste On-Site Consultation Program (Georgia Tech Research Inst.)	Technical assistance	$220,000[b]	10-15
Illinois: Hazardous Waste Research & Information Center	Research Technical assistance	$1.3 million	10
Minnesoat: Minnesota Waste Management Board	MnTAP Research grants, Governor's Award	$235,000	25
New York: Industrial Materials Recycling Act Program (NY State Environmental Facilities Agency)	Technical assistance Industrial financing	$494,000	<25
North Carolina: Pollution Prevention Pays	Technical assistance Challenge grants	$590,000	<50
North Carolina Board of Science & Technology	Research and Education grants		
Governor's Waste Management Board	Governor's Award		
North Carolina Technical Development Authority	Financial assistance		
Pennsylvania: PennTAB (operated by Penn State University; funded by Department of Commerce)	Technical assistance	$150,000[c]	<25
Tennessee: Safe Growth Cabinet Council	Governor's Award	$1.8 million[c]	<25
Department of Economic and Community Development	Technical assistance		
Center for Industrial Service (University of Tenn.)	Hazardous Waste Extension research		
Waste Management, Research & Education Institute (Univ. of Tennessee)	Engineering research and development, policy research		
Wisconsin: Bureau of Solid Waste	Information out-reach Research grants, Tax exemptions	$850,000	<25

<u>Environmental Protection Agency Waste Minimization Program (2)</u>

While no program within EPA is specifically geared to source reduction and recycling, the Office of Solid Waste (OSW) has become the focus of activities with regard to waste minimization since the passage of the RCRA amendments of 1984. Two other programs within EPA that also influence waste minimization are the Office of Water (OW), and the Office of Research and Development (ORD).

° Office of Solid Waste. The Office of Solid Waste (OSW) is charged with developing and implementing RCRA and its amendments, which promote waste minimization directly by mandating:

- The inclusion on hazardous waste manifests of the generator's certification that waste quantity and toxicity are reduced to the maximum degree economically practicable;
- The inclusion of descriptions of the generator's efforts to reduce waste volume and document actual reductions achieved in biennial reports to EPA;
- The requirement that generators certify annually that they are minimizing waste quantities and toxicity to the extent feasible, as a condition of all treatment, storage, and disposal permits issued after September 1, 1985; and
- The preparation of a Report to Congress on the desirability and feasibility of instituting performance standards, management practices, or other actions to "assure such wastes are managed in ways that minimize present and future risks to human health and the environment."

° Office of Water. The Office of Water (OW) has controlled waste-water pollutants by requiring in-plant (source) reductions in several industries. Under the Clean Water Act (CWA), effluent limitations guidelines and standards are issued to control discharges of pollutants from industrial facilities (or point sources). The bases for the limitations in some of the guidelines and standards are chemical use minimization or substitution and water use reductions, which in turn reduce pollutant discharges. Although the wastewater itself is not a RCRA hazardous waste, sludges from wastewater treatment often are; thus, the effluent guidelines may serve to reduce some RCRA hazardous wastes associated with wastewaters.

° <u>Office of Research and Development.</u> Activities of ORD in waste minimization include a small business/small quantity generator research program in ORD's Office of Environmental Engineering and Technology (OEET), and waste reduction research conducted by the Hazardous Waste Environmental Research Laboratory (HWERL).

- The OEET Small/Business Quantity Generator Research Program provides financial support for research and information efforts of agencies or associations working directly with small businesses. Its current efforts include two main focuses: (1) supporting the research efforts of State technical assistance programs (providing financial support

to North Carolina's PPP program) and (2) providing funding
to the Governmental Refuse Collection and Disposal Association to set up a clearinghouse to furnish information on
waste management options to small businesses.

State Programs

Several of the states within the U.S. have become very active in
promoting waste minimization. In fact, in many ways the States have
been more involved than the Federal government in promoting waste
minimization.

Table 6 lists some of the state programs involved with promoting
waste minimization.

Expanded EPA Program (2)

In its recently submitted report to Congress, EPA stated that
generators should continue to determine which waste minimization
techniques are economically practicable at this time and that EPA
should not specify command and control regulations for waste minimization. (2)

The EPA did state however that work would continue to be done at
the Agency to reassess this determination and if necessary, additional
authority from Congress to do so in its report. EPA, however proposed
to report back to Congress on this issue in December 1990, the earliest
date at which it believes a recommendation on the need for a mandatory
waste minimization program could be made.

In the interim, EPA does plan however to expand its waste minimization efforts as discussed below in the context of proposed three-
point waste minimization strategy.

The elements of this strategy are:

1. Information Gathering: Detailed data on industry's response
    to the land disposal restrictions program and other existing
    waste minimization incentives must be gathered in order to
    make a final determination on the desirability and feasibility
    of performance standards and required management practices.

2. A Core Waste Minimization Program: During the interval when
    the new provisions of HSWA are taking effect, EPA will
    launch a strong technical assistance and information transfer
    program through the States to promote voluntary waste mini-
    mization in industry, government, and the non-profit sectors
    of the economy. It will also work with Federal agencies to
    encourage procurement practices that promote the use of
    recycled and reclaimed materials.

3. Longer Term Options: Based on an analysis of the new data
    gathered under (1) above, performance standards and other
    mandatory requirements can be imposed, if necessary, once

the new legislation has taken full effect and its impact on
waste generation has been determined.

Summary

There are many waste minimization activities currently being
undertaken by U.S. industry. However, there is the potential for
significantly more reduction of wastes if all available options and
processes are implemented. The EPA estimates the future reduction
potential to range from 17 to 33%. (2) The great unknown regarding
what waste reduction will occurrr in the absence of increased regulations is the result of the land disposal prohibition for selected
hazardous waste now just beginning to be implemented. Incentives for
waste minimization in the U.S. are strong. Between now and 1990 it
will become much clearer as to what role waste minimization will play
as a major environmental improvement strategy in the U.S. (1)

References

1. S. K. Stoddard, etal. Alternatives to the Land Disposal of
   Hazardous Wastes: An Assessment For California. Prepared by The
   Toxic Waste Assessment Group, Governor's Office of Appropriate
   Technology. 1981
2. Minimization of Hazardous Waste: Report to Congress. Office of
   Solid Waste U.S.EPA, 1986. PB-87-114-336 and PB-87-114-334.
3. Congressional Budget Office. Hazardous Waste Management: Recent
   Changes & Policy Alternatives. Prepared for Senate Committee in
   Environmental & Public Works.
4. Waste Minimization Issues & Options. Volume 1. 1986.
   EPA/530/SW-86-041. Office of Solid Waste. U.S. EPA.
5. Serious Reduction of Hazardous Waste, U.S. Congress Office of Technology Assessment, OTA-ITE-317 (Washington, D.C.: U.S. Government
   Printing Office, September 1986).

PRODUCTION, TREATMENT AND DISPOSAL OF ACID TAR WASTES

By S. Haverhoek, Shell Internationale Petroleum Maatschappij,
Health, Safety and Environment Division, The Hague

ABSTRACT

Acid tar has been generated by the benzole refining, white oil refining
and waste oil rerefining industries in the past and still is being
produced on a small scale today.
Maximum production of acid tar occurred during the 1950's and early
1960's coïnciding with peaks in benzole refining and white oil
production, prior to the existence of stringent controls on the disposal
of hazardous wastes and the availability of improved refining
technologies. In line with accepted practices at that time acid tar
wastes were usually deposited in lagoons on or adjacent to production
sites. A number of these disposal sites still exist today as legacies of
past practices and may require remedial action in the near future as a
result of the physical danger they often represent.

Some experience exists about the on-site restoration of acid tar dumps,
mainly using lime as fixation agent. With respect to clean-up,
incineration of acid tar has been attempted in the past but proved to be
impracticable due to the corrosive properties of the material and the
massive sulphur-dioxide emissions.
Today incineration of acid tar in advanced equipment is being attempted
again taking account of the possibilities of recovering its calorific
value.
The state of the art with respect to clean-up is presented in this
paper including consideration of estimated associated costs.

1. INTRODUCTION

Acid tar is a hazardous waste produced by several industries which use
or have used concentrated sulphuric acid (oleum) for the treatment of
hydrocarbon materials.
It is a black, acrid-smelling, corrosive substance with very aggressive
persistent and toxic characteristics.
Currently there is growing concern about the past waste disposal
practice applied for acid tar, namely direct landfill. Problems being
encountered with existing acid tar dumps comprise: odour emissions,
break-out of acid tar from open and covered dumps, physical danger
(quick sand) due to viscosity and low pH, groundwater contamination by
phenols where spent caustics were co-disposed.
As many millions of tonnes of acid tar have been produced, which were
subsequently landfilled throughout the industrialized world, clean-up
and/or restoration of acid tar ponds could present a massive undertaking
with high associated costs.
It is against this background of concern that an inventory of facts on
current and past treatment and disposal practices of acid tar has been
undertaken in order to contribute to the most practicable longterm
solution to this waste problem.

2. ACID TAR PRODUCTION PROCESSES

Acid tars are produced by three industrial processes which all involve the use of concentrated sulphuric acid to purify an organic material. These three processes are benzole refining, white oil production and waste oil rerefining.

2.1. Benzole refining

Crude benzole is a by-product of coal-carbonization, a process to produce metallurgical coke and domestic, smokeless, solid fuel. Coal carbonization was also carried out in the past to produce "town gas" but this activity was generally terminated since the introduction of natural gas some decades ago.

The crude benzole was treated with concentrated sulphuric acid to produce a purified BTX (Benzene, Toluene, Xylene) fraction which could be distilled.
The residual acid tar from this treatment consisted of unwanted sulphur-containing compounds, unsaturated hydrocarbons, sulphonated aromatic hydrocarbons and heterocyclic compounds and free sulphuric acid.

The composition of the acid tar and the quantities produced can vary considerably and depend amongst other things on the type of coal used, carbonization temperature, product specification, degree of acid treatment.

Benzole refining reached its peak e.g. in the U.K. around 1960. After this the production of acid tar declined, mainly due to a decrease in coke production and the introduction of alternative refining methods.

2.2. White oil production

The use of concentrated sulphuric acid in oil refining used to be a widespread activity including treatment of kerosine, paraffins, lubricants and white oils for the removal of colour, mainly caused by the presence polycyclic aromatic compounds and naphthenes.
Particularly luboil refineries have produced large quantities of acid tar before the change-over to solvent extraction processes in the early 1960's in Western Europe. There is evidence that in some older refineries outside Western Europe and North America the oleum treatment process is still in use.

In Western Europe at present, acid treatment is restricted to the production of "white oils" for medicinal use and specialized lubrication applications and this is done in a very small number of refineries. The product is usually "polished" by treatment with Fuller's earth, which was previously disposed of along with the acid tar but is now usually incinerated in cement kilns or solidified prior to landfill. Acid tar production is low nowadays due to the precise luboil feedstock specifications achieved at present in the refinery by extraction of the aromatic hydrocarbons (by furfural and sulpholane) and the unsaturated hydrocarbons (by MEK/toluene and hydrotreating).

The small quantities of freshly produced acid tar are nowadays either incinerated for sulphuric acid recovery or solidified with lime prior to landfill disposal.

In the past it was common practice for refineries with acid treatment processes to dispose of their acid tars in lagoons on or near the refinery sites, considerable efforts to develop alternative disposal methods having largely failed. Most of these acid tar ponds are more than 25 years old and may require restoration or clean-up due to the physical risk (quick sand) they may pose to human and animal life. It should be noted that many dumps are still viscous with low pH and more or less accessible. The size of an acid tar pond usually varies between 10.000 and 100.000 m^3.

2.3. Oil rerefining

Oil rerefining is used to regenerate spent lubricants either for direct re-use or for blending. The classical rerefining process entails:

    a. the removal of solids and debris by screening
    b. oil/water/sludge separation in a holding tank, aided by heating or alkali addition
    c. Steam stripping of the oil layer for removal of light fuels and solvents followed by oil/water separation
    d. after cooling the oil is pumped into tanks where concentrated sulphuric acid is added and thorough mixing is applied
    e. the treated oil is further purified by Fuller's earth filtration
    f. the acid tar and spent Fuller's earth are disposed of together
    g. the separated water and other sludges are further treated before disposal.

The volume and composition of the acid tar can vary considerably due to the oil quality and the extent of the treatment applied.

Other more sophisticated rerefining processes are based on solvent treatment and distillation or possibly hydrofinishing. With these types of treatment the generation of acid tar is avoided, but plants of this type are scarce.
The production pattern in oil rerefining is rather uncertain.
In the U.K. and the FRG there was much rerefining during the second world war to maintain supplies of lubricants for military purposes.
The second upswing can be identified after the two oil crises where it became economically attractive to reclaim luboils. However in a number of countries waste oils are only upgraded to second hand fuel oils as rerefining is not permitted. Following the collapse of the oil price since the beginning of 1986 some of the existing oil rerefiners are ceasing activity.
The contribution of the oil rerefining to the acid tar problem is small compared to benzole and white oil acid tar production.

3. CHARACTERISTICS OF ACID TARS

3.1. Physical characteristics

Acid tar is a dark-coloured liquid with a persistent, penetrating and
noxious odour.
The consistency of acid tar varies with the originating process and
depends also on the amount of acid used. Thin tars, indicating an
excessive use of acid, were usually regenerated for sulphuric acid
recovery, a practice which is still applied to day. The more viscous
tars with still a varying degree of mobility are found in acid tar
lagoons. The viscosity is both shear and temperature dependent.
The leacheability of acid tars is very limited. Groundwater
contamination is only observed in cases where other chemicals were
co-disposed with the acid tar. E.g. phenol from spent-caustic.
The presence of other materials, drums, trees, wildlife corpses etc.
could hamper the clean-up.

3.2. Chemical characteristics

The chemical characteristics of acid tars are difficult to define due
to the complex nature of the hydrocarbons present whose composition
varies between the different sources.

Acid tars have been described as being "strongly acid", highly
reactive, corrosive substances with a pH of usually less than 1., due to
the presence of concentrated sulphuric acid.

Most of the work on chemical characterization of acid tars has
consisted of analysis of simple group parameters namely water, acid,
organic and ash content. Table 1 gives typical results.

Table 1. Composition of acid tars (in % w/w)

	H_2SO_4	organic	water	solids
benzole refining	35–60	40–50	5–10	0
white oil production	25–90	5–70	5	0– 5
oil rerefining	10–30	40–60	10–40	5–10

The organic part of the acid tar is a complex mixture of sulphonated
paraffins and naphthenes, sulphonated polycyclic aromatics and
asphaltenes and small amounts of sulphonated mono and bicyclic
aromatics. Sulphonated heterocyclic aromatics can also be present.
In acid tar from oil rerefining heavy metals like lead and chromium can
be present.

4. TREATMENT AND DISPOSAL

1. Historical practices

The following treatment and disposal methods have been used for acid tar:
a. Dilute sulphuric acid and oil recovery by steam or water treatment and dumping of tarry residue.
b. Ammonium sulphate production using solvent (creosote), water and heat and neutralization with ammonia. Recycling of oily solvent. Dumping of tarry residues.
c. Neutralization (with lime, caustic) prior to disposal.
d. Utilization as boiler fuel by mixing with cokes (1:3-8), crude coal tar, or refinery fuel.
e. Dumping in slag heaps, lime pits, disused coal mines, quarries.
f. Incineration of thin acid tars for sulphuric acid recovery.

The petroleum refining industry (ref. 1) has devoted considerable effort to developing economic and environmentally acceptable methods for acid tar treatment, mainly aiming at utilizing its calorific value. These attempts largely failed to produce a reliable process. The introduction of solvent extraction methods has ultimately superseded the oleum treatment process thus eliminating the acid tar problem in petroleum refining. The majority of the acid tars produced were deposited in lagoons, as being the best available solution although there is some evidence that acid tars were added piecemeal to boiler fuel. The benzole refining industry and the oil rerefining industry basically relied on "landfill" methods.

2. Current practices

Today's practices on acid tar disposal mainly from oil rerefining operations are basically not different from the historic practices although there is a clear trend that freshly produced acid tar is disposed of by incineration. The current quantities of acid tar produced are only a fraction of those in the past at least in the Western world due to the fact that benzole refining is nearly phased out and refineries have ceased acid tar generation. There is evidence that e.g. in Eastern Europe considerable quantities of acid tar are still being produced in refinery operations and that acid tar dumping in lagoons even in conjunction with spent caustic is still common practice.

Acid tar generation from white oil manufacturing is small nowadays due to the relative purity of the luboil feedstocks used. The acid tar is usually regenerated by sulphuric acid manufacturers. The saturated Fuller's earth used to remove the last traces of aromatics from the white oil after the acid tar separation is usually incinerated in cement kilns, as controlled landfill is increasingly prohibited due to the high residual oil content (10-15%).

5. ACID TAR CLEAN-UP

The presence of old acid tar dumps has been tolerated and sometimes ignored for several decades. However, with improvement in scientific understanding of the potential environmental impact of such dumps, with increasing public attention and with developing soil protection legislation, this situation is likely to change. A number of acid tar dumps which posed an immediate danger have been cleaned-up using on-site stabilization techniques. The know-how obtained from these cases will be a useful basis for establishing an acceptable treatment method for acid tar dumps. This might also result in improvement of some present disposal practices mainly related to oil rerefining.

5.1. Lime stabilization

Experience with the clean-up of acid tar dumps is rather limited but in all known cases involves the in-situ application of slaked and/or quick-lime with or without any other additions.

The method involves the use of a drag-line to remove the viscous acid tar sludge from the lagoon and spread it on a mixing area.
With a crane and grab quick lime is added and thoroughly mixed with the sludge.
During this neutralization and fixation process a considerable amount of heat and steam is produced. The quick lime addition and mixing is repeated several times until the sludge is transformed into a virtually odourless, light brown, pulverulent product. Site restoration and compacting is carried out with bulldozers and rototillers.
This process is capable of transforming an objectionable and hazardous acid tar pond into a dune of inert earth-like product. The acid tar/quick lime ratio can be about 1:3 depending on the acidity of the waste and on the level of homogenation, so a considerable volume increase has to be taken into account.

The leachability of the endproduct is claimed to be limited, due to the favourable stability and the good compacting and water repellent properties of the material.
A French company which applies the process claims a leachate quality of COD 250 ppm after 3 months of immobilization.

From a chemical point of view this process is likely to be effective provided that thorough homogenation is applied. The calcium oxide reacts exothermically with any water present but also with the sulphonic acid groups, thus providing neutralization and linking of the sulphonated compounds (chemical transformation):

$$R_1-SO_3H + CaO + HSO_3-R_2 \qquad R_1-SO_3-Ca-SO_3-R_2 + H_2O$$

Any excess of sulphuric acid will result in the formation of gypsum $CaSO_4$ aq. which is also known to have stabilizing and fixation properties.

5.2. Incineration

Acid tars represent a considerable calorific value. This was recognized
many years ago as illustrated by the attempts by refineries to use acid
tar as boiler fuel. However technological developments both in corrosion
abatement and in flue gas treatment have advanced considerably since the
1950's. Moreover toxic waste incinerators (rotary kilns) have appeared
on the scene which are nowadays equipped with waste heat boilers and
highly developed flue-gas cleaning devices, and in which almost any
waste can be destroyed.

It may therefore be assumed that in a rotary kiln acid tars can in
principle be incinerated satisfactory. An important requirement is
however a suitable feed preparation method, because acid tar is neither
solid nor pumpable and is very sticky.
Further aspects to be taken into account are: transportation, flue-gas
desulphurization and land restoration after excavation of the acid tar.
The flue gas desulphurization would require considerable amounts of lime
in the dry-scrubbing process and also would yield substantial quantities
of gypsum to be disposed of.
From a waste management point of view incineration would have a slight
preference as potentially dangerous constituents of acid tar are
destroyed completely whereas with lime stabilization the wastes are
chemically bound rather than eliminated.

5.3. Other alternatives

Russian researchers have investigated the re-use of acid tar for
asphalt manufacture. The ultimate results are not known but given the
very strict specifications of asphalt materials this outlet is not
likely to be developed on a large scale due to the variability in acid
tar composition (ref. 2).

Belgian researchers are currently investigating the utilization of acid
tar as a support fuel in a fluid bed steam generation system based on
powder coal. The acid tar would have to be pretreated with lime and coal
fines in order to produce a pelletized material which can be fed to the
fluid bed incinerator proportionately to the base coal fuel.
Costs are not known. The concept might be attractive, but the
technology has yet to be proven with respect to the high calcium
sulphate levels present.

5.4. Clean-up cost estimates

On site lime stabilization costs for acid tar can vary between 50 and
100 USD per tonne of acid tar depending on the amount of lime to be used
to achieve a suitable end product. The acid tar to lime ratio depends on
the acid tar composition and on the efficiency of homogenation of the
mixture. Additives to facilitate the lime dispersion and therefore the
homogenation might give savings on the amount of the lime required but
on the other hand will add to the reagent cost per tonne. Further cost
factors are site investigation, laboratory work, labour, civil work
equipment and site restoration.
If incineration is the preferred route to dispose of acid tar, feed
preparation would be required in order to be able to feed the toxic

waste incinerator in a controlled manner. This would entail a similar
process to that described for lime stabilization and hence comparable
costs would have to be incurred in addition to incineration costs
proper. As incineration costs of high calorific wastes as support fuel
in licenced toxic waste incinerators are in the order of 50 USD/tonne
including benefits from steam generation, the ecomonics of this route
are not likely to compare favourably with the in-situ stabilization.
Transportation and handling costs and flue-gas treatment provisions will
make the overall incineration economics even less favourable.
Moreover it is likely that only 10% of the capacity of a toxic waste
incinerator would be dedicated to acid tar treatment in order not to
lose other regular customers. Consequently total treatment of an acid
tar dump by incineration would be a very costly and time consuming
operation.

6. CONCLUSION

Lime stabilization of acid tar might be an environmentally acceptable
treatment method for the clean-up of acid tar dumps provided longterm
leaching behaviour is acceptable. The economics of this method compare
favourably with incineration, which would involve extensive additional
treatment and handling operation. Although costly, incineration of acid
tar from dumps is slightly better from an environmental point of view
provided adequate flue-gas treatment is performed.

7. REFERENCES

1. Institute of Petroleum Journal vol. 38, No. 337 January 1952,
   pp. 1-57.
2. Frolov, A.F.; Aminov, A.N.; Timrot, S.D.
   Composition and Properties of acid tar and asphalt produced from
   acid tar, Chem. & Techn. Fuels & Oils 17 (1982)5-6, 284-288

Chapter 3

PHYSICAL TREATMENT METHODS

IN SITU VITRIFICATION
FOR DECONTAMINATION OF SOILS CONTAINING PCBs

R. F. Battey
Bechtel National, Inc.
San Francisco, CA 94119
USA

and

J. T. Harrsen
General Electric Company
Schenectady, NY 12345
USA

ABSTRACT

An emerging thermal treatment process known as in situ vitrification
(ISV) is being developed to immobilize radioactively contaminated soils.
ISV is a permanent remedial action that destroys solid and liquid organic
contaminants, and incorporates radionuclides and heavy metals into a
glass and crystalline form. The essential elements of the process
consist of an electrical power system supplying the energy to melt the
contaminated soil, a hood to contain gaseous effluents, an off-gas
cooling and treatment system, and a process control station. This study
presents the results of a hypothetical remedial action on PCB
contamination at a relatively shallow site. The remedial work plan
involved excavation of a vitrification trench to a depth of 18 feet and
backfilling with soil from four of the six contaminated areas on the
site. The plan called for two deeper contaminated areas inside a
building to be vitrified in place. A detailed list of activities was
developed and costs estimated for each. Total costs of ISV for this
hypothetical (but typical) site were estimated at $210-220/ton of soil
vitrified. The most significant cost elements were electricity and
labor. Alternative total project costs for disposal of the same quantity
of soil in a landfill about 200 miles distant would be about $290/ton.

INTRODUCTION

Battelle Memorial Institute, at the Pacific Northwest Laboratory at
Richland, Washington, is developing an in-place stabilization technique
for radioactively contaminated soils. The process, called in situ
vitrification (ISV), converts contaminated soils into a stable glass and
crystalline waste form that has approximately the same chemical durability
properties as granite. With some additional development, ISV could be
applied to chemically hazardous waste sites. The process is being funded
by the U.S. Department of Energy and has been under development since
1980. It has been described in a number of papers and reports [1-4].

IN SITU VITRIFICATION

In situ vitrification is a thermal treatment process that converts
contaminated soil into a chemically inert and stable glass and
crystalline product. Figure 1 illustrates how the process operates.
Four molybdenum electrodes are inserted into the ground in square array
to the desired treatment depth. Because soil is not electrically
conductive once moisture has been driven off, a conductive mixture of
flaked graphite and glass frit is placed among the electrodes to act as
the conductive starter path. An electrical potential is applied to the
electrodes, which establishes an electrical current in the starter path.
The resultant power heats the starter path and surrounding soil up to
3600°F, well above the initial soil melting temperatures of 2000 to
2500°F. The graphite starter path is eventually consumed by oxidation,
and the current is transferred to the molten soil, which has become
electrically conductive.

As the molten or vitrified zone grows, it envelops nonvolatile
hazardous elements, such as heavy metals, and destroys organic components
by pyrolysis. The pyrolyzed by-products migrate to the surface of the
vitrified zone where they combust in the presence of oxygen. A stainless
steel hood placed over the area being vitrified directs the gaseous
effluents to an off-gas treatment system. Remaining ash (noncombustible
material) dissolves or becomes encapsulated in the molten soil. Natural
convective currents within the molten soil help distribute the stabilized
materials uniformly. The molten soil cools to a durable glass and
crystalline form resembling natural obsidian.

PNL began developing ISV technology in 1980 under the support of the
U. S. Department of Energy. Since that time, numerous experimental tests
under a variety of conditions and waste types have been conducted.
Table 1 shows the different scales of testing units that PNL used in
developing ISV technology. The successful results of these tests have
proven the feasibility of the process. Also, economic studies have
indicated that large economies of scale are attainable with the ISV
process.

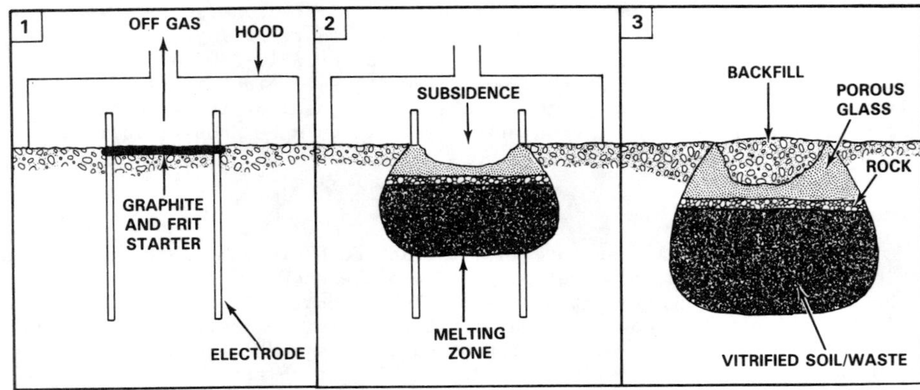

Figure 1. In situ vitrification process

TABLE 1
Testing units for developing in situ vitrification technology

Equipment Size	Electrode Separation (m)	Block Size
Bench Scale	0.11	1 – 2 kg
Engineering Scale	0.23 – 0.36	0.05 – 1.0 t
Pilot Scale	1.2	10 – 50 t
Large Scale	3.5 – 5.5	400 – 800 t

LARGE-SCALE EQUIPMENT

The large-scale process equipment for in situ vitrification as developed by Battelle is show in Figure 2. Controlled electrical power is distributed to the electrodes, and special equipment contains and treats the gaseous effluents. The process equipment required to perform these functions can be described most easily by dividing it into four major components: off-gas treatment system, off-gas hood, electrical power supply, and electrodes.

Except for the off-gas hood, all the components are contained in three trailers. They consist of an off-gas trailer, a process control trailer, and a support trailer. The off-gas hood and off-gas line, which are installed on the site for collection of gaseous effluents, are dismantled and placed on flatbed trailers for transport.

Off-Gas Treatment

The ISV process was developed expressly to immobilize buried inorganic low-level radioactive wastes. Test work has shown that organic, combustible materials are pyrolyzed at the temperatures of the molten soil (up to 3000°F) and essentially destroyed. However, off-gases from the Battelle on PCB-contaminated soils test contained small amounts of PCBs, in excess of that permitted by EPA regulations, and four orders of magnitude smaller amounts of furans and dioxins.

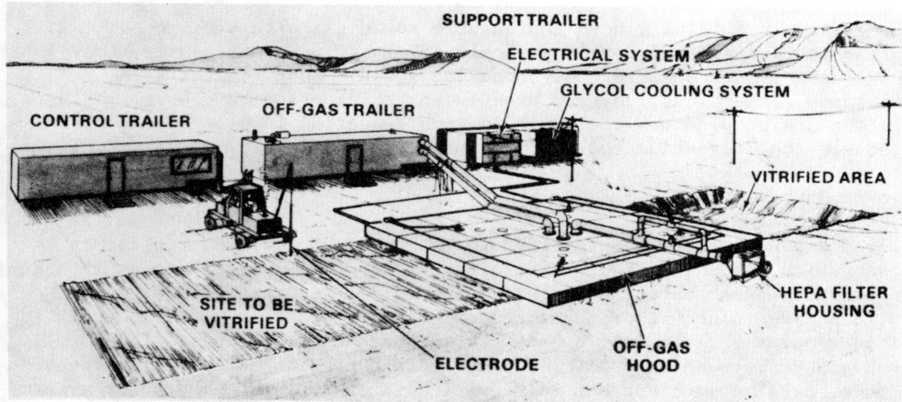

Figure 2. Large-scale ISV setup

The process of vitrification generates inorganic fumes and results in the formation of fine particulates. These particulates can be expected to adsorb most of the PCBs and similar organic compounds that enter the off-gas treatment section. The off-gas treatment system developed by Battelle for ISV of radioactive, transuranic wastes is designed to contain particulates entrained in the off-gas from the vitrified mass and has a removal efficiency of 99.999%. The off-gas treatment system consists of the following steps: indirect cooling with circulating glycol-water coolant, direct quench, two-stage high pressure drop venturi scrubbing, indirect cooling and condensation using glycol-water coollant, electrical reheating, and HEPA filtration. The small amount of PCBs not adsorbed on fine particulates will be removed with activated carbon. Particulates (with adsorbed PCBs) and soluble PCBs can be removed from the scrubbing water by pumping a side stream through a cartridge filter and an activated carbon filter.

Off-Gas Hood

The hood, which contains the gaseous effluents for the large-scale system, is constructed of 16-gauge stainless steel panels supported by trusses and beams (see Figure 3). Panels are bolted and gasketed together to relieve stresses due to nonuniform thermal expansion. Soil is backfilled along the outside of a flexible skirt around the lower edge to form a seal and minimize in-leakage of air. It is designed for a vacuum of 6 inches of water. The entire hood weights about 22 tons.

In recent field tests, Battelle has used a 2-inch thick layer of porous insulation material, which rests directly on the molten soil during a run to reduce the temperature and thermal stresses the hood is exposed to.

Battelle has recently designed a much lighter hood, weighing only about 5 tons for the large-scale system. It utilizes an umbrella-tent construction with external supports, from which is hung a flexible, heat-resistant metallized cloth.

Electrodes

Battelle initially used nonreusable solid graphite electrodes, 4 inches in diameter, for large-scale ISV tests. More recently, they have used 2-inch diameter molybdenum rods, in 6-foot lengths with threaded connections, covered by a 1-inch thick graphite sleeve. This latter design promises to result in some reduction in electrode cost because the molybdenum rod sections may be reusable.

Power Supply System

The power system for the ISV process uses a Scott-Tee transformer connection, shown in Figure 4. This converts 3-phase input power into a balanced 2-phase output configuration, and was selected over several alternatives since it produces a nearly square block of vitrified soil. This results in a minimum overlap between adjacent settings. The resistance between each pair of electrodes decreases as the melt size grows, causing more current to flow at a fixed voltage. The power system

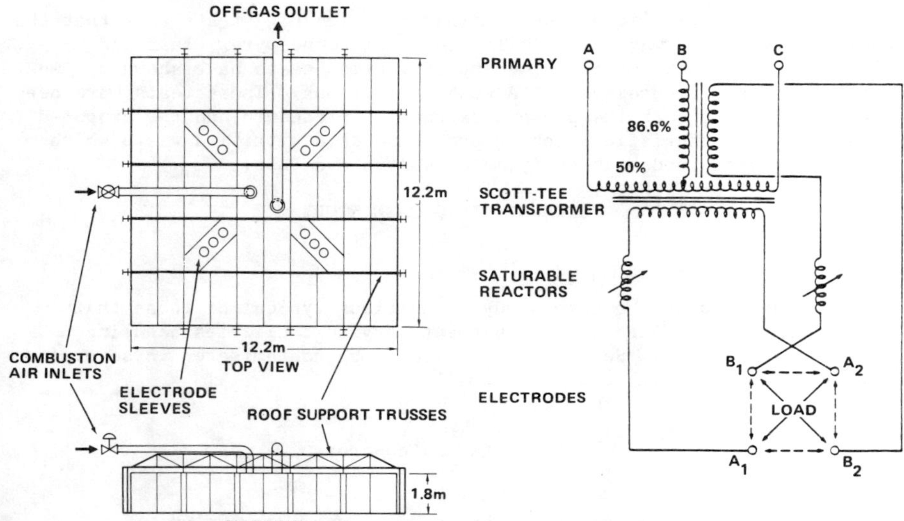

Figure 3.   Large-scale ISV
            off-gas hood

Figure 4.   Large scale ISV Scott-
            Tee transformer design

uses multiple intermediate voltage taps and saturable reactors in order
to operate near the maximum power and thereby minimize run times. The
power supply is discussed more fully in reference [2].

ISV PROCESS CHARACTERISTICS

Physical Limitations

The range of electrode separations for the large-scale unit varies
from 3.5 to 5.5 m. The unit has been designed to produce a vitrified
depth of as much as 13 m at 3.5 m electrode spacing. Tests at Bettelle
have shown that the ISV system can successfully vitrify sites containing
metal objects, combustible organics, void volumes, and moisture. If the
soil moisture is replenished during vitrification due to surface runoff
or the combination of a high water table and a high soil permeability,
the additional energy use may become prohibitive.

PCB Tests

An engineering-scale ISV test at Battelle with PCB-contaminated soil
resulted in small quantities of PCBs, furans, and dioxins in the
untreated off-gas, but none in the vitrified mass. The majority of
samples of the adjacent soil showed no detectable PCB content, but a few
samples directly adjacent to the block contained measurable
concentrations up to 0.7 ppm PCBs. This is below the lower limit usually
set by EPA for a cleanup action.

Waste Form Quality

The vitrified waste mass is physically and chemically similar to
obsidian, which has a reported natural weathering rate of 1 mm in
10,000 years. Distribution of elemental species is quite uniform

throughout the vitrified mass. Comparative leaching tests show that the
durability of ISV waste is similar to granite and pyrex glass and
superior to that of marble. Recent extraction tests have shown the waste
to be nontoxic according to EPA RCRA definitions. These tests have been
performed using both the present extraction procedure and the proposed
toxicity characteristic leaching procedure on a vitrified waste which
initially contained both heavy metal wastes and PCBs.

SITE-SPECIFIC CASE STUDY

Contaminated Areas

In developing the case study, conditions typical of those that
might be found at electrical equipment service facilities handling
PCBs were selected. Soils were assumed to be contaminated in several
areas as shown in Table 2.

TABLE 2
PCB-contaminated areas

Area	Dimensions (ft)	Depth (ft)	Max Conc. (ppm)	Volume (yd^3)
1	70 x 110	5	300	1426
2	83 x 113	3	300	1042
3	30 x 30	10	1000	333
4	35 Diam.	30	5000	1069
5	35 Diam.	30	5000	1069
6	44 x 76	10	1000	1239
Total				6178

For the purposes of this study, areas 1 and 2 are assumed to be
storage areas for electrical equipment containing PCBs and
PCB-contaminated materials. Areas 3 and 6 are assumed to be outside
service areas. Areas 4 and 5 inside the service building are assumed to
be dry wells receiving floor drainage and equipment washings.

Work Plan

Battelle's experience with tests, supported by a computer model of
vitrification, is that depths of 10 to 20 feet show the greatest energy
efficiencies for the large-scale equipment. Extremely shallow depths,
particularly areas 1 and 2 with 3- and 5-foot depths, would result in low
utilization of off-gas equipment and site personnel. Run times would be
about as short as changeover times between runs. A typical run at 12-18
feet depth takes about 60-120 hours, compared to a 20-hour change over
time. The selected approach was to excavate an 18-foot deep vitrification
trench and backfill it with contaminated soil from the shallow outside
areas for vitrification.

Areas 4 and 5 inside the building would be vitrified in place to the
30-foot contaminated depth because of the difficulties of excavating
inside the building. Clean soil initially excavated from the
vitrification trench would be stockpiled and later used to backfill the
excavated contaminated areas. Due to settlement during vitrification, an

additional amount of backfill would be required from off-site to restore all areas to the original grade. The mobilization and final cleanup and closure activities typical for this type of remedial action would also be required. The vitrification trench would require 18 separate runs, each melting an area about 24 ft x 24 ft x 18 ft deep. Each run would require 111 hours, plus 20 hours for change over. For the two in situ melts inside the building, Battelle estimates a vitrification time of 320 hours each to reach a depth of 30 feet and a width of 35 feet. The total calendar time for vitrification for all 20 melts would be 3038 hours or about 4.2 months. From Battelle's simulation runs, the average vitrification power requirement would be 3000 kW for both inside and outside locations.

The sequence of activities for the case study site campaign would involve three phases: mobilization, vitrification, and demobilization. Thirteen tasks have been identified and are listed in Table 3. Each vitrification melt, or setting, would require specific support activities in addition to operation of the electrical supply and off-gas processing equipment. These activities are listed in Table 4.

ECONOMIC EVALUATION

The capital cost of an ISV unit specifically designed for treating PCB-contaminated soils has been estimated, based on Battelle's costs for their large-scale unit designed for treating buried radioactive wastes. Operating costs for the case study have been estimated for two variations of electrical power supply: electricity purchased from a local utility company, and the rental of diesel generator sets and the purchase of fuel. In addition, operating costs have been estimated for excavation and off-site landfill at the nearest facility permitted to dispose of PCB wastes.

Estimated Capital Cost

The estimated cost for a large-scale unit, adapted for use on PCB-contaminated soils from Battelle design is $2.09 million. This is about 25% less than the latest estimate for the Battelle large-scale unit for use on radioactive wastes [5].

Estimated Operating Costs

The estimated operating costs for the case study site campaign are shown in Table 5. The method of estimation of some of the components is discussed below. Five year, 9%/yr financing was assumed. The annual cost was multiplied by 7/12. It should be noted that the ISV equipment would only be in service at the site approximately 4.5 months. If consecutive sites were prepared in advance, the ISV equipment could be kept busy for a larger percentage of the time. This is particularly important for short-duration campaigns, but less of a factor for longer duration sites.

TABLE 3
Work plan tasks

Phase 1 - Mobilization

1-1	Establish exclusion area
1-2	Construct decontamination pad
1-3	Remove concrete from areas 3, 4, 5, and 6
1-4	Excavate vitrification area
1-5	Excavate contaminated areas 1, 2, 3, and 6
1-6	Support building foundations in area 3
1-7	Backfill areas 1, 2, 3, and 6
1-8	Mobilize vitrification equipment

Phase 2 - Vitrification

2-1	Vitrify soil in trench
2-2	Vitrify soil in building, Areas 4 and 5

Phase 3 - Demobilization

3-1	Demobilize vitrification equipment
3-2	Pour concrete in areas 4 and 5
3-3	Demobilize exclusion area

TABLE 4
Vitrification support activities

Prior to Vitrification

1	Drill holes and set casing for electrodes
2	Move unused hood from previous to next scheduled melt area (area may not be adjacent to current melt area)
3	Pull casing and set electrodes in place
4	Seal hood opening around electrodes
5	Lay graphite-glass starter path
6	Lay Kaowool insulation over melt area under hood
7	Assemble off-gas piping
8	Backfill and level previously vitrified areas

At End of Vitrification Run

9	Pull electrodes as soon as possible (for ease of pulling and for verification of melt depth)
10	Disconnect off-gas piping
11	Move trailers (may not be necessary for every run)
12	Connect off-gas piping to previously placed hood off-gas line
13	Connect electrical leads
14	Perform miscellaneous maintenance on off-gas equipment

TABLE 5
Estimated site costs

	$1000	$/ton(1)	$/yd^3(2)
Capital, prorated for 7 months	313	28	51
Electricity			
Case A – Purchase @ $0.063/kWh	646	57	105
Case B – Rent generator @ $76,000/mo	380	34	62
Buy fuel @ $0.475/gal	349	31	56
Electrodes @ 235 $/ft	372	33	60
Site preparation & closure	280	25	45
Vitrification support	479	43	78
Operations	190	17	31
Maintenance and operating supplies	74	17	12
Waste disposal	30	3	5
Total:			
Case A – Purchase electricity	2384	212	386
Case B – Rent equipment, buy fuel	2467	219	399

---

(1) Total soil vitrified = 11,270 tons
(2) Total contaminated soil originally in ground = 6,178 yd^3

Purchased electricity would be available at an average of 6.29 ¢/kWh
including both energy and demand charges. No allowance was made for a
high voltage line to the site. Alternatively, diesel generator sets
could be rented in locations where power is not readily available.
Rental prices for four 1250 kW units, plus controls, plus a fuel tank
would be $76,000/month.

Site work and vitrification support costs include equipment, labor,
and supervision (site manager and site engineer). The vitrification
support crew would be staffed in 3 shifts, 7 days a week, even though the
time required for their duties accounts for only 40% of this time. Wage
rates from the current union labor agreements were used. Direct labor
rates were $17.47 to $25.05 per hour. A contractor's overhead and fee of
18.4% was added. Local equipment rental and lease rates were used for
cranes, etc. Transportation costs for equipment for 1000 miles to the
next job and relocation costs for the site manager and engineer were
included in site work costs.

It was assumed that two operators would be on duty 3 shifts, 7 days a
week during Phase 2. Operator direct labor rates were $21.00 to
22.77 per hour. An allowance for contractor's overhead and fee of 18.4%
was added. An allowance for periodic visits by a safety inspector to the
site, including an initial training period, was included.

Disposal of the residual materials noted in Section 4 would be at the
nearest PCB–permitted landfill. The largest quantity would be the
material from the decontamination pad — costs were estimated at $27,000
for five truckloads of bulk materials. Miscellaneous drummed waste
disposal was estimated at $3,000.

Landfill Alternative

For comparison, costs were estimated for excavation, transportation, and disposal of all PCB-contaminated soil at the nearest landfill permitted to dispose of these materials. Rates were assumed as follows for bulk material. Disposal in a PCB area of the landfill would be $230 per ton, and transportation for 220 miles is $865 for a 30-yd^3 truck load (equivalent to $43.25 per ton). Costs are summarized in Table 6.

TABLE 6
Landfill alternative estimated costs

	$1000	$/ton(1)	$/yd^3(2)
Site preparation and closure	179	16	29
Purchase backfill	31	3	5
Disposal, including transportation	3080	273	499
Total	3290	292	533

(1) Total excavated = 11,270 tons
(2) Total contaminated soil in ground originally = 6,178 yd^3

CONCLUSIONS

The ISV process are developed by Battelle has been demonstrated on a variety of scales on soils containing both metals and organics, including PCBs. Costs for a large-scale system designed for PCB-contaminated soils have been estimated at $210 to 220 per ton. This is about 3/4 of the cost to dispose of the same quantity of soil in a nearby landfill.

REFERENCES

1. Oma, K. H., Brown, D. R., Buelt, J. L., FitzPatrick, V. F., Hanley, K. A., Mellinger, G. B., Napier, B. A., Silviera, D. J., Stein, S. L., and Timmerman, C. L., "In Situ Vitrification of Transuranic Wastes: Systems Evaluation and Applications Assessment." PNL-4800, Pacific Northwest Laboratory, Richland, Washington. 1983.

2. Buelt, J. L., FitzPatrick, V. F., and Timmerman, C. L. "Electrical Technique for In-Place Stabilization of Contaminated Soils." Chemical Engineering Progress, 81 (3), pp. 43-48. March 1985.

3. Buelt, J. L. and Carter, J. G. "Description and Capabilities of the Large-Scale In Situ Vitrification Process." PNL-5738, Pacific Northwest Laboratory, Richland, Washington. 1986.

4. Timmerman, C. L. and Oma, K. H. "An In Situ Vitrification Pilot-Scale Radioactive Test." PNL-5240, Pacific Northwest Laboratory, Richland, Washington. 1984.

5. Timmerman, C. L., Pacific Northwest Laboratory. Personal Communication. May 1986.

6. FitzPatrick, V. F., Pacific Northwest Laboratory. Personal Communication. Jan. 1987.

IN-SITU RECOVERY OF HYDROCARBON CONTAMINATED SOIL
UTILIZING THE INDUCED SOIL VENTING PROCESS

by

George Edward Hoag
Department of Civil Engineering
Uinversity of Connecticut
Storrs, Connecticut 06268, USA

and

Arthur L. Baehr
Water Resources Division
United States Geological Survey
Reston, Virginia 22092, USA

and

Michael C. Marley
Department of Civil Engineering
University of Connecticut
Storrs, Connecticut 06268, USA

ABSTRACT

Induced soil venting is a process for removing immiscible
contaminants, such as gasoline, from the unsaturated zone. Air is
withdrawn from the unsaturated zone which results in hydrocarbon vapor
movement along the induced flow path toward a collection pipe. Factors
affecting process performance are vapor pressures of the contaminants, air
permeability of the soil and the ability to develop a flow field which
encompasses the immiscible contaminants. A synthesis of laboratory
experiments, modeling and full-scale implementation of the soil venting
process to remove gasoline from the subsurface will be discussed in this
paper.
Results of laboratory studies using soil columns at field capacity
moisture content, reveal that 100 ± 1 percent of the gasoline at residual
saturation can be removed by soil venting. A detailed analysis will be
made of the mechanisms affecting system performance, considering physical
and chemical factors based on laboratory and field data. Application of
the process to remediate soil contaminated with other volatile
contaminants will also be discussed.

INTRODUCTION

The transport and fate of solvents and oils once introduced to the subsurface environment is a complex phenomena because of the various phases present (i.e. solid (soil), liquid (water), and gas (soil air)) in the unsaturated zone where underground storage tanks are predominantly located. Upon a solvent entering the unsaturated zone, it can partition (i.e. a thermodynamic reversible reaction) between the solvent and solubilized states and between the solvent and gaseous states. Adsorption onto the solid phase may occur from the solvent, solubilized and gaseous states. The potentiometric flow of the solvent phase is complex because it is predominantly immiscible with water and in a system with potentiometric flow of water and gases simultaneously taking place.

The three major environmental health and safety problems which result from the discharge of gasoline hydrocarbons and solvents to the subsurface environment are groundwater pollution, explosion hazards and exposure to toxic vapors.(1) Groundwater contamination is the result from the solubilization of normal and substituted alkane, alkene, and aromatic hydrocarbons. Explosion and toxic vapor hazards are caused by volatilization of hydrocarbon components to soil-air (i.e. gas phase) and subsequently entering confined gas phases (i.e. building basements and excavation sites by diffusion and advection).

Although the phenomena of subsurface dissolution of aromatic hydrocarbons from a solvent (e.g. gasoline) phase to a aqueous phase is recognized as a significant evironmental problem, little has been done to investigate the problem. Most solvent spill recovery procedures only address the removal of the solvent product (i.e. recoverable mobile phase of the solvent) from groundwater systems by depression well or trenches and gasoline skimming pumps.(2,3) The authors generally agree, however, that groundwater contamination by water soluble aromatic hydrocarbons is usually a much more extensive and long-term threat to public health and the environment. The particular reasons for this can be readily understood from the mechanisms which control the transport of all solvent phases following a spill. Once introduced to the unsaturated zone, the mobile phase of the solvent becomes immobilized in the soil interstices and capillaries. This capacity is termed the residual soil saturation by the solvent phase. Because the solvent phase residually saturates a soil, remedial actions that only address the recovery of the mobile product in typical depression wells frequently leave behind the majority of a spilled solvent. Thus, effective remedial actions must address the removal of the immobilized solvent phase on soil.

LABORATORY RESEARCH OBJECTIVES

The objectives of this research were to develop an experimental apparatus and procedure to study the mechanisms involved in induced soil venting of gasoline hydrocarbons from a residually saturated soil. The purpose of studying induced soil venting was to investigate the feasibility of using this technique as a possible remedial action at hydrocarbon spill sites. Additionally, it was proposed to develop a model to predict the soil venting process and to evaluate the model against actual experimental data.

METHODS AND MATERIALS

Soil venting experiments were conducted in plexiglass column reactors. The columns had a diameter of 89 mm, a length of 650 mm and a cross-sectioned area equal to 6200 mm^2. To allow a uniform distribution of air throughout the porous media in the columns, a 100-mesh brass screen sandwiched between two 6 mm thick plexiglass spacers was fixed 45 mm from the reactor base. The soil venting apparatus is shown in Figure 1.

27 experimental runs were performed. To achieve residual saturation of the sand gasoline was allowed to saturate the column, the sand column was then allowed to drain freely to its residual state.

In the experimental runs with moisture content at field capacity, the soil was first saturated with distilled water and then allowed to drain freely before the gasoline was introduced.

Air loading rates used during the study varied from 16.1 $cm^3 \cdot (cm^2 \cdot min)^{-1}$ to 161 $cm^3 \cdot (cm^2 \cdot min)^{-1}$. These rates are considerably higher than the 3 $cm^3 \cdot (cm^2 \cdot min)^{-1}$ rate used by Thornton and Wootan (1982), however, they are consistent with achievable air loading rates recommended by Yu(1985).(4,5)

Hydrocarbon samples were taken from the reactor at various time intervals from a 12 mm diameter sampling outlet located at the top of the reactor. Samples were taken with Hamilton 10 ml glass air-tight syringes with teflon plungers equipped with Minert valves. Sampling frequency varied, depending on air flow rate. The range of sampling frequency varied from 15 seconds to several hours. The range of time required to complete each run was from less than one day to four days. The hydrocarbon vapor samples were analyzed by gas chromatography/mass spectrometry (GC/MS).

A Perkin-Elmer Sigma 1 gas chromatograph and microprocessor with an FID detector was used for routine qualitative and quantitative separation of the gasoline vapor hydrocarbon components. A Hewlett-Packard Model HP 5985 mass spectrophotometer with Model HP5840A gas chromatograph was used for compound identification. The packed GC column used was a 3 m x 3.2 mm stainless steel column with 5% AT-1200 + 1.75% Bentone 34 on Chrom-W-AW 100/120 mesh. The temperature program used was: initial temperature, 40°C for 5 min; increase temperature to 160°C at a rate of 5°C/min, hold at 160°C for 7 min. Injector and detector temperatures were 140°C and 180°C, respectively. Helium (ultra carrier grade) was used as the carrier gas for the GC/FID analysis and was mass flow controlled at 15 $cm^3 \cdot min^{-1}$. Air (high purity) and hydrogen (high purity) were used for the FID at 210 kPa and 154 kPa, respectively. Helium (ultra carrier grade) was used at a flow rate of 18 $cm^3 \cdot min^{-1}$ for the GC/MS analysis.

Samples were analyzed as soon as possible. To insure precision, a 100 $\mu\ell$ sampling loop attached to a gas sampling valve was used for sample injection.

The packed GC column method was specifically developed at the University of Connecticut for routine gasoline analysis. In Figure 2 a chromatogram of the gasoline used in the venting study is shown. An identification of the 52 individual peaks separated by the column is presented in Table 1. In Table 2 molecular weights, vapor pressures and mole fractions in gasoline of the 52 peaks are presented.

Individual hydrocarbon and total hydrocarbon concentrations were calculated based upon area intergration units using standard curves based upon the response of pentane. Because of the extensive number of samples per run and peaks per samples, the data was entered into an IBM-3081D computer and a FORTRAN program was written to calculate hydrocarbon concentration.

A mass balance on the gasoline in the reactor was then conducted. Mass loss rates for total and individual hydrocarbon components were calculated for comparison to theoretically predicted values.

RESULTS AND DISCUSSION

Soil Residual Saturation by Gasoline

One of the objectives of this research was to evaluate the quantity of gasoline held at residual saturation in quartz sand, over a range of soil conditions. The variable soil parameters studied were: sand particle size, sand density and sand moisture content. In each of the 27 experimental runs, the quantity of gasoline retained at residual saturation in 1831 cm^3 of sand, was recorded. Table 3 presents the sand particle size, dry density, moisture state and quantity of gasoline retained at residual saturation in each experimental run. Included in Table 3 are the results of three further gasoline retention experiments conducted by the authors under identical experimental conditions, but prior to the soil venting study.

It can be seen from Table 3 that for the dry state as the particle size decreases the quantity of gasoline at residual saturation increases. This would be expected as decreasing the particle size increases the available surface area for contact and reduces capillary size thereby increasing capillary forces. The particle size effects are lessened when the sand is initially water wetted. This, again, is as expected as the water will tend to fill up the smaller capillaries prior to the gasoline addition.

Induced Soil Venting

For each of the 27 experimental runs, GC analysis of the vapor samples produced discrete values of concentration for the gasoline as a single parameter and for each of the 52 composite peaks used to simulate the overall gasoline composition.

The individual hydrocarbon and total hydrocarbon concentrations, calculated by gas chromatographic analysis, were used in conjunction with the sampling time intervals and induced air flow rates to calculate approximate values of individual and total hydrocarbon mass loss and mass loss rates from the soil columns.

Cumulative mass loss of gasoline, as calculated from vapor analysis in each of the experimental runs, was compared to the gravimetrically measured gasoline mass loss in each run. A mass balance based on GC analysis was used to evaluate the precision of the experimental and analytical techniques. In 14 of the 27 experimental runs, the mass balance was within ± 5%. In 21 of the 27 runs, the mass balance was within ± 10%. For the remaining six runs, the mass balance varied between ± 10 to ± 26%. Under the given experimental conditions and analytical technique, a deviation of ± 10 percent in the mass balance was considered acceptable. Inconsistencies observed in the air flow monitoring equipment may have been a source of error in the experimental runs with greater than 10% deviation in the mass balance. In all 27 experimental runs, the total gravimetrically measured mass loss of gasoline was 100 ± 1% of the initial mass of gasoline present at residual saturation in the soil.

In a multi-component mixture, such as gasoline, the equilibrium concentration of a particular compound in the gas phase is constantly changing as vapors are removed. It was assumed that the components of gasoline act as individual ideal substances (i.e. activity coefficients = 1.0), because the literature indicates that a mixture of chemical components with similar structure (such as gasoline) exhibits the behavior of an ideal mixture.(6) To calculate the vapor phase equilibrium concentration $(Z(t)_i)$ at any time t, of a component i, the following equation is used:

$$Z(t)_i = (Vp(T)_i \; x(t)_i \; V \; MW_i)/R \; T \tag{1}$$

where,

$Z(t)_i$ = vapor phase concentration of component i in equilibrium with the liquid phase in gas space V cm^3, at anytime t, (mg),

$Vp(T)_i$ = vapor pressure of pure component i, at temperature T, (atm),

$x(t)_i$ = liquid mole fraction of component i at time t, (mole/mole),

V = volume, equal to $1,000 \; cm^3$ for calculating vapor concentration in $mg \cdot \ell^{-1}$,

MW_i = molecular weight of component i, $(mg \cdot mole^{-1})$,

R = universal gas constant = 82.04 $(atm \; cm^3 \cdot mole^{-1} \; {}^o K^{-1})$,

T = system temperature $({}^o K)$.

To calculate the total vapor phase equilibrium concentration $(Z(t))$ the individual component saturation concentrations are summed as follows:

$$Z(t) = \sum_{i=1}^{i=52} (V_p(T)i \; x(t)_i \; V \; MW_i)/R \; T \tag{2}$$

where, $Z(t)$ is the total vapor phase concentration of the mixture in equilibrium with the liquid phase in the gas space V cm^3, at time t, (mg).

The mass loss rate of an individual component at any time t, is the product of the component vapor phase equilibrium concentration at time t, and the air flow rate through the reactor as follows:

$$M(t)_i = Q \; Z(t)_i \tag{3}$$

where,

$M(t)_i$ = component mass loss rate to the vapor phase at time t, $(mg \cdot min^{-1})$,

$Z(t)_i$ = vapor phase concentration of component i, at time t, in 1 liter of gas space, $(mg \cdot \ell^{-1})$,

Q = induced air flow rate $(\ell \cdot min^{-1})$.

The overall mass loss rate at any time t, is the product of the total hydrocarbon concentration at time t, and the air flow rate through the reactor:

$$M(t) = Q \; Z(t) \tag{4}$$

where,

M(t) = overall hydrocarbon mass loss rate from the soil column at time t, $(mg.min^{-1})$,

Z(t) = total hydrocarbon concentration at time t, in 1 liter of gas space, $(mg \cdot \ell^{-1})$.

The total hydrocarbon mass loss within a time increment Δt is calculated as:

$$ML (\Delta t) = M(t) \; \Delta t \tag{5}$$

where,

$ML(\Delta t)$ = total hydrocarbon mass loss within time increment Δt, (mg),

Δt = time increment between successive vapor phase concentration calculations, (min).

After a time increment of Δt, a mass of $ML(t + \Delta t)_i$ of component i has been lost and is approximately equal to:

$$ML(t + \Delta t)_i = M(t)_i \; \Delta t \tag{6}$$

With this mass loss, the components mole fraction within the remaining liquid phase changes to $x(t + \Delta t)_i$ resulting in a new component vapor phase equilibrium concentration $Z(t + \Delta t)_i$. The new liquid mole fraction of component i, $x(t + \Delta t)_i$, at time $t + \Delta t$, is calculated from:

$$x(t + \Delta t)_i = \frac{(MC(t)_i - ML(t + \Delta t)_i)/MW_i}{\displaystyle\sum_{i=1}^{i=52} (MC(t)_i - ML(t + \Delta t)_i)/MW_i} \tag{7}$$

where $MC(t)_i$ = mass of component i remaining in the liquid phase at time t, (mg).

The mass $MC(t = 0)_i$ was obtained by GC analysis of the liquid gasoline used in the experiment and from the measured mass of gasoline, held at residual saturation, in the reactor at the start of the experiment. Substituting the values of the new mole fractions from Equation (7) into Equations (2) and (4) will give the new vapor phase equilibrium concentration and hydrocarbon mass loss rate at time $t + \Delta t$.

This calculation technique is an iterative process where the dynamics of a changing vapor phase equilibrium is simulated by incremental changes. A computer model to evaluate the changing vapor phase saturation concentration, hydrocarbon mass loss rate and cumulative hydrocarbon mass loss with time has been developed. Because of the constraints of the experiment, the saturated vapor equilibrium model is considered over the entire length of the reactor, as opposed to a local equilibrium at discrete points within the reactor (7).

Plots were developed to compare the theoretically derived model with the observed data. Figure 3 presents a superimposition of the calculated total hydrocarbon mass loss rate versus time and the theoretical total hydrocarbon mass loss rate versus time, for run #2. The curves show good agreement, except for the hydrocarbon mass loss rates in the initial several minutes. The calculated total hydrocarbon mass loss rate was higher than the theoretical hydrocarbon mass loss rate in the initial time period, this was true of all 27 experimental runs. A possible explanation of this phenomenon would be the presence of low molecular weight (high vapor pressure) components that were not accurately identified by the GC/MS study. If these components were properly identified, and the appropriate parameters (i.e. higher vapor pressure) were input to the model, a closer match of theory and real curves might exist over the entire length of the plot. A subamient cooling accessory, not available at the time of the study, would have allowed a lower initial oven temperature to be used in the GC analysis. Increased separation of the light ends (higher vapor pressure) components would result from lowering the initial oven temperature.

Overall, the theoretical curve agreed well with the observed curve and the theoretical model evidently predicts the real (observed) soil venting process accurately.

CASE STUDY, IN-SITU GASOLINE SPILL RECOVERY

The objectives of the case study are to present sampling of the unsaturated zone using prototype gas sampling probes permanently installed in the ground and determining the effectiveness of the soil venting process to remediate a gasoline spill. Vadose zone probes consisted of screened Sch.80 PVC pipe, 1.5" diameter, 8" long and slotted (0.020") over the middle 4" of the probe. Filter fabric was placed along the inside diameter of the probe to minimize the aspiration of silt and clay sized particles during sampling. The probes were capped at both ends. The upper end of the probes had a threaded teflon fitting and ferrule nuts to form an air tight seal for the placement of an 1/8" Teflon tube which extended to a Luer fitting at the surface of the ground. The Teflon tube was placed in a 5/8" Sch.40 PVC pipe, which also extended to the ground surface for protection purposes. A 2' steel casing street box was used for access to the Luer fitting for sampling. Sampling consisted of aspirating 300 cm^3 of air from the vadose zone probe with a syringe, followed by taking a 10 cm^3 sampling with a Hamilton 10 cc gas-tight syringe with a Minert valve and a Luer fitting. Samples were stored at ambient temperatures to minimize condensation and were stored in the dark to reduce photooxidation. Samples were injected into a GC with a 0.100 cm^3 sampling loop on a 10 port Valco valve. Samples were analyzed within 10 hours of acquisition.

Approximately 400 to 500 gallons of gasoline were spilled at this site based on inventory records. A total of 80 gallons were removed either by bailing or by a groundwater depression and gasoline skimming system. Thus the remainder of gasoline left in the soil was 320 to 420 gallons. These numbers are only best estimates.

Examination of the service station site plan in Figure 4 indicates that the property is approximately 125' x 150', with a service station building and two pump islands. There are 5 groundwater monitor wells located on the property and a 6 inch diameter recovery well with a depression pump. Hydrocarbon vapor concentrations in the unsaturated zone were sampled with 10 gas sampling probes located 15 feet below grade (approximately 3 feet above the water table). To remove hydrocarbon vapors from the soil, 3 - 6 inch diameter wells were placed in the unsaturated zone with vacuum pumps connected to each well.

A cross-section of the site shown in Figure 5 illustrates the relative locations and characteristics of the vapor phase monitoring and soil venting systems. The soil venting wells were each driven by 1.5 HP vacuum pumps with discharges 5 feet above the service station roof. Gas samples were taken periodically from the soil venting system to measure individual and total hydrocarbon concentrations in the vented vapors.

Gasoline Plume Characterization

Gas samples were taken from the vadose zone gas sampling devices and analyzed for gasoline range hydrocarbons by gas chromatography. Total hydrocarbon concentration was calculated for each gas sample. Using this method of vapor sampling and analysis a vapor plume diagram was developed for the site based on samples collected one hour prior to when the venting system was started up. Data plotted in Figure 4 were in close agreement with samples collected two months earlier from the same vadose zone gas sampling probes. It can be assumed from the data that vapor phase hydrocarbon concentrations greater than 50 mg/ℓ represent the hydrocarbon vapor envelop directly above the spilled gasoline.

Groundwater elevations and product thicknesses were measured in the groundwater monitoring wells at the site. In Figure 6 these data from Monitor Well A are presented from December 1984 through October 1985 including the period of soil venting. Up to 2 feet of product was present in the wells, however the product thickness was not corrected for capillary rise in the soil, which will reduce the actual gasoline thickness on the water table by at least 25 percent. Similarly, groundwater elevations were not corrected for depression caused by the product layer on top of the groundwater table. Elevations of the product in the Figure are not meant to be quantitative for estimating the volume of product spilled. It can be seen, however, that the groundwater elevation fluctuated about four feet during the monitoring period reported here.

Performance of the Soil Venting Process

The soil venting system with its 3 vacuum pumps operated for about 90 days. In this period of time 364 gallons (1376 ℓ) of gasoline was recovered using the system. Of this quantity of gasoline removed by the soil venting method, 90 percent was removed in the first 40 days of operation. In Figure 7, the cumulative gasoline recovered in each of the three vacuum wells is plotted versus time of operation. It can be seen that after 60 days of operation only a very small amount of additional gasoline was recovered. This is confirmed by examination of Figure 6 for groundwater monitor well A during the venting period. It can be seen that after about 40 days of venting, only a skim of gasoline remained in the groundwater monitor well. By the time the venting system was taken out of operation, there were no detectable gasoline layers or films present in any of the groundwater monitor wells. The initial gasoline recovery rate was about 26 liters/day (7 gallons/day) and this steadily declined for the duration of system operation.

In addition to the above interpretations for the efficiency evaluation of the venting system, additional vapor samples were taken from the 10 gas sampling probes and analyzed for hydrocarbons by gas chromatography. All soil hydrocarbon vapor readings were less than 0.1 mg/ℓ at the time of sampling.

CONCLUSIONS

The process of induced venting of gasoline residually saturated soil has been investigated. Under the experimental conditions studied, 100± 1% of the gasoline was removed. Vapor phase equilibrium saturation of the air, used for the venting, was apparent. This was evidenced by the close correlation between the actual experimentally derived data and data produced from a phase equilibria saturation model developed to describe the process.

The study suggests that induced soil venting would be an effective and predictable remedial action at hydrocarbon spill sites, particularly through the reduction of the overall time of possible contaminant input. This is achieved by the removal of hydrocarbons from the residually saturated soil in contact with the fluctuating water table and infiltrating rainwater.

Furthermore, controlled venting of the hydrocarbon contaminated soil will reduce the changes of fires and explosions by directing the vapors to a chosen location. If discharge of the vented hydrocarbons into the atmosphere is not permitted, the vapors may be passed through an activated carbon filter or other treatment process.

ACKNOWLEDGMENTS

The work upon which this publication is based was supported in part by funds provided by the Department of the Interior, Washington, D.C. as authorized by the Water Resources Research Act of 1984 (P.L. 98-242). The authors thank the University of Connecticut, Institute of Water Resources, and Civil Engineering Department for the kind support of this work.

For GC/MS work, we thank, the University of Connecticut, Institute of Material Science, and particularly Mr. Gary Levine for his effort and interest in this work.

Table 1.   COMPONENTS OF REGULAR (LEADED) MOTOR GASOLINE
IDENTIFIED BY GC/MS.

Peak Number	Retention Time (min)	Compound Names (synonym in parentheses)
1	1.54	Air
2	1.68	2-methyl propane (isobutane), terbutyl alcohol
3	1.81	unknown
4	2.30	n-pentane and isopentane
5	2.58	2-pentane
6	3.10	1-pentane
7	3.83	2-methyl pentane
8	4.20	3-methyl pentane
9	4.74	unknown(molecular weight = 86) methyl tertbutyl ether
10	5.34	2-ethyl-1-butene
11	6.00	methylcyclopentane
12	7.64	unknown (molecular weight = 100)
13	8.06	3-ethyl pentane
14	8.73	2,2,4-trimethyl pentane (isooctane)
15	9.25	n-heptane
16	9.80	1-methyl-1-cyclohexane
17	10.79	benzene
18	11.96	unknown
19	12.52	unknown (molecular weight = 114)
20	12.88	unknown (molecular weight = 114)
21	13.71	1,2-dimethyl cyclohexane
22	14.15	2,4-dimethyl heptane
23	14.89	unknown (molecular weight = 124)
24	15.92	toluene
25	16.53	4,4-dimethyl-3-ethyl-2-pentane
26	17.04	4-methyl octane
27	17.34	4-n-propyl heptane
28	18.57	n-nonane
29	19.84	ethylbenzene (phenylethane)
30	20.59	p-m-xylene,1,2 dibromoethane and phenylendiamine
31	21.13	3,4 dimethyl heptane
32	21.62	O-xylene
33	22.20	unknown
34	22.66	2,6-dimethyloctane
35	23.49	n-propyl benzene
36	24.08	methylethyltoluene
37	24.90	isopropyl benzene (cumene)
38	25.76	1,2,4-trimethyl benzene and O-ethyl toluene
39	26.48	vinyl-2-ethyl hexyl ether
40	27.18	m-styrene and n-butyl benzene
41	27.92	dimethyl ethyl benzene
42	28.81	diethyl benzene
43	29.65	1,2,4,5-tetramethyl benzene (durene)
44	30.02	n-dodecane
45	30.65	1,1-dimethylethyl benzene
46	31.12	ethyl styrene
47	31.71	2,6-dimethylstyrene
48	32.84	1,6-dimethyl indan
49	33.72	dimethyl isopropyl benzene
50	34.51	2,5-dimethyl undecane
51	35.91	unknown
52	37.71	naphthalene

TABLE 2. REGULAR LEADED GASOLINE COMPONENT FACTORS AT 21°C

Peak	Molecular Weight (g/mole)	Vapor Pressure (Atm)	Gasoline mole Fraction
2	58.12	3.0634	0.00969
3	86.18	0.1676	0.05482
4	72.15	0.6000	0.06249
5	70.13	0.5622	0.05978
6	70.13	0.7276	0.01356
7	86.18	0.2366	0.05866
8	86.18	0.2117	0.03120
9	86.18	0.2200	0.04640
10	85.17	0.2000	0.00832
11	84.16	0.1524	0.03533
12	100.20	0.0500	0.03911
13	100.20	0.0632	0.03100
14	114.30	0.0537	0.02156
15	100.20	0.0495	0.02436
16	96.18	0.0500	0.00825
17	78.11	0.1043	0.04435
18	110.20	0.0300	0.00491
19	114.30	0.0200	0.02158
20	114.30	0.0200	0.01484
21	110.20	0.0090	0.00576
22	128.16	0.0040	0.01555
23	124.20	0.0060	0.00465
24	92.14	0.0305	0.06428
25	124.20	0.0050	0.00000
26	126.24	0.0050	0.00842
27	142.28	0.0020	0.00619
28	126.24	0.0043	0.00774
29	106.17	0.0100	0.01789
30	106.17	0.0087	0.05424
31	130.27	0.0040	0.00143
32	106.17	0.0068	0.02116
33	120.19	0.0050	0.00084
34	144.30	0.0020	0.00347
35	120.19	0.0034	0.00599
36	134.22	0.0025	0.02608
37	120.19	0.0054	0.01647
38	120.19	0.0023	0.02746
39	120.19	0.0022	0.00229
40	126.20	0.0014	0.01982
41	134.22	0.0015	0.00534
42	134.22	0.0011	0.01756
43	134.22	0.0005	0.00133
44	170.34	0.0001	0.00416
45	134.22	0.0005	0.01211
46	132.20	0.0005	0.00741
47	144.21	0.0005	0.01358
48	146.23	0.0005	0.00895
49	148.25	0.0005	0.00595
50	184.34	0.0001	0.00411
51	148.25	0.0010	0.00371
52	128.17	0.0003	0.01000

Table 3. GASOLINE RETENTION AT RESIDUAL SATURATION

Run #	Air Flow Rate (ℓ/min)	Average Particle Diameter $\times 10^{-2}$ (cm)	Dry Absolute Density (g/cm^3)	Moisture State ++	Gasoline Retention (g/Kg Sand)	Gasoline Retention (ℓ/m^3 sand)
1	1.425	8.90	1.44	Dry	46.64	86.1
2	3.000	8.90	1.44	Dry	35.27	65.1
3	7.000	8.90	1.44	Dry	42.47	78.4
4	3.000	21.90	1.47	Dry	33.62	63.4
5	3.000	2.25	1.47	Dry	122.36	230.6
6	1.425	8.90	1.55	Dry	41.23	81.8
7	3.000	(MX)	1.74	Dry	68.42	152.6
8	3.000	8.90	1.55	Dry	41.58	82.6
9	7.000	8.90	1.55	Dry	42.28	84.0
10	3.000	21.90	1.57	Dry	34.89	70.2
11	3.000	2.25	1.62	Dry	97.77	203.2
12	1.425	8.90	1.71	Dry	40.88	89.6
13	3.000	(MX)	1.85	Dry	64.36	152.6
14	3.000	8.90	1.71	Dry	42.16	92.4
15	7.000	8.90	1.71	Dry	37.37	81.9
16	3.000	21.90	1.67	Dry	32.05	68.6
17	3.000	2.25	1.72	Dry	92.36	203.6
18	1.425	8.90	1.48	Field	36.90	70.0
19	3.000	(MX)	2.00	Dry	54.61	140.0
20	3.000	8.90	1.48	Field	28.41	53.9
21	7.000	8.90	1.48	Field	31.00	58.8
22	7.000	8.90	1.57	Field	28.18	56.8
23	1.425	8.90	1.57	Field	24.06	48.4
24	1.425	8.90	1.71	Field	28.11	62.6
25	3.000	8.90	1.57	Field	29.57	59.5
26	3.000	8.90	1.71	Field	26.51	58.1
27	7.000	8.90	1.71	Field	27.15	59.5
28*		2.25	1.50	Field	44.19	85.0
29*		2.25	1.62	Field	33.72	70.0
30*		2.25	1.72	Field	39.59	87.3

*Runs 28 through 30 were conducted by the authors, prior to the venting study, for purposes of measuring residual saturation.
(MX) indicated mixed sand 33 1/3% by wt of fine, medium and coarse sands
++ Field State = Moisture content varying from 2% to 10% by weight.

REFERENCES

1. Hall, P.L., and Quam, H., "Countermeasures to Control Oil Spills in Western Canada.", Groundwater, 14, 3, 1976.

2. McKee, J.E., Laverty, F.B., and Hertel, R.M, "Gasoline in Groundwater," Journal of the Water Pollution Control Federation (JWPCF), 44, 2, 1972.

3. Williams, D.E., and Wilder, D.G., "Gasoline Pollution of a Ground Water Reservoir-A Case History.", Groundwater, 9, 6, 1971.

4. Thornton, J.S. and Wootan, W.L. 1982. "Venting for the removal of hydrocarbon vapors from gasoline contaminated soil", J. Environ. Sci. Health. A17:31-45.

5. Yu, L. 1985. Study of Air Flow through Porous Media, Masters Thesis, University of Connecticut, Civil Engineering Department.

6. Reid, R.C., Prausnitz, J.M., Sherwood, T.K., The Properties of Gases and Liquids, 3rd edition, McGraw-Hill Book Company, New York, 1977.

7. Baehr, A.L., Immiscible Contaminant Transport in Soils with an Emphasis on Gasoline Hydrocarbons., Ph.D Dissertation, University Delaware, November 1984.

All Dimensions in millimeters

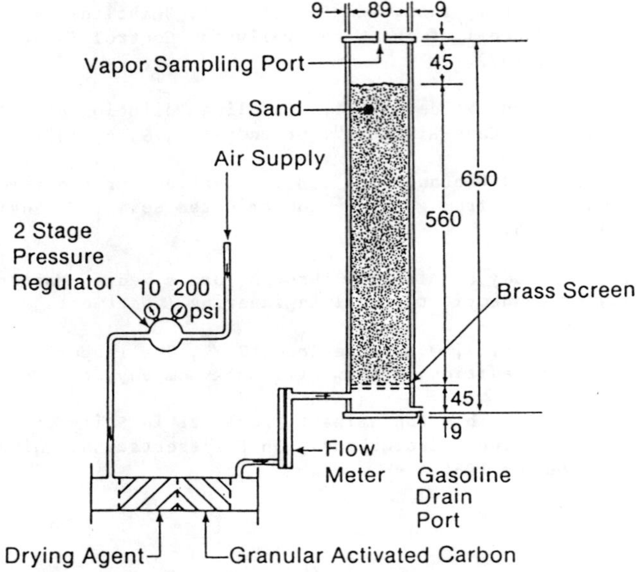

Figure 1 : <u>SOIL VENTING TEST APPARATUS</u>

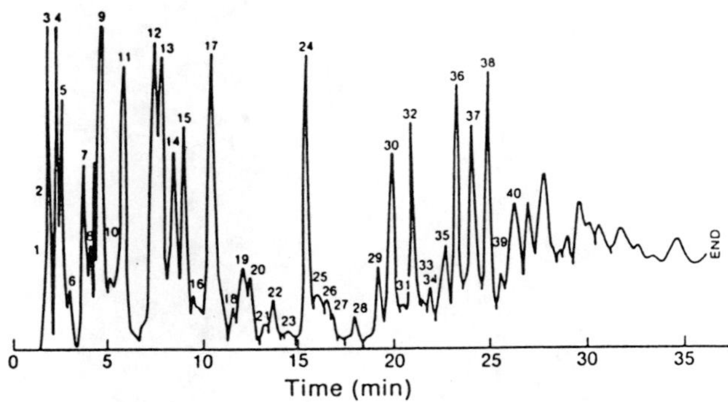

Figure 2 : <u>GC/FID CHROMATOGRAM OF REGULAR LEADED GASOLINE</u>

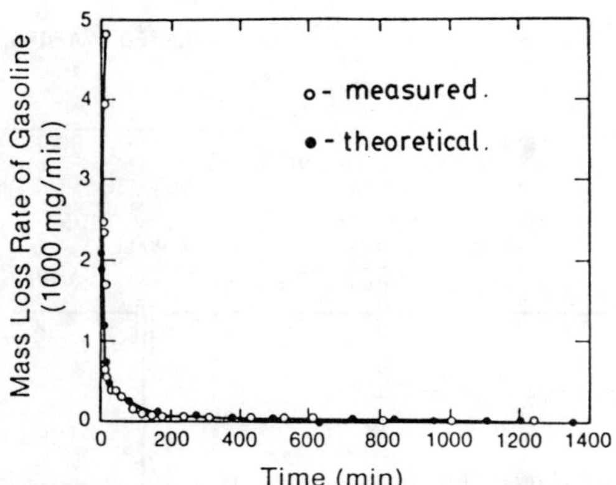

Figure 3: <u>COMPARISON OF THEORETICAL AND OBSERVED MASS LOSS RATES OF GASOLINE, RUN 2.</u>

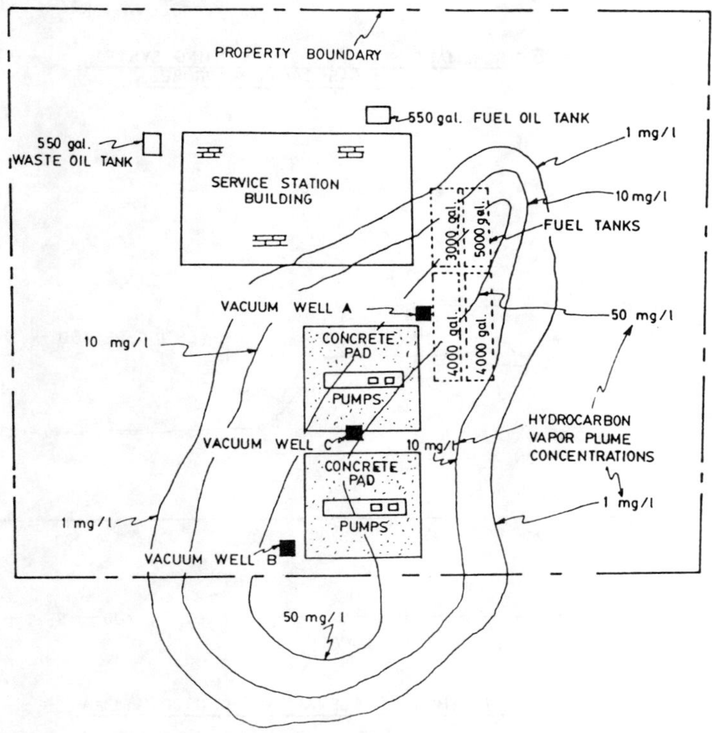

Figure 4: <u>SITE PLAN AND HYDROCARBON PLUME</u>

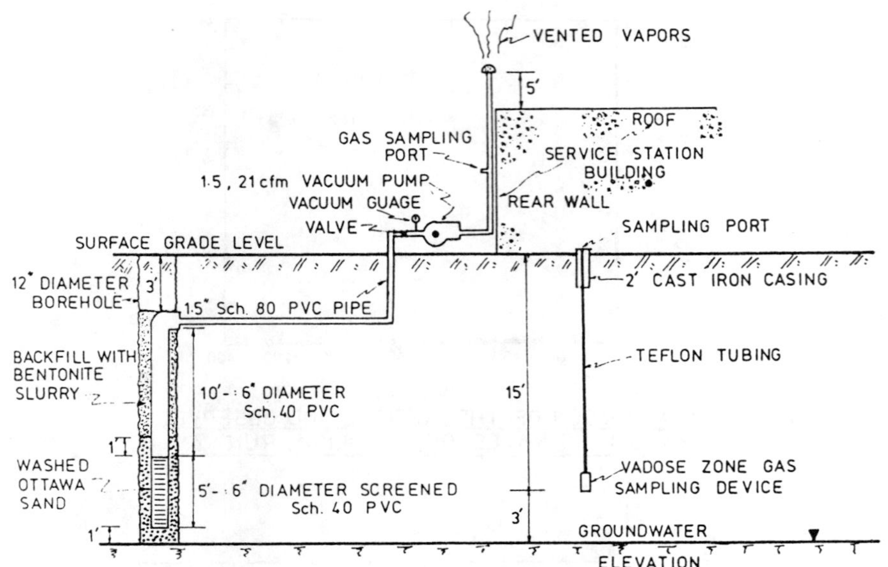

Figure 5 : SCHEMATIC PLAN FOR SOIL VENTING SYSTEM WITH GAS SAMPLING PROBES

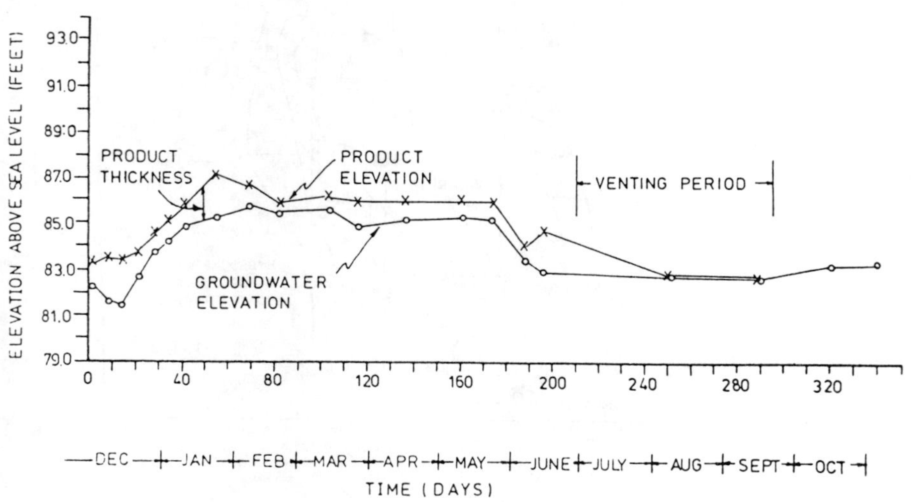

Figure 6: GROUNDWATER AND PRODUCT ELEVATIONS-MONITOR WELL A

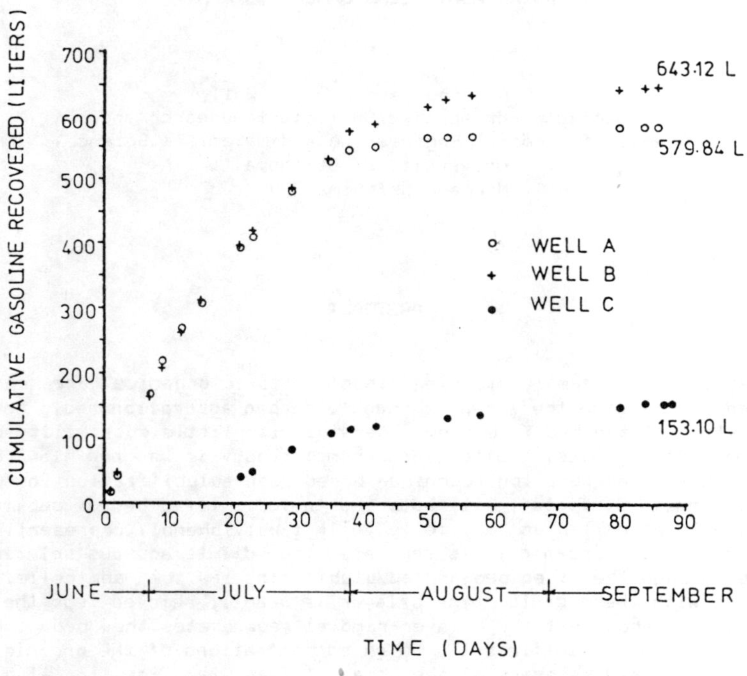

Figure 7: SOIL VENTING GASOLINE RECOVERY

A NEW WASTEWATER TREATMENT TECHNOLOGY
USING ADMICELLAR CHROMATOGRAPHY

C.R. Grout and J.H. Harwell
Institute For Applied Surfactant Research and
School of Chemical Engineering and Materials Science,
University of Oklahoma,
Norman, Oklahoma 73019,
USA

ABSTRACT

Wastewater streams containing dissolved toxic organics are commonly purified by passing the stream through a carbon adsorption bed; however, regeneration of the bed is energy intensive with little possibility of recovering most organics. Admicellar chromatography is a new low energy separation and regeneration technique based upon solubilization in surfactant systems. Surfactant is adsorbed onto a fixed bed of packing and forms a bilayer called an "admicelle". Tert-butylphenol, representing the class of toxic organics, is separated from dilute aqueous solutions by passing through the fixed bed and adsolubilizing in the admicelle. When saturated with the organic, the bilayer is readily removed from the fixed bed. Results show that the bilayer removal regenerates the bed and the effluent contains sufficiently high concentrations of the organic (200% increases) to make recovery of the organic feasible. Results also show that the organic effluent concentration can be above the saturation point suggesting a possible two phase separation between the surfactant and the organic solution.

INTRODUCTION

Wastewater streams containing dissolved toxic organics are a common problem in chemical, petroleum, and other industries. Process streams from these industries, including synfuel plants and metal plating plants, may also contain heavy metals. Both pollutants are usually at levels above permitted emmision levels and must be removed before the wastestream is discharged to the environment or reused in the process. In some instances, it would be advantageous to recover the organic for reuse or resale, if this could be done economically.

Current treatment methods that can recover organic pollutants include solvent extraction and adsorption. The principle problems with these methods are the high energy and materials costs during the downstream recovery stage and the limited types of pollutants that can be recovered. Activated carbon adsorption beds are low energy systems, but are dependent upon the regeneration costs of the carbon once it is saturated with the organic.

Carbon regeneration methods that are capable of recovering organic pollutants include hot-gas and solvent regeneration [1-2]. As in the case of solvent extraction, these methods have high energy and downstream equipment costs and are also limited in the types of pollutants that can be recovered.

The purpose of this paper is to introduce a new low energy separation and regeneration process: Admicellar Chromatography. Admicellar chromatography is based upon adsolubilization which is defined as the incorporation of compounds into surfactant aggregates adsorbed on surfaces, which compounds would not be in excess at the solid/solution interface in the absence of surfactant [3]. Several advantages over existing separation and regeneration techniques will be discussed.

Background

The ability of metal oxides to adsorb single component ionic surfactants has been studied exstensively [4-7]. Oxides such as alumina can be viewed as surfaces which contain patches with different adsorption energies. These patches chemically adsorb H_2O molecules which result in a surface of close packed oxygens, some of which carry hydrogen atoms. This configuration allows the surface to be charged by protonation or deprotonation of the adsorbed water [8-9]. Because of this property, the charge on the surface can be altered by manipulating the solution pH. Such a surface is called amphoteric. The pH at which the surface has a net zero charge is called the point of zero charge (PZC). By adjusting the pH either above or below the PZC, the surface can be made to be oppositely charged compared to the surfactant that is to be adsorbed. Reversing this procedure will force the surfactant to desorb from the surface.

At low concentrations, surfactant adsorption obeys Henry's Law as single surfactant molecules (monomers) adsorb on each patch. As the surfactant concentration increases, surfactant aggregates begin to form on the most energetic patches. These aggregates are micelle-like structures called admicelles [7]. The concentration where admicelles first form on the most energetic patches is called the critical admicelle concentration (CAC). Adsorption reaches a plateau value where surfactant begins to form micelles in the bulk solution. This concentration is known as the critical micelle concentration (CMC).

The proposed separation and regeneration process is based upon the phenomenon of adsolubilization. Adsolubilization can be considered a surface phenomenon analogous to micellar solubilization, a widely studied property of surfactant systems. Solubilization is defined as "the spontaneous dissolving of a substance (solid, liquid, or gas) by reversible interaction with the micelles of a surfactant in a solvent to form a thermodynamically stable isotropic solution with reduced thermodynamic activity of the solubilized material" [10]. Since the surfactant aggregate is adsorbed onto a solid substrate, the system is not isotropic, but the mechanisms of solute incorporation in the admicelle are identical to those in solubilization.

Process Overview

The proposed process is divided into three steps (Figure 1). In Step 1, the Bed Preparation Step, a dilute surfactant solution (Stream A) is equilibrated with a fixed bed of adsorbate with an amphoteric surface, such as chromatographic alumina. For industrial applications, an active alumina such as Alcoa H-151, Kaiser KA-201, or Pechiney Alumina A would be suitable. Even though the stream is dilute, the concentration of the stream is above the CMC to minimize bed preparation time. The pH of the stream is adjusted to allow the surfactant to adsorb onto the surface. Stream B (bed effluent before surfactant breakthrough) can be recycled into the process or emitted directly into the environment since it is initially void of surfactant.

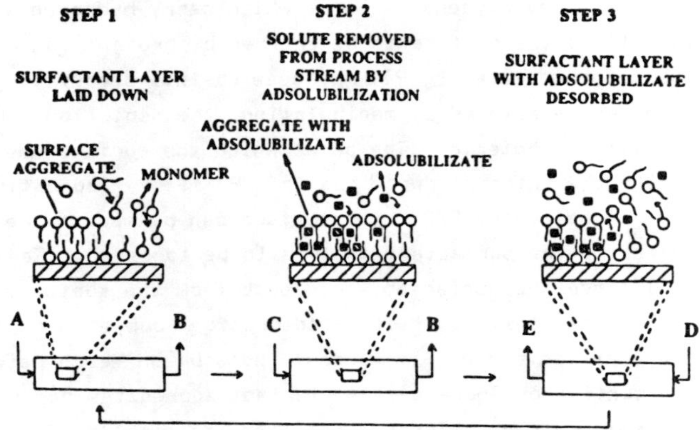

Figure 1. Schematic of Admicellar Chromatography Process.

After the surfactant wave has propogated sufficiently through the bed, Step 2, the Adsolubilization Step, is started by introducing the stream containing the solute that is going to be recovered (Stream C). Stream C may also contain surfactant at the system CMC to maintain the equilibrium between the bilayer and surfactant in the bulk phase, preventing the premature desorption of the admicelles. As in the beginning of Step 1, Stream B is void of surfactant, but sometime the process, either late in Step 1 or early in Step 2, surfactant will begin to appear in the effluent as the wave forming the bilayer reaches the outlet. This stream can be recycled to prepare a second bed. Stream B also initially contains no solute, but as the admicelles become saturated, the solute will begin to appear at the column outlet.

When solute breakthrough occurs, Step 3, the Bed Stripping Step, is begun by injecting a stream at a pH which will cause the surfactant to desorb (Stream D). As the surfactant desorbs, the solute is removed with the surfactant, but at higher concentrations. Since the solute is at a higher concentration, any downstream removal or recovery equipment can be scaled down and operating costs lowered. This result could be very significant in the overall economics of removing or recovering the solute. If the solute is not to be recovered, the adsolubilization step serves as a substitute for a carbon adsorption bed. The bed stripping step can be thought of as a low energy regeneration process that does not require dumping the bed. In either application, when the concentration of the solute in Stream E falls below the point where it is no longer economical or necessary to continue recovering it, the influent stream is switched back to Stream A, the column begins re-equilibrating with surfactant, and a new bilayer forms.

The proposed process has several advantages over conventional techniques. The ease at which the bed can be regenerated and the ease of recovering the solute are advantages over carbon bed adsorption. Fixed-bed adsolubilization can be used for high and low molecular weight compounds, a definite advantage over membranes, which usually do not work well on low molecular weight solutes.

Because of the wide availability of commercial surfactants and suitable substrates, a surfactant/substrate system can be designed to fit a specific pH range or a specific solute. Commercially available surfactants are also relatively inexpensive for the concentrations required (mM) and are biodegradable. In addition to substrates previously mentioned, aluminosilicates, acid-treated clays, silica gels, or a diatomaceous earth would be suitable materials.

Besides the removal of organic pollutants, heavy metals which are prevalent in some waste streams can be treated with this process. Theoretically, it should be possible to recover a number of target molecules, including trace metals and organics, simultaneously.

There are potential disadvantages to the process. Depending on the column porosity and the surface area of the adsorbent, phase boundaries may be crossed during the bed stripping step. Precipitate formation, emulsion formation, or a phase separation within the bed are all possible. If any of these developments occur and create a highly viscous phase, the column might be plugged, forcing the bed to be dumped.

MATERIALS AND METHODS

To test the technical feasibility of the process, two runs were performed which recovered 4-_tert_-butylphenol (TBP) from dilute aqueous solution by adsolubilization in sodium dodecyl sulfate (SDS) admicelles. The SDS was used as received and the TBP was recrystallized from a mixture of water and ethanol and dried at low heat then stored in an evacuated container. HCl and NaOH, 0.01N-1.0N, were used to adjust the pH; water was distilled and deionized.

The column used for each run was a 6mm X 250mm glass column dry packed with a low surface area alumina (PZC 9.5). Each run consisted of four phases. The initial phase was to equilibrate the alumina bed with a stream of water at pH 8.4-8.6. After equilibrating the column, a 14.0-18.5 mM SDS solution at pH 8.4-8.6 was injected into the column (phase 2). The effluent SDS concentration was monitored until the concentration approached the inlet SDS concentration. When this condition was observed, the inlet stream was switched to a new stream at pH 8.0-8.1 containing SDS at 8.0 mM and TBP at 1.0-2.75 mM (phase 3). In Run I, the concentration of the SDS and TBP in the effluent were monitored until the TBP concentration approached the inlet TBP concentration. The inlet stream was then switched to the final stream containing water at pH 12.0-12.2 (phase 4). The concentrations of the SDS and TBP were again monitored until they approached zero. In Run II, regeneration began when the amount of TBP injected equaled only half the TBP removed in Run I. Table 1 summarizes the experimental conditions for each run.

Burets were used to measure the amount of each stream injected, and the pH was measured using a flow cell equipped with a pH electrode. SDS concentrations were measured using High Performance Liquid Chromatography (HPLC) and TBP concentrations were measured with a gas chromatograph (GC). A schematic of the experimental set-up is shown in Figure 2.

RESULTS AND DISCUSSION

Concentrations for each run were taken at the outlet of the bed and plotted as a function of pore volumes of solution injected. A pore volume is the void space in the column or the space not taken up by the packing. A pore volume is also the volume of liquid needed to move a slug of solution from the inlet to the outlet of the column; approximately one pore volume of solution needs to be injected before any noticeable change in outlet concentration is detected.

TABLE 1
Summary of Run Conditions

RUN I	Stream A	Stream C	Stream D
[SDS] (mM)	16.20	7.96	-----
[TBP] (mM)	-----	2.54	-----
pH	8.44	8.00	12.01
Pore Vol. Injected*	9.02	9.63	9.29
Flowrate (ml/min)	.15	.15	.15

*Pore Volume= 5.73 ml

RUN II	Stream A	Stream C	Stream D
[SDS] (mM)	18.22	6.90	-----
[TBP] (mM)	-----	2.55	-----
pH	8.51	8.00	12.07
Pore Vol. Injected*	8.17	3.38	7.42
Flowrate (ml/min)	.15	.15	.15

*Pore Volume= 5.57 ml

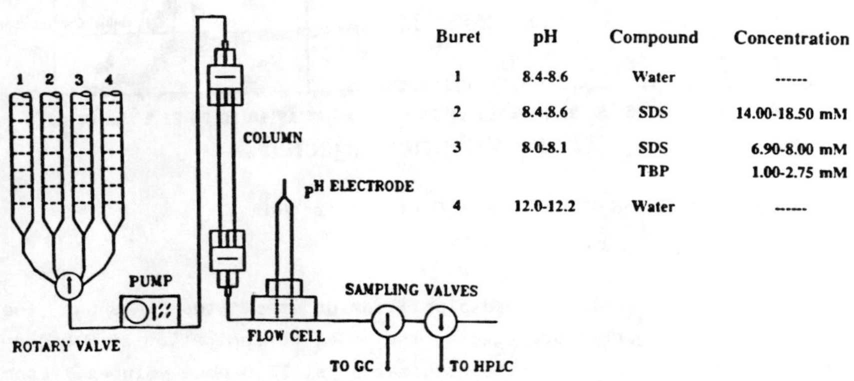

Buret	pH	Compound	Concentration
1	8.4-8.6	Water	------
2	8.4-8.6	SDS	14.00-18.50 mM
3	8.0-8.1	SDS	6.90-8.00 mM
		TBP	1.00-2.75 mM
4	12.0-12.2	Water	------

Figure 2. Schematic of Experimental Apparatus.

To check for any loss of SDS or TBP due to irreversible adsorption on the column, mass balances were calculated for each run. This procedure was also done to find out how much of each component could be recovered. By intergrating the area under each component curve, the amount coming out of the column could be calculated. Subtracting these values from the amounts injected gave the total SDS and TBP mass balances for the process.

The SDS and TBP mass balances all closed within 3%.

The SDS and TBP profiles for Run I are shown in Figure 3. The horizontal lines depicts the inlet SDS and TBP concentrations of each stream. The vertical lines split each run into the three steps of the process. In Step 1, SDS was not detected until 2.0 pore volumes had passed through the column. After this point, the SDS concentration rose sharply and approached the inlet concentration after 3.8 pore volumes.

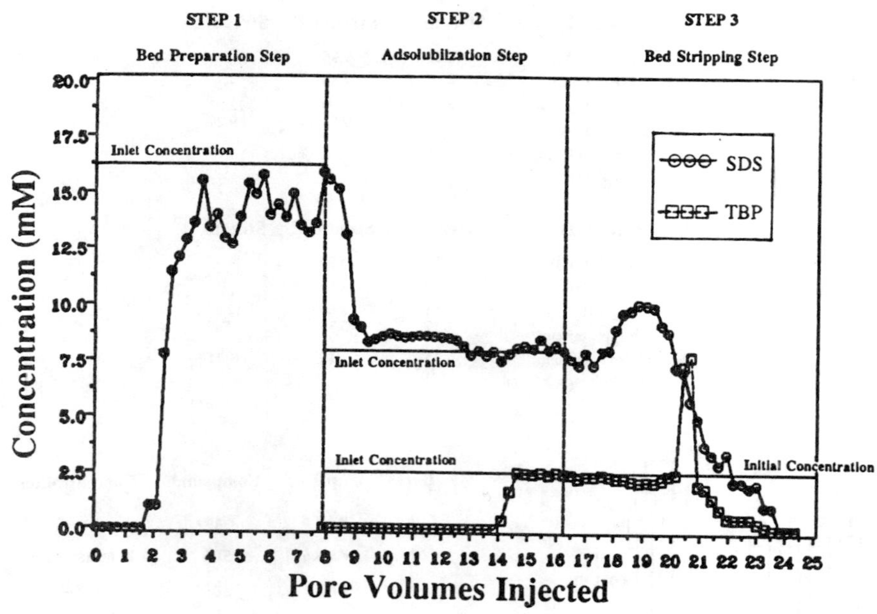

Figure 3. SDS and TBP Effluent Profiles for Run I.

The SDS profile during the adsolubilization step (Step 2) shows the SDS concentration began to decrease to the inlet concentration at 8.5 pore volumes and reached the inlet concentration at 13.0 pore volumes. From 9.4 to 13.0 pore volumes there was a steady concentration of 8.1 mM and then the concentration dropped to the inlet concentration of 7.96 mM.

TBP was completely removed from the wastestream over a range of 6.5 pore volumes. After the TBP was detected, there was a sharp concentration rise to the inlet concentration similar to the SDS profile.

During the bed stripping step (Step 3), the SDS concentration stayed close to the inlet concentration of the adsolubilization stream for 2.0 pore volumes and then the effects of the desorbing SDS were detected. As shown, the SDS concentration rose to 10.0 mM and maintained this concentration over a range of 3.0 pore volumes before dropping off sharply and approaching zero concentration after 24.0 pore volumes.

The desorbing TBP concentration also maintained the inlet concentration from the adsolubilization step, but at 20.0 pore volumes there was a sharp rise in the TBP concentration from 2.47 mM to 7.74 mM. This represents an over 200% increase compared to the wastestream concentration of 2.54 mM. The concentration dropped sharply at 21.0 pore volumes and approached zero concentration after 24.5 pore volumes.

During the bed stripping step, the pressure in the column rose from 70 psig to approximately 300 psig at 20.0 pore volumes. The pressure increase corresponds with desorption of TBP from the column suggesting an increase in solution viscosity.

Figure 4 represents the concentration profiles for Run II. Compared to the first run, there is little difference in the general shapes of the curves or process performance. The only difference noted is the shorter time needed to recover the TBP.

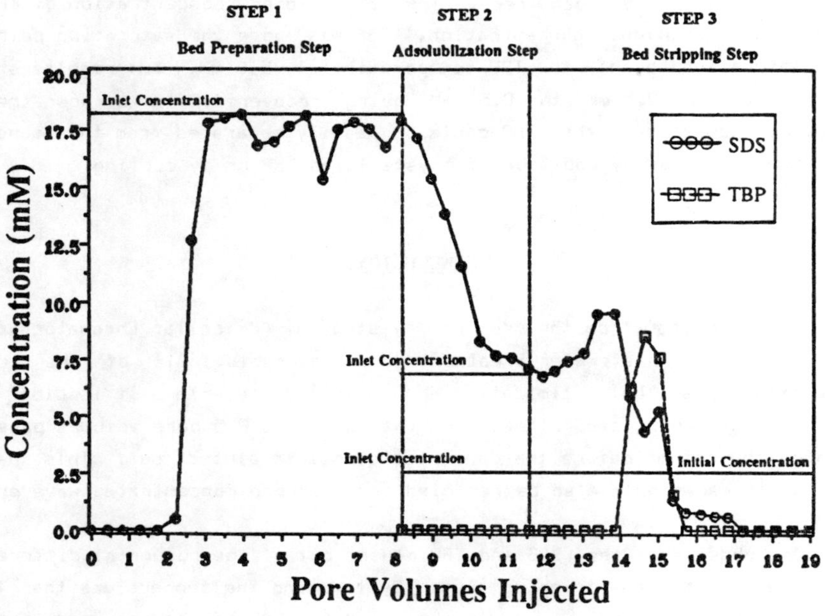

Figure 4.   SDS and TBP Effluent Profiles for Run II.

Results for the adsolubilzation step show that the SDS did not drop to the inlet concentration as quickly as in the previous run. Because of the introduction of the bed stripping stream at a much earlier time in the process, there could be interactions between the different chromatographic waves within the column which would result in the slower concentration decrease.

The results for the bed stripping step were slightly different than in Run I. The SDS profile had the same general rise in concentration, but desorbed at a much quicker rate and was at zero concentration after 6.0 pore volumes compared to 8.0 pore volumes during Run I. The TBP profile was also different in Run II compared to the first run. The TBP stayed at zero concentration for a range of 2.25 pore volumes and then desorped in a sharp peak. The concentration increased to 8.5 mM which resulted in an over 225% increase comparec to the inlet concentration of 2.55 mM.

From the results of the two runs, it is clear that this process can effectively remove organics from wastestreams. Increases of over 200% were consistently acheived with minimal loss of SDS or TBP. Two other points are also significant in this process. First, the TBP desorbed in a very sharp peak in Run I and II. This portion of the stream could be treated again by recycling the first 2.5-3.0 pore volumes of the bed stripping step and then recovering the higher concentration of TBP. Second, at this higher concentration, TBP was _above_ the saturation point. The saturation point for TBP is approximately 5.0 mM, but results show concentration of 7.5 mM to 8.5 mM being recovered. In a downstream recovery operation, the TBP could be readily separated from this supersaturated solution by addition of crystallized TBP or by cooling.

CONCLUSIONS

The results from the preliminary study of Admicellar Chromatography as a new form of wastewater treatment were favorable. All of the runs were single pass, but optimization should include recycle. It is clear in the bed stripping step of each run that at least 2.5 pore volumes passed through the column before the concentrated solute eluted out. This part of the stream should also be recycled and a second concentrated wave produced.

As observed, the SDS and TBP eluted out of the column at different times during the bed stripping step. By lowering the temperature the TBP could be separated out in a pure phase and the SDS can be returned to the process. Overall, only pH changes are needed to make the process work, which is a significant energy savings over conventional adsorption bed regeneration methods.

It was clear that the psuedo-industrial application of Run II decreased the run time considerably from 25.0 to 17.5 pore volumes. This application also resulted in a more concentrated peak of organic in a much shorter time, both results advantageous for the practical applications proposed. It should also be noted that in each run a large excess of SDS was injected in Step 1. Given the delay in the TBP wave, perhaps only a half pore volume of concentrated SDS would be needed.

Future work will study the economic feasibility of the process compared to carbon adsorption beds. These studies will include the design of a typical industrial application with recycle. Effects of recycle will also be considered along with the effects of pH, solute concentration, and optimum choice of surfactant and packing material.

REFERENCES

1. Hutchins, R.A., Activated-Carbon Systems for Separation of Liquids. In Handbook of Separation Techniques for Chemical Engineers, ed. P.A. Schweitzer, McGraw-Hill, New York, 1979, pp. 415-445.

2. Sutikno, T. and Himmelstein, K.J., Desorption of Phenol from Activated Carbon by Solvent Regeneration. Ind. Eng. Chem. Fundam., 1983, 22, 420-425.

3. Scamehorn, J.F. and Harwell, J.H., Surfactant-Based Treatment of Aqueous Process Streams. In Surfactants and Chemical Engineering, eds. D.T. Wasan, D.O. Shah, and M.E. Ginn, Marcel Dekker, New York, 1986, In Press.

4. Gaudin, A.M. and Fuerstenau, D.W., Quartz Flotation with Anionic Collectors, Min. Eng., 1955, 7, 66-72.

5. Cases, J.M. and Mutaftsheiv, B., Adsorption Et Condensation Des Chlorhydrates D'Alkylamine 'A L'Interface Solide-Liquide, Surf. Sci., 1968, 9, 57-72.

6. Scamehorn, J.F., Schechter, R.S., and Wade, W.H., Adsorption of Surfactants on Mineral Oxide Surfaces from Aqueous Solutions. I. Isomerically Pure Anionic Surfactants, J. Colloid Interface Sci., 1982, 85, 463-478.

7. Harwell, J.H., Hoskins, J.C., Schechter, R.S., and Wade, W.H., A Pseudophase Separation Model for Surfactant Adsorption: Isomerically Pure Surfactants, Langmuir, 1985, 1, 251-262.

8. Yopps, J.A. and Fuerstenau, D.W., The Zero Point Charge of Alpha-Alumina, J. Colloid Interface Sci., 1964, 19, 61-71.

9. James, R.O. and Parks, G.A., Characterization of Aqueous Colloids By Their Electrical Double Layer and Intrinsic Surface Chemistry Properties. In Surface and Colloid Science Vol. 12, ed. E. Matijevic, Plenum Press, New York, 1982, pp. 119-216.

10. Rosen, M.J., Surfactants and Interfacial Phenomena, Wiley and Sons, New York, 1978, pp. 123.

MODELING METAL PLATING INDUSTRY
FOR INTEGRAL POLLUTION CONTROL STRATEGIES

J.P.M. Ros
Laboratory of Waste and Emissions,
National Institute of Public Health
and Environmental Hygiene,
Bilthoven, the Netherlands

ABSTRACT

This paper describes a model for metal plating industries, which has been developed by the RIVM. This model can be used to assess new technologies for recycling process chemicals in an integral pollution control strategy. It can be concluded, that for most types of metal plating industries a combination of process integrated measures, mostly drag-out-tanks and recovery technologies, and end-of-pipe treatment is the best solution. The discharge of heavy metals in wastewater and sludge is considerably lower as compared with end-of-pipe treatment only, whereas costs remain about the same. The paper shows the influence of specific circumstances in a metal plating shop as well as effluent standards on the choice of a particular combination of technologies.

INTRODUCTION

During the last ten to fifteen years many technologies have been developed for the recovery of process chemicals in the metal plating industry [1]. Many of them were presented as the answer to the environmental problems of these firms, offering profits as well. In reality, any recovery technology can only be a part of a complete package of measures for pollution control.

In a metal plating industry there are always many sources of pollution, because the process exists of many treatment steps. The simplest way of pollution control is treatment of all rinsing water and concentrates in one purification system. In this case, however, the simplest solution is not the best. Such a system could only be a DND (Detoxification,

Neutralization, Dewatering), which transfers the problem of the heavy metals from wastewater to hazardous waste (sludge).

An integral pollution control should be more than end-of-pipe treatment. It should include process integrated measures like drag-out tanks and recovery technologies, but there is no standard solution, because every plating shop is different. The complexity of the firms makes it difficult to decide on the right alternative. Therefore a computer model has been developed for the metal plating industry (or even, for surface treatment of metals in aquaous systems). The model has three objectives:
- to find the best applicable combination of technologies to reduce the discharge of heavy metals and cyanide with wastewater
- to judge the possibilities of new technologies
- to provide a basis for a policy aimed at the prevention of hazardous waste.

This approach for the metal plating industry could serve as an example for other industries.

DESCRIPTION OF THE MODEL

Every metal plating shop consists of one or more lines with different treatment tanks, like degreasing tanks, acid dips and tanks for galvanizing and passivation. The products are dipped into the aquaous solutions in these tanks and rinsed afterwards. Two sources of environmetal problems can be distinguished, rinsing water and polluted concentrates. There is a drag-out of process solution with the products to the rinsing tank, so that the rinsing water gets polluted. The process solutions themselves are polluted too, mainly because their function is to absorb polluting substances, like in degreasing. They have to be renewed after a while. So, metal plating creates wastewater and waste, both polluted with heavy metals.

Before starting model calculations first model shops have to be designed, which consist of process steps mentioned above. Every process tank is defined by a series of parameters, given in table 1.

The effect of process integrated measures depends on these parameters, the configuration of the drag-out tanks and the efficiency of recovery technologies. This is described mathematically, based on mass-balance figures on especially the last drag-out tank [2]. The effect is calculated separately for every process tank.

TABLE 1
Parameters defining the characteristics of every
treatment tank in a model metal plating shop

Parameter	Dimension
Volume tank	l
Drag-out (Qov)	l/h
Evaporation loss (Qr)	l/h
Hand/automatic	
Dilution factor for rinsing	
Working time (d)	h/y
Costs for transport/discharge of concentrates	Dfl/l
concentration (heavy metals, CN)	g/l
input impurities	g/h

The computer model itself describes all possible flows of water and
waste with their components and the influence of technologies on them.
Figure 1 gives a schematic view of all possible flows.
Some general data (like water costs, efficiencies) are kept in files
separately from the main program. Before starting the calculations, the
program is loaded with these general data, data of the model industry
and the package of measures. The package includes a configuration of
drag-out tanks and -if chosen- a recovery technology and/or a system for
rinsing water treatment for every process tank, the total wastewater
treatment system (end-of-pipe) and the treatment of concentrates. Table
2 gives a survey of all possibilities.

TABLE 2
Measures and technologies included in the
model metal plating industry

Process integrated (coupled at one process tank)	Treatment of total stream
Drag-out tanks (none, one or two continuously or batch)	Wastewater treatment - none
Cascade rinsing	- ion exchange
Recovery technologies	- DND with sedimentation
- electrolysis	- DND with ultrafiltration
- reversed osmosis	- neutralization
- evaporation	Treatment of concentrates
- electrodialysis	- transport for treatment elsewhere
Rinsing water treatment	
- ion exchange	- DND (batch)
- ultrafiltration	- treatment in the wastewater
- chemical rinse	treatment system

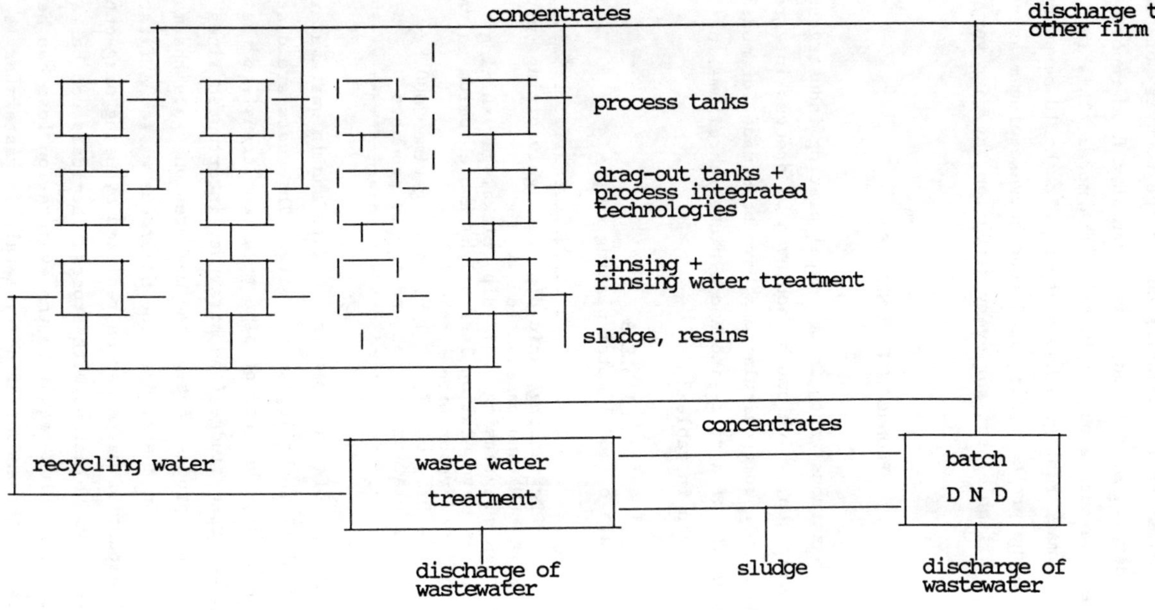

Figure 1. Representation of flows of heavy metals in a metal plating firm

The output of the model includes data on the distribution of heavy
metals in wastewater, sludge or concentrates and of the yearly costs
(including water costs and the benefits due to the recovery of process
chemicals). More details of the model are given in a RIVM-report [3].
Reduction of the drag-out is not included in the model. This is
considered to be a matter of good-housekeeping. It should be pointed
out, that many of the process integrated good-housekeeping measures are
very simple and profitable, but for every situation they are very
specific.

PRODUCTION OF SLUDGE

The management of hazardous waste in the metal plating industry should
imply primary prevention, which can be achieved by process integrated
measures. As an example some calculations have been made for model
shop A, which consists of a zinc-line and a nickel/chromium-line. More
specific data are given in table 3.

TABLE 3
Data of model metal plating shop A

Zinc-line:	Degreasing, acid-dip, zinc-galvanizing (with CN) and passivation
Nickel/chromium-line:	Degreasing, acid-dip, nickel-galvanizing (evaporation loss 5 l/h) and chromium-galvanizing (evaporation loss 4 l/h)
For both lines:	Drag-out 2,5-3 l/h, working-time 2000 h/y, plating of iron and copper products

In figure 2 the amounts of heavy metals in waste and in wastewater for
several combinations of technologies are given. The costs are discussed
later. It will be clear, that end-of-pipe treatment transfers the heavy
metals from wastewater to sludge (the straight, interrupted lines in
figure 2). Without any process integrated measures one gets the maximum
amount of heavy metals in the sludge, the hazardous waste. A first
reduction of the amount of waste can be realized by using drag-out
tanks. From these tanks solutions with process chemicals can be returned
to the process tank. A volume equal to the evaporation loss can be
returned. In this way the amount of heavy metals in wastewater is
reduced without the production of waste.

However this reduction is limited. A further reduction of the discharge
with wastewater can be achieved by renewing the solutions in the drag-
out tanks. The higher the frequency of renewing these solutions, the
smaller the discharge with wastewater will be, but the amount of
concentrates increases more than linear. This is clearly shown in
figure 2.

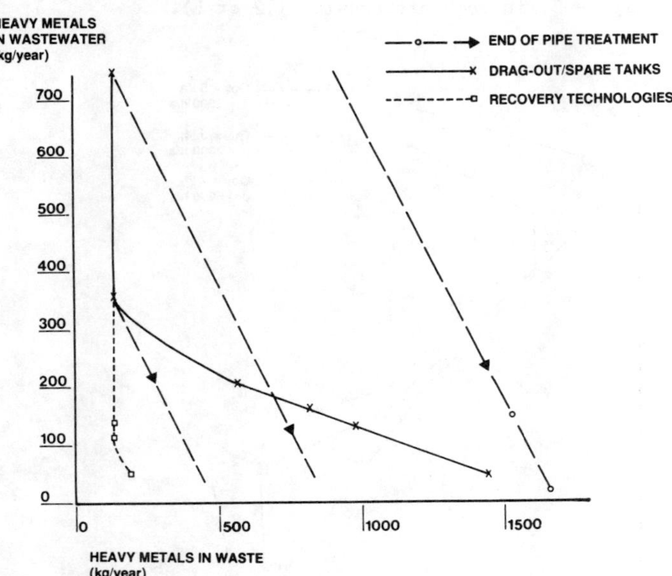

Figure 2. Heavy metals in wastewater and sludge for different
combinations of pollution control measures for
model shop A

The only way a further reduction of heavy metals in wastewater and in
waste can be achieved is by applying recovery technologies. It can be
concluded for model shop A, that the reduction of waste amounts to 60 %
as compared with the combination of drag-out tanks and end-of-pipe
treatment producing the smallest amount of waste and to more than 90 %
as compared with the application of only end-of-pipe treatment. It
should be noted that the exact figures are different for each situation.

FACTORS INFLUENCING THE STRATEGY

The influence of certain factors on the choice of a particular package
of pollution control measures is explained by means of calculations on a

very simple model shop (B), consisting of one nickel-tank with an
evaporation loss of 5 l/h. Three alternatives have been regarded, in
which the drag-outs have been varied. They have been chosen to be 5, 3
and 1 l/h. The number of applicable technologies has been limited to one
or two drag-out tanks (presented with a 1 or 2 in figure 3),
electrolysis (E in figure 3) with or without DND as end-of-pipe
treatment (□ in figure 3, in combination with 1,2 or E).

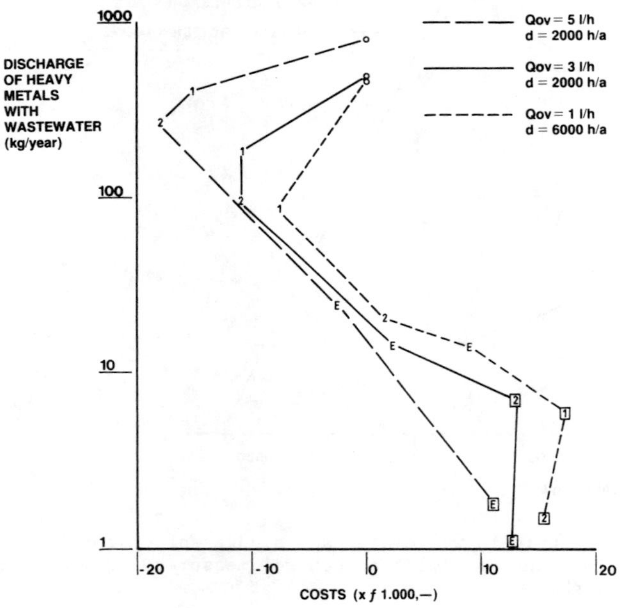

Figure 3. Results of calculations for model shop B

It is clear that the use of drag-out tanks offers the cheapest way of
working, but the optimal number of tanks is dependent on the drag-out.
In practice, a firm is faced with effluent standards (concentrations and
amounts of heavy metals) and in most cases drag-out tanks alone will not
be sufficient. Figure 3 makes it obvious that the allowed amount of
heavy metals in wastewater determines the choice, which may be different
for each of the three examples. In case of a high drag-out, electrolysis
will be the most advantageous choice. In case of a low drag-out,
however, the use of drag-out tanks will be more profitable.

Limitation of the concentration in the effluent leads to the application
of DND, because process integrated measures prevent the process
chemicals from going to the rinsing tank and less rinsing water will be
needed. In this simple case, the process integrated measures only reduce
the water flow and the amount of heavy metals in it, but not the
concentration of nickel. Thus in some cases the limitation of
concentrations prohibit the introduction of recovery technologies.
Other factors, which may be of importance are water costs, costs of the
discharge of heavy metals, costs of drag-out tanks, reduced production
(this is included in the model) and of course specific data on the
firm itself.

COSTS

In the end, costs will be the main factor. As far as the metal plating
industry is concerned, costs are more important than the reduction of
waste or sludge, as long as the industry can transport it somewhere at
reasonable costs. To gain some insight into the costs of different
packages of measures calculations have been made for eight model
shops, all with Zn, Ni, Cr and Cu-plating facilities.
For each of these three combinations have been studied, all three with
DND as end-of-pipe treatment. This does not mean DND is the only
possible choice. In fact, in the Netherlands ion-exchange in combination
with a DND-batch installation for the treatment of regenerates and
concentrates and process integrated measures is regarded as the best
practicable technology for large metal plating firms. For small plating
shops drag-out tanks and sometimes simple recovery technologies (like
elektrolysis) with neutralization of the wastewater only and DND-batch
for the concentrates are best practicable. Results of calculations for
these combinations are presented in a RIVM-report [4].
The three combinations, of which the results are shown in figure 4, are:
a. only DND
b. drag-out tanks in combination with DND
c. drag-out tanks and recovery technologies in combination with DND.
Figure 4 shows the total costs for every kg of heavy metal removed from
wastewater and concentrates and the final discharge with wastewater.
These results are dependent on the size of the shop expressed in the
total amount of heavy metals in wastewater and in concentrates without
any measures or treatment system (except good-housekeeping). It should

be noted, that the lines in figure 4 are a simplification of the reality, even for model shops. There is a variation in costs and in discharged heavy metals, which cannot be neglected. The lines present a general tendency.

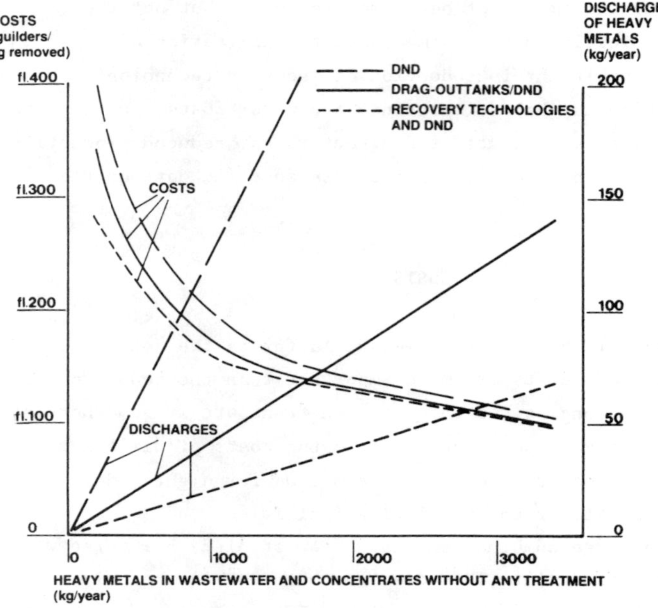

Figure 4. Costs and results of DND as end-of-pipe treatment
with and without process integrated measures for
different model plating shops (Zn, Cu, Cr, Ni).

It can be stated, that drag-out tanks and even more clearly recovery technologies lead to considerably smaller amounts of heavy metals in wastewater at roughly the same or even lower costs as compared with the results of only DND.

The question remains, why these recovery technologies have not yet been introduced on a large scale. There are several reasons for this. In the past many firms have chosen for end-of-pipe treatment to comply with the standards. Introduction of new recovery technologies is therefore not (yet) under discussion. The firms are not interested in the amount of sludge they produce. Sometimes the investments are too high. Finally it must be stated, that many of the metal plating shops know too little about these rather new recovery technologies, they do not believe them to be profitable and especially the smaller firms are afraid that they will not be able to use them properly.

REFERENCES

1. OECD, Emission control costs in the metal plating industry, Paris, 1983.

2. Ros J.P.M., Procesgeintegreerde maatregelen in de galvano-industrie, Pt Procestechniek, 39 (1984) 12, 48-51 (in dutch).

3. Ros J.P.M., van der Plaat J., Opzet van het model galvanisch bedrijf, RIVM-report 851401001, 1986 (in dutch).

4. Ros J.P.M., Berekeningsresultaten met het model galvanisch bedrijf, RIVM-report 851401002, 1986 (in dutch).

HEAVY METALS PRECIPITATION FROM SMELTER WASTE WATER
AND OBTAINABLE SLUDGE QUALITIES

Björn Lindquist
Metallurgical Technology
Boliden Contech AB
932 00 Skelleftehamn
Sweden

ABSTRACT

Background

Boliden Metall AB (Rönnskär Works) has copper and lead smelters as well as plants for recovery of by-products in the form of nickelsulphate, silver, gold, selenium, arsenic, sulphuric acid, and liquid SO_2.

According to the environmental permit from the Swedish government in 1975 all of the process waste water as well as rain and flush water had to be handeled in a central waste water treatment plant before being emitted to the Gulf of Bothnia. The plant had to be built within three years. It was taken into operation in the beginning of 1978, half a year before the deadline.

Waste Water

Waste water which is fed to the purification plant has two origins:

1. Process water, mainly from copper and lead smelter gas cleaning systems.

2. Rain and flush water from the whole smelter area.

The water is caracterized by a wide variation in flow rate, sludge amount and concentration of soluble metals such as Cu, Pb, Zn, Cd, As, and Hg. Its level of acidity varies between 1.0 and 3.0.

Treatment Process

The treatment process has been developed with the following aims:

1. Metals should be precipitated and form a sludge which can easily be recycled to the smelter. That means a high metal content.
2. Other constituents of the waste water which can not be emitted to the environment should be precipitated and form a sludge which can easily be stored without environmental problems. Examples of these are sulphates, acids, and fluorine.

3. The overall metal recovery should be high.

The Process Steps

Sum total a maximum amount of 200 m^3 of water per hour can be treated. The incoming water is fed into two buffer tanks. The pH-value is controlled with either sulphuric acid or caustic soda dependent upon the incoming water analysis. Thereafter sulphide is added to get a predetermined pH-value.

After that the sulphide precipitate is pumped into a thickener where floccuation agents are added. The sludge from the thickener underflow is taken out to a centrifuge. The clear water is then treated with a lime solution to get a pH-value of 12. The precipitate which is hereby formed is pumped into a small pond from which the clear water is pumped to the recipient.

Dosage of chemicals are automatically controlled by pH and analyses.

INTRODUCTION

General Smelter Description

The main products of the Boliden's Rönnskär Works smelter are copper and lead. The copper production is more than 100 000 metric tons, and the lead production is 65 000 metric tons per year. A large part of the activities at the Rönnskär Works are concentrated on the production of by-products. Such as: sulphuric acid, liquid sulphur dioxide, white and metallic arsenic, selenium and selenium compounds, gold, silver, nickel sulphate, zinc clinker, and so on. Enclosure No 1 shows the material flow in the smelter.

Process and Wash Water Sources

There is an extensive piping system at the Rönnskär Works for both incoming water and wastewater. For the incoming clean water, there are two piping systems: one for ordinary tap water, used mainly for drinking and pump sealing, and another for brackish water, used for process applications and washing. For wastewater there are several piping systems. Sewage water is separated and led to the municipal sewage treatment plant. Cooling water from indirect cooling applications is discharged directly into the plant effluent receiver. The rest is piped to the sulphide precipitation plant for treatment. Runoff water and cleaning water from the smelter also goes to the sulphide precipitation plant.

The following process sources are major contributors to the influent of the sulphide precipitation plant:
1)   Purge water from the central gas cleaning plant.
2)   Purge water from the scrubbers that treat gases from the deleading furnace in the slag fuming plant.
3)   Purge water from the gas cleaning unit on the fluidized solid roaster.
4)   Purge water from the scrubber after the lead Kaldo furnace.
5)   Condensed water out of gas ductwork from distant areas to the central gas washing plant.
6)   Drainage water from process gas fans.
7)   Water from washing wet electrostatic precipitators, mainly those for gases going to the sulphuric acid and SO_2 plants.
     Surface and wash water sources are:
1)   Rain water from the smelter, including runoff from roofs.
2)   Wash water from cleaning floors and equipment.

3)   Water from the washing of vehicles.
4)   Water from the laundry which cleans protective clothing.
5)   Condensed water from arsenic refinery

The composition and flow rates of these streams are given in Enclosure No 2. The composition is given for elements in solution.

Survey of Swedish Effluent Guidelines

One of the goals of Swedish legislation for environmental protection is to prevent contamination of clean waterways by industrial wastewaters. This legislation recognizes that total protection is not feasible, but that contamination should be prevented as far as possible. Permission must be obtained to operate any process which can pollute the environment. The Licensing Board for Environmental Protection (or Franchise Board) is the authority which examines permit applications. This Board specifies conditions and environmental controls which must be satisfied; it takes into account primarily the technical feasibility. The financial impact of control requirements is also considered. In addition, public and private interests are taken into consideration. Similar examinations in accordance with the Environment Protection Act are also made by the National Environment Protection Board and, in certain cases, the County Administration.

In Sweden, both the examination of applications for permits and the monitoring of pollutant levels are done on a case by case basis. This means that there are no general, binding emission standards. However, the Environment Protection Board has issued general guidelines for both examination of applications and monitoring of pollutant levels. General guidelines for air pollution control are particularly well documented.

DEVELOPMENT WORK

Background

The main purpose of the test work was to find an efficient process to reduce the heavy metal content in the previous mentioned wastewater. A secondary goal was that the process should produce a sludge containing heavy metals in a form suitable for return to the smelting processes. This was important since the regulations for the storage of heavy metal sludge make alternate disposal methods expensive.

Another condition had to be met for recirculation purposes: the sludge was not to contain fluoride. Boliden did not intend, however, to release fluoride-containing waters into the plant effluent receiver; fluoride was to be removed separately and not returned to the process. These circumstances added another demand on the process to be developed, namely that fluoride should be removed separately.

Test Program

The following parameters were to be evaluated:
* Precipitation agent
* pH levels
* Contamination levels
* Temperature
* Sedimentation and sludge dewatering properties

A summary of the types of experiments performed is presented in Enclosure No 3. Included in this table are the precipitation agents and dosages, pH levels, temperatures, and sedimentation and dewatering techniques which were tested.

Results

General

All of the fundamental work was done with synthetic solutions. The base for the solutions was distilled water, and the heavy metals were added mainly in the form of chlorides and sulphates. Because most of the water used for process purposes is brackish, sodium chloride was also added. The actual wastewater also contains SO_2, which also had to be added to the synthetic water.

In the second test series, the aim was to find a better process control scheme. All these tests were performed with real waters.

Most of the tests were made at room temperature. Two precipitation tests with a combination of sodium carbonate and sodium sulphide were performed at $40^{\circ}C$.

Only methods for the concentration and separation of sludge from sodium hydroxide – sodium sulphide precipitation were studied.

Precipitation of heavy metals as sulphides is a well known analytical chemistry procedure. Precipitation of heavy metal hydroxides is also well known. Equilibrium constants can easily be found in the literature. Unfortunately, these data are only for single elements in neutral solutions, at low salt concentrations. By comparison, the wastewaters teated are acidic and quite complex in composition. In spite of these limitations, the literature data are useful in predicting the relative ease of precipitating a metal.

The solubility products for the metals of interest are given in Enclosure No 4.

Precipitation

In Enclosure No 5 are shown some examples of precipitation results. All the analysis reported were made on filtered samples.

Reaction time for lime precipitation was one hour, for sodium hydroxide and sodium carbonate, 15 minutes, and for sodium sulphide, 20 to 30 minutes.

Sedimentation

Sedimentation tests were done both with and without the addition of polyelectrolyte. Enclosure No 6 is an example of settling curves for precipitate both with and without the addition of polymer.

In industrial applications of precipitation reactions, overdosage of the precipitating agent occasionaly occurs. Normally, this is not a problem, but with sodium sulphide the overdosage leads to change in the precipitate. The amount of small flocs tends to increase, perhaps due to a change of charge and the formation of elemental sulphur when sodium sulphide reacts with the sulphur dioxide in the water. The resulting sedimentation properties are poor. The influence of polyelectrolytes on the sedimentation properties in the case of a sulphide overdosage were studied. It was found that the polymer dosage had to be doubled in order to achieve good sedimentation.

A limited number of flotation tests were performed. The experiments showed that flotation is possible. However, the process is more sensitive to small changes in sludge characteristic than is normal sedimentation. In addition, presedimentation is a must for flotation because it is not possible to float heavy material like metallurgical dust or sand which is present in the incoming water.

A filter leaf test procedure was used to simulate drum filter performance. Only the filtration properties of a sludge precipitated with sodium hydroxide-sodium sulphide were studied. Variables in these types of tests were filter cloth material, filter aid type, and type and dosage of poly-electrolyte.

PROCESS SELECTION AND DESIGN

Considerations Leading to Process Selection

The following criteria were used to choose the water treatment process: the new plant was to treat scrubber blow-down, wash water from the washing of floors and equipment, and rain water from the heart of the smelter area. All of these waters contain heavy metals, and the water from the gas scrubber also contains fluoride. The heavy metal and fluoride content in the discharge water was to be minimized through treatment in the wastewater treatment plant. Because many of the metal constituents are harmful to the environment, they must att be reduced as fas as possible. The heavy metals separated from the water are preferably to be in a form that would permit their return to the metallurgical processes.

The best precipitation results were reached with sodium sulphide after preneutralization using sodium hydroxide, sodium carbonate or lime. Lime could not be used as a preneutralizing agent because the heavy metal sludge would also contain fluoride, making it unsuitable for return to the smelter furnaces. Sodium hydroxide was chosen for neutralization: This was an economic decision, because sodium carbonate could also have been used.

A special treatment step was necessary for fluoride precipitation. This had to come after the heavy metal sludge had been separated from the water. Lime was chosen as the fluoride precipitation agent.

Design Basis

Flow rates

The waters to be treated in the sulphide precipitation plant were:
* Process water
* Surface water
* Wash water
The compositions for these different streams are given in Enclosure No 2.

The flow rates for individual incoming flows was chosen as:
* Process water: maximum flow 75 m^3/h, with a possible increase to 100 m^3/h. Minimum flow 25 m^3/h.
* Wash and rain water: maximum flow 125 m^3/h. Minimum flow 25 m^3/h.
The flow rate after mixing the two streams is:
* Maximum flow 200 m^3/h.
* Minimum flow 50 m^3/h.
The flow rate of the rain water can be large on occasions of heavy downpour. The surge capacity of the rain water pipe system was calculated to 1 300 m^3. To further dampen the flow to the plant surge tanks were installed. Both factors were considered sufficient to guarantee a reasonably constant flow rate.

PLANT DESCRIPTION

Equipment

Flow-sheet of the plant is shown in Enclosure No 7.
The incoming water streams are taken to their surge tanks. The tank for

process water has a volume of 140 m^3 and the one for rain and wash water 280 m^3. During very high rainfall it is also possible to pump rain and wash water directly to the fluoride precipitation tank. During these occasions the level of impurities is so low that it does not influence the composition of the fluoride sludge.

By means of level controls in the surge tanks the valves after the pumps control the flow rate. The flow rates of the process water and the sum of rain/wash water is monitored. Based upon the results of a sample analysed upon it is decided to change the pH-level to somewhere between 1.8 and 3.0. This is done by adding sulphuric acid or caustic soda directly into the stream in the pipe-line. As source for sulphuric acid, spent acid from the evaporation plant of the tank house in the copper refinery is used.

Precipitation is then done with a solution of sodium sulphide to reach a pH level of 4, which is controlled in the reaction tank by means of a pH electrode. Addition of sulphide solution is done in the pipe line before the tank. The pump for sulphide solution is speed controlled.

Volume of the reaction tank is 55 m^3. It has an impeller. Water flows by gravity directly to a pump tank of 50 m^3, from which the water is pumped to the thickener. Before entering the thickener the flocculation agent is added into the same line. The thickener has a total volume of 680 m^3 and a diameter of 15 m.

Water transparency in thickener is controlled manually and used for determining the working time of the centrifuge.

Outgoing underflow from the thickener is pumped to a buffer tank. The pump is controlled by the buffer tank level indicator. Solid content in the underflow is 3 - 11 %.

Flow rate to the centrifuge is set manually to obtain a certain pressure. By means of this pressure the centrifuge can control its own differential rotation speed. It is equipped with a microprocessor. Capacity of the centrifuge is 400 kg solids per hour. Solid content of the sludge is 25 - 30 %. Reject water quality is very good as sludge content is less than 5 mg/l. Water is recycled back to pump tank before the thickener.

Thickener overflow flows by gravity to a supply tank and to the fluoride precipitation tank, in which slaked lime is added to a pH-level of about 12. The fluoride precipitation tank has a volume of 124 m^3. Water is then by gravity sent to a basin for sedimentation of calcium fluoride together with gypsum. The clear outgoing water is sampled continously before being emitted.

The plant is equipped with two parallell multilayer filters to filtrate the thickener overflow. These filters consists of anthracite sand, gamet sand and bearing materials. As overflow water quality is normally very good it has been observed that practically no improvement in water quality has been obtained by letting the over-flow water pass them. The multilayer filters are therefore seldom used.

Most of the materials in the plant which are in contact with the water are made of epoxi plastics. The centrifuge is made of stainless steel as well as the equipment for polyelectrolyte preparation.

All tanks are evacuated directly to the plant chimney. There are four warning devices for H_2S. There is, however, practically never any liberation of H_2S.

Process

The sulphide sludge is transported in containers to the copper smelter fluid-bed-roaster. The sludge is then fed by means of a pump to the charging chute of the fluid-bed-roaster and thereby entering the roaster the same way as copper concentrates.

Fluoride precipitate is pumped from the basin to a deposit area and is used for landfilling purposes.

A waste product from the Swedish pulp industry, which contains 90 - 100 g Na_2S per liter is used as a source of sodium sulphide. The plant is, however, also equipped for utilisation of solid sodium sulphide. The flocculation agents in the thickener and in the centrifuge are Magmafloc 2027 respectively Magmafloc 919.

Incoming water is analysed by x-ray every 3rd minute for As, Cu, Zn, Pb and Cd after filtration of the sample.

Overflow after thickener is also analyzed with x-ray twice an hour for the same elements.

PROCESS PERFORMANCE

General

During commission of the plant in 1978 some problems had to be taken care of and some changes have also been undertaken. The biggest change was to replace the original drum filter with the centrifuge. Nowadays the plant is working in a very effcient manner.

Environmental

A balance of the distribution of the metals in the process is shown in enclosure No 8.

It is based upon average figures during one year's operation. The average incoming flow-rate was 125 m^3/h. Amount of process water was 45 m^3/h and rain and wash water 80 m^3/h.

As can be seen metal extraction to the sulphide precipitate is very high, or, for

Cu	99.9	%
Zn	99.1	%
Pb	99.96	%
As	98.3	%
Cd	99.9	%
Hg	99.97	%

The sulphide sludge has normally following composition:

Au	5.7	g/t
Ag	506	g/t
Cu	11	%
Zn	10.6	%
Pb	12.6	%
As	17.0	%
Se	0.4	%
Cd	0.81	%
Hg	0.65	%

Amount produced per year is 1 000 tonnes dry weight.

Fluoride precipitate amounts to 1 300 tonnes per year dry weight. It consists mainly of fluorin and sulphur combined with lime.

Ingoing water quality is given in Enclosure No 9. Outgoing water quality is seen in Enclosure No 10.

Technical

Consumption figures are as follows:

Na$_2$S	1 250	tonnes per year
NaOH	2 250	-"-
CaO (90 %)	3 000	-"-
Magmafloc, total	6	tonnes per year
Electricity	2 085	MWh per year

The process has now been working successfully for about 9 years. Problems encountered during commissioning have been solved and the plant is accessable practically 100 % of the time.

Maintenance work is very limited. Practically all of it is lubrication and some preventive maintenance based upon observations of the crew.

The plant is manned during day-time. In the evenings and nights operational information of the plant is shown in the sulphur dioxide plant control room. Such information is flow-rates, pH-levels, tank levels and some analytical values.

Flowsheet – Rönnskär works

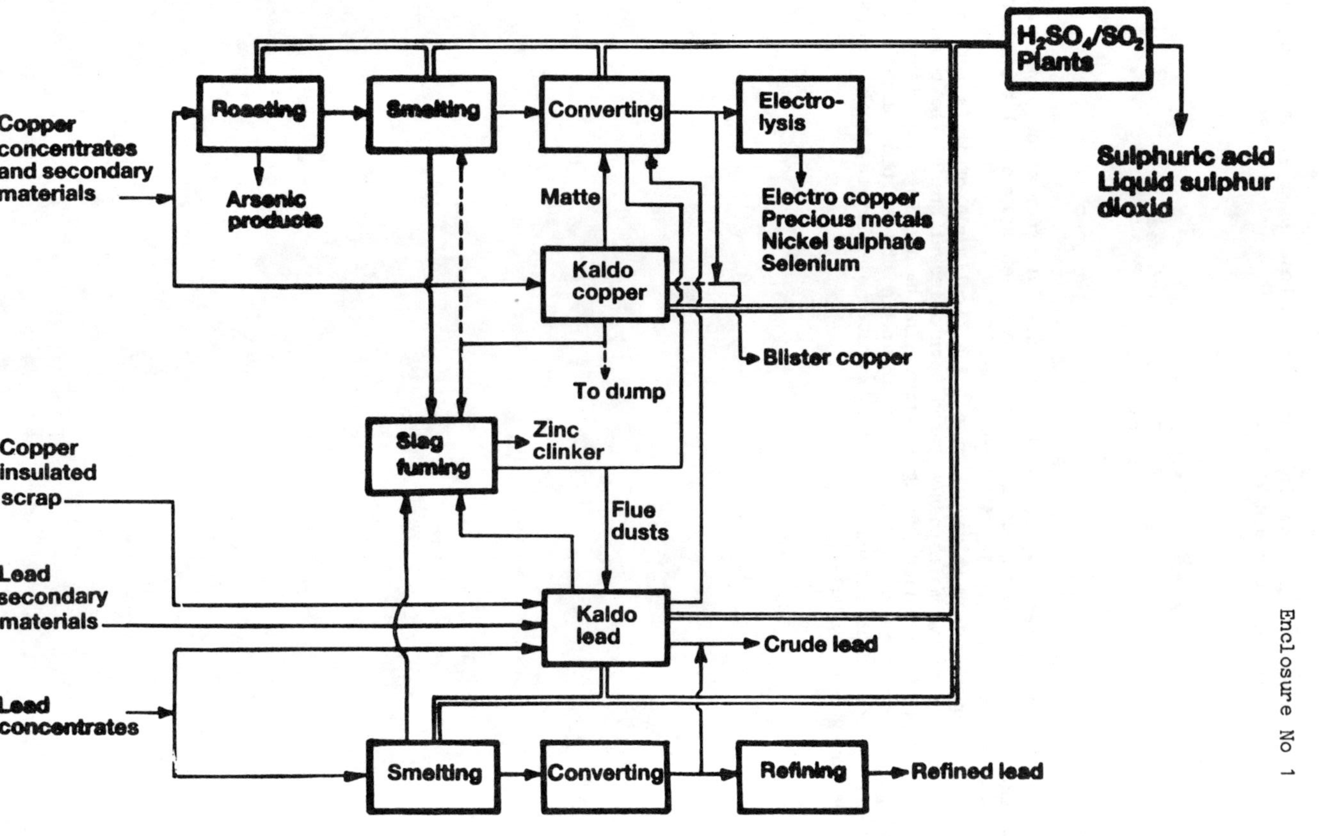

INCOMING STREAMS TO THE WASTEWATER TREATMENT PLANT

	Flow rate m^3/h	Liquid Phase Analyses, mg/l														
		Cu	Zn	Pb	Cd	Bi	Sb	As	Se	Hg	Ci	Cl	F	SO_4	SO_2	pH
Process water from																
Central gas cleaning plant	20	70	450	10	10	2	5	400	5	5	3	5000	500	6500	1000	1.6
Scrubber for deleading furnace	15	0.4-1	50-100	0-30	2-5			50-100				1700	1600-2000			2
Gas cleaning unit on the fluo solid washer	3	1	30	40	2			1500		0.02		3000	150			1-2
Wet electrostatic pre-cipitators	5	20-50	200-500	0-30				50-500	3-8	1-10						2
Surface water from the central industrial area																
Wash water from	40-150	1-50	1-220	0-2	1-10			10-100		0.001		1650	1-100			3-7
Cleaning of floors and equipment																
Washing of vehicles																
Laundry																

Enclosure No 2

SURVEY OF EXPERIMENTAL CONDITIONS IN THE BENCH SCALE EXPERIMENTS

Precipitation agents	Lime	Lime + Na$_2$S		Na$_2$CO$_2$	Na$_2$CO$_3$ + Na$_2$S		NaOH	NaOH + Na$_2$S	
Dosage	By pH	By pH	2.6 mM/ℓ 5 10 20	By pH	By pH	By pH	By pH	By pH	By pH
pH-levels	5.5 7 8.5 10.0 11.5	5.5		5.5 7 8.5 10	4-8	4.5-10.5	10.5	3-6.5	3-6.7
Temp ^{o}C	25°C	25°C		25°C	25°C and 40°C		25°C	25°C	
Sedimentation properties	Temp 25°C				25°C and 40°C		25°C	25°C and 40°C different polyelectrolytes	

Separation of sludge								microflotation
								thickening
								separation with:
								  centrifuge
								  drumfilter[x]
								  precoatfilter
								  filterpress
								
								filter aids:
								  lime
								  sawdust
								  celite
								  polyelectrolyte	

x) leaf test

SOLUBILITY PRODUCTS OF HYDROXIDES AND SULFIDES

Metal	Hydroxide	Sulfide
Cu^{2+}	-19.8	-35.9
Fe^{2+}	-16.3	-18.8
Fe^{3+}	-38.6	-
Hg^{2+}	-25.4	-52.2
Ni^{2+}	-15.3	-21.0
Pb^{2+}	-19.9	-28.1
Zn^{2+}	-16.1	-24.5
Bi^{3+}	-31	-98.8
Cd^{2+}	-14.3	-28.9

PRECIPITATION WITH $Ca(OH)_2$, $NaCO_3$, NaOH AND Na_2S. EXAMPLES OF RESULTS.

| | Precipitation Agent | | | | | | | | Analyses mg/l | | | | | | | | | | |
| | $Ca(OH)_2$ | | $NaCO_3$ | | NaOH | | Na_2S | | | | | | | | | | | | |
Test	Addition g/l	pH	Addition g/l	pH	Addition g/l	pH	Addition g/l	pH	Cu	Zn	Pb	Cd	Se	As	Bi	Sb	Hg	F	SO_4
Raw Water									37.7	290	2.1	10.8	2.2	57	1.3	8.8	0.014	19.8	3110
1	1.35	5.56							31.0	241	0.28	5.1	2.2	48	0.08	5.4	0.020	15.6	2890
2	2.17	7.57							2.14	37.6	<0.02	1.9	2.2	9.2	<0.02	3.2	0.010	9.0	1970
3	2.58	8.68							0.08	0.54	0.04	0.29	2.1	2.0	<0.01	2.5	0.009	16.8	1930
4	2.96	10.0							<0.02	0.02	0.02	0.01	2.1	1.50	0.03	2.3	<0.001	17.6	2140
5	3.71	11.54							<0.02	0.76	<0.02	0.01	1.3	1.72	0.01	2.0	<0.001	16.4	2030
6	1.3	5.5					0.195	5.5	0.18	142	0.04	0.04	1.9	42.4	0.01	3.5	<0.001	17.2	3050
7	1.3	5.5					0.39	5.46	<0.04	13.2	0.04	0.02	1.7	32.0	0.03	3.2	<0.001	16.6	3160
8	1.3	5.5					0.78	6.19	0.44	5.7	<0.02	0.14	1.4	1.9	0.06	2.5	<0.001	14.0	2530
9	1.3	5.5					1.56	11.04	0.04	0.60	<0.02	0.01	2.1	18.7	0.02	2.9	<0.001	14.0	2140
10*	1.3	5.48					0.65	6.48	0.10	3.38	0.04	0.02	1.9	1.20	0.03	2.3	<0.001	15.4	2370
11			1.08	5.56					30.0	222	0.12	7.7	2.2	39.0	0.10	5.0	0.017	15.4	3320
12			4.30	7.35					7.0	139	<0.02	2.5	2.0	26.2	0.04	4.8	0.015	15.6	4040
13			6.12	8.55					3.30	1.54	<0.02	0.40	2.0	12.3	0.03	4.4	0.014	16.0	4120
14			17.0	9.96					2.94	0.62	<0.02	0.02	2.2	16.0	0.02	4.4	0.012	7.6	6600
15			44.4	10.61					6.32	1.52	<0.02	0.08	2.1	32.0	0.04	5.4	<0.003	16.0	7130
16			1.24	5.71			1.01	6.3	0.22	2.72	<0.02	0.07	1.8	0.66	0.03	2.9	<0.001	10.6	3100
17					0.78	5.46	0.86	6.36	0.80	1.68	<0.02	0.06	1.7	0.20	0.04	2.9	<0.001	12.8	2990

* Removed lime precipitate before sulphide addition.

Suspended solids in clear phase as a function of surface loading

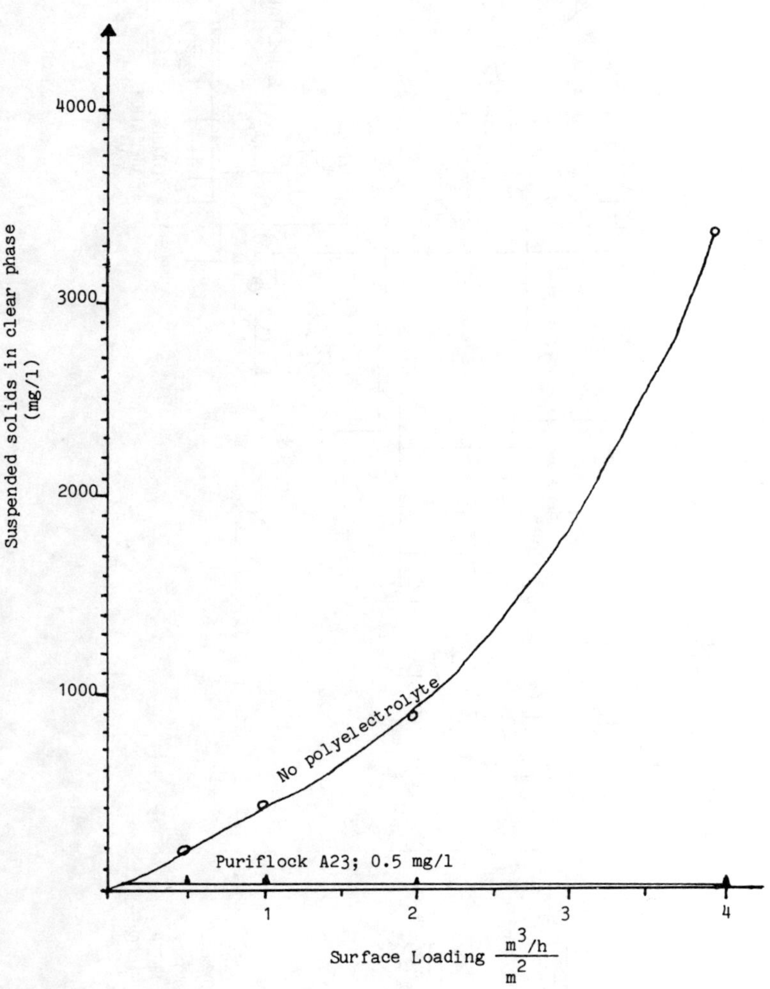

FLOW-SHEET OF WATER TREATMENT PLANT

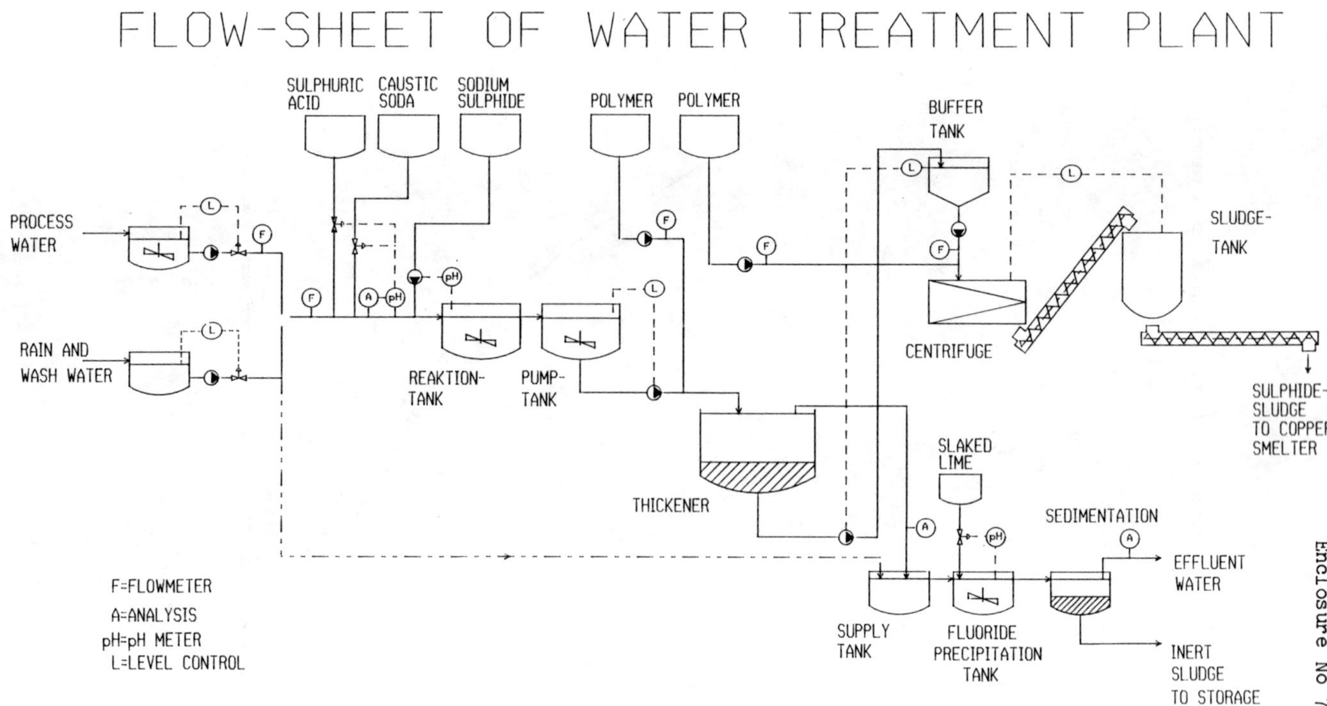

Metal balances in waste water treatment process, 1 year

	Amount dry weight (tonnes)	Cu (tonnes)	Zn (tonnes)	Pb (tonnes)	Cd (tonnes)	As (tonnes)	Hg (tonnes)
Incoming water		110.1	107	126.05	8.105	173	6.502
Sulphide sludge	1 000	110	106	126	8.1	170	6.5
Thickener overflow		0.1	1	0.05	0.005	3	0.002
TOTAL		110.1	107	126.05	8.105	173	6.502
Outgoing water		0.09	0.1	0.04	0.002	0.5	0.001
Fluoride precipitate	1 300	0.01	0.9	0.01	0.003	2.5	0.001
TOTAL		0.1	1	0.05	0.05	3	0.002

Ingoing water quality to water treatment plant.

Total contents, mg/l

Cu	110	mg/l
Pb	106	-"-
Zn	126	-"-
Cd	8	-"-
As	175	-"-
Hg	6	-"-
F	500	-"-
Ag	0.5	-"-
Se	0.5	-"-

Outgoing water quality, mg per litre of metals.

Total content, sum of solids and in solution.

	Normal level	Recommended level by Franchise Board
Cu	0.09	0.1
Pb	0.04	0.1
Zn	0.1	2.0
Cd	<0.002	0.05
As	0.5	10
Hg	0.001	not given

A METHOD FOR THE RELIABLE PREDICTION OF THE QUALITY OF CEMENT-BASED SOLIDIFIED HAZARDOUS WASTES AT THE TIME OF PREPARATION

Baldwin G, Rushbrook P E, Dent C G
Waste Research Unit
Harwell Laboratory
DIDCOT
Oxfordshire OX11 ORA
UK

ABSTRACT

Solidification is a process used for treatment of many hazardous wastes. However, predicting the physical and leaching characteristics and hence the "acceptability" of a particular solidified waste mix at the time of production is difficult at present. This paper outlines work funded by the UK Department of the Environment and conducted by the Waste Research Unit at Harwell to determine a practical method by which such a prediction can now be made for the cement-based solidification process. The method defines "acceptable" compositions of the waste mixes in terms of the proportions of cement, liquid wastes and solid wastes present. The ultimate goal is to provide "quality control" charts as a tool to indicate the waste/cement combinations most likely to give rise to solidified products meeting all legal and other criteria.

INTRODUCTION

Solidification is a treatment process utilised in the UK and elsewhere to precondition and allow certain liquid or solid hazardous wastes to be landfilled. These wastes are "retained" (although the mechanisms are not fully understood) within a cement or polymer matrix and thereby prevented from entering freely into the environment. Solidification is a treatment option purported to be suitable for a wide range of inorganic industrial wastes (1-6). A properly managed solidification process should not be viewed so much as a method of total containment of waste but rather as giving rise to a product from which toxic components are released into the surrounding strata at a controlled rate which will not give rise to any adverse environmental effects.

Various solidification techniques are cited in the literature   (7) and some of these are listed below:

i)   Cement-based techniques.

ii) Silicate-based techniques.

iii) Pozzolanic processes (not containing cement).

iv) Thermoplastic techniques (including the incorporation of wastes in bitumen, paraffin and polyethylene.

v)   Organic polymer techniques (including urea-formaldehyde and unsaturated polyester).

vi) Surface encapsulation (jacketing).

vii) Vitrification processes.

Within the UK, only the cement-based solidification process is currently operating on a commercial scale. Prior to incorporation into the final mix, some wastes require pre-treatment; this may be chemical (e.g. oxidation of cyanides) or physical (e.g. particle size reduction for solid wastes). Whatever the process, the aims of the pretreatment are to render a waste more amenable to the solidification process and further reduce any risk to the environment following landfill disposal of the processed product.

Recently in the UK, concerns have been expressed as to the acceptability of the final product when certain industrial wastes have been subjected to the process. Instances have been reported (8,9) where the quality of the product has not achieved the quality expected by the licensing authority. Unfortunately, present methods of testing are not standardised and at best the quality of the product can be evaluated only many days after landfilling. This situation is considered unsatisfactory and as a consequence, the UK Department of the Environment (DoE), as part of a comprehensive programme, commissioned the Waste Research Unit (WRU) at the Harwell Laboratory to undertake a major laboratory study to identify the physical and leaching characteristics of cement-based solidified wastes. The objective of this work was to investigate in detail for a whole range of wastes those proportions of constituents (cement, wastes and fillers) which could be expected to produce, with a high degree of certainty, a satisfactory product. The proportions of components considered in the prepared samples were designed to reflect those which might be expected to be encountered in the process when operated by commercial companies responding to the strong influence of market forces. In addition to cement as a major constituent, some treatment plant operators have also used pulverised fly ash (PFA) in their mixes. PFA is plentiful, serves as a useful bulking agent and is claimed to assist the rate of setting by virtue of its pozzolanic properties. Consequently, the study largely confined itself to assessing cement/PFA/waste mixtures.

This paper presents a brief overview of the research programme carried out by WRU and introduces the concept of "quality control charts" for use by treatment plant operators as a technique to indicate the suitability of waste mixes at their time of production. In essence, boundaries are established for mix composition to define parameters to be met in order to achieve pre-defined and agreed properties of the solidified product.

LABORATORY PROGRAMME

Before any quality assurance method could be devised, a better understanding of the physical characteristics and leaching behaviour of representative solidified waste mixes was required. To achieve this, an extensive laboratory programme was designed and undertaken to obtain selected measurements and characteristics for a variety of mix compositions (see Section 3). The work was divided into two stages. The first involved the production and assessment of poorly prepared waste mixes which had been produced with a minimum of pre-treatment. The intention was to create the conditions which might prevail in a poorly-run disposal operation. This stage provided data to establish the proportions of constituents that would be expected to produce unacceptable physical and leaching characteristics.

The second stage of the study involved the preparation of further mixes, this time produced under more rigorously controlled conditions and where necessary after pretreatment.

Mix Components

Wastes

The wastes incorporated into the experimental mixes were chosen as being representative of large volume arisings which currently are processed by cement-based solidification operators in the UK, ie:

i)   Cyanide waste, liquid;
ii) Cyanide waste, solid;
iii) Mixed inorganic acid waste;
iv) Caustic waste solution;
v)   Metal-bearing slags.

Cement

Ordinary Portland cement (10) was used for all mixes produced in both stages of the study.

Pulverised Fuel Ash

In the first stage of the study, PFA from a local power station was chosen for use in the mixes. In the second stage, for reasons of variability in the origin and quality of the coal used, PFAs from several sources were incorporated into the trial mixes.

Preparation of Mixes

First Stage

In the first stage of the work, pretreatment of the wastes i.e. oxidation of cyanide (treatment with sodium hydroxide and sodium hypochlorite at pH values above 10.5), neutralisation of caustic and acidic wastes to pH 7.0, and size reduction of metal slags, was considered to be the _minimum_ believed necessary prior to mixing with cement. A sequence of operations which would be expected to be followed at a full-size treatment

plant was used. This involved pretreatment, mixing of wastes, addition of
PFA, further mixing, addition of cement and then final mixing. The product
was then poured into a suitable container; further details of the
composition of the trial waste mixes are given in Table 1.

Second Stage

In the second stage, wastes were again pretreated before admixing with
cement and PFA although more rigorously controlled conditions were used on
this occasion. The products were cast into moulds and left to set in an
environment not subject to extreme ranges of temperature or excessive
disturbances. The choice of proportions of liquid wastes, solid wastes,
cement and PFA was derived from the results of the experiments conducted in
the first stage. Attention was focussed specifically on areas of
uncertainty which had been identified in the first stage around the
boundary between 'acceptable' and 'unacceptable' mixes, ie mixes in the
range of 4 to 12% cement. This boundary was defined on the basis of the
results of the physical tests performed on the solidified products.
Details of individual mixes are given in Table 1.

PHYSICAL AND LEACHING CHARACTERISTICS MEASURED

Parameters Measured

The primary objective of both stages of the laboratory programme was
to measure specific parameters for each of the trial mixes, and to relate
the results for each parameter to the mix compositions. In the first stage
the parameters considered were:

i)   rate of solidification;
ii) rate of supernatant production;
iii) heavy metal concentrations in leachates (i.e. Fe, Pb, Zn, Cd, Ni, Cr)
     from leaching tests.

These parameters were supplemented in the second stage by the following:

iv) concentration of arsenic in leachate from leaching tests;
v)   compressive strength after setting;
vi) hydraulic conductivity.

Rate of Solidification

Previous work at Harwell Laboratory has shown that the strength of concrete is directly related to the development during the setting process of the internal calcium alumino-silicate matrix. A similar phenomenon is assumed to account for the strength of cement-based solidified wastes.

The rate of setting of each waste mix was measured daily using a cone penetrometer. In this study, all measurements were made with an 80g cone which conforms to the specifications in BS 1377: 1975 (11). The cone was found satisfactory for penetration measurements over a wide range of waste consistencies including those for soft and unconsolidated, hard and brittle products.

Supernatants

Supernatant liquids are known to arise in landfills receiving solidified wastes and these were initially present in most of the laboratory trial mixes. With some mixes, the liquids were rapidly re-absorbed into the setting product, whereas with others, they were retained for several days before being lost through evaporation. To ensure consistent results, samples were left to set in a cool, stable environment. In this respect, the period of retention of supernatants was not comparable with that to be expected at a landfill, where conditions are more variable.

Permeability Measurements

Permeability or 'hydraulic conductivity' is defined as the rate of movement of a fluid through a porous medium in response to a pressure gradient. Within the programme, measurements of hydraulic conductivity, based on Darcy's Law, were made for the solidified samples. The general procedure employed followed a standard method reported in the literature (12). Since the basic technique was originally designed for soils rather than solidified waste, some modifications were necessary. Considerable difficulty was encountered in obtaining reproducible results.

Each newly-prepared waste mix was poured directly into a container and left to solidify. Measurements of hydraulic conductivity were then made after 7 and 28 days. The permeates obtained were analysed for the presence of cadmium, chromium, copper, iron, nickel, lead and zinc.

Leaching Tests

Leaching tests, using the standard Waste Research Unit method (13), were performed on the solidified waste mixes 28 days after preparation. Distilled water was used as the leaching fluid. 25g portions of each mix, in the form of several small pieces, were shaken for successive two-hour periods to generate leachate samples. These were filtered and subsequently analysed for cadmium, chromium, copper, iron, nickel, lead and zinc. Samples from the second stage were also analysed for arsenic.

Compressive Strength

Five cubes of material (approximately 25mm square) were cut from most of the solidified waste mixes produced in the second stage. The exceptions were Mixes F1 and F3 which were too weak to cut, and Mixes F5 and F7 where only one and two samples respectively survived during handling and loading into the test machine.

Testing was performed using a test machine that conformed to the British Standard for the compression testing of hardened cement paste and concrete (14). Each sample was stressed at the recommended rate of 15 MPa min⁻ It should be noted that because of the need to conserve sufficient sample for other tests, the dimensions of each cube of solidified waste were less than those commonly used in the British Standard test (BS dimensions are 100mm along all sides). No attempt has been made to apply a correction factor to the results obtained.

QUALITY CONTROL CHARTS

In general the physical and leaching characteristics of the waste mixes improved with increasing cement and decreasing liquid waste compositions. A detailed discussion of the laboratory results will not be presented in this paper but it was evident that different proportions of cement, waste (liquid and solid) and PFA could be identified which produced solidified products with characteristics likely to be "acceptable", "unacceptable" or "marginally acceptable" to regulatory authorities within the UK.

The results obtained from the second stage of the laboratory programme were plotted onto triangular graph paper (e.g. Figure 1), from which those proportions conforming to acceptable environmental criteria can be easily identified. By interpolation, other cement, waste and PFA proportions can be traced onto the graph and an indication provided as to whether they are likely to produce solidified wastes in the "acceptable" or "unacceptable" zones of the chart.

It is apparent that a clear understanding of the chemical and physical processes occurring within mixes undergoing solidification is only just emerging (15,16,17,18,19). The quality assurance method presented here is based upon using "external" measurements such as setting rate, compressive strength, hydraulic conductivity, leachate and supernatant quality (heat of reaction and porosimetry measurements could perhaps be incorporated at a later stage). This is suggested as a sensible and practical approach until such a time as a much fuller understanding of the internal processes is available.

In the UK, the setting of physical and leaching criteria to determine "acceptable" and "unacceptable" waste mixes is established independently by the operator and local regulatory authority on a site-by- site basis. Elsewhere this could be determined at regional or national level. The quality control procedure introduced here enables the quality of waste mixes to be assessed during mixing rather than after 28 days (or a similar length of time) as at present. This offers a significant advantage to both regulators and operators. The implementation of the procedure is site-specific. It involves three stages; the first two stages involve the

necessary analytical work and the third stage involves the production of a
"quality control chart".

The first stage would involve conducting physical tests on sample
mixes using the expected bulk waste stream. It is envisaged that 20 to 30
mixes would be prepared, representing a broad range of different
compositions of cement, waste and PFA. These tests could include setting
rate, compressive strength, supernatant retention and hydraulic
conductivity tests (although difficulties may be encountered in achieving
representative measurements of conductivities). The results would then be
plotted onto triangular diagrams (e.g. Figure 1) and any discernible trends
noted.

The second stage would involve carrying out leaching tests on those
solidified samples which satisfy the physical criteria acceptable to the
regulatory authority. Analyses would include those for heavy metal
concentrations, together with any other parameters required by the
regulatory authority.

The third stage would be the preparation of a control chart. Minimum
values to which an acceptable mix must conform would be set by the
regulatory authority. A control chart similar to Figure 1 would then be
drawn up using the results from the previous stages. The chart would
indicate compositions falling above and below the minimum values. The
composition of all mixes produced by the treatment plant would then be
required to fall within the "zone of acceptability", where values for all
the physical tests would be above the minimum constraints and in which the
leachate from the resultant mixes would be below prescribed concentrations.
It is recommended that the control chart would then become a requirement of
the operator's site licence or permit. However, these charts should be
subject to continued review and redrawn using the results from new sample
mixes in the following situations:

i)   when the composition of any major constituent in the bulk waste
     stream, or composition of cement or PFA, varies by more than
     predetermined values (e.g. as an illustration, 10% for wastes and 2%
     for cement or PFA);

ii)   significant modifications are made to the pretreatment or
      solidification plant;

iii) if new information becomes available and the licensing authority
      wishes to alter the minimum values;

iv)   every 2 or 3 years to ensure the existing control chart is still
      reliable.

Once a treatment operation is underway, the regulatory authority
should regularly visit the site. Reference would then be made to the
control chart to determine if the mixes being made up are likely to produce
an acceptable product once landfilled. Proper monitoring would require an
operator to divulge details of the proportions of these constituents in
use. Obviously, the procedure would be most usefully employed where
co-operation exists between the operator and the authority. It is also
suggested that an "observation cell" is set aside in the landfill where
mixes whose composition is in dispute (between the operator and authority)
can be placed and left to set for 28 days. If it transpires that the mix
was unacceptable, it would then become the responsibility of the operator
to ensure alternative disposal via an approved route.

To verify the operator's information on compositions, it might be
possible to install security-locked flow meters or weighing equipment on
input flow lines, or the mixer itself could be mounted on a weighing
device. In addition, it is recommended that every 3 to 6 months (or more
regularly if problems have been found) the regulatory authority should
collect and analyse samples of the waste streams and mixes under
preparation. This would ensure the control chart was still valid for
existing operations. Failure to link the results from these samples to
that predicted by the control chart might lead the authority to suspect
that the information on proportions supplied by the operator was incorrect,
and would warrant further investigation.

CONCLUSIONS

The quality assurance procedure proposed for solidified wastes is not foolproof but is presented as a method worthy of further consideration and refinement. This would involve verifying the laboratory-based results with full-scale mixes in a treatment plant. The objective of control charts is to switch from the current practice of monitoring whereby the decision on the acceptability of a mix is made many days after preparation, to a situation enabling a decision to be made at the time of preparation. The legal conclusiveness of any monitoring test must also be reviewed very carefully, although collaboration rather than confrontation between operator and regulatory authority would be the most satisfactory way forward.

ACKNOWLEDGEMENT

This work has been funded by the UK Department of the Environment (DoE). The authors wish to thank members of the DoE's Land Wastes Division for their advice and assistance.

REFERENCES

1. United Kingdom Patent, No 148525 (1973).

2. United Kingdom Patent, No 2016438A (1978).

3. United Kingdom Patent, No 2040712A (1979).

4. European Patent, No 0013822 (1979).

5. United States Patent, No 4230568 (1978).

6. United States Patent, No 4154307 (1985).

7. Guide to the Disposal of Chemically Stabilised and Solidified Waste. United States Environmental Protection Agency. US Army Engineer Waterways Experiment Station, Mississipi, 1980, SW872.

8. Anon, Getting to Grips with Waste Solidificaton. ENDS Report (120) (January 1985), pp 11-13.

9. Hazardous Waste Management - An Overview. First Report of the UK Hazardous Waste Inspectorate, June 1985.

10. British Standards (BS) 12 (1958); 4027 (1966), 4248 (1968), 146 (1968), 4246 (1968), 1370 (1958), 915 (1947). British Standards Institution.

11. British Standard (BS) 1377 (1975). British Standards Institution.

12. American Society of Agronomy. Methods of Soil Analysis. Ed. Black, C A
    Part 1, pp 520-531.

13. The Testing of Hazardous Wastes to Assess their Suitability for
    Landfill Disposal. Harwell Report, AERE-R10737, November 1982.

14. British Standard (BS) 4550: Part 3: Section 3.4 (1978). British
    Standards Institution.

15. Eaton, H C, Tittlebaum, M E, and Cartledge, F K. Techniques for
    Microscopic Studies of Incineration and Treatment of Hazardous Waste,
    Eleventh Annual Research Symposium, Cincinnati, Ohio, 29 April-1 May
    1985, pp 135-142.

16. Clark, A I, Poon, C S, Perry, R. The Rational Use of Cement-Based
    Solidification Techniques for the Disposal of Hazardous Wastes.
    Proceedings of New Frontiers for Hazardous Waste Management,
    International Conference, Pittsburgh, 15-18 September 1985, pp 339-347.

17. Brown, T M, Bishop, P L. The Effect of Particle Size on the Leaching
    of Heavy Metals from Stabilised/Solidified Wastes. Proceedings of new
    Frontiers for Hazardous Waste Management International Conference,
    Pittsburgh, 15-18 September 1985, pp 356-363.

18. Walsh, M B, Eaton, H C, Tittlebaum, M E, Cartledge, F K and Chalasani D
    The Effect of Two Organic Compounds on a Portland Cement-Based
    Stabilisation Matrix in _Hazardous Waste and Hazardous Materials_, 1986,
    Vol 3, No 1, Mary Ann Liebart, Inc, Publishers, pp 111-123.

19. Poon, C S, Clark, A I, Perry, R. Permeability Study on the
    Cement-Based Solidification Process for the Disposal of Hazardous
    Wastes. _Cement and Concrete Research_, 1986, Vol 16, Pergammon Press
    Ltd, pp 161-172.

TABLE 1
Composition of each sample waste mix

Stage of study	Sample number	Cement	Liquid waste				Solid waste		
			Oxidised solid cyanide waste (in liquid form after oxidation)	Oxidised liquid cyanide waste	Neutralised acid/alkali	Total liquid waste	PFA	Metal slags	Total solid waste
Stage 1	A 1	16	22	23	32	77	-	7	7
"	A 2	4	22	23	32	77	12	7	19
"	A 3	25	26	26	17	69	-	6	6
"	A 4	11	17	17	17	51	33	5	38
"	A 5	33	19	20	22	62	-	6	6
"	A 6	16	19	20	22	62	16	6	22
"	B 1	11	34	21	20	75	-	14	14
"	B 2	5	24	16	15	53	32	10	42
"	B 4	14	36	21	20	72	-	14	14
"	B 6	17	32	19	19	70	-	13	13
"	C 1	27	17	17	10	44	27	2	29
"	C 2	4	26	29	16	71	21	4	25
"	C 3	7	9	9	5	23	69	1	70
"	D 1	2	36	28	9	71	12	15	27
"	D 2	4	24	19	7	50	36	10	46
"	D 3	5	24	19	7	52	33	10	43
"	E 1	3	12	24	31	67	30	0	30
"	E 2	6	13	27	37	77	17	0	17
"	E 3	11	6	13	17	36	53	0	53
"	E 4	8	9	20	25	54	38	0	38
Stage 2	F 1	2	23	21	21	65	17	16	33
"	F 2	2	18	16	16	50	24	24	48
"	F 3	2	14	13	13	40	29	29	58
"	F 4	2	10	10	10	30	34	34	68
"	F 5	4	19	18	18	55	21	20	41
"	F 6	4	15	15	15	45	26	25	51
"	F 7	4	13	11	11	35	31	30	61
"	F 8	4	9	8	8	25	36	35	71
"	F 9	7	20	20	20	60	17	16	33
"	F10	7	18	16	16	50	22	21	43
"	F11	7	14	13	13	40	27	26	53
"	F12	7	10	10	10	30	32	31	63
"	F13	9	19	18	18	55	18	18	36
"	F14	9	15	15	15	45	23	23	46
"	F15	12	18	16	16	50	19	19	38
"	F16	12	14	13	13	40	24	24	48
"	F17	14	23	21	21	65	11	10	21
"	F18	14	19	18	18	55	16	15	31
"	F19	14	15	15	15	45	21	20	41
"	F20	17	18	16	16	50	17	16	33

All percentages are rounded to the nearest whole number.

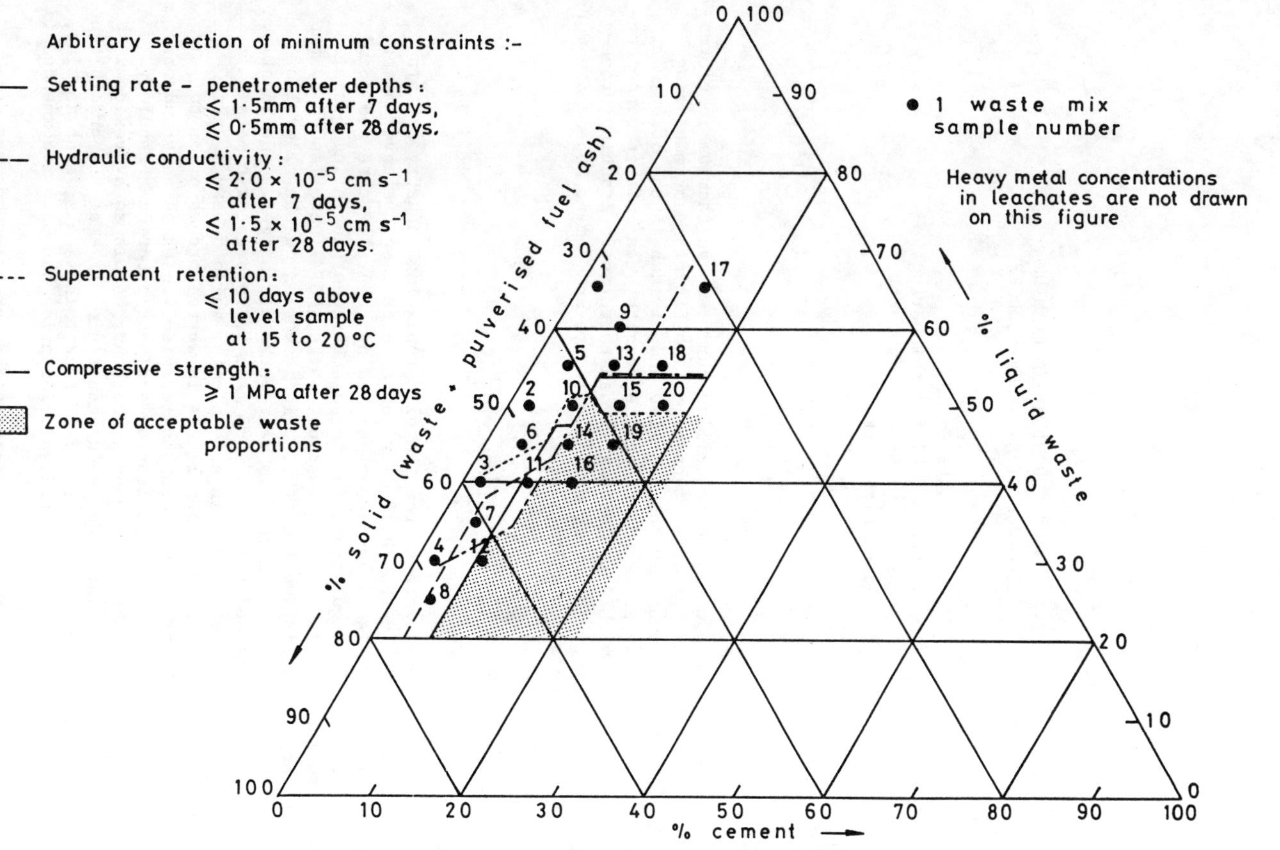

Figure 1. An example of a solidification control chart with a suggested "zone of acceptable waste proportions"

FIXATION OF CHROMIUM AND CADMIUM
IN ELECTROPLATING WASTE SLUDGES

Mriganka M. Ghosh
Department of Civil Engineering,
Pennsylvania State University,
University Park, PA 16802,
USA

ABSTRACT

Studies were conducted to immobilize chromium and cadmium in elctro-
plating waste sludges. The effectiveness of two fixation schemes using
sodium silicate and cement in one and lime and fly ash in the other were
compared. Both methods produced acceptable results with extract concentration of metals well within EPA's toxicity chatacteristic leaching procedure (TCLP) guidelines. However, sodium silicate and cement appear to work
better for electroplating waste sludges.

INTRODUCTION

The process wastewaters from the electroplating industry contain
many toxic heavy metals that have detrimental effect on the environment.
Consequently, their discharges are regulated by federal, state and local
ordinances. In 1978, regulations for the discharge of toxic industrial
wastewaters into publicly owned treatment works (POTW) were issued by the
EPA. For the electroplating industry, this necessitated the removal of
toxic heavy metals and cyanides from wastewater discharges. Consequently,
manufacturing process modifications were made to remove toxic metals
entirely from the waste streams in some plants, while others sought
alternative methods to concentrate metals in waste streams prior to
disposal. One of the treatment technologies widely used already and more
likely to be installed is neutralization and precipitation technology
which destroys the cyanide and removes the heavy metals as hydroxides.
Precipitation of metal hydroxides using caustic or lime as the precipitant
is most common. However, for the removal of certain metals, such as zinc,
copper, cadmium and mercury, sulfide precipitation has been recommended (1).

With the promulgation of new regulations for hazardous wastes,
precipitation alone often fail to meet the discharge criteria for many of
the toxic metals. The Resource Conservation Recovery Act (RCRA) of 1976
led to further revision of the regulations for the management of hazardous
wastes. A waste is deemed hazardous if it exceeds the parameters set by the

EPA in the categories of ignitability, reactivity, corrosivity and toxicity. Wastes containing heavy metals fall into the last classification.

Sludges containing heavy metals often used to be dumped into unsecured pits, ponds or landfills, while another portion incinerated. Stringent regulations for land disposal have led to the segregation of such wastes, with sludges containing toxic metals requiring disposal in approved hazardous waste landfill sites. Even these secured sites have come under close scrutiny since metals can leach due to changing environmental conditions and report to the surface runoff or the groundwater or both. An alternative to landfilling is to isolate the waste material into a solid matrix that can be safely disposed of by conventional landfilling. The liquids and sludges are solidified into a structurally sound material suitable for landfill. However, solidification alone does not go far enough to meet RCRA guidelines. The hazardous constituents are rendered immobile by using "fixative" agents. This latter process is called fixation or stabilization (2,3). Three commercially successful methods are listed below.
1. Calcilox process by Dravo Lime Co., Pittsburgh, PA (4). This is a cementitious material derived from blast furnace slag.
2. Poz-O-Tec, marketed by IU Conversion Systems, Inc., Philadelphia, PA, uses lime, fly ash and other additives (5).
3. Chemfix process, marketed by National Environmental Controls, Metairie, LA (6). This process converts metal hydroxide sludges to a solidified product using soluble silica gel and a setting agent, cement, for example. The solid product is not affected by changes in the aquatic environment. However, sulfate, chloride and sodium are not bound by the solid matrix and may leach if the material is reworked from its cast-in-place condition.

Other methods for the immobilization of neavy metals that have been used with varying degree of success involve thermoplastic techniques, organic polymerization and encapsulation (7,8). Questions have arisen as to the long term integrity of the solid mass formed by the solidification/ stabilization processes and health hazards resulting from the release of toxic metals. The purpose of the present study is to determine if chemically fixed hazardous inorganic wastes could be disposed safely in unlined landfills. In particular, the performance of two fixation methods, namely silicate-cement and lime-fly ash, in immobilizing chromium and cadmium in hydroxide sludges was compared.

BACKGROUND

Chemical Precipitation

Fixation of sludges to immobilize metals depends largely on the sludge characteristics. The moisture content and the soluble metal concentration are important considerations. Therfore, the nature of the chemical precipitants and the dewatering method used are equally important factors in the fixation process.

The amphoteric nature of the hydrated metal oxides makes them soluble at either low or high pH levels. The Chemfix process converts metal hydroxides into very insoluble silicates which presumably are not affected by pH fluctuations. Sulfide precipitation of metals has received considerable attention in recent years (1,9). High degree of removal can be obtained over a broad pH range. Excellent removal of certain metals, such as As, Cu, Cd and Hg even at low pH levels, high reactivity and therefore

smaller reaction times, low solubility, and the feasibility of selective
metal recovery make sulfide precipitation attractive. A comparison of the
solubilities of hydroxides and sulfides of certain metals are given in Table
1.

TABLE 1
Comparison of solubilities of metal hydroxides and sulfides (10)

Metal	Hydroxide Solubility, mg/l	Sulfide Solubility (pH=8), mg/l
Cadmium (II)	2.3×10^{-5}	6.7×10^{-10}
Chromium (III)	8.4×10^{-4}	no precipitate
Lead (II)	2.1	3.8×10^{-9}
Mercury (II)	3.9×10^{-4}	9.0×10^{-20}
Zinc (II)	1.1	2.3×10^{-7}

Fixation Process

Fixation, for the most part, has become synonymous with solidification/
stabilization. It produces a monolithic solid with reduced surface area and
hence low leachability. Toxic materials are immobilized by chemical fixing
agents to form insoluble compounds and entrapping them in an inert solid
matrix. The final solid is impermeable, res-stant to chemical and biological
attack and has high resistance to freeze-thaw cycles (7). Although fixation
does not normally immobilize contaminants completely, it slows the rate of
their release sufficiently as not to affect the environment adversely and
permit more flexible disposal options to be considered. The degree of
leaching from the solid mass depends on its permeability and the solubility
and concentration of entrapped contaminants. Chemical stabilization may
affect the solubility of the contaminants by altering the pH of the waste,
the form of the contaminant (dissolved versus insoluble), or by seques-
tering the contaminant within the solid matrix. Solubility of the contaminants determine the leaching process in most situations (11). Leaching
occurs by diffusion from the solid surface to the water in contact with it.
Internal leaching of contaminants depends on their solubility. Minimal
leaching should occur if the solid matrix should maintain its structural
integrity. If the mass should crack or deteriorate in any manner, internal
leaching will accelerate (12,13). Numerous results from fixation studies
are reported that satisfy U.S. EPA's EP-Toxicity test as to the non-
hazardous nature of the solid mass (3,8,12). Heavy metal concentration in
the leachates are often much lower than the permissible limits. Field
studies of leaching at a few chemically stable waste disposal sites that
have been in operation for a short period of time, seem to indicate no
contamination of the groundwater table (14,15).

A brief description of the two fixation methods tested in the present
study follows.

Silicate-cement process. This method employs a soluble silicate and a
setting agent, portland cement, that react with polyvalent metals to
produce a solid matrix or a pseudo-mineral that become soil-like in

appearance with subsequent weathering. The formation of the matrix involves three types of reactions: 1) between metal ions and silicates; 2) between soluble silicates and the reactive components of cement; and 3) a series of hydration, hydrolysis and neutralization reactions between cement and water. Some water remains uncapsulated and may appear as runoff. An inverse correlation exists between the solids content of the raw sludge and the amount of additives required. The metal-silicate matrix is based on tetra-hedrally coordinated silicon atoms alternating with oxygen atoms to form the backbone of a linear chain. Charge side groups (O⁻) react with metal ions to form strong ionic bonds between adjacent chains producing a three-dimensioned matrix as in pyroxene minerals.

Lime-fly ash process. The performance of the lime-fly ash process depends on the composition of the fly ash, lime dosage and the moisture content of the sludge. The constituents of importance in the fly ash are silicate, lime, alumina and unoxidized carbon. Lime and fly ash undergo possolanic type reactions with intermediate products similar to those formed by the reactions of lime with water. However, the reactions occur at a much slower rate. Upon mixing, the material behaves somewhat like a silty clay which undergoes cementitious reactions during curing and develops the characteristics of a low strength concrete-like material (16). The required amounts of ingredients in the mix varies with raw sludge characteristics. But a 2:1 fly ash to sludge ratio with about 5% lime by weight is typical, with a resultant solids content of about 75% in the solidified product.

MATERIALS AND METHODS

Electroplating waste sludge

The waste sludge from an electroplating firm was used in this study. An on-site treatment facility produced a metal hydroxide sludge with about 5% solids which was further dewatered to about 25% solids by belt press. For silicate-cement fixation, raw sludge was used while dewatered sludge was used for lime-fly ash stabilization. A typical analysis of the sludge is given in Table 2.

TABLE 2
Analysis of test sludge

Metal	Concentration (mg/1)	EPA Hazardous No.	EP-Toxicity Level (mg/1)
Arsenic	0.8	D 004	5
Barium	<5	D 005	100
Cadmium	2.8	D 006	1
Chromium	5.6	D 007	5
Mercury	<0.01	D 009	0.2
Selenium	<0.05	D 010	1

The pH of the fixatives used were, 11.7 for liquid sodium silicate, 12.1 for cement, 10.9 for fly ash (Wyoming coal), and 9.8 for commercial lime. Initially, fly ash from both Wyoming and Missouri coal were used. However, Missouri fly ash failed to produce an acceptable solid mass. Consequently, only Wyoming fly ash was used in all experiments. Apparently, low lime, alumina and silicate contents of the Missouri fly ash were responsible for its poor performance. Wyoming fly ash had 36% SiO_2, 17% Al_2O_3, 28% CaO, 5% MgO, 6% Fe_2O_3, and traces of other oxides.

Procedure

Design mixes were prepared to simulate optimum and worst conditions. For the silicate-cement process, a constant volume of raw sludge with approximately 5% solids was used in all formulations to which nine different concentrations of sodium silicate, ranging from 2% to 12% of sludge volume were added in 1% increments in different experiments. The additions for cement were 4, 8, and 12% of sludge volume for each of the silicate concentration used. A total of 27 mixtures were tested.

In the lime-fly ash process, for each of the four weight percents of the raw sludge used, namely 20.5%, 31.7%, 39.6%, and 47.7%, four weight percents of lime, 2, 4 , 6 and 8%, were used. The fly ash content in each mixture was so varied as to yield a total of 100%. A total of 16 formulations were tested using the lime-fly ash method. The final solids contents of the formulations immediately after mixing was directly related to the weight percent of fly ash used and relatively insensitive to the amount of sludge.

The samples from various formulations were cured in 2 in. diameter by 6 in. long aluminum cylinders under controlled conditions of humidity and temperature for different lengths of time. The compressive strength of the cured samples was tested on a Young testing machine. The samples were then pulverized and passed through a No. 4 (4.76 mm) U.S. standard sieve. To determine the rate of leaching of metals from the stabilized sludge, 200 g of the pulverized samples were placed in large plastic bottles with 2 l of distilled deionized water. The U.S. EPA EP-Toxicity test procedure was used where the pH of the slurry was lowered to and maintained at 5 at all times during the test using 0.5N acetic acid. During this 24-hr leaching test the slurry was kept well mixed using a rotary shaker. At no time during extraction the aggregate amount of acid added was more than 4 ml per g of sample or an allowable maximum of 800 ml for the 200 g samples tested. The extract was filtered through 0.45 micron membrane filter, pH measured (in the range of 6 to 8 for all samples), and preserved at a pH of 2 with nitric acid. The samples were analyzed for Cr and Cd using atomic absorption spectroscopy.

Leaching by rainwater was simulated in a different column leaching test proposed by Jackson (17). Only four out of a total of forty-three available samples of stabilized sludge were tested by this method. Pulverized samples were placed in a 75 mm diameter fritted glass funnel through which deionized distilled (pH=6.8) was drained slowly by applying a small vacuum. To simulate a rainfall of 40 in. per year, typical of the site where the electroplating firm was located, approximately 50 ml of distilled water were through the column each for five weeks. Only weekly composites of column leachates were analyzed for metals.

In analyzing for a specific metal in a mixed metal extract, the

method of standard additions is used to correct for any interferences (18).
In this procedure, to analyze each sample, four identical aliquots of the
extract are taken three of which are spiked with a known concentrations of
the test metal. The volume of the spikes are same as that of the sample
aliquots. An identical volume of distilled water is added to the fourth
aliquot of the extract which serves as the blank.

RESULTS AND DISCUSSION

The average concentrations of metals in the raw sludge tested are
shown in Table 2. The leaching tests yielded highly variable concentrations of Cr and Cd in the leachates. Table 3 compares typical results
obtained for chromium by the two fixation methods while similar data for
Cd are presented in Table 4. The concentration of metals are well within
EPA's EP-Toxicity criteria although complete immobilization of metals
cannot be obtained. Inappropriate fixative agent to sludge ratio,
insufficient mixing, and high moisture content of sludge may all contribute to incomplete immobilization. Further, no inference should be made as
to the long term effects of environmental factors on leaching. In similar
studies Bishop and Gress (19) observed that the use of 17N glacial acetic
acid instead of 0.5N acid in an otherwise identical EP-Toxicity test
caused greater amounts of metals to leach from the cementitious matrix.
Apparently, metals are bound to the matrix by pH-sensitive mechanisms and
thus their release is subject to the total acidity as well as pH. The
difference in leachate metal concentrations in the two fixation methods
obtained by the EP-Toxicity test is more pronounced for Cr than Cd. The
results clearly favor the silicate-cement as better fixatives. Further,
the leachability of either of these metals was inversely related to the
amount of silicate used within the range of silicate dosages tested. On
the other hand, in the lime-fly ash method, lime seems to be the key
ingredient controlling the quality of the stabilized mass. In the column
leaching test, metals extracted with simulated rainwater (pH=6.8) were
insignificant. Again, somewhat higher metal concentrations could be
leached from the lime-fly ash stabilized sludge (Table 5).

Structural integrity of the stabilized sludge is one of the key
objectives of any fixation method. High compressive strength of the
stabilized sludge may allow its use as a road subbase or in other similar
constructions. Also, high-strength composites tend to be less porous with
greatly reduced permeabilities, and, consequently, reduced leachability.
The unconfined compressive strength of the fixed sludges varied from 2.8
to 5.2 T/sq ft, with silicate-cement matrices yielding much higher values.
Again, the ingredients that contributed to higher strengths in these two
methods were silicate in the silicate-cement method and lime in the lime-
fly ash method. The highest permeability, 5.2×10^{-5} cm/sec, was obtained
in the solid matrix produced by 8% silicate and 8% cement.

TABLE 3

Leaching of chromium by EP-toxicity test

Silicate-Cement

Ingredients	%Vol	Chromium (mg/1) at Indicated Time (day)		
		30	60	90
Sodium silicate:	2			
Cement:	4	0.238	0.2	0.423
	8	0.18	0.095	0.456
	12	0.081	0.165	0.053
Sodium silicate:	8			
Cement:	4	0.208	0.122	0.168
	8	0.1	0.167	0.145
	12	0.16	0.066	0.147

Lime-Fly Ash

Sludge wt: 20.5%
Final solids: 85.2%

Fly ash (%wt)	Lime (%wt)			
71.5	8	0.113	0.184	0.191
73.5	6	0.233	0.238	0.348
75.5	4	0.222	0.305	0.468
77.5	2	0.233	0.281	0.446

Sludge wt: 47.7%
Final solids: 65.6%

Fly ash (%wt)	Lime (%wt)			
44.3	8	0.183	0.2	0.188
46.3	6	0.221	0.319	0.45
48.3	4	0.285	0.236	0.305
50.3	2	0.289	0.228	0.364

Weight of stabilized sludge solids = 200 g
Chromium in raw sludge = 5.6 mg/1

TABLE 4

Leaching of cadmium by EP-toxicity test

Silicate-cement

Ingredients	%Vol	Cadmium (mg/l) at Indicated Time (day)		
		30	60	90
Sodium silicate:	2			
Cement:	4	0.06	0.017	0.018
	8	0.1	0.12	0.02
	12	0.1	0.01	0.008
Sodium silicate:	8			
Cement:	4	0.02	0.018	0.015
	8	0.068	0.016	0.015
	12	0.008	0.01	0.006

Lime-Fly Ash

Sludge wt: 20.5%
Final solids: 85.2%

Fly ash(%wt)	Lime (%wt)			
71.5	8	0.038	0.068	0.096
73.5	6	0.028	0.038	0.116
75.5	4	0.065	0.033	0.112
77.5	2	0.091	0.043	0.089

Sludge wt: 47.7%
Final solids: 65.6%

Fly ash (%wt)	Lime (%wt)			
44.3	8	0.059	0.072	0.1
46.3	6	0.048	0.058	0.1
48.3	4	0.036	0.076	0.088
50.3	2	0.051	0.035	0.08

Weight of stabilized sludge solids = 200 g
Cadmium in raw sludge = 2.8 mg/l

TABLE 5

Extraction of metals by column leaching test

Metal	Week				
	1	2	3	4	5
Silicate-cement					
Chromium (mg/l)	0.12	0.043	0.013	0.012	0.015
Cadmium (mg/l)	0.039	0.029	0.015	0.021	0.012
Lime-fly ash					
Chromium (mg/l)	0.27	0.19	0.15	0.1	0.13
Cadmium (mg/l)	0.029	0.046	0.039	0.037	0.043

Rate of application of pH 6.8 water = 50 cc/day

REFERENCES

1. Bhattacharyya, D., Jumwan, Jr., A.B. and Grieves, R.B., Separation of toxic heavy metals by sulfide precipitation. Separation Sci. Technol., 1979, 14, 441-52.

2. Pojasek, R., Stabilization/solidification of hazardous wastes. Env. Sci Technol., 1978, 12, 382-6.

3. Christensen, D. and Wakamiya, W., A solid future for solidification, Toxic and Hazardous Waste Disposal, Vol. 4, ed. R. Pojasek, Ann Arbor Science Publishers, Inc., Ann Arbor, MI, 1975, pp. 75-89.

4. Selmeczi, J.G., Stabilization of sludge slurries, U.S. Patent No. 3.930.795, November 18, 1975.

5. Anon. Poz-O-Tec process for economical and environmentally acceptable disposal of scrubber sludge and ash, IU Conve-sion Systems Brochure, Undated.

6. Anon., Leaching tests for solids developed from treatment of industrial wastes, Chemfix Data Sheet 105, Undated.

7. Survey of Solidification/Stabilization Technology for Hazardous Industrial Wastes, EPA-600/2-79-056, July 1979.

8. Roberts, B. and Smith, C., Microencapsulation: Simplified hazardous waste disposal, Toxic and Hazardous Waste Disposal, Vol. 1, ed. R. Pojasek, Ann Arbor Science Publishers, Inc., Ann Arbor, MI, 1980, pp. 55-66.

9. Control and Treatment Technology for the Metal Finishing Industry: Sulfide Precipitation, EPA Technology Transfer Report No. EPA-625/8-80-003, April 1980.

10. _Lange's Handbook of Chemistry_, 12th Edition, McGraw-Hill Book Co., New York, 1979.

11. Landreth, R. and Mahloch, J., Chemical fixation of wastes, _Ind. Water Engr._, 1977, _14_, 16-19.

12. Johnson, J. and Lacione, R., Assessment of processes to stabilize arsenic-laden wastes. In _Disposal of Hazardous Wastes_, Proc. Sixth Annual Res. Symp., EPA-600/9-80-010, 1980, pp. 181-6.

13. Goodbee, H. and Joy, D., Assessment of the loss of radioactive isotopes from waste solids to the environment. Part 1: Background and theory, Oak Ridge National Laboratory, ORNL-TM-4333 (1974).

14. Myers, T, Francingues, N., Thompson, D. and Malone, P., Chemically stabilized industrial wastes in a sanitary landfill environment. In _Disposal of Hazardous Wastes_, Proc. Sixth Annual Res. Symp., EPA-600/9-80-010, 1980, pp. 223-41.

15. Jones, L., Malone, P. and Myers, T., Field investigation of contaminant loss from chemically stabilized sludges. In _Disposal of Hazardous Wastes_, Proc. Sixth Annual Res. Symp., EPA-600/9-80-010, 1980, pp. 187-202.

16. State of the art of FGD sludge fixation, Michael Baker, Inc., Report No. EPRI FP-671, Janualry 1978.

17. Jackson, K., Comparison of three solid waste batch testing methods and a column leach test method. Proc. Hazardous Solid Waste Testing: First Conference, ASTM, January 1981.

18. Test methods for evaluating solid waste physical/chemical treatment methods. EPA Waste Characterization Branch, Office of Solid Waste Management, U.S. EPA, May 1980.

19. Bishop, P.L. and Gress, D.L., Leaching from stabilized/solidified hazardous wastes. Proc National Conference on Environmental Engineering, American Society of Civil Engineers, Minneapolis, MN., 1981, pp. 423-29.

TREATMENT OF HAZARDOUS AND TOXIC WASTES USING SURFACTANT-BASED SEPARATIONS PROCESSES

Jeffrey H. Harwell and John F. Scamehorn
Institute for Applied Surfactant Research and
School of Chemical Engineering and Materials Science
The University of Oklahoma,
Norman, Oklahoma 73019
USA

ABSTRACT

Surface active molecules, surfactants, exhibit a number of unique properties which can exploited in the area of separations. Though having promising potential applications in several areas, these separations seem especially well suited for treating dilute aqueous solutions, an area of growing importance as wastewater emission standards become more stringent. This paper reviews several surfactant-based separations techniques which appear particularly promising for development in the area of wastewater treatment and in the control of fugitive emissions from painting process: micellar-enhanced ultrafiltration, surfactant-enhanced carbon regeneration, foam flotation, admicellar chromatography, surfactant liquid membranes, and absorption into emulsions.

INTRODUCTION

Surfactants are one of the most unique and versatile classes of chemical compounds [1]. One of the most useful aspects of their behavior is their tendency to spontaneously aggregate at interfaces and in aqueous solutions. A surfactant molecule is formed by joining a hydrophobic moiety that is immiscible in aqueous solution with a hydrophilic moiety which readily dissolves in water. A common surfactant which may be considered typical is sodium dodecylsulfate, in which an alkane (dodecane) is chemically bonded to a salt (sodium sulfate). At very low concentrations surfactants form monolayers at liquid/liquid or air/liquid interfaces, they form micelles (spherical or cylindrical aggregates of 50-200 molecules) or vesicles in aqueous solution, and admicelles or hemimicelles at solid/liquid interfaces. At higher concentrations, surfactants can induce a phase separation in which a dilute aqueous surfactant phase is in equilibrium with a concentrated aqueous surfactant phase called a coacervate phase, or an emulsion or a microemulsion in which two immiscible phases are dispersed one inside the other (Figure 1).

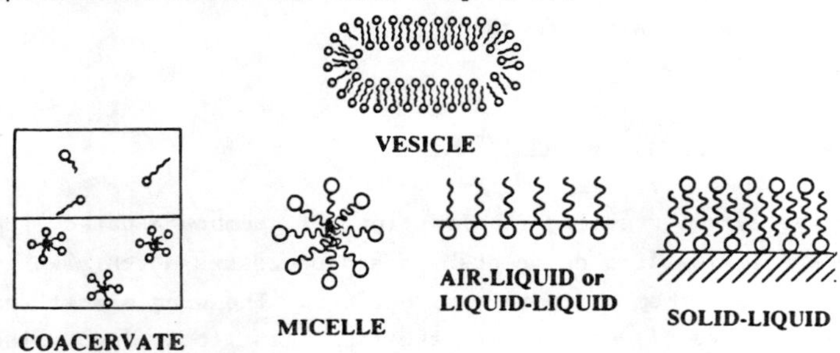

Figure 1. Some surfactant aggregates.

While surfactants have long been important in industries as varied as cleaning and pharmaceuticals or paints and food processing, over the last decade and a half there has been a spurt of interest in applying surfactants to separations [2]. The two principal motivations for this activity have been the demanding new separations required for the recovery of delicate biological molecules from fermentation broths and the increasingly demanding purifications required by the growing world-wide awareness of the need to control environmental pollution.

In this paper we would like to briefly survey some on the new surfactant-based separations which seem to us to have great potential for application to the separations demands of wastewater treatment technology. Our purpose is to increase interest in the investigation and development of surfactant-based separations as an important new class of separations techniques for water treatment. For convenience we will divide our presentation of these techniques into those which are micellar-based, those based on adsorbed surfactant aggregates, those which are emulsion-based, and those which are foam-based.

SEPARATIONS BASED ON MICELLES

It is sometimes suggested that a micelle may be thought of as a droplet of organic liquid 3-5 nm in diameter covered with an ionic coat. While this is certainly an oversimplified model it does accurately convey some of the potential of micelles in separations technology: the oil-like core can act like an organic solvent in solvent extraction, and the ionic-coat can act like an ion exchange medium, while the small size of the aggregates suggests an enormous area of contact between the aqueous phase and the "phase" into which the target compounds are being extracted. Surfactant-based separations which may also be classed as based on micelles [3]are micellar-enhanced ultrafiltration, extraction into reverse micelles, micellar chromatography, and surfactant-enhanced carbon regeneration. We will only discuss the first and last of these as techniques with potential for wastewater treatment.

Micellar-Enhanced Ultrafiltration (MEUF™)

Despite the ever increasing application of membrane-based separations, there are limitations on membrane-based processes which make it uneconomical to apply them in certain situations. The single most serious limitation in the area of wastewater treatment is that low molecular weight compounds and ions such as trace metal ions cannot be rejected by available ultrafiltration membranes.

If surfactant is added to the process stream, however, low molecular weight organics will become solubilized into the oil-like core of the micelles while multivalent ions will become bound to the ionic coat of the micelle by electrostatic attraction. Since the micelle may contain several hundred molecules under these conditions, this permits removal of the target compounds with an ultrafiltration membrane with large pores permitting high flux with excellent separation efficiency (Figure 2).

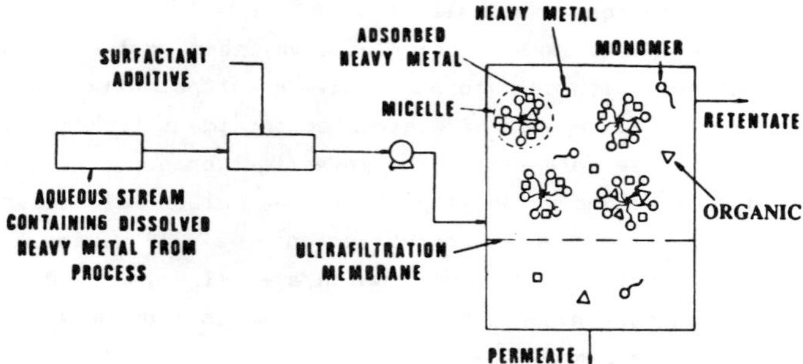

Figure 2. Schematic of the micellar-enhanced ultrafiltration process.

Both dissolved organics and multivalent metals can be effectively removed by MEUF™. Dunn et al. [4] demonstrated that 4-tertbutyl-phenol (molecular weight 150) could be separated from water with membranes of pore sizes as high as 50,000 with rejections as high as 98 percent. Similarly, Scamehorn et al. [5] elsewhere demonstrated that Zn^{+2} could be removed from aqueous solution with rejections of over 99.8 percent.

While there will be some leakage of surfactant through the membrane, it is generally at very low concentrations (<100 ppm), and it should be remembered that many nontoxic, biodegradable surfactants are commercially available, and that most municipal waste treatment facilities are already designed to deal with them, as witnessed by the detergents emitted daily into municipal sewage systems. Finally, if necessary the leaked surfactant can be removed for recovery and recycle by other surfactant-based process, such as admicellar chromatography, discussed below.

Surfactant-Enhanced Carbon Regeneration

Activated carbon adsorption is one of the most important water treatment techniques. A principal disadvantage of carbon adsorption is the need to periodically regenerate the bed. Fixed-bed adsorption is an intrinsically unsteady state process, and when the bed becomes saturated with contaminants, it must either be replaced or regenerated. Unfortunately, conventional regeneration techniques tend to be high energy (such as _in situ_ thermal regeneration with high pressure steam) and may also require dumping the bed (such as burning off the contaminant in a furnace). Such techniques also generally preclude the possibility of recovering the target compound for recycle or other use.

Surfactant-enhanced carbon regeneration (SECR) is a low-energy process with great potential for recycle of the target compound [7]. In SECR, when the bed must be regenerated, a concentrated solution of surfactant is passed through the saturated bed, The micellar "phase" can have a very high capacity for solubilizing the organic adsorbed on the carbon. As shown in Figure 3, the solute desorbs from the carbon surface

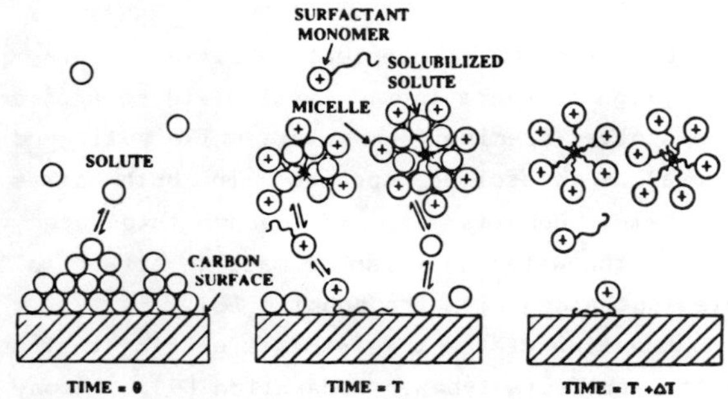

Figure 3. Carbon surface regenerated by solubilization of solute in micelles.

and solubilizes in the micelles. The surfactant is then washed out of the bed with water and the regenerated bed placed back on-stream (Figure 4).

In a demonstration of SECR a carbon bed saturated with tert-butylphenol was regenerated with a concentrated stream of a cationic surfactant. Only 0.05% of the original volume of water was required to regenerate the bed, with 70% removal of the solute. The final concentration of tert-butylphenol in the regenerant solution was 1000 times its concentration in the original process stream. These results show the great potential for SECR as a carbon regeneration techniques [7].

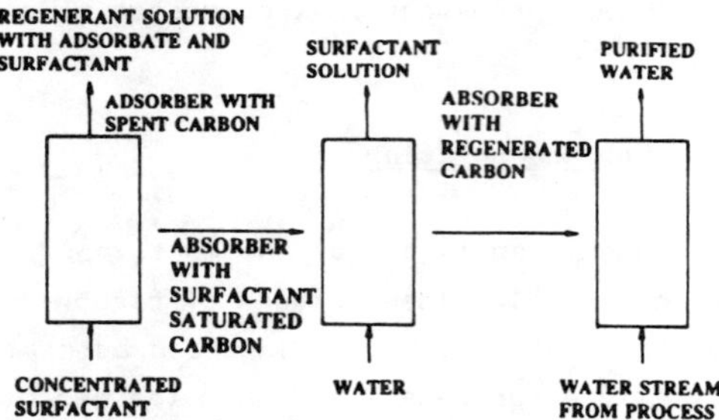

Figure 4. Application of SECR to regenerate a bed of carbon.

SEPARATIONS BASED ON ADSORBED SURFACTANT AGGREGATES

Surfactant aggregates at solid/liquid and air/liquid interfaces exhibit the micelle-like attributes of being able to exchange monovalent for multivalent ions and to have their solvent-like cores act to solubilize or adsolubilize organic compounds. This capability introduces still other potentials for the use of surfactants in separations.

Foam Separations

Foam separations can be conveniently divided into particulate separations and colligend separations. Particulate separations deal with the removal of undissolved species from water, while colligend separations deal with the removal of dissolved species. In both cases the basic approach is the same: Bubbles of air sparged into water float to the surface; anything in the water that can be made to attach to the bubbles can then be floated out along with the bubbles [8].

There is no question at the time of this writing that ore flotation is the most important surfactant-based separation [9]. It may serve as a suitable model of particulate flotation, although undissolved materials have been successfully separated by foaming, including bacterial spores, algae, and precipitates. In particulate flotation, the goal is to induce the particle to adhere to the surface of the rising bubbles. This may be accomplished by adding to the water to be treated a surfactant which will adsorb on the surface of the particles. When the particles collide with with a bubble, the hydrocarbon moieties of the surfactants are in a lower free energy state when removed from the aqueous solution; since the surfactant is also adsorbed onto the particle surface, the particle adheres to the bubble and floats out of solution (Figure 5). There has been a great deal of expertise built up in the

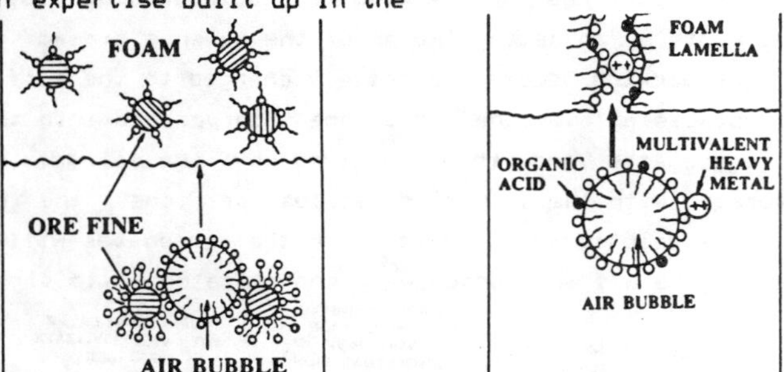

Figure 5. Ore flotation (left), a particulate flotation process, and heavy metals removal by foam flotation (right), a colligend flotation process.

technology of ore flotation, and its implementation in the area of wastewater treatment should be pursued [10]. It seems especially promising as an alternative to some filtration processes.

Colligend flotation is based on adsorbed surfactant aggregates if the aggregate in question is the monolayer at the air-water interface. In this variation of the flotation concept, surfactant adsorbs at the air/water interface of rising bubbles. Multivalent ions, such as heavy metal ions, dissolved in solution exchange for the counterions associated with the charged headgroups of the surfactants in the monolayer (Figure 5). Since the surfactant must stay with the foam, the counterions must stay with the foam or a separation of electrical charge would occur. Strontium, for example, can be essentially completely removed from solution at concentration as low as 6×10^{-5} M. Additionally, uncharged or even like-charged organic materials dissolved in the solution may become solubilized between the hydrocarbon tails of the surfactant monolayer and also be floated out of solution. Dyes, organic acids (including phenolics), humic acids and fatty acids have all been successfully removed by this approach. Additionally, design procedures for continuous flow foam separation units are available in the open literature [8-10].

Admicellar Chromatography

Aggregates of adsorbed surfactant on a packing with an amphoteric surface make an attractive alternative to both adsorption on activated carbon and to ion exchange with conventional ion exchange resins. As previously mentioned, the adsorbed aggregates (called admicelles or hemimicelles) retain the micelle's ability to incorporate organic compounds into a solvent-like core and to exchange monovalent counterions for multi-valent counterions. These aggregates can be induced to form on the surface of inorganic packings such as alumina, acid treated clay, silica, or diatomaceous earth by adjusting the pH of the aqueous stream so that the surface of the packing becomes oppositely charged to the surfactant. Organics or ions passing over the bed become incorporated into the adsorbed surfactant aggregates and thus retained by the column. When the bed becomes saturated with the target molecules or ions, the bed can be flushed with a solution at a pH at which the aggregates will desorb from the packing (Figure 6). This produces a concentrated waste stream of much

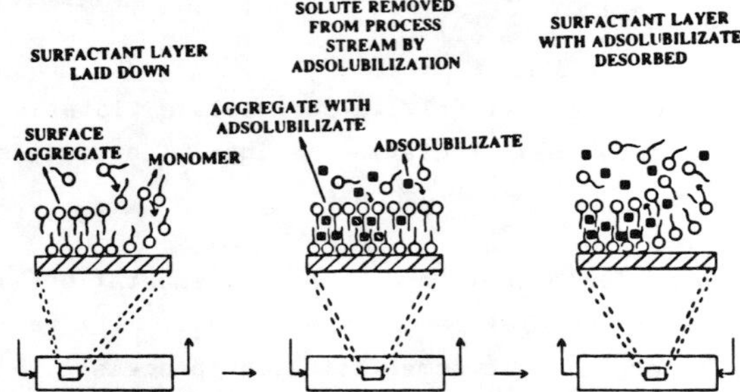

Figure 6.   The basic operating scheme of an admicellar chromatograph.

lower volume than that of the original stream while simultaneously regenerating the bed. The technical feasibility of this process has been demonstrated with the recovery of tert-butylphenol using sodium dodecyl-sulfate adsorbed on chromatographic alumina. Concentration increases of over 200% per pass have been obtained, including supersaturated solutions of the organic [11-12].

SEPARATIONS BASED ON EMULSIONS

Not only may the core of the surfactant aggregates act as a "solvent" phase for target compounds, but the surfactant can also be used to disperse an actual solvent phase within an aqueous phase. Adaptations of this phenomenon to separations can produce interesting advantages.

Surfactant Liquid Membranes

In a liquid membrane separation, an emulsion is formed between two immiscible phases. The emulsion is then dispersed into a third phase which is immiscible with the external phase of the emulsion. Components to be removed from the third phase then diffuse through the external phase of the emulsion and into the phase encapsulated in the droplets (Figure 7).

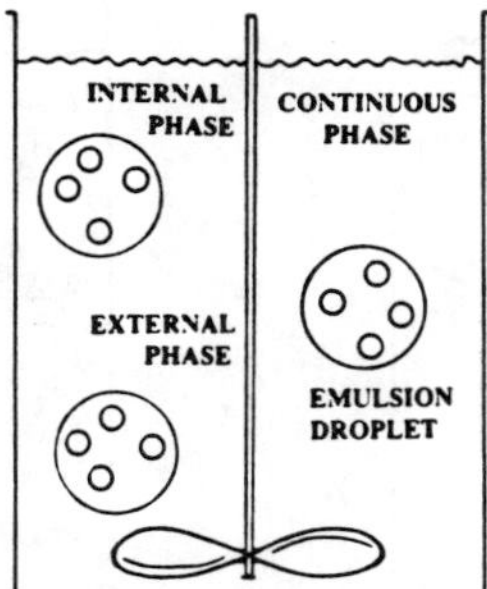

Figure 7. Dispersed emulsion droplets in a liquid membrane separation.

This separation technique is especially useful for enhancing the selective separation of one component of the continuous phase from another component which might also be soluble in the encapsulated phase; the external phase of the emulsion acts like a membrane, retarding the transport of the compound least soluble in the external phase. It is also especially useful when the encapsulated phase is a reagent phase, concentrating the products of the reaction in the emulsion droplets, as, for example, in the extraction of cupric ions from water [13-14].

Absorption into Emulsions

It has recently been suggested that fugitive emissions from spray painting processes can be controlled by absorption in an emulsion [15]. Vapor from the spray chamber is sparged into an oil-in-water emulsion stabilized by surfactant. Water soluble components of the vapor dissolve in the continuous water phase; oil soluble components dissolve in the surfactant stabilized oil droplets. When the emulsion is broken after one or the other phase becomes saturated, a concentrated sludge is formed containing the components of the vapor. Residual components of the vapor in the oil phase are the lower molecular weight components and may be removed, for example, by vacuum distillation. Then the emulsion can simply be reformed and recycled into the absorption unit. (Figure 8)

CONCLUSIONS

The combination of the ability to treat streams containing extremely low toxin cocentration efficiently and the low toxicity/environmental impact of the surfactant itself give surfactant-base separations unique advantages over many conventional techniques. Research in this area should expand rapidly over the next decade.

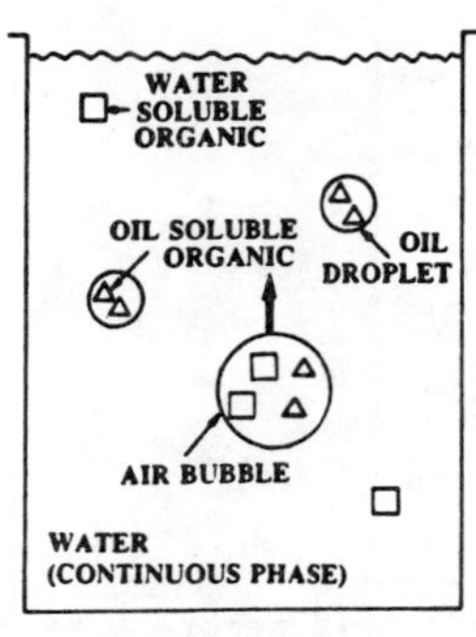

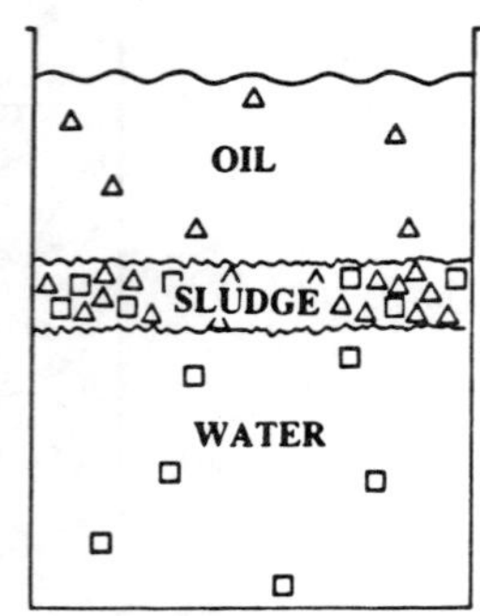

Figure 8. Left, capture of fugitive emissions from spraying process in an emulsion. Right, concentration in sludge after emulsion is broken.

ACKNOWLEDGEMENTS

Aspects of the work of the authors which was reported here have been supported by the United States National Science Foundation (CBT-8318864) and the Office of Basic Energy Sciences of the Department of Energy (DE-A505-8TER13175), and by the Oklahoma Mining and Mineral Resources Research Institute.

REFERENCES

1. Rosen, M.J., <u>Surfactants and Interfacial Phenomena</u>, Wiley and Sons, New York, 1978.

2. Rosen, M.J., (ed.), <u>The Role of Surfactants in New and Emerging Technology</u> Marcel Dekker, New York, In Press.

3. Scamehorn J.F. and Harwell, J.H., An overview of surfactant-based separation processes, in <u>The Role of Surfactants in Emerging Technologies</u>, (Rosen, M.J. ed.), Marcel Dekker, in press.

4. Dunn Jr.,R.O., Scamehorn,J.F., and Christian, S.D., Use of micellar-enhanced ultrafiltration to remove dissolved organics from aqueous streams, <u>Sep. Sci. Technol.</u>, <u>20</u>: 257 (1985).

5. Dunn Jr., R.O., Scamehorn, J.F., and Christian, S.D.,Concentration polarization effects in the use of micel'ar-enhanced ultrfiltration to remove dissolved organic pollutants from wastewater <u>Sep. Sci. Technol.</u>, In Press.

6. Scamehorn, J.F., Ellington, R.T., Christian, S.D., Penney, B.W., Dunn Jr., R.O. and Bhat, S.N., Removal of multivalent metal cations from water using micellar-enhanced ultrafiltration. _AICHE Symp. Ser._, 1986, _250_, 48-58.

7. Blakeburn, D.L. and Scamehorn, J.F., Surfactant-enhanced carbon regeneration, in _Surfactant-Based Separation Techniques_ (Scamehorn, J.F. and Harwell, J.H. ed.), Marcel Dekker, New York, In Press.

8. Okamoto, Y. and Chou, E.J., Foam separation processes, in _Handbook of Separation Techniques for Chemical Engineers_, (P.A. Schweitzed, ed.), McGraw-Hill, New York, 1979, Chap. 2.5.

9. Fuerstenau, M.C. (ed.), _Flotation. A.M. Gaudin Memorial Volume_. Am. Inst. Min. Met. Petrol. Eng., New York, N.Y., 1976.

10. Lemlich R. (ed.), _Adsorptive Bubble Separation Techniques_, Academic Press, New York, 1972.

11. Fitzgerald, T.P. and Harwell, J.H., Admicellar chromatography: A new low energy separation process. _AICHE Symp. Ser._, 1986, _250_, 142-153.

12. Grout, C.R. and Harwell, J.H., Fixed-bed adsolubilization: A new wastewater treatment technology. _Ind. Engr. Chem. Res._, in preparation.

13. Hatton, T.A.. Lightfoot, E.N., Cahn, R.P. and Li, N.N. _Ind. Eng. Chem. Funda._,1983, _22_, 27.

14. Li, N.N., Wasan, D.T. and Gu, Z., Surfactant liquid membranes, in _Surfactants and Chemical Engineering_ (Wasan, D.T. , Shah, D.O. and Ginn, M.E., ed.), Marcel Dekker, New York, In Press.

15. Wetlaufer, D.B., Surfactants in novel separation techniques, in _The Role of Surfactants in New and Emerging Technology_ (Rosen, M.J., ed.), Marcel Dekker, New York, In Press.

ADSORPTION OF ARSENIC ON HYDROUS OXIDES

Jill Ru-hsing Yuan
Department of Environmental Engineering
National Chung-Hsing University
Taichung, Taiwan
R.O.C.

and

M. M. Ghosh and R. S. Teoh
Department of Civil Engineering
Pennsylvania State University
University Park, PA 16802
USA

ABSTRACT

Adsorption studies of arsenic were made using ALCOA F-1 alumina. The adsorption is greatly affected by the pH of the solution. High adsorption capacities of arsenate are achieved at $pH < pH_{zpc}$, with maximum adsorption occuring around pH 5. Maximum adsorption of arsenite occurs around pH 8. The adsorption behavior of MMA is very similar to that of arsenate. Adsorption of arsenic are enhanced by the presence of cations. Experiments with fixed alumina beds indicate that arsenic can be effectively removed for long periods of time at regular EBCT's and MLR's. Successful regeneration of spent alumina is feasible using 0.5% NaOH followed by dilute acid rinse.

INTRODUCTION

Arsenic (As) is ubiquitous in the environment. For most potable waters, the contamination of As are mostly due to natural sources. However, ground waters have been reportedly contaminated with arsenic as a result of mining [3] and waste disposal [4]. Smelting, petroleum-refining, herbicide manufacturing and other industrial activities may also introduce significant quantities of this toxic element to the environment locally.

Five forms of As, namely monomethyl arsonate (MMA), dimethyl arsinic acid (DMAA) or cacodylic acid (CA), trimethyl arsine oxide

(TMAO), arsenite and arsenate were found in the contaminated soil. All forms of organoarsenicals are believed to form arsenate in soil, possibly by microbial demethylation. The toxicity scale of arsenic is as follows: arsine >arsenite >arsenate >alkyl arsenic.

In aquatic system, arsenic is stable in four oxidation states (+V, +III, 0, -III) under different redox conditions. In oxygenated waters, arsenate species are stable. At E_h values characteristic of mildly reducing conditions, arsenite species become stable [5]. Hydroxides of iron, chromium and aluminum strongly adsorb or form insoluble precipitates with arsenite and arsenate. This may play an important role in controlling arsenic concentration in unpolluted waters. In the presence of sulfides, realgar (AsS) and orpiment (As_2S_3) may be present as as stable solids at pH less than about 5.5 and E_h of about 0 volt. $HAsS_2$ and AsS_2^- are the predominate aqueous species. Under extremely reducing conditions, arsine may occur in traces being only slightly soluble. The most commonly used forms of methylated arsenic in herbicides are $CH_3AsO(OH)_2$ (MMA) and $(CH_3)_2AsO_2H$ (CA). They are both highly soluble in water.

Many methods have been used to remove arsenic from aqueous solutions. Precipitation with lime and ferric salts and adsorption onto a coagulant floc is the most frequent method used for arsenic removal [4-6]. Sulfide precipitation [6], adsorption on activated carbon [7,8], and adsorption on activated alumina [9-11] have also been used for removing arsenic. Ion exchange has been used with limited success [7,12].

The present paper deals with the feasibility of removing arsenic by adsorption on the hydrous alumina surface. The specific objectives are to establish removal capacities and kinetics of adsorption under different conditions of pH and competition from other contaminant ions. The effect of mass loading rate (MLR) and empty bed contact time (EBCT) were studied in fixed bed adsorption systems. Further, spent alumina beds were regenerated to determine reuseable capacity and material loss during regeneration in an effort to evaluate process economics.

MATERIALS AND METHODS

Activated Alumina

All adsorption experiments were conducted using a commercial grade of activated alumina, ALCOA F-1 alumina. The alumina contained 92% $r-Al_2O_3$ and some minor impurities. The BET surface area of the 28x48 mesh alumina was 218 m^2/g as measured using a Quantasorb surface area analyzer. The test alumina for any adsorption experiment was prepared by washing it with 0.01 N NaOH solution and rinsing with deionized water (DI) till the pH and conductivity of the supernatant became constant. It was then dried overnight at $105^{\circ}C$ before use.

Alkalimetric Titration

The test alumina suspensions were prepared by suspending prewashed alumina (2g/l), aged for two weeks in contact with DI water, in stock NaCl solution so that final ionic strength of 10^{-3} M was obtained. The titrants used were 0.1 N HCl and 0.1 N NaOH. Titrations were carried out at ionic strengths of 10^{-3}, 3×10^{-3}, 10^{-2}, and 10^{-1} M. Blank titrations of 2 g/l alumina suspensions preaged for 24 hours at identical ionic strengths were carried out on filtered supernatants. The titrant volumes for blanks were subtracted from those for alumina suspensions to obtain net titration for the alumina surface. The surface charge, σ_o, was computed as follows:

$$\sigma_o = F[(M \; Vt)/(V \; Ca \; S)] \tag{1}$$

where F = Faraday constant (coul/mol); M = concentration of acid or base (mol/l), Vt = net amount of acid or base used (ml); V = volume of sample (ml); S = specific surface area of alumina (m^2/g); Ca = concentration of alumina (g/ml).

Equilibrium Adsorption Experiments

All adsorption isotherms were obtained using batch adsorption experiments. Predetermined amounts of washed alumina (1 to 5 g/l) were place in 150 ml plastic bottles. The ionic strengths in all bottles were maintained at 10^{-1} M and the final sample volume was adjusted to 125 ml. The concentration of sorbate was varied from 1 to 20 mM depending on the arsenic species being adsorbed. The pH was adjusted using 1 M NaOH or 1 M HCl. The bottles were continuously shaken at toom temperature $(20 \pm 1^{\circ}C)$ for attainment of equilibrium. Limited experiments at other temperatures were also conducted. To determine kinetics of removal, samples were removed at specified time intervals, pH measured, filtered through 0.45 um Millipore filter, and arsenic concentration in the filtrate determined using atomic absorption spectrophotometry (AAS) following hydride reduction of arsenic.

Column Studies

The studies for fixed bed adsorption of arsenic on alumina were conducted by passing arsenic solution through 0.9 cm diameter, 76 cm long glass minicolumns containing alumina supported by Whatman No.1 filter paper. Different amounts of alumina (1 to 5 g) were used to vary the empty bed contact time (EBCT) in different experiments. The rate through the column was held constant by a reciprocating piston pump or a masterflex pump. Generally, an experiment was terminatd when the adsorption capacity of alumina was exhaused. In certain cases, experiments were terminated if no breakthrough occurred for a prolong period (more than 30 days).

Regeneration of spent alumina was performed by applying NaOH solutions of varying strengths, 0.5%, 1% and 3% to determine optimum regenerant concentration based on alumina loss and sorption capacity restored. Hydraulic rates ranged from 68-79 m^3/m^2-da. Regenerants were applied in an upflow mode.

RESULTS AND DISCUSSION

Surface Acidity of Alumina

Figure 1 shows the surface charge density distribution of alumina as a function of pH. The pH_{zpc} of F-1 alumina was found to be 7.92.

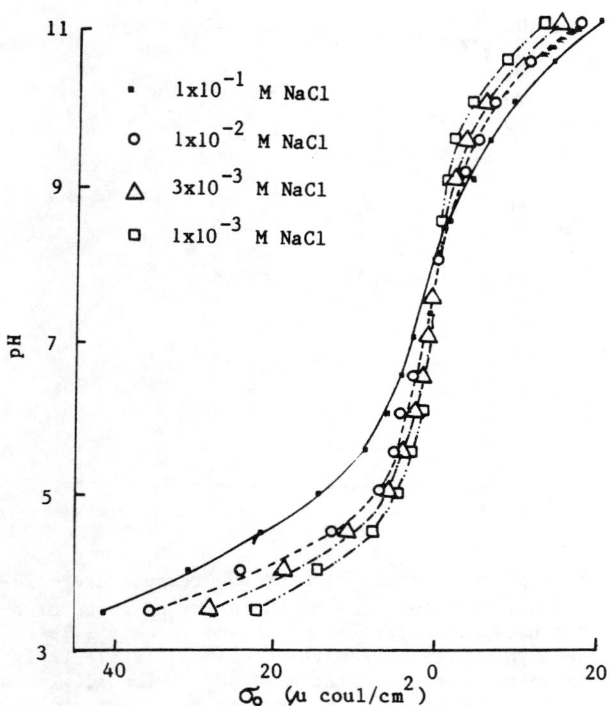

Figure 1. Surface charge distribution on alumina.

This value may vary depending on the origin, aging, doping by
impurities, and the degree of hydration. Values in the range of 7.4 to
9.0 for r-Al$_2$O$_3$ have been reported. Although alumina surface can be
treated as a diprotic acid, the acidity of each surface group is
affected by the neighboring ones. The two intrinsic acidity constants
and the total number of surface sites were calculated from Fig. 1 with
the method developed by Huang [13]. The constants were calculated to be
6.44 and 9.4 at an ionic strength of 10^{-1} M. The total number of
exchangeable sites was found to be 1.67 mmol/g or 4.6x10^{14} sites/cm^2.
This number is about six times higher than that reported by Huang [14],
but it is close to the value reported by Davis [15]. This method works
well at high ionic strengths as used in this study. A pH$_{zpc}$ of 7.92,
which is the same as obtained from Fig. 1, was determined analytically
by taking an average of the two acidity constants.

<u>Removal Kinetics</u>

Typical kinetic data for the adsorption of arsenate on alumina are
shown in Fig. 2. The removal was rapid in the first 24 hours and then
slowed considerably as the reaction approached equilibrium. The
significant effect of pH on arsenate adsorption is apparent. At a pH of
6.5 or less, 95% of the maximum adsorption was obtained in less than 24
hours. At a surface loading of 67 μmol/g, arsenate concentration

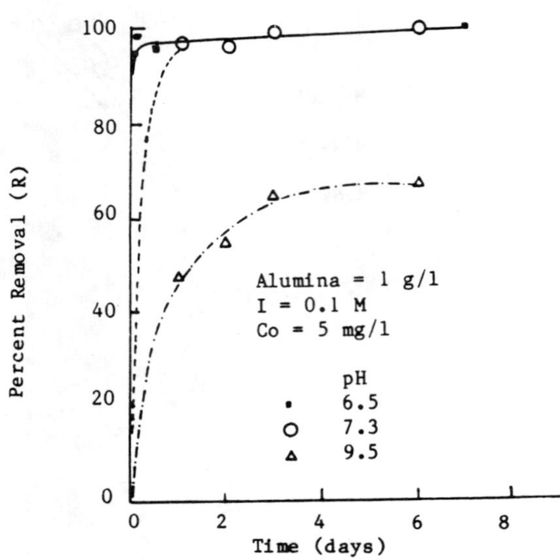

Figure 2. Effect of pH on arsenate adsorption (temperature = 20°C)

decreased from 5 to 0.1 mg/l in a few hours in the pH range of 6 to 7. At a pH of 9.5, arsenate concentration decreased to only 1.75 mg/l even after 6 days. The high degree of removal at low pH levels imply that the adsorption may be non-specific. At pH $<$ pH$_{zpc}$, the alumina surface is positively charged and coulombic interaction with arsenate is feasible. At pH $>$ pH$_{zpc}$, the surface carries a net negative charge. Adsorption at these pH's must occur by specific interactions inspite of any coulombic repulsion. The effects of temperature on the adsorption of arsenate are shown in Fig. 3. The results for Fig. 3 were obtained

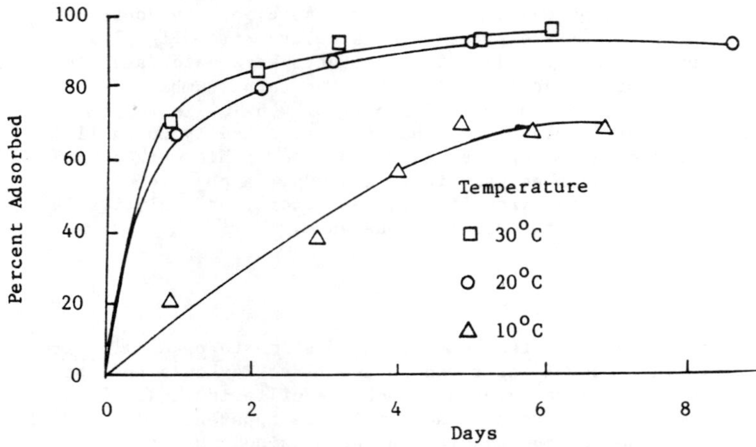

Figure 3. Effect of temperature on arsenate adsorption (alumina = 1 g/l; As(V) = 5 mg/l; I = 10^{-1} M; pH = 8.5).

at pH 8.6 and an ionic strength of 10^{-1} M. In all experiments the concentrations of alumina and arsenate were held constant. At 20°C or

higher, the rate of adsorption remained relatively unchanged. Based on
the results presented in Figs. 2 and 3, equilibrium adsorption was
assumed to occur in 6 days regardless of temperature and pH.
   Arsenite, on the other hand, exhibited different adsorption
behavior (Fig. 4) which will be discussed in the next section.. The

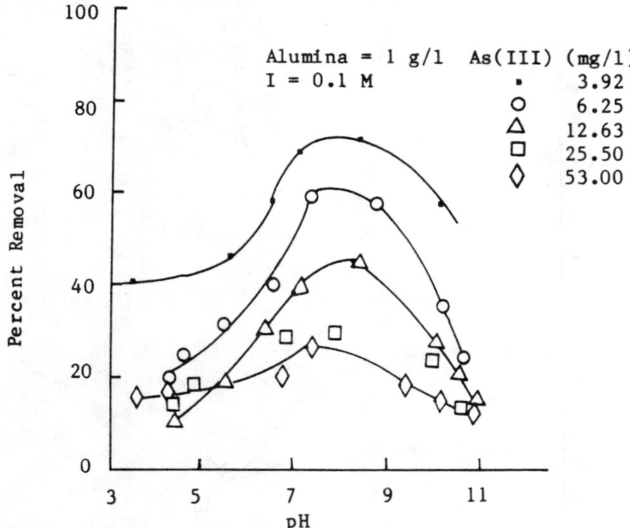

Figure 4. Effect of pH and adsorbate concentration on adsorption of
arsenite.

rate of approach to equilibrium was much slower than that for arsenate,
although a two-step removal similar to arsenate, was observed. Similar
results for the removal of arsenite on amorphous $Fe(OH)_3$ have been
reported by Pierce and Moore [16].

Equilibrium Adsorption

   Typical data for the adsorption of arsenate are shown in Fig. 5.
The initial concentration of As(V) was varied from 0.013 to 0.7 mM in
these experiments while the amount of adsorbent was held constant.
Maximum adsorption was obtained between pH 4 and 5. Except at high
concentration, adsorption appears to be independent of pH at pH < pH_{zpc}
and decreased with increasing pH. At increased initial concentrations,
the optimum pH of adsorption decreased due to allowering of pH_{zpc}. In
the pH range of 3 to 6, $H_2AsO_4^-$ is the predominant species, and,
apparently, the major species being adsorbed. In this pH range alumina
surface acquires a net positive charge and adsorption is facilitated by
coulombic interaction. Hingston [17] suggested that the optimum
adsorption pH depends on the pK of the weak acid. However, for arsenate
adsorption the optimum pH was almost two units lower than pK_2 (6.98) for
H_3AsO_4. The adsorption of arsenite increased to a maximum at pH 8,
then decreased with increasing pH (Fig. 4). Gupta and Chen [9] observed

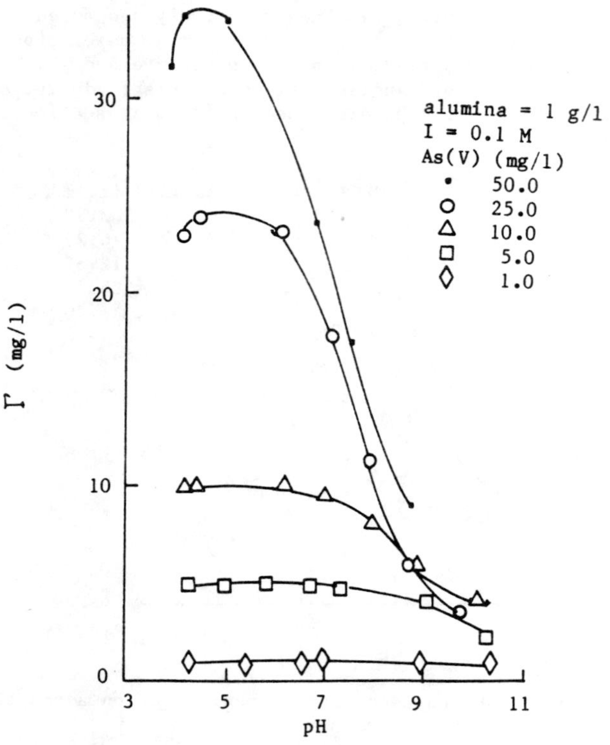

Figure 5. Effect of pH and adsorbate concentration on the adsorption of arsenate.

only slight variation in the adsorption of As(III) in the pH range of 4 to 9 with a sudden drop beyond pH 9. Frost et al. [18] observed an increase in As(III) adsorption on kaolinite in the pH range of 3 to 9 followed by a decrease. The adsorption of MMA and DMAA exhibited pH dependence very similar to that of arsenate. However, the adsorption of DMAA was much less than that of MMA or arsenate.

Competitive Interactions

The effects of common ions such as sulfate, silicate, bicarbonate and calcium ions, on the adsorptions of arsenic were studies at various pH's and concentrations. For most experiments, an initial adsorbate concentration of 0.1 mg/1 was used and the concentrations of competing ions varied. The result suggests that the adsorption of arsenate and arsenite may be affected adversely slightly by the presence of these anions and enhanced by cations.

Adsorption in Fixed Beds of Alumina

Fixed bed adsorbers were used to evaluate the effects of operational parameters, such as empty bed contact time (EBCT) and mass loading rate (MLR) on the adsorption of arsenic on activated alumina. The effect of MLR is shown in Fig. 6 for an EBCT of 5 minutes and a pH

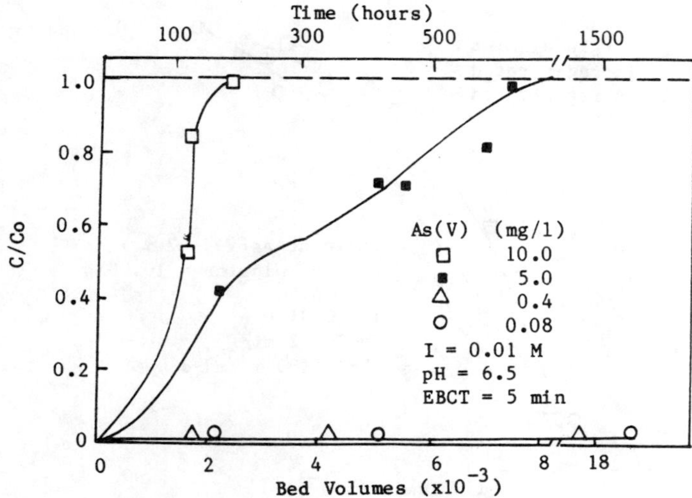

Figure 6. Effect of mass loading rate on adsorption arsenate in fixed alumina beds.

of 6.5. As the effluent history profiles indicate, MLR is a significant factor in determining the life of an adsorption bed. At an initial concentration of 0.08 mg/1 (1.1×10^{-3} mM), typical of many natural waters, 30,000 bed volumes (BV) could be treated before breakthrough of arsenic occurred. Columns were operated at EBCT's of 3, 5, and 8 minutes, all receiving an As(V) concentration of 5.3×10^{-3} mM at pH 6.5. All columns could be operated for 1500 hours or longer before breakthrough occurred. An EBCT of 3 minutes or longer was needed for satisfactory operation.

The economics of adsorption processes depend largely on the ability to rejuvenate spent beds. Reportedly, NaOH works well for desorbing arsenic from alumina [10]. Limited studies were made to determine the optimum concentration of NaOH that would produce good desorption of arsenic without large loss of alumina. Result indicates that spent alumina beds could be successfully regenerated using dilute solutions of NaOH followed by rinsing with dilute HCl and, finally, with deionized water. A total of 50–60 BV of 0.5% NaOH solution followed by 10–15 BV of 0.1 percent HCl resulted in almost complete desorption of arsenate. At these dosages, the loss of alumina was about 3 percent by weight per regeneration. The loss in adsorption capacity was negligible after several regenerations (Fig. 7).

SUMMARY AND CONCLUSION

Adsorption of arsenic on activated alumina is greatly affected by pH. High adsorption densities of arsinate and organoarsenicals can be obtained at pH less than the pH of the zero point of charge. Maximum adsorption of arsenate occurs at pH 5 or less. The specific adsorption of arsenate, MMA and DMAA can be described well by the Langmuir isotherm, while arsenite adsorption could not. Arsenite adsorption was unsatisfactory. Of the two organoarsenicals, MMA seems to be better

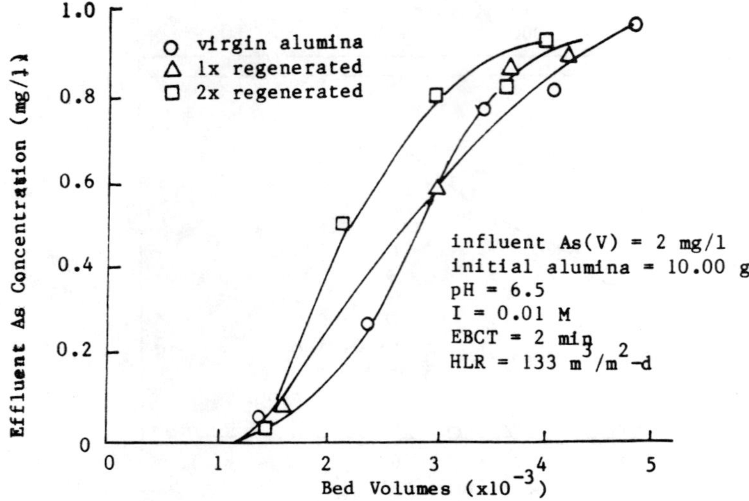

Figure 7. Effect of multiple regenrations on capacity of alumina.

adsorbed on alumina than DMAA.

Experiments with fixed beds of alumina indicated that arsenic can be effectively removed for long periods of time at an EBCT of 3 minutes or longer and at low MLR's in the pH range of 4 to 7. Results also indicate that removal is best accomplished in the arsenate form. Therefore, oxidation of arsenite and other forms of arsenic to the arsenate form may be desirable. Successfull regeneration of spent alumina beds to desorb arsenate is feasible with NaOH at concentrations as low as 0.5 percent; but a much stronger caustic solution (3%) is needed for arsenite desorption. No appreciable loss in adsorption capacity occurs due to regeneration. The loss of alumina can be held at 3 percent per regeneration.

ACKNOWLEDGEMENTS

This study was funded by a research grant from the Office of Exploratory Research and Development, U.S. EPA, Grant No. R 809425010. Mention of trade names does not constitute endorsement either by the author or EPA. The authors wish to thank Ron J. Schlicher who was involved in this study.

REFERENCES

1. Bottomley, D. J., Origins of some arseniferous groundwaters in Nova Scotia and New Brunswick, Canada, J. Hydrol., 1984, 69, 223-257.

2. Holm, T. R., Anderson, M. A., Iverson, E. G., and Stanforth, R. R., Heterogeneous interactions of arsenic in aquatic systems, A.C.S. Symposium Series 93, 1979, pp. 711-736.

3. Ferguson, J. F. and Gavis, J., A review of the arsenic cycle in

natural waters, Water Research, 1972, 6, 1259.

4. Shen, Y. S., Study of arsenic removal from drinking water, Jour. Amer. Water Wks, Asscn., 1973, 65, 543.

5. Gulledge, H. and O'Connor, J. T., Removal of arsenic(V) from water by adsorption on aluminum and ferric hydroxides, Jour. Amer. Water Wks. Asscn., 1973, 65, 548.

6. Logsdon, G. S., et al, Removal of heavy metals by conventional treatment, Proc. 16th Water Quality Conference, University of Illinois, Urbana, IL, 1974, Feb. 13-14, pp 111-133.

7. Lee, J. Y. and Rosehart, R. G., Arsenic removal by sorption process from wastewaters, Canada Minining Metallurgical Bull., 1972, 65, 11-33.

8. Huang, C. P. and Fu, P. L. K., Treatment of arsenic(V)-containing water by activated carbon process, Jour. Water Poll. Control Fed., 1984, 56, 233.

9. Gupta, S. K. and Chen, K. Y., Arsenic removal by adsorption, Jour. Water Poll. Control Fed., 1978, 50, 493.

10. Rubel, F. and Williams, F. S., Pilot study of fluoride and arsenic removal from potable water, EPA-600/S2-83-107, 1984, Feb..

11. Rosenblum, E. and Clifford, D., The equilibrium arsenic capacity of activated alumina, EPA-600/S3-83-107, 1984, Feb..

12. Schlicher, R. J. and Ghosh, M. M., Removal of arsenic from water by physical-chemical treatment, AIChE Symp. Ser., 1985, 243: 81, 152-164.

13. Huang, C. P., The surface acidity of hydrous solids, in Adsorption of Inorganics at Solid-Liquid Interfaces, 1981, Ann Arbor Science Publishers, Inc., Ann Arbor, MI.

14. Huang, C. P. and Stumm, W., Specific adsorption of cations on hydrous r-Al_2O_3, Jour. Colloid Interface Sci., 1973, 43, 409.

15. Davis, J. A. and Leckie, J. O., Surface ionization and complexation at the oxide/water interface: III. Adsorption of Anions, Jour. Colloid Interface Sci., 1980, 74, 32.

16. Pierce, M. L. and Moore, C. B., Adsorption of arsenite on amorphous iron hydroxides from dilute aqueous solutions, Env. Sci. Technol., 1980, 14, 214.

17. Hingston, F. J., A review of anion adsorption, Adsorption of Inorganics at Solid-Liquid Interfaces, Ann Arbor Science Publishers, Inc., 1981, Ann Arbor, MI.

18. Cabrera, F., Madrid, L. and DeArambarri, P., Adsorption of phosphate by various oxides: Theoretical treatment of the adsorption envelope, Advances in Chemistry Series No. 189, Amer. Chem. Soc., 1980.

MEASUREMENT OF THE TOXICITY OF KCN AND SOME ORGANIC COMPOUNDS FOR ACTIVATED SLUDGE USING THE WAZU-RESPIRATION METER

Henri Spanjers and Abraham Klapwijk
Department of Water Pollution Control
Wageningen Agricultural University
De Dreijen 12
6703 BC Wageningen
The Netherlands

ABSTRACT

A device is described for continuously monitoring the respiration rate of activated sludge. Activated sludge is pumped from the system being monitored through a respiration vessel. In the inflow and outflow of the vessel dissolved oxygen concentration is measured. With these data an on-line computer calculates the respiration rate of the activated sludge.

The investigation on toxicity of potassium cyanide and some organic compounds on activated sludge is carried out in a batch reactor as well as a treatment plant on laboratory scale. The effect of these compounds on the respiration rate is measured. For potassium cyanide a dose response curve is derived from the batch as well as the continuous experiments for endogenous oxygen uptake and for nitrification.

The technique has been proven to be valuable for quickly obtaining important information on toxicity and biological treatability of organic and anorganic compounds.

INTRODUCTION

Activated sludge plants are capable to cope with considerable variations in flow rate and concentration of contaminants in the influent. Some compounds however may reach concentrations by which the biological purification process becomes severely inhibited. This may result in poor effluent quality and loss of active biomass. Therefore it is of great importance to have the disposal of a device which enables one to survey the influent on toxic chemicals and to perform laboratory tests on toxic or suspicious chemicals.

Various methods have been described in the literature to determine the toxicity of chemicals to activated sludge. For example methods based on the measurement of chemical parameters, specific glucose uptake or respiration rate.
In the first method the effect of chemicals on the efficiency of the process is considered by means of measurements of chemical parameters, like COD and TOC, in the influent and effluent [1]. Unfortunately there is a certain time lag before the information is available as a result of the time needed for chemical analysis. Then the process may already be disturbed for some time. Moreover, the method is not suitable for automatic surveillance.
The second method is based on the determination of the inhibition of glucose uptake [2]. However, this method is unsuitable for the automation and on-line toxicity monitoring. Moreover the method is aspecific so that results are hard to interpret.
The third method is the determination of the inhibition of respiratory activity, expressed as respiration rate, of activated sludge. The respiration rate is an important indication of the state of the activated sludge. Therefore this parameter is a widely used activated sludge parameter. An OECD test for toxicity to activated sludge is based on the determination of the effect on the respiration rate [3]. However there is a lack of universal, easy to use and reliable respiration meters for continuous application.

In this paper a new respiration meter for continuously measuring the respiration rate of activated sludge is presented. This device is suitable for use in batch as well in continuous operation. The device differs from previous flow-through designs in that the activated sludge flows continuously through the system, whereas in other systems, the sludge is kept in a chamber.
Besides toxicity studies the device is also applicable for monitoring the sewage quality and control of an activated sludge plant [4, 5].

MATERIALS AND METHODS

Description of the apparatus

The Wazu-respiration meter (figure 1) is composed of a closed respiration chamber through which activated sludge is pumped from the system being monitored [4, 5]. Dissolved oxygen is measured periodically in the inflow and in the outflow of the vessel with one oxygen sensor. This is realised in alternately changing the flow direction using a reversible pump.
Assuming the vessel can be concerned as a continuously-stirred system, the respiration rate is calculated from the mass balance of oxygen:

$$V.(dc/dt) = Q.c_i(t) - Q.c(t) \qquad (1)$$

where: c = oxygen concentration in the outflow $(g.m^{-3})$
     c_i = oxygen concentration in the inflow $(g.m^{-3})$
     V = volume of the vessel (m^3)
     Q = activated sludge flow rate through the vessel $(m^3.h^{-1})$
     r = respiration rate $(g.m^{-3}.h^{-1})$

For digital implementation of the respiration measurement equation (1) is integrated over one sampling interval T, assuming that c and c_i are constant during this interval.

$$c(t) = A.c(t-1) + (1-A).c_i(t-1) - (1-A).r(t-1)/a \qquad (2)$$

where: $a = Q/V$
$A = e^{-aT}$

The respiration rate is numerically solved in the calculation programme.

The Wazu-respiration meter is connected to a modular microprocessor system (Siemens SMP) which collects the dissolved oxygen concentration from a dissolved oxygen meter (WTW OXY 91), calculates the respiration rate and controls the activated sludge flow. In this work we used the respiration meter in combination with a batch reactor as well as a treatment plant both on laboratory scale (Figure 1).

a.

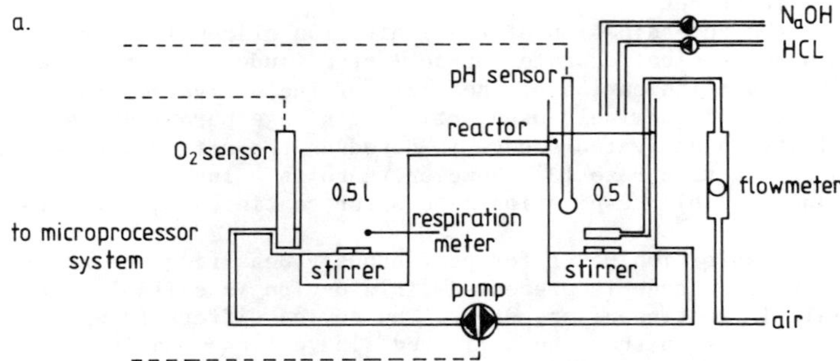

b.

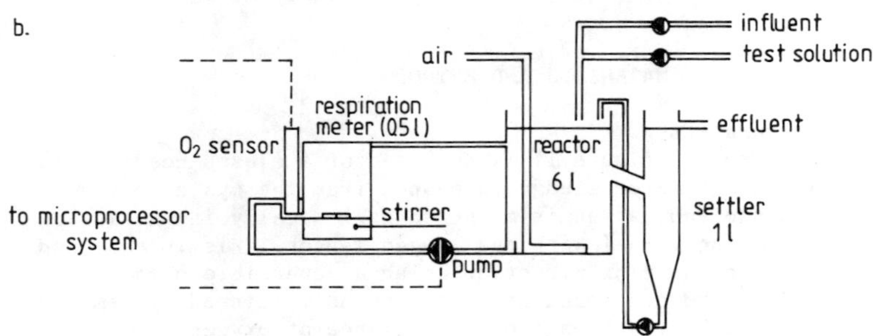

Figure 1: Schematic diagrams of the experimental unit with the WAZU-respiration meter for:
a. batch experiments and
b. continuous experiments.

The activated sludge samples were taken from a sequencing batch reactor which was discontinuously loaded with presettled domestic sewage (from Bennekom, the Netherlands) maintaining an average sludge load of 0.15 kg COD per kg MLSS per day. In this reactor the pH was automatically kept at 7.5 ± 0.15 by addition of HCL or NaOH. The biomass concentration was adjusted to 4 kg MLSS.m^{-3}.

Batch experiments

The activated sludge samples were taken from the sequencing batch reactor.
This sludge is concerned to be endogenous in the sense that the concentration of soluble organic substrate was minimal. Experiments were done
with endogenous sludge and with nitrifying sludge.
After the endogenous respiration rate was constant during at least 15 minutes a certain amount of the chemical to be tested was added. The course of
the respiration rate was recorded for at least 60 minutes after addition of
the chemical.
In order to assess the effect of a chemical on nitrifying activated sludge
an excess of ammoniumchloride (80 $g.m^{-3}$) was first added to the sludge.
This excess was sufficient for a maximum nitrification rate during one hour.
The chemical to be tested was added after a constant respiration rate was
reached.

Continuous experiments

The reactor was fed with presettled domestic sewage (from Bennekom, the
Netherlands) maintaining a sludge load of 0.15 kg BOD.kg $MLSS^{-1}.day^{-1}$.
To dispose of constant influent quality the sewage was tekan from a stock
at 4°C. After a constant respiration rate was reached, a small flow of test
solution was mixed with the sewage. The respiration rate was measured
during 12 to 24 hours.

RESULTS AND DISCUSSION

Potassium cyanide (batch experiments)

Figure 2 shows the course of the specific endogenous respiration rate in
the batchreactor after addition of various doses of KCN. Within 60 minutes
the activity drops to a constant level and the decrease is proportional to
the dose.

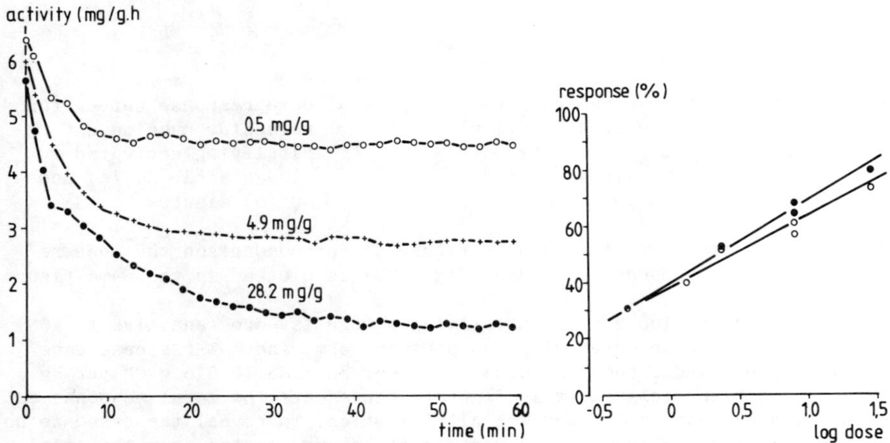

Figure 2. Specific respiration rate over
time for three different doses
of potassium cyanide.

Figure 3. Dose response curves
for potassium and
endogenous activated
sludge after 30 (o)
and 60 (o) minutes

From these results the percentage inhibition (response) was calculated from equation (3) below:

$$\text{response} = (Ro - R).100\%/Ro \qquad (3)$$

where: Ro = (specific) respiration rate before addition
       R = (specific) respiration rate a certain time after addition

In figure 3 the response is plotted against the dose. Concerning the range there is a fairly linear relationship between response and log dose. From figure 3 the 50%-response dose (that is the dose that leads to 50% reduction in specific oxygen uptake rate) can be derived as: 2.5 g KCN per kg MLSS (1.0 g CN per kg MLSS).

Figure 4 shows the response of nitrifying activated sludge to different doses of KCN. In this case the measured oxygen consumption is the sum of the endogenous respiration rate and the nitrification rate. From the figure it can be seen that for low doses the respiration rate returns from a lower level within 60 minutes.

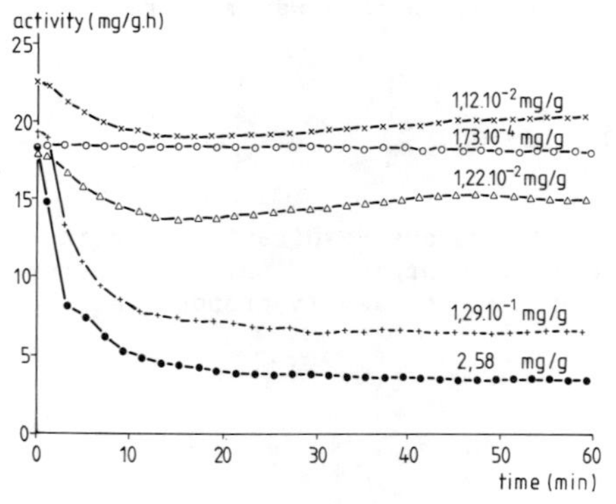

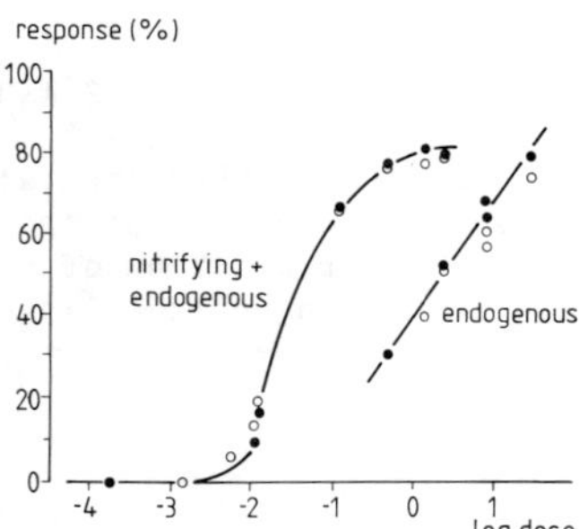

Figure 4. Specific respiration rate against time of nitrifying sludge for different doses of potassium cyanide.

Figure 5. Dose response curves for potassium cyanide and nitrifying activated sludge after 30 (●) and 60 (o) minutes.

The dose response curve is given in figure 5. For comparison the dose response curve for endogenous sludge (figure 3) is plotted in the same figure.

As can be seen in figure 5 the nitrifying sludge is more sensitive to KCN and the dose response curve displays another form. The 50%-response dose for nitrifying sludge comes to 0.04 g KCN per kg MLSS (0.016 g CN per kg MLSS). In fact the left curve in figure 5 represents the total response on the endogenous respiration and the nitrification. To reveal the response on the nitrification the endogenous curve must be substracted from the total response curve. In figure 5 this is only possible for a small range of doses. It appears that the nitrification is being inhibited for 95-96% in the range 0.5 - 2.6 g KCN per kg MLSS.

Others [6] found a 50% inhibition of acetate oxidation for 0.038 g CN per kg MLSS for not adapted activated sludge and 0.13 g CN per kg MLSS for adapted sludge. A 80% response dose of 0.1 g CN per kg MLSS was also reported [7]. However the authors did not communicate the type of sludge (endogenous or exogenous). In our experiments the 80% response dose was 9.1 g CN per kg MLSS for endogenous sludge and 0.37 g CN per kg MLSS for nitrifying sludge after a contact time of 60 minutes. A 66% inhibition of acetate oxidation of endogenous respiration was found [8] for 6.6 g CN per kg MLSS.

Potassium cyanide (continuous experiments)
The effect of continuously adding KCN on the respiration rate in the pilot plant is shown in figure 6. Table 1 summarizes the most relevant process parameters.

TABLE 1

Process parameters for continuous experiments with potassium cyanide

	experiment		
	1	2	3
influent flow ($1.h^{-1}$)	0.338	0.350	0.386
KCN concentration influent ($g.m^{-3}$)	0.29	1.0	8.4
suspended solids ($kg.m^{-3}$)	4	2.9	4.2
total volume activated sludge (1)	5.80	5.80	6.34
return sludge flow ($1.h^{-1}$)	0.36	0.36	0.36

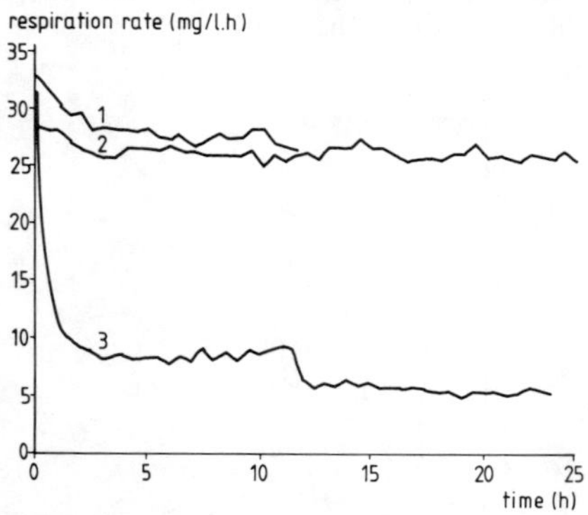

Figure 6. Respiration rate in the pilot plant after starting the dosage of three different concentrations of potassium cyanide.
For explanation of numbers, see table 1.

The respiration rate is decreasing more or less depending on the concentration of KCN in the influent. Assuming that the reactor can be concerned as a completely mixed system and that the cyanide is neither subject to adsorption nor breakdown and volatilization, the theoretical concentration of KCN can be calculated as a function of time. Then from figure 6 a dose response curve can be derived (figure 7).

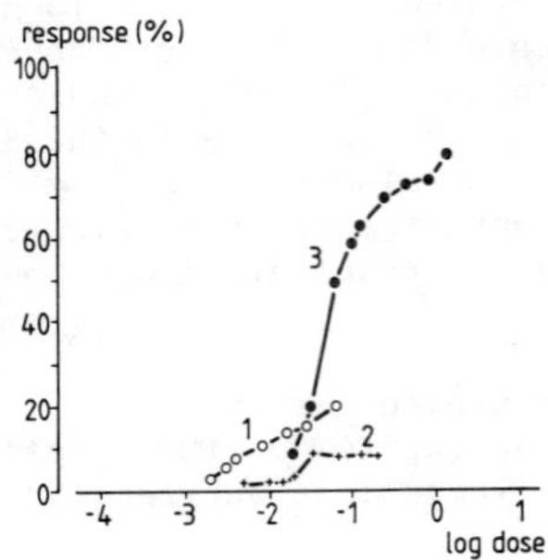

Figure 7. Dose response curves for potassium cyanide and activated sludge derived from continuous experiments. For explanation of numbers, see table 1.

Comparison of figures 5 and 7 reveals that approximately batch and continuous experiments lead to the same dose response curves. This leads to the conclusion that the continuously fed reactor experiments reflect the effect on the nitrifying activated sludge as found in the batch experiments.

The course of the respiration rate after ending the cyanide dosage is presented in figure 8. The recorded respiration rate indicates the recovery of the activated sludge. This affirms the theory that the mechanism behind the inhibitory effect of KCN is reversible [9].

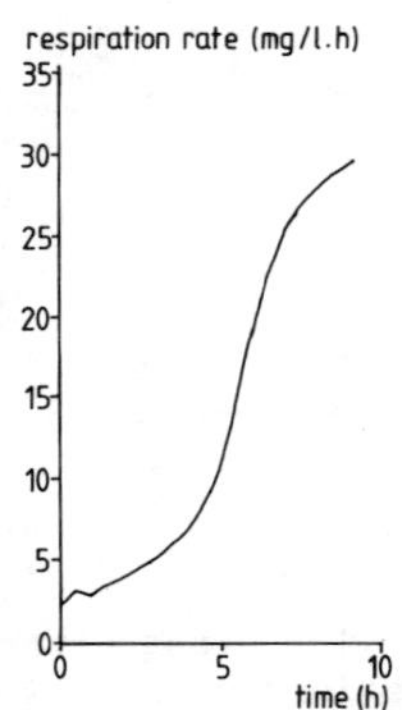

Figure 8. Respiration rate in the pilot plant after ending the cyanide dosage.

Some organic compounds

Batch experiments are carried out in order to trace the effect of 2-nitropropane on endogenous as well as nitrifying activated sludge. The results are presented in figure 9.

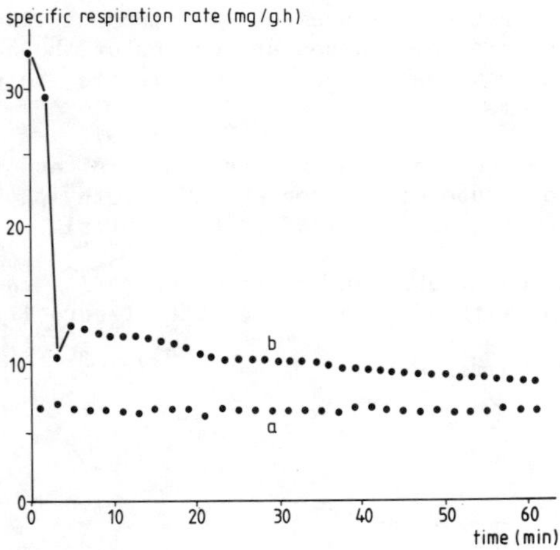

Figure 9. Specific respiration rate of activated sludge over time after
addition of 2-nitropropane.
a. Endogenous sludge. Dose 74 g per kg MLSS
b. Nitrifying sludge. Dose 84 g per kg MLSS

A dose of 74 g per kg MLSS has no effect on the endogenous respiration. In
other experiments (not shown in figure 9) it appeared that even a dose up
to 247 g.kg^{-1} had no significant effect on the endogenous respiration. For
nitrifying sludge however, 2-nitropropane appears to be considerably toxic
(curve b in figure 9). In this experiment the endogenous level of the specific respiration rate (measured before addition of ammonium chloride) is
8.6 g.kg^{-1}.h^{-1}. This leads to the conclusion that the nitrification is
inhibited for almost 100% after a contact time of 60 minutes.

Figure 10 shows the effect of a dose of 6 g per kg MLSS of triethylphosphate on endogenic as well as nitrifying activated sludge in batch
experiments. Subtraction of the endogenous curve from the total effect
curve gives the response of the nitrifying organisms. This is also shown in
figure 10.

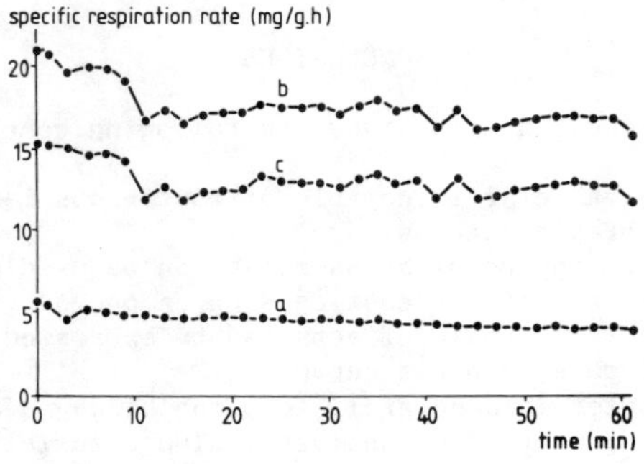

Figure 10. Specific respiration rate of activated sludge against time
after addition triethylphosphate. Dose: 6 g per kg MLSS.
a: endogenous respiration; b: respiration nitrifying sludge
c=b-a: oxygen uptake for nitrification

In discontinuous respiration measurements a component like trethylphosphate gives rise to serious problems caused by foaming of the activated sludge after addition of the component. The WAZU-respiration meter however, can cope with these problems.

From the figure it can be seen that the responses of endogenous respiration and nitrification to a dose of 6 g per kg MLSS trethylphosphate only differ slightly (28% and 20% respectively) after 60 minutes).

The effect of a non-toxic substance on the respiration rate is shown by a continuous experiment with addition of acetate (figure 11).

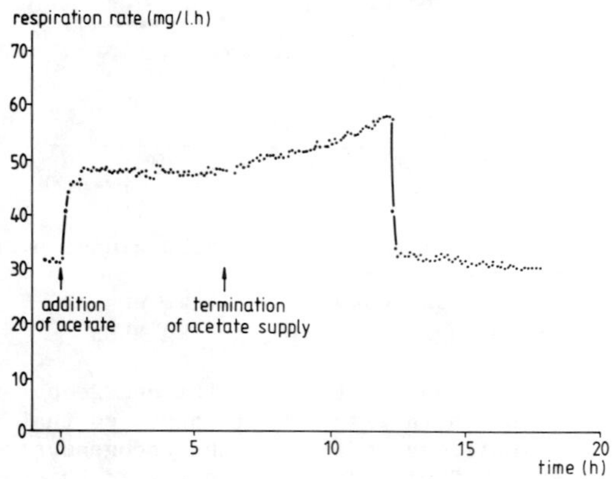

Figure 11. Respiration rate in the pilot plant. Addition of acetate during 6 hours.

The continuous addition of acetate to the influent causes an important increase of the respiration rate due to the biological oxydation of the compound. After a certain period of constant respiration rate, a gradual increase, probably due to adaptation, is observed. After termination of the acetate supply the respiration rate remains on a higher level due to the oxydation of accumulated acetate. After the acetate is removed from the solution the respiration rate returns to the steady state level.

CONCLUSIONS

Based on the experiences in this study the following conclusions can be drawn:
1. The Wazu-respiration meter is capable of continuous measurement of the respiration rate of activated sludge.
2. In that application the respiration meter can be used as an instrument to determine toxic effects of chemicals on endogenous and on nitrifying activated sludge. These toxic effects can be expressed in quantitative terms by means of dose-response curves.
3. The respiration meter is a helpfull tool for biodegradation tests.
4. There is a need for a standard activated sludge toxicity test based on continuous measurement of the respiration rate.

Based on the assumption that the inhibition of the respiration rate is to be regarded as a toxicity parameter, we believe that the respiration meter has two major fields of application:
1. In the research laboratory it can be used for rapid screening of toxicants. The examination of long term respiration rate trends helps to distinguish acute and chronic toxicity.
2. At treatment plants it is applicable as a toxicity monitor for the incoming waste water. The monitor must be situated upstream to provide sufficient time for control action. The activated sludge in the monitor has to be of the same composition as the sludge in the treatment plant.

ACKNOWLEDGEMENTS

The work has been supported mainly by the Dutch Institute for Inland Water Management and Waste Water Treatment. We thank Dimitri van den Akker of the Department of Physics and Meteorology (group Measurement and Control Engineering) and Mark Feenstra.

REFERENCES

1. Henney, R.C., Fralish, M.C. & Lacina, W.V. (1980). Shock loads of chromium (VI). J. Wat. Pollut. Control Fed. 52, 2755-2760.

2. Larson, R.J. & Schaeffer, S.L. (1982). A rapid method for determining the toxicity of chemicals to activated sludge. Water Res. 16, 675-680.

3. OECD (1981). Guidelines for testing of chemicals.

4. Klapwijk, A., Spanjers, H. & Van den Akker, D. (1987). Control of an activated sludge plant with a Wazu-respiration meter. Paper to be published at the EWPCA Symposium on May 19th - 23rd, 1987 in Munich.

5. Klapwijk, A. & Spanjers, H. (1986). Respiratiemeter bewaakt biologische zuivering van afvalwater. PT/Procestechniek 41, 47-51.

6. Blok, J. (1973). Inhibitie van actief slib systemen door cyanaat en cyanide. H_2O 6, 524-526.

7. Blok, J. & Ten Berge, W.F. (1975). Toxiciteitstesten voor afvalwater t.a.v. biologische zuivering. H_2O 8, 328-332.

8. Vandebroek, R. (1986). Study and development of a microcomputer controlled sensor for the determination of the biodegradability and the toxicity of wastewaters: the Rodtox. Thesis, State University of Ghent.

9. Schlegel, H.G. Allgemeine Mikrobiologie (4 Drucke). Thieme-Verlag.

REMOVAL OF PHOSPHORUS FROM WASTEWATER BY
ALUMINUM CHLORIDE/POLY-ALUMINUM CHLORIDE BLENDS

D.D.Joye
Department of Chemical Engineering
Villanova University
Villanova, PA 19085
USA

and

J.M. Courtney, III
Courtney Industries, Inc.
Baltimore, MD 21224
USA

ABSTRACT

The phosphorus removal effectiveness of blends of aluminum chloride and poly-aluminum chloride was studied by obtaining turbidity reduction and phosphorus content in jar tests on real clarifier influent from industrial and municipal sources. The blends of aluminum chloride and poly-aluminum chloride showed a strong synergistic effect for phosphorus removal from various wastewaters. Values in the range of .07 - .08 mg P/L were achieved in jar tests from wastewaters initially containing 5 - 20 mg P/L. The best blend ratios seem to be in the area 40 - 60% AC by weight and about 70 - 80% of the optimum dosage value for AC. However, in some wastewaters this synergism was not evident. Possible causes are being explored. The present work shows that aeration does not negatively affect the performance of the blends, and that pH and calcium ion have an effect on phosphorus removal.

INTRODUCTION

Removing phosphorus from industrial and municipal wastewaters is extremely important for preventing various undesirable biological effects in the aquatic ecosystem, particularly rivers and lakes which act as sinks for treated wastewater discharge.

Aluminum chloride and poly-aluminum chloride are relatively new chemical coagulants in the USA, and very little is known about their ability to remove phosphorus. Previous work (1,2) has shown that aluminum chloride, poly-aluminum chloride and blends of the two are unusually effective inorganic coagulants for turbidity reduction (clarification) of wastewater streams. In particular, the blends are extremely efficient and capable of achieving turbidities in the drinking water range (less than 1 NTU) in some wastewaters. Although economically attractive for producing very low effluent turbidity, they seem to be generally more advantageous for potable water uses than for industrial wastewater applications. In the case of phosphorus removal, however, extremely low effluent levels, e.g. .02 -.03 mg P/L, may become the new standards in the near future. Very efficient coagulants such as the aluminum chloride/poly-aluminum chloride (AC/PAC) blends have recently (3) been shown to achieve these levels of phosphorus removal, but not consistently in all cases. Thus, the objective of this ongoing research is to evaluate the effectiveness of the AC/PAC blends for phosphorus removal from real treatment plant influent, and establish the parameters that may affect its use. In this work we investigate the effects of aeration and pH adjustment with lime, two widely practiced treatment procedures which may have an effect on the performance of the blends.

LITERATURE SURVEY

The whole phosphorus cycle has been the subject of intense study for the last decade or more. Chemical and biological methods for reducing phosphorus have been reviewed (4). Phosphorus levels of 6 mg/L are typical for many treated municipal wastewaters, although levels of 1 -2 mg/L were reported in some plants (5). Detergent bans have been extremely effective in reducing phosphorus in rivers and lakes but cannot by themselves solve the phosphorus problem. Phosphorus limits, for example 1.0 mg/L, have been used in many areas with positive results (6).

Careful chemical coagulation using one of the familiar flocculants, e.g. alum, ferric chloride, lime, or ferrous sulfate, can reduce phosphorus levels to about 1.0 mg/L or slightly below (7,8,9). However, standards below .03 mg/L are being used as new target figures. From the standpoint of phosphorus in the ecosystem, these new standards are reasonable enough, because no detrimental effects in the ecosystem are observed at phosphorus levels below about .03 mg P/L. Removal by biological treatment has not been able to achieve these levels in practice. If new standards of .02 -.03 mg P/L are to be mandated, more effective removal technologies must be brought to light. Only the recently reported AC/PAC blends (3) have shown levels in this region, and this provides the reason for present research in this area.

DESCRIPTION OF THE EXPERIMENT

Standard jar-testing apparatus and procedures were used for the experimental phase of the present study. Wastewater samples were obtained from local treatment plants, both municipal and industrial. Total dissolved solids were typically about 300 -600 mg/L for the municipal sources with influent phosphorus levels in the 5-20 mg/L range. Sampling was done at two locations in the treatment process, before and after the aeration basin. The industrial source had total solids in the 40,000 mg/L range and phosphorus content of 58 mg P/L. Coagulation was carried out in a Phipps-Bird six-place paddle mixer with a 1-liter sample volume per place. Turbidity of the supernatant after 8-10 minutes settling was taken with a Hach Turbidimeter 2100. The general procedure is reported in previous publications (1-3). In the procedure, the polymeric coagulant was added second, after about three minutes agitation with the inorganic, and the agitation intensity reduced to minimize shear degradation of the polymer. The pH of the samples was measured using a Fisher Accumet, model 230, pH/ion meter with a combination electrode. Initial pH was adjusted to 6-7 range to start, but no further adjustments were made. This follows typical industrial practice.

Analysis for phosphorus (in the supernatant) was an involved, but well-standardized procedure using a Spec-20 spectrophotometer. Several methods for phosphorus analysis are available, but we used the ascorbic acid method with digestion (total conversion of all forms of phosphorus to orthophosphate) described in the equipment literature (10). The specific, step-by-step procedure for phosphorus analysis can be found in a U.S. EPA report (11). Basic procedure is to (a) convert all phosphorus to orthophosphate by persulfate digestion, (b) add combined reagent color developer including ascorbic acid, and (c) use spectrophotometer to determine phosphorus. Phosphorus standards are made up fresh for each day's run.

The treatment chemicals were a flocculant-grade aluminum chloride containing 10.7% aluminum as Al2O3, and poly-aluminum chloride with 10.5% alumina content. The flocculants were added in solution form. The aluminum chloride solution contained 28% by weight aluminum chloride, and the poly-aluminum chloride solution contained 22.5% poly-aluminum chloride solids by weight.

Dosages were varied in the normal manner to find the optimum dose. Blends of aluminum chloride/poly-aluminum chloride (AC/PAC) in various proportions were studied with and without pH adjustment.

RESULTS AND DISCUSSION

Earlier work from reference (3) showed that optimum dose for turbidity removal coincided roughly with that for phosphorus removal. Also addition of organic polymers following common practice with single inorganic flocculants showed that improvements on the order of 30% for phosphorus were achieved, but no dramatic reductions were obtained. On several municipal wastewaters the AC/PAC blends showed consistent synergism achieving levels of phosphorus in the .02-.06 range. In one heavily loaded industrial wastewater, the synergistic effect was not noted; in this case, the best phosphorus removal was equivalent to the AC alone.

The results of the present study are shown in Tables 1 and 2, which represent municipal wastewater sampled before and after the aeration basin. Table 1 and 2 show the synergism of the blends in both cases. The results are typical for what we've come to expect, except that phosphorus levels do not quite make the .02-.06 region obtained previously. In virtually all cases, blending the coagulants yields better results than either used alone. The best blend ratios are located in the lower, center-right region of the tables. Blending is of particular interest for economic reasons, since PAC is expensive as Using less of this and more of the AC, which is much cheaper, would be of very high interest, if the same removal levels could be attained. The tables show much better removal with the blends.

TABLE 1

Phosphorus Removal from Municipal Influent (Unaerated)
Using AC/PAC Blends

Turbidity (NTU)

mg AC/L mg PAC/L	0	6.4	12.8	19.2	25.6	31.9	38.3
0	44	28	10	5.6	4.8	5.3	3.8
6.1	30	3.0	2.2	2.1	2.8	3.3	12.0
12.2	4.9	2.4	1.9	1.8	2.4	1.9	2.2
18.3	2.4	2.3	1.6	1.3	1.4	2.4	4.2
24.4	2.4	2.2	1.4	1.0	1.1	1.2	4.9
31.0	2.6	2.3	1.1	1.1	1.4	1.9	4.4
36.6	2.6	1.7	.9	1.0	1.2	4.9	–

Phosphorus (mg P/L)

	0	6.4	12.8	19.2	25.6	31.9	38.3
0	5.6	2.3	.43	.43	.41	.37	.33
6.1	2.5	1.5	.20	.20	.19	.23	.38
12.2	.78	.26	.20	.20	.26	.15	.22
18.3	.20	.26	.13	.13	.20	.16	.26
24.4	.27	.26	.17	.17	.17	.12	.33
31.0	.29	.12	.11	.11	.12	.08	.34
36.6	.20	.13	.11	.11	.14	.24	–

TABLE 2

Phosphorus Removal from Municipal Influent (Aerated)
Using AC/PAC Blends

Turbidity (NTU)

mg AC/L mg PAC/L	0	3.2	6.4	9.6	12.8	16.0	19.2
0	66	2.9	2.5	1.9	1.6	1.6	1.4
3.1	1.9	2.3	1.1	1.1	0.9	1.3	2.1
6.1	1.4	0.8	0.8	1.0	1.0	1.1	1.4
9.2	1.2	.6	1.0	0.7	1.1	0.9	1.4
12.2	1.3	.4	1.2	.9	1.4	.9	1.3
15.3	1.0	.4	1.0	1.3	0.9	1.0	1.4
18.3	0.9	.4	1.1	.7	.5	1.8	1.9

Phosphorus (mg P/L)

	0	3.2	6.4	9.6	12.8	16.0	19.2
0	19	1.6	.96	.53	.61	.64	.17
3.1	1.6	.49	.13	.17	.18	.20	.21
6.1	1.06	.21	.32	.12	.19	.20	.20
9.2	.45	.18	.11	.22	.19	.08	.11
12.2	.38	.09	.21	.10	.11	.18	.35
15.3	.18	.20	.17	.09	.22	.09	.20
18.3	.16	.24	.23	.09	.22	.09	.15

Tables 1 and 2 also show the effect of aeration on the treatment method. Aeration is commonly practiced in the USA. It generates a kind of biological floc which is recycled into the treatment process. It was originally thought that this floc would interfere with the coagulating action of the blends, but the data above show that aeration actually improves both turbidity and phosphorus removal slightly. More importantly, it leads to reduced amount of treatment agent for the same removal. And in any event it does not interfere with the blends at all. Thus, whether the wastewater had been aerated or not should have no negative effect on the blends, and could not explain why some wastewaters do not show the synergism of the blends.

That pH affects coagulation is well known. How pH affects the performance of the blends is shown below for a few selected cases. Tap water was used to make up phosphorus-containing water by adding potassium dihydrogen phosphate. Initial phosphorus was 44 mg P/L.

TABLE 3

pH Effects in Phosphorus Removal from Synthetic, Phosphorus-Containing Solution for AC, PAC and a Blend of the Two

AC dose = 143.7 mg/L

lime dose, ml	0	5	10	40
pH	3.5	6.0	8.0	11.5
phosphorus, mg/L	32.7	21.6	11.3	0.97

PAC dose = 110 mg/L

lime dose, ml	0	3	8	30
pH	3.5	6.0	8.1	11.4
phosphorus, mg/L	30.9	20.2	19.1	1.35

AC/PAC blend (AC = 72 mg/L, PAC = 55 mg/L)

lime dose, ml	0	4	9	35
pH	3.5	6.0	8.1	11.3
phosphorus, mg/L	29.8	15.0	6.9	0.97

AC/PAC blend (AC = 72 mg/L, PAC = 55 mg/L)
(pH Adjusted by NaOH dose instead of saturated solution of Ca(OH)2)

pH	3.5	6.0	8.1	11.3
phosphorus, mg/L	30.4	20.7	31.5	39./

Calcium Hydroxide only

lime dose, ml	–	2.0	7.5	39.0
Ca++/P	–	.04	.14	.77
pH	–	6.0	8.1	11.5
phosphorus, mg/L	–	40.5	31.5	4.1

These results show several important and interesting phenomena. First, the blends are affected by pH as suspected. Second, when the pH is adjusted with a phosphorus-neutral agent such as NaOH, the phosphorus removal effectiveness is altered. The best pH seems to be around 6.0, or between 3.5 and 8.0. The removal efficiency gets worse at more acidic pH and also at more basic pH. This is consistent with previous work in aluminum ion chemistry (12). Third, lime itself, with no added coagulant, is also effective at removing phosphorus, but the removal is not very significant until the very high pH of 11.5. This is very close to the solubility limit for calcium ion. Fourth, the blend sample shows enhanced phosphorus removal at all pH levels over either single component. All seem to work best at the extremely high pH of 11.4, probably as a result of the lime becoming more effective as pH is increased (as pH increased, lime concentration increased). Therefore phosphorus removal is enhanced by lime addition and high pH.

Table 4 shows turbidity and phosphorus removal data for an industrial wastewater of very high initial turbidity and phosphorus content. This water originated from a milk-processing plant, and because of stringent cleanliness requirements in the process, contained a high degree of phosphorus from detergent cleaners.

TABLE 4

Performance of the AC/PAC Blends in the Treatment of a
Heavily Loaded Industrial Wastewater

Turbidity (NTU)

mg PAC/L \ mg AC/L	0	72	108	144	180	270
0	360	400	410	540	56	20
137	475	6.7	8.6	5.9	46	46
206	87	7.6	5.9	6.1	59	82
275	21	7.5	6.6	23	14	80
343	6.4	8.6	9.0	22	25	82
412	19	8.0	8.9	8.5	32	110

Phosphorus (mg P/L)

mg PAC/L \ mg AC/L	0	72	108	144	180	270
0	58	–	–	–	43	43
137	44	36	27	50	28	39
206	27	13	24	26	37	31
275	24	34	30	34	36	34
343	19	23	22	34	35	34
412	25	34	27	25	55	45

This particular wastewater is a troublesome one. Synergism of the blends is evident, with the best blend ratios in the lower central portion of the grid. Turbidity reduction is not spectacular, but reasonable. The phosphorus reduction is not particularly good. We suspected that 10 minutes of settling in the jar-test procedure was not enough for adequate settling, so some samples were allowed to sit overnight, and the phosphorus of the supernatant re-evaluated. The blend 144 mg/L AC and 206 mg/L PAC showed an improvement to 7.4 mg P/L. This was a typical result and a much better, though not spectacular, number. The turbidity in the supernatant sample has a strong effect on the phosphorus analysis, because most of the phosphorus (that which is removed) is in the floc. To get a better measure of phosphorus in the supernatant, one must first be able to get a clear supernatant, either by filtering or by settling for longer periods. In order to get a better phosphorus and turbidity removal in the wastewater above, a two-step process might be more effective. Another reason why performance of the blend was relatively unspectacular was that very high dosages were required to coagulate. This resulted in a very low pH, typically 2.5-3.5. We did not pH adjust, but it would seem to be advantageous, particularly if lime were used to do this.

CONCLUSIONS

1. The aluminum chloride/poly-aluminum chloride blends reported in this
   study were shown to have increased effectiveness over the use of
   either separately in turbidity reduction and phosphorus removal.

2. Aeration treatment does not negatively affect the performance of the
   blends, indeed it actually enhances removal efficiency; less dose is
   needed to achieve the same removal.

3. AC/PAC blends are capable of achieving phosphorus levels of .10 mg/L
   and below in laboratory tests on wastewaters used in the present
   study.

4. The efficiency of the AC/PAC blends is not negatively affected by changes in pH of solution from about 5 up to about 12. When lime or calcium containing species is used, performance is actually enhanced.

5. In heavily loaded industrial wastewater, the AC/PAC blends were capable of removing about 90% of influent phosphorus in one treatment, but a second treatment would be needed to achieve levels below about 1 mg/L.

REFERENCES

1. Courtney, J.M., III and D.D. Joye, "Experimental Study of the Effectiveness of Aluminum Chloride for Clarifying Wastewater", in Toxic and Hazardous Wastes, M.D. LaGrega and D.A. Long (eds.) Proc. 16th Mid-Atlantic Industrial Waste Conf., Technomic Publ. Corp., Lancaster, PA, 1984. pp.343-350.

2. Kisciras, R.P., J.M. Courtney, III and D.D. Joye, "Comparative Performance of Aluminum Chloride, Poly-Aluminum Chloride and Blends in Wastewater Clarification", in Toxic and Hazardous Wastes, I.J. Kugelman (ed.), Proc. 17th Mid-Atlantic Industrial Waste Conf., Technomic Publ. Corp., Lancaster, PA, 1985. pp. 156-165.

3. O'Neill, M.J., D.D. Joye and J.M. Courtney, III, "Phosphorus Removal From Municipal and Industrial Wastewater by Aluminum Chloride and Derivatives", in Toxic and Hazardous Wastes, G.D. Boardman (ed.), Proc. 18th Mid-Atlantic Industrial Waste Conf., Technomic Publ. Corp., Lancaster, PA, 1986. pp.598-606.

4. Ryczak, R.S. and R.D. Miller, "A Review of Phosphorus Removal Technology", U.S. Army Technical Report 7706, NTIS Doc. No. ADA 040 802, 1977.

5. Gakstatter, J.H., M.O. Allum, S.E. Dominguez and M.R. Crouse, "A Survey of Phosphorus and Nitrogen Levels in Treated Municipal Wastewater", Jour. WPCF, 50(4), 718 -722 (1978).

6. Hartig, J.H., "Progress in Municipal Phosphorus Control", Water/Eng. and Mngmt, March 1985, pp. 32 -33.

7. Switzenbaum, M.S., J.V. DePinto, T.C. Young and J.K. Edzwald, "A Survey of Phosphorus Removal in Lower Great Lakes Municipal Treatment Plants", Jour. WPCF, 52(11), 2628 -2633 (1980).

8. Van Dam, D., "Economical and Efficient Phosphorus Control at a Domestic Industrial Wastewater Plant", Jour. WPCF, _53_(12), 1732 -1737 (1981).

9. Cullinane, M.J. and R.A. Shafer, "Reduce Lagoon Algae Problems with Coagulants", Water and Wastes Eng., June 1980, pp. 19 -21.

10. Bausch & Lomb Spectrophotometer Literature, "Phosphorus - Ascorbic Acid", Bausch & Lomb, Inc., 1974.

11. "Methods for Chemical Analysis of Water and Waste", U.S. EPA Report No. EPA-600-4-79-020, Environmental Monitoring and Support Lab, Cincinnati, OH, March 1979.

12. Diamadopoulos, E. and A. Benedek, "The Precipitation of Phosphorus from Wastewater Through pH Variation in the Presence and Absence of Coagulants", Water Res., _18_(9), 1175-1179 (1984).

TREATMENT AND DISPOSAL OF INDUSTRIAL
WASTE BY SOLIDIFICATION

P. J. Howard
Leigh Environmental Limited
Brownhills, Walsall, WS8 7BB
UK

ABSTRACT

Industrial waste disposal by solidification techniques has been
successfully practiced over a period of time particularly with
inorganic waste materials. The paper describes one such system
employing the SEALOSAFE technology giving practical experiences and
indicating the monitoring neccessary to operate the treatment plant.
The paper also discusses new developments in the solidification of
organic waste and briefly describes uses of the solidified product and
the process economics.

Industry, particularly manufacturing industry, exists for the
purpose of isolating or purifying various substances. All these
substances which are useful to and used by industry do not exist in a
natural state but need to be extracted. The processes of extraction
and purification produce a waste which if not reusable or easily
reclaimed
requires disposal.

Thus over the decades the generation of industrial wastes and
their disposal has become a fact of life. Over the years legislation
has developed to cater for the safe disposal of such wastes although
the emphasis of the legislation will vary from country to country.
Throughout the world the following disposal methods can be seen in
operation:

> sea dumping
> land filling
> discharge to underground strata
> discharge to watercourse
> discharge to sewer
> discharge to atmosphere
> biological purification
> incineration
> chemical treatment/degradation
> solidification and encapsulation

One such concept in the listed alternatives is that of waste solidification and encapsulation.

The greatest problem facing industry as far as waste disposal is concerned is the safe and effective disposal of its liquids and sludges. Solid wastes can generally be landfilled at suitable sites but free flowing materials present many problems when considering this traditional outlet. Solidification processes are designed to overcome the inherent difficulties experienced in the disposal of liquids and sludges in that the material is converted to a solid mass prior to ultimate disposal.

One such solidification process employing the Sealosafe technology has been operating in the United Kingdom for the past 12 years. The process has safely handled and treated a wide variety of industrial waste products. The technology has also been successfully applied in the countries of Canada and Japan.

The process involves the treating of hazardous industrial waste, capable of being contained in an aqueous slurry, by the addition of calcium containing cement typically (CaO, 64.0%; SiO_2, 20.5%; Al_2O_3, 5%) and a powder consisting of aluminium silicate or aluminosilicate, (typically S^iO_2, 49.0%; Al_2O_3, 24.8%; Fe_2O_3, 10.2% plus traces of C. and S.) so forming a flowable slurry which sets to a rigid mass. The ratio of calcium containing cement to aluminium silicate or aluminosilicate is in stoichiometric and for any given water content over the range 50:1 and 1:50.

The quantity of water should not be less than 20% as this is insufficient to hydrate the product and the flowable slurry is not formed. However, the greater the concentration of water in the slurry the longer will be the setting time. Additionally increasing the concentration of any particular hazardous waste will reduce the initial and ultimate compressive strength of the rigid mass.

If the hazardous waste material is neutral or alkaline, the silicate or aluminosilicate and the cement are mixed and the hazardous waste added. If the hazardous waste material is acidic it is desirable to make it alkaline with a lime bearing material before the addition of the reagent powders.

The structure of the final rigid mass has been debated extensively following the inception of Sealosafe technology. Although a definitive structure has not been established it is most likely that the hazardous waste is bound into the rigid mass by chemical reaction and physical forces related to the formation and molecular structure of complex calcium silicates and aluminosilicates.

As in good sound environmental practice prior to acceptance of any waste at a solidification plant operating the Sealosafe technology a sample is submitted for laboratory analysis, evaluation, formulation and testing. The waste sample is throughly screened for suitability and a sample product made and tested. The test work will include leaching, permeability and compressive strength studies and determinations. When the waste product has achieved suitability in all parameters is it then only cleared for processing.

The formation determined to treat the waste in this initial stage is used as a reference basis for treatment at the plant. The process used involves the blending of wastes in pre-determined sequences and quantities and high speed mixing to form a homogeneous waste stock.

The stock is then blended in a paddle mixer and this stage of the process is termed the "polymerisation" step.

After a predetermined contact time in the mixer, the waste is fully reacted and is both safe and suitable for transport to a disposal site. At this stage, the polymerised product is in the form of a slurry which can be pumped or transported to a remote disposal site or poured for casting purposes for up to 24 hours.

There are no elevated temperatures or pressures involved in the processes although the settling time of the polymerised product can be reduced by operating at higher temperature. The whole waste is solidified during reaction and no by-products are formed requiring separate disposal.

In about three days the polymerised product sets to a rigid mass. Hardening continues for 28 days at which time the product will have reached its maximum compressive strength, setting may continue for six months after initial product formation. As far as can be ascertained over the period of the existing operational units the property of compressive strength is sustained.

The principal areas of interest especially as far as environmental significance is concerned are:

> permeability
> leaching
> compressive strength

The test schedule reproduced in Appendix A is that adopted by Sealosafe plant laboratories in order to ensure reproducibility and conformity to laid down parameters.

Table 1 gives typical values of coefficient of permeability in cms per second .

Table 2 gives typical compressive strengths obtained in MN/m .

Tables 3 to 5 give typical leachate values.

From this data it will be seen that solidification techniques are effective in rendering waste environmentally inert.

TABLE 1

Coefficient of permeability in cms per second[2].

Clean Gravel	1.0 and greater
Clean Coarse	1.0 to 0.01
Sand Mixture	0.01 to 0.005
Fine Sand	0.005 to 0.001
Silty Sand	0.002 to 0.0001
Silt	0.0005 to 0.00001
Clay	0.0000001 and smaller
Typical Concrete	0.000001
SEALOSAFE Polymers	0.0000001 to 0.0000002

TABLE 2

Compressive Strengths in MN/m^2.

	After 3 days	After 7 days	After 28 days
An arsenical waste		390	750
An aqueous effluent	193	330	610
A chromic waste		108	220
A chromium waste		155	310

There are currently over 2,000 different types of wastes formulated for Sealosafe treatment and these represent wastes produced by a large section of modern industry.

Solidification has proved effective in stabilising aqueous wastes with a maximum organic concentration of around 10%. Anything above this level and the organic constituents tend to delay the setting, reduce the final compressive strength and leach significantly.

TABLE 3

One day E.L.T. and two day A.S.T.M. leaching data in mg/l

| | 1 day E.L.T. on product | | | | 2 day A.S.T.M. | | |
	Ni waste	Cr waste	Cu waste	Zn waste	Sludge	Raw Zn waste	Product
Cadmium	–	–	–	<0.0005	–	0.04	<0.001
Chromium	–	0.09	–	0.3	0.07	0.08	0.06
Copper	<0.002	–	0.02	–	–	–	–
Iron	–	–	–	0.01	0.005	10.1	<0.005
Lead	–	–	–	<0.005	0.01	0.05	0.03
Nickel	<0.05	<0.005	–	<0.005	–	50	<0.005
Zinc	–	<0.005	–	<0.005	<0.005	226	<0.005
T.O.C.	<1.0	<1.0	3.0	2.9	78	4.4	8.2
C.O.D.	<25	<25	<25	97	806	47	67
Cyanide	0.03	–	0.04	–	0.03	–	–

TABLE 4

Analysis of Rainwater Collecting In Land Reclamation Site
U.K. Water Pollution Control Authority

	mg/l		
P.V. 4 hour	1.8	–	–
B.O.D.	1.6	3.0	1.8
Suspended Solids	63	133	85
Ammonial Nitrogen	0.7	0.5	0.1
Nitrite Nitrogen	0.1	0.1	–
Nitrate Nitrogen	3.3	4.1	2.4
Chloride	99	170	51
Cadmium	–	–	–
Chromium	0.04	0.02	0.02
Copper	0.05	0.05	0.02
Nickel	0.02	0.01	0.12
Zinc	0.08	0.02	0.06
Arsenic	0.26	0.01	0.00
Mercury	0.02	0.01	0.03

TABLE 5

Japan Environmental Protection Agency Leaching Test

	Electro-plating waste mg/l	Humus sludge mg/l	Paint sludge mg/l	Latex sludge mg/l	Acid tar mg/l
Cyanide	<0.007		–	–	<0.007
Cadmium	<0.01	<0.01	<0.01	<0.01	0.02
Chromium	<0.01	–	<0.01	<0.01	<0.01
Copper	<0.3	<0.3	<0.3	<0.03	<0.3
Lead	<0.3	<0.3	<0.3	<0.3	0.37
Nickel	<0.1	–	<0.1	<0.1	0.15
Total Mercury	–	<0.0005	<0.0005	<0.0005	–
Organic Phosphorous	–	<0.02	–	–	–
Arsenic	<0.001	<0.001	<0.001	<0.001	<0.001

Experience with the technology of Sealosafe has shown it to be able to accomodate substantial organic contaminants in wastes accepted for treatment including:

> latex
> acid tars
> oily wastes
> pharmaceutical wastes

The methods employed to achieve this vary depending on the nature and concentration of the organic material in the waste.

One such system involves the solidification of an oily waste produced from the cleaning and descaling of aero engines. The treatment includes for the reduction of chromate, neutralisation, the removal of metals in solution and solidification of the remaining oily solids. After separation the oily solids are mixed with an oil absorbant which is a complex material formed by mixing olefinic polymer with a fine inorganic powder obtained from soil. The "encapsulated gel" so formed is then fed to a mixer wherin are added the reagent powders as already defined. The "polymerised" product is then allowed to set to a rigid mass in its normal manner.

Other organic constituents have been treated using the Sealosafe technology by using intermediates such as ceramic cements, polybutadiene polymers and some forms of polyethylene.

The advantages of the polymerised product can be summarised as:

low permeability and leaching
considerable compressive strength
non inflammable
non biodegradable
non odourous
environmentally acceptable
cost effective

The product is at the present time being used in a land
reclamation project in which an abandoned clay quarry, some 30 metres
deep, is being filled. The use of the product will make possible the
release of the new landform for immediate development. The first
phase of reclamation is now complete. The area concerned was capped
with a thin layer of clay some 18 months ago since when the land has
shown no signs of instability and has quickly returned to its original
natural habitat including a re-emergance of wildlife.

An active research and development programme continues into
other ways of product re-use. Investigations have been carried out
into its use as bulk fill, cavity fill, industrial grout, sub-base,
soil stabilisation and embankment work.

The Sealosafe process is versatile and by varying the amounts
and ratios of the reagents used in the treatment, a product with a
guaranteed specification can be acheived. This is essential for any
re-use of the product.

As indicated initially it is the liquid wastes from industry
that pose the greastest problems for safe disposal. Many countries
operate sites that accept liquid wastes as part of normal, landfill
operations with the safeguard of a butyl rubber, polythene, puddled
clay or chalk or concrete liner to the site. However, even with the
safeguard of a liner, these sites are generally equipped with
longterm, probably infinite monitoring boreholes, and invariably
include leachate collection systems. Ensuing leachate has to be
treated on-site and then removed or taken away in its raw state for
treatment eleswhere.

Concrete bunkers, storage tanks, liquid lagoons, drum stores and
a host of other methods are employed for liquid waste "disposal". All
short term solutions.

Sealosafe technology provides a system producing solids which
can easily be contained and effectively isolate industrial wastes from
the environment.

ACKNOWLEDGMENTS

The author wishes to thank the Directors of Leigh Interests PLC
for permission to publish this paper. The views expressed by the
author are personal and may not necessarily represent totally those of
Leigh Interests PLC.

APPENDIX A

Test Schedule for "polymerised" product.

____

Permeability
____

Permeability is measured by passing water under pressure through
a specimen clamped in a specially designed holder and determining the
water flow rate. The co-efficient of permeability is defined as the
rate of flow of water per unit area of specimen under unit hydraulic
gradient, the dimensions being cms sec.

The product test is cast as a slurry into a greased ring mould
to produce a squat cylinder of 5cm diameter and 1.5cm depth. The
sample should be vibrated or tapped to remove entrained air. Each
mould is placed in an incubated environment at 20°C +/-2°C at
equilbrium humidity for three days, after which the samples can be
removed from the moulds. When removed, the samples are placed over
water in a closed container at 20°C+/-2°C for a further 25 days
curing.

The test samples can then be clamped into the permeability
apparatus and sealed in place by compression of the elastomeric
insert. Figures 1A and 1B illustrate the test apparatus.

The space above the product is filled with water by opening taps
T and C so that water enters the space from aspirator A and air, and
eventually water, is expelled through outlet tube D. Tap C is then
closed and the vertical tube B and capillary measure S allowed to fill
completely. The apparatus is allowed to stand for several hours to
ensure that the product is fully saturated. For the test, tap E is
opened to allow water to pass through the specimen and the rate of
flow is measured using the capillary tube and scale.

The test should be repeated on the same sample. The two
replicate results will be the same only when the specimen is fully
saturated and if the two measurements are different the test should be
repeated until constant permeability is measured.

Permeability is calculated from the expression:

$$k = 2.3 \frac{al}{At} \cdot \log 10 \frac{hl}{h2}$$

where a	=	cross section area of water column
l	=	length (thickeness) of specimen
A	=	cross section area of specimen
t	=	time for water to fall between hl and h2
hl	=	initial height of column
h2	=	final height of column

Figure 1A

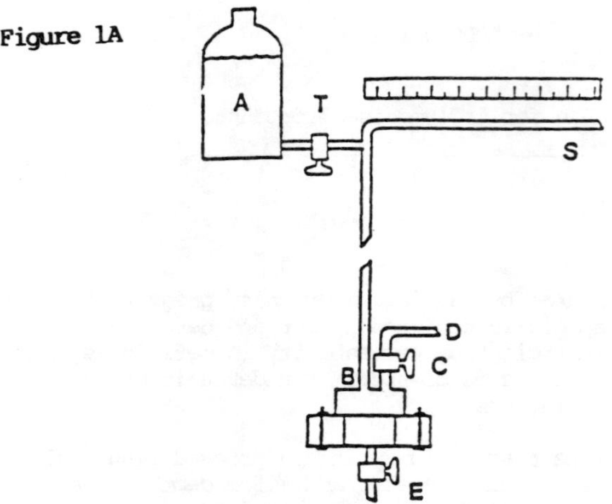

Figure 1B

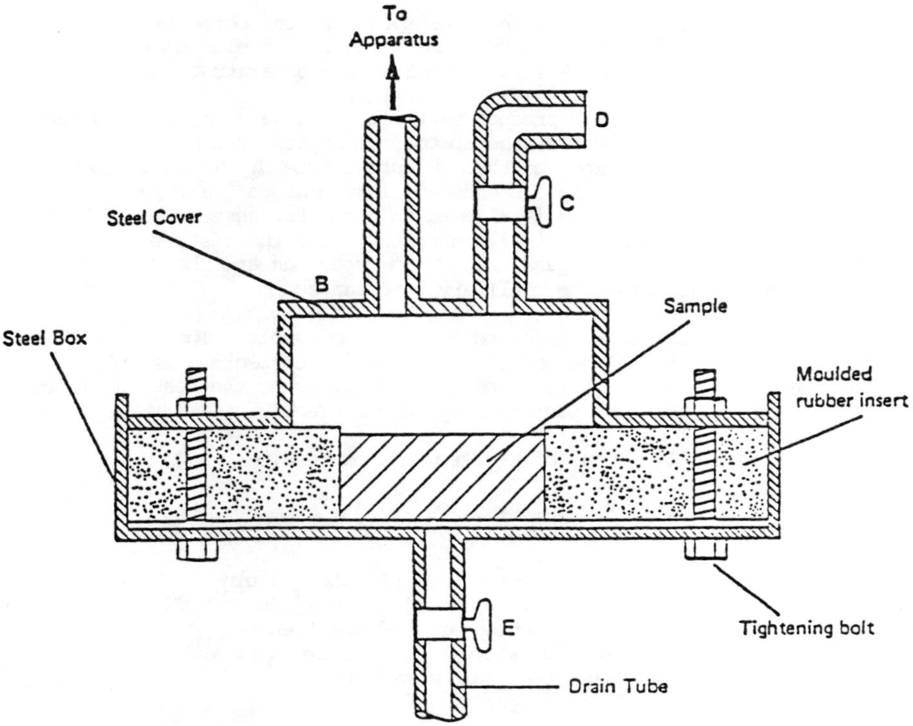

Leaching
____

In the absence of any definitive test method for determining
leaching characteristics, Sealosafe laboratories have adopted the
American Society for Testing and Materials (ASTM) proposed method for
leaching of Waste Materials (D.19:12, Method A). It is recognised
that the proposed method is gaining widespread acceptance in U.S.A.
and elsewhere and is considered to be the best option available.

The test, known as the water shake extraction procedure,
comprises the shaking of a known weight of specimen with water of a
specified quality and separation of the aqueous phase for subsequent
analysis. The method is intended to determine collectively the
immediate surface washings from the specimen and the time-dependent
diffusion controlled contributions to leachings from the specimen.

Apparatus required is a reciprocating platform shaker operated
at between 60 and 70 strokes of 25mm per minute.

The first part of the test is the determination of solids
content incubation at 104 °C +/-2 °C for 18 hours +/- 2 hours.

The second part of the test is the shake procedure.. Specimen
containers are chosen, such that the specimen plus water equal to four
times the specimen weight occupies between 80% and 90% of the
container volume. Specimens of 700g weight are preferable and the
container should then be a 4 litre wide necked bottle. The container
is weighed, or tared, the specimen introduced and re-weighed. Water
is then added equivalent to four times the weight of the specimen.
The container is closed.

The container is then agitated continuously on the shake
platform for 48 hours+/- 0.5 hours at 20 °C +/- 2 °C. After this, the
container is opened and the aqueous phase decanted off, centrifuged or
filtered through paper as appropriate. The aqueous portion is then
vacuum filtered through a 0.45 μm membrane and the filtrate analysed.

Analytical results obtained in mg/l of filtrate are converted to
mg leached per g of dry sample (L) as follows:

$$L = \frac{C}{250\ S}$$

where C = Concentration of constituent in mg/l

       S = Solids content in g/g

Compressive Strength

The unconfined compressive strength of a product sample may be determined by employing one of two types of machine:

i)   A compression testing machine of the lever, self indicating or proving ring type that has been verified as having a grade B accuracy complying with Section 2 of British Standard BS 1610:1964. It should be capable of providing a uniform rate of deformation of approximately 1mm/min.

ii)   One conforming to Section 15 of the ASTM Methods E4, Verification of Testing Machines. It must apply the load continously, rather than intermittently, and without shock.

The product under test is cast as a slurry into one of three types of mould:

i)   A tapered mould, having two steel plugs, of 50mm mean diameter and 100mm height with suitable ejecting plunger and displacing collar conforming to BS 1924:1975, Test 10.

ii)   Split mould with parallel bores of the above dimensions.

iii) Split cube moulds of 50mm or 100mm side.

iv) Mould of inside diameter 71mm and height 229mm for producing specimens of diameter 71mm and height 142mm as detailed in ASTM D1632 - 63.

The mould should be smeared with light oil or releasing fluid. The sample should be vibrated or tapped to remove entrained air. The specimen is cured for three days at an equilibrium humidity in an incubator at 20 °C +/- 2 °C, then removed from the mould and coated with paraffin wax. Curing at 20° C +/- 2° C is continued for 25 days.

After curing, the wax is removed from the specimen. If the ends are not plain they are either trimmed to a tolerance of 0.05mm or capped with gypsum plaster or cement paste to a tolerance of 0.05mm. The length and cross section area of the specimen are measured and the specimen placed centrally on the lower platten of the machine.

The force is applied to both ends of the specimen such that the rate of deformation is uniform and approximately 1mm/min. The maximum force exerted during the test is recorded.

The unit compressive strength of the specimen is calculated as maximum per unit area.

IONIZING WET SCRUBBERS
ADVANCED TECHNOLOGY FOR AIR POLLUTION CONTROL
FOR HAZARDOUS WASTE INCINERATORS

Dipl.-Ing. M. Griem
Geschäftsführer
CEILCOTE Korrosionstechnik GmbH
Am Brunnenweg
6083 Biebesheim/Rhein
Germany

ABSTRACT

Ionizing Wet Scrubbers (IWS) represent a new and advanced technology of low energy
gas cleaning. They combine the advantages of electrostatic precipitators and wet
scrubbers in a unique way. Noxious, corrosive and odorous gases, as well as coarse
and extremely fine solid and liquid particulate are removed with high efficiency
in one unit simultaneously. Details of this effective collection principle are followed
by a selection of formulas describing the design and obtainable removal efficiency.

Due to its simplified design and low pressure drop, operating energy costs of the
IWS are only a fraction of those for alternative air pollution control equipment.
The low overall cost and high removal efficiency for a variety of pollutants with
fluctuating loadings make the Ionizing Wet Scrubber ideal for hazardous waste in-
cincerator applications.

INTRODUCTION

Modern air pollution control technology offers more than just high collection efficiencies for dust, noxious gases or odors. Advanced systems are especially designed
to handle at a high level of operation reliability simultaneously various types of
pollutants with fluctuating concentrations. This can be done with a minimum of
overall energy usage and investment cost. One such system, the Ionizing Wet Scrubber
(IWS), has proven to be an excellent choice for cleaning exhaust gases from liquid
and/or solid waste incinerators. Air pollution control engineers select more often
the IWS-systems relative to Venturi scrubber / packed bed absorber combination
or other conventional systems. This is particularly true for large incineration facilities which generate high concentrations of extremely fine particulate (< 1 µm)
in the combustion process.

IONIZING WET SCRUBBER (IWS)

The IWS is a packed crossflow scrubber preceded by a high voltage ionizer
(Figure 1). The principles of electrostatic particle charging, image force attraction, inertial impaction and gas absorption are here combined in one unit to
collect coarse particles, aerosols and noxious gases at the same time.

The air flows horizontally through the scrubber, while the scrubbing liquid is
recycled from the top downwards to the integrated sump. In the first section
of the IWS the particles are electrostatically charged passing through a serie
of negative electrode wires and grounded electrode plates, which are only 25o
to 3oo mm wide. These 'mini-plates' are continuously flushed with liquid to
prevent build-up of a resistive layer of particles. After the particles are charged
they enter an electrically neutral packed bed which is constantly irrigated
with scrubbing liquid.

INERTIAL IMPACTION

Large particles (3 μm and larger) are removed by inertial impaction. Particles
impinge on a Tellerette packing or scrubbing liquid surface are captured and
flushed away. Tellerette packing is used almost exclusively within the IWS.
The unique geometrical design configuration of the Tellerette packing (Figure
2) makes it an excellent impingement target and highly efficient collector
due to its target size and large number of target areas. Greater particulate
collection efficiency, higher gas absorption rates and greater flow capacities
are characteristic of Tellerette packing as opposed to other types of packing
materials and design.

IMAGE FORCE ATTRACTION

Fine particles (3 μm and smaller) are collected in the IWS by the mechanism
of so-called image force attraction. This attraction takes place when an electrostatically charged particle comes within the boundary (approximately 1 mm)
of a Tellerette packing filament's or scrubbing droplet's neutral surface.

The particle induces an electrostatic charge of opposite polarity at the surface, which causes the particle to be attracted to the surface. This attraction is similar to that which would exist between a charged particle and its mirror image of equal distance behind a neutral surface (see Figure 3). Because of the great number of filaments and the large area of packing material per volume unit (12o - 18o m^2/m^3), Tellerettes are ideal for high collection efficiencies. Due to its open geometrical configuration dust loads up to 5 ooo mg/m^3 can be tolerated without plugging problems. Larger dust loads can be handled in using a prescrubber to remove the large particles before the charging section.

GAS ABSORPTION

Tellerette packing was originally designed for improved gas absorption. The packed bed of the IWS is an excellent absorber for soluble or reactive gases such as HCl, HF and SO_2. Efficiencies up to 99,9% and more can be obtained by selecting the right scrubbing liquid and packing depth. If scrubbing shall take place in two or more steps no additional equipment is required. This can readily be achieved with the crossflow scrubber in one body because its flow geometry permits the use of two or more packed beds in one unit and allows different scrubbing liquids to be used without mixing.

The scrubbing liquid is recirculated by a pump. Fresh liquid is added to balance evaporation losses and to keep absorbed gases in the recycled liquid. Overflow or blowdown of fluid from the scrubber is kept to a minimum.

COLLECTION EFFICIENCY

The IWS combines the advantages of electrostatic precipitators and wet scrubbers in a unique way. Within one device gaseous compounds are removed by highly efficient packing material, coarse particulate matter is collected by inertial impaction on scrubbing droplets and fixed internals and simultaneously the troublesome very fine particulates are captured with the help of electrostatic forces. The collection efficiency (η_E) required to meet regulations is calculated as follows:

$$\eta_E = \frac{c_{IN} - c_{OUT}}{c_{IN}} \qquad [1]$$

The general formula for collectors is:

$$\eta = 1 - e^{-NTU} \qquad [2]$$

NTU or the number of transfer units represents for a specific case how difficult the removal of contaminants is and describes the size and complexity of the collection device. Typical gaseous contaminants in the flue gases are HCl, HF, SO_2 and SO_3. The required NTU is commonly calculated as

$$NTU_G = \frac{C_{IN} - C_{OUT}}{\Delta C_{IN} - \Delta C_{OUT}} \, \ln \frac{\Delta C_{IN}}{\Delta C_{OUT}} \qquad [3]$$

The (ΔC)-values represent the concentration or pressure difference between partial pressure and vapour pressure above the scrubbing liquid at equilibrium. A length of 1,5 m of 2" Tellerette packing provide approximately NTU_G = 5 for HCl-gas which represents a removal efficiency of 99,3%. A packing depth of 3,o m achieves efficiencies above 99,9%.

Particulates are removed either by inertial impaction or by electrical forces. The NTU for E-filters is calculated in accordance with the so called Deutsch-Equation

$$NTU_E = w \, \frac{A_E}{\dot{G}} \qquad [4]$$

The efficiency depends on the relationship of electrical collection surface (A_E) to gasrate ($\dot{G}$) and the effective migration velocity (w). The w-value can change considerably with filter design and operation conditions.

Typical values for full scale units vary from o,o2 to o,2 m/s. The formula for mechanical separators can be expressed similar to the Deutsch-Equation as follows:

$$NTU_M = w_o \, \frac{A}{\dot{G}} \, \eta_A \qquad [5]$$

The value (w_o) represents the speed of the particle at which it hits its collection target. This speed is approximately the same as the gas velocity. Typical values are 1 to 2 m/s for low energy scrubbers. The ratio collection area (A) to gas rate ($\dot{G}$) can be much smaller and thus the equipment size and investment cost. The limiting factor, however, is the target efficiency (η_A). If a droplet, a filament or any other type of impingement target) with the size (D) is placed in a flue gas stream dust particles cannot follow the streamlines due to their mass inertia. All particles within the distance (b) in figure 4 hit the target and will be removed from the gas stream. The target efficiency is defined as:

$$\eta_A = \left(\frac{b}{D}\right)^2 \qquad [6]$$

The value (b) and, therefore, the target efficiency (η_A) depends on various parameters described in the Stoke-Number (K_T)

$$K_T = \frac{Cu \cdot \varsigma_p \cdot d^2 \cdot w_o}{18 \cdot \mu \cdot D} \qquad [7]$$

The number is proportional to the particle density (ς_p), the gas velocity (w_o) and the square of the particle size (d). If a given dust shows a mean diameter (d) of about lo μm the target efficiency can be as high as o,95 and the removal is almost complete after the first layer of targets. If the particle size is reduced to 1 μm the ratio w_o/D must be increased by the factor of hundred to obtain the same result. To pass small targets at high gas velocities a high energy input is required. A conventional low energy wet scrubber will not do the job, all particles follow the gas streamlines and flow around the targets. This can be changed dramatically if the particle is charged. The particles become attracted even if they travel on a gas streamline outside the target dimension (D). The distance (b) is now larger than (D) and the target efficiency will be larger than one. The removal efficiency will increase for particles of all sizes but especially for the very fine ones to a remarkable high level. The target efficiency for electrostatically augmented wet dedusters can be calculated as follows:

$$\eta_A = \left(\frac{5 \cdot Cu \cdot Q_p^2}{4 \cdot \pi \cdot \mu \cdot d \cdot w_o \cdot \epsilon_o \cdot D^2}\right)^{0,4} \qquad [8]$$

The actual efficiency will be higher since the above equation does not consider inertial impaction followed by condensation and agglomeration effects.

In common electrostatic precipitators the efficiency drops in the o,2 to o,6 µm size range. With fabric filters efficiency decreases sharply when particles are o,5 micron or smaller. The collection efficiency of an IWS decreases only slightly as fine particles become smaller.

The IWS is an 'allround collector'. A single-stage IWS will remove a fairly constant percentage of incoming particles regardless of particle size distribution and dust load. Where a higher collection efficiency is required, a second stage or even a third stage will be added and will have approximately the same collection efficiency as the first IWS-stage despite the reduction in loading.

ENERGY SAVINGS

Pressure drop through a single-stage IWS is only 35o to 5oo Pa and is controlled primarily by the pressure drop through the wet scrubber section. Additional small amounts of energy are required for high-voltage charging and for the recycle pump. In most incineration applications a two or sometimes three stage IWS will give the required collection efficiency for gas and dust. Power consumption for a two stage IWS is approximately 1,o Watt per 1 m³/h flue gas. This is similar to or lower than what is required for a costly combination of electrostatic precipitator and packed counter-current scrubbers for the acid gas collection.

For years when the cost of electrical power was relatively low for industrial users, the lowest investment cost approaches in handling fine particulate control was to use Venturi scrubbers. Pilot tests have demonstrated, however, that to equal the performance of the IWS in the collection of fine particle matter, venturis must operate at a pressure drop of 1o ooo to 25 ooo Pa. Multiple, high-head capacity fans and exorbitant power consumption are required for such applications. Multistage IWS systems with pressure drops of less than 1 ooo Pa offer substantial savings in energy consumption and operating cost. The difference in initial cost between a venturi/packed bed absorber system and an IWS installation can usually be offset in two years or less for many large incineration facilities and possibly for smaller plants, depending on the specific application.

The American Environmental Protection Agency (EPA) has performed last year an air pollution control cost study for hazardous waste incinerators and found that the IWS system is the most competitive air pollution control system in overall cost.

CONSTRUCTION AND CORROSION RESISTANCE

The IWS shell and most internal parts are fabricated of Duracor fiber-glass reinforced polyester (FRP) and other polymer materials. The use of plastic assures corrosion-free operation in the presence of corrosive gases as HCl, HF, SO_2 and SO_3. Additionally, structural support requirements are minimized by the unit's light weight. The only metal required is for the electrical components employed within the ionizing section. The electrode wires are manufactured of Hastelloy C. The welted electrode plates are either of Hastelloy or conductive FRP. The IWS is light in weight, inside and outside corrosion resistant and require only little maintenance. The unit incorporates various stages in one body and thus being very compact in design and easy to install.

HAZARDOUS WASTE INCINERATOR APPLICATIONS

The first commercial installation of an IWS system was used in the USA at the General Electric Silicone Products plant to remove high quantities of HCl and high loadings of very fine SiO_2 particulate from the effluent gases of an incinerator burning chlorinated solvents and chlorinated siloxis and silanes. In this application over 8o% of the particulate matter is o,2 um or below. This prototype has been in operation for lo years. HCl gas is reduced from 7o ooo mg/m³ to only 2o mg/m³. Dust removal is about 8oo mg/m³ down to 2oo mg/m³ and in a second step down to 5o mg/m³. The gas stream is 29 ooo m³/h. The same company has installed four more IWS at different locations to handle similar waste incinerator off-gases with capacities ranging from lo ooo m³/h to 85 ooo m³/h handling dust loads up to lo ooo mg/Nm³.

Other applications for the use of IWS systems are contract incineration operations. IWS equipment cleans flue gases generated from rotary kiln and afterburner incinerating solid and liquid wastes of all types.
One system is in operation since 1983. Both IWS installations handle 14o ooo m³/h effluent gas each in two parallel gas streams (Figure 5 and 6).

The first installation in Europe was a one-stage IWS removing HCl, SO_2 and aerosols from a hazardous waste incinerator in a chemical plant in Germany. The unit is handling 2o ooo m³/h since 1979 and will be upgraded soon with a second stage to meet the new German '86 TA-Luft' regulation.

Another IWS system in Italy is successful in operation since 1982. Liquid and solid chlorinated solvents and silanes are burned to full destruction. High loads of HCl and very fine SO_2 dust is removed by a two-stage IWS. In this case the body of the IWS is extended to incorporate a third stage if future regulations should make it necessary to add a third stage. The addition can be done without any changes on the outside of the system and with a minimum of downtime. (Figure 7)

SUMMARY

Altogether in the last 6 years about 2o IWS systems have been ordered and installed to clean effectively at low cost off-gases from all types of hazardous waste incinerators. The characteristics of the described system can be summarized as follows:

- The IWS is capable to collect coarse as well as very fine particles of o,1 µm and smaller to meet stringent regulations such as the TA-Luft in Germany.
- Collection efficiency varies little with load or particle size over a wide range.
- The packed crossflow scrubber can simultaneously absorb gases at no extra expense.
- The IWS body and most internals are constructed of fiberglass reinforced polyester. The unit is light in weight and corrosion resistant inside and outside.
- Energy consumption is very low and thus total operating cost.
- The IWS is compact in design, easy to install and to maintain. Operation is fully automated.
- Existing systems can be easily upgraded with the addition of a compact one-stage unit.

In general it can be stated that the growing demand for low cost high efficient removal devices for very small particles in a corrosive gas stream will increase the usage of Ionizing Wet Scrubbers in controlling off-gases from hazardous waste incinerators. The IWS has excellent collection capabilities and is highly energy-efficient.

NOMENCLATURE

A	Collection area in scrubbing zone
A_E	Collection area in electrostatic precipitator
b	Distance between critical particle path lines
C_{IN}	Concentration of contaminants before the equipment
C_{OUT}	Concentration of contaminants after the equipment
Cu	Cunningham Correction Factor
D	Collector target diameter or width
d	Particle diameter
$\dot{G}$	Volumetric flow rate of gas
K_T	Stoke-Number
NTU	Number of Transfer Units
NTU_E	Number of Transfer Units in electrostatic precipitators
NTU_G	Number of Transfer Units in gas absorbers
NTU_M	Number of Transfer Units in wet scrubbers
Qp	Charge on particle
w	Effective migration speed of particles in E-Filters
w_o	Relative speed particle to collection target (gas velocity)
ε_o	Permittivity of free space
μ	gas viscosity
η_A	Target efficiency
η_E	Collection efficiency
ρ_p	Density of the particle

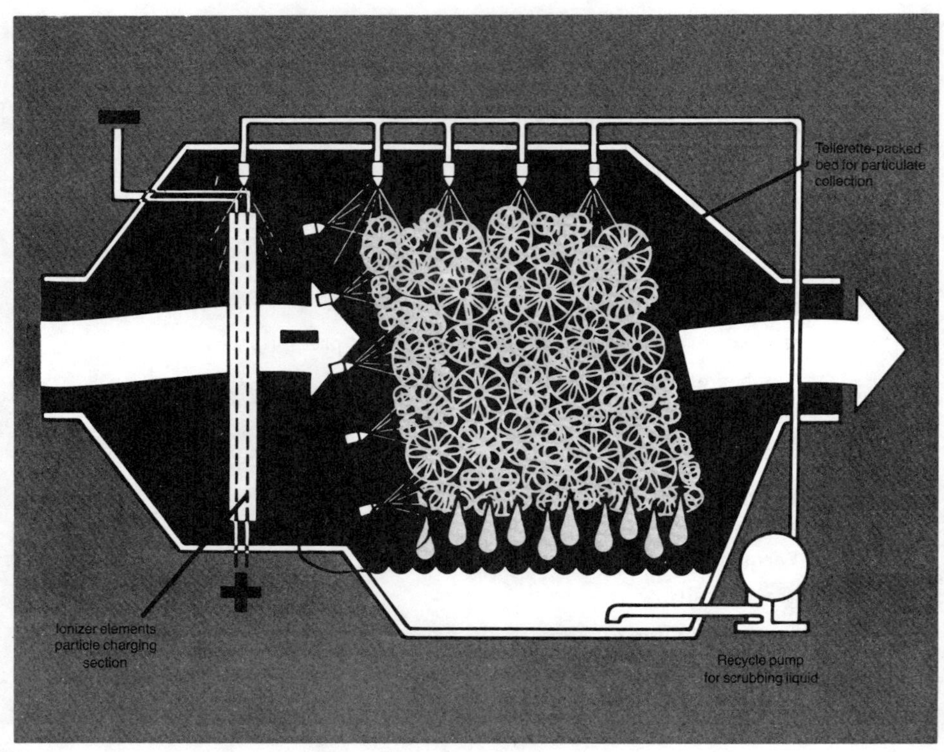

Figure 1:   Simplified Cross Section of IWS

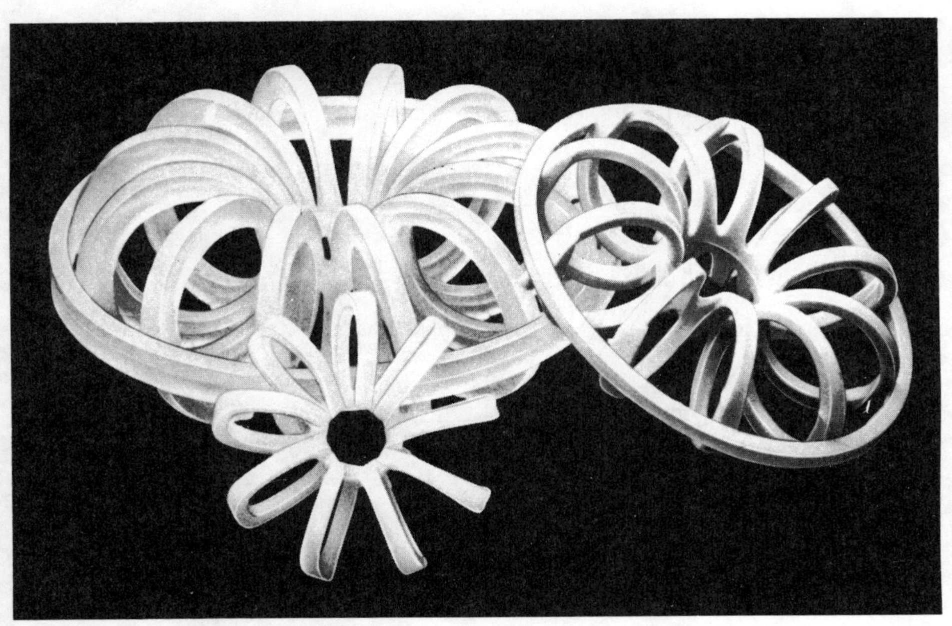

Figure 2:   Tellerette Packing (1", 2", 3" size)

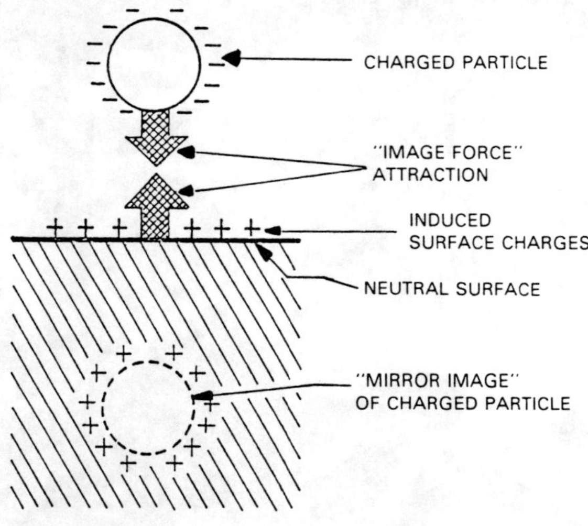

Figure 3:   Image Force Attraction

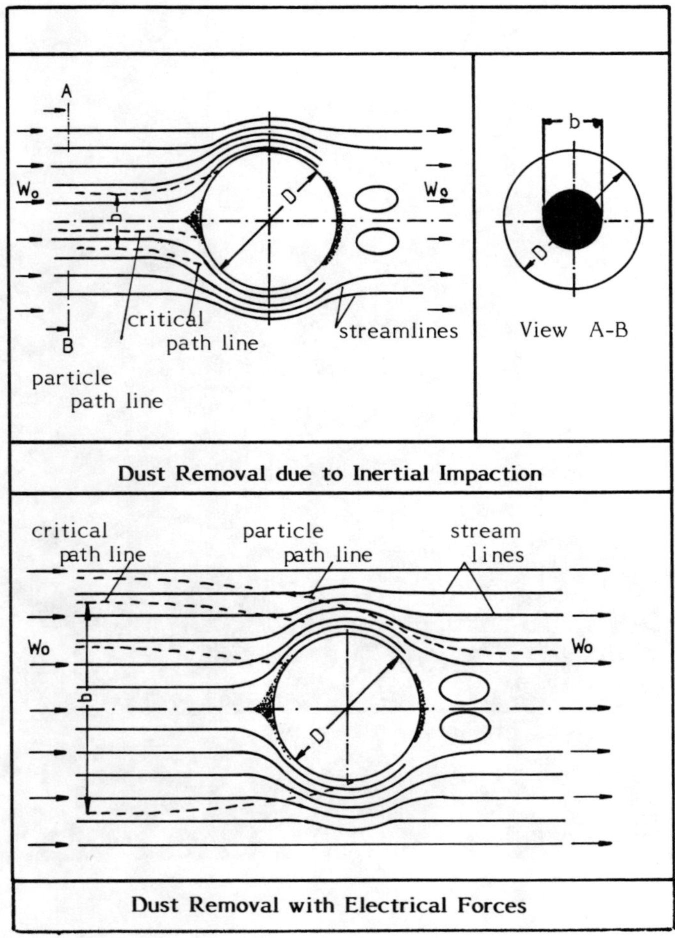

Figure 4:   Dust Removal on a Round Target

Figure 5:   The IWS removes acid gases and submicron particulate matter from hazardous waste incinerator emissions

Figure 6: Looking down the liquid waste feed pipe rack,
the secondary combustion chamber is connected to the
rotary kiln to the right and pollution control equipment
to the rear. Two IWS systems handling 14o ooo m³/h

Figure 7: Removal of HCl gas and submicron SiO_2 from solid and liquid waste incineration with an IWS system

Chapter 4

BIOLOGICAL TREATMENT METHODS

NEW APPROACHES AND APPLICATIONS OF BIOLOGICAL PROCESS TECHNOLOGY FOR THE TREATMENT OF TOXIC ORGANIC WASTES[*]

Alan F. Rozich
University of Delaware
Civil Engineering Department
Newark, DE 19716

and

Anthony F. Gaudy, Jr.
University of Delaware
Civil Engineering Department
Newark, DE 19716

ABSTRACT

Biological systems have the potential to offer cost-effective technologies for treating toxic organic waste streams and restoring hazardous waste sites. Although it has been shown that many industrial pollutants are amenable to treatment via biodegradation, pollution control engineers are often somewhat hesitant regarding the application of biological processes for toxic waste treatment. The purpose of this paper is to present a predictive modelling approach which has been developed over the last few years for use in the design and operation of activated sludge systems which are treating toxic, or inhibitory, wastewaters. Additionally, the application of this technology for developing strategies for managing joint municipal/industrial treatment situations will be discussed. Finally, process technology applications such as modified extended aeration systems and high-rate biological processes for toxic waste treatment will also be presented.

LIST OF SYMBOLS

D = dilution rate; ratio of flow, F, to volume of liquor, V, in aeration tank, time^{-1}.

F = rate of flow of incoming substrate or wastewater, $\text{vol} \times \text{time}^{-1}$.

F_R = recycle sludge flow rate, $\text{vol} \times \text{time}^{-1}$.

k_d = specific decay rate, time^{-1}.

K_i = the inhibitor constant; used in the Haldane and other model equations for inhibitory substrates to account for the effect of concentration of the inhibitory substrate on the specific growth rate, mgl^{-1}.

K_s = the saturation constant; the shape factor in the rectangular hyperbola form of the Monod relationship; it is defined as the concentration of the limiting substrate at which $\mu = 0.5\ \mu_{max}$, mgl^{-1}.

S = soluble substrate concentration, mgl^{-1}.

S_e = soluble substrate concentration in the clarifier effluent; it is equal to the soluble substrate concentration in the reactor for a completely mixed reactor, mgl^{-1}.

S_i = concentration of substrate in inflowing feed to a biological reactor, mgl^{-1}.

S_R = soluble substrate in the biological solids recycled to the reactor, mgl^{-1}.

$\bar{t}$ = nominal aeration tank detention time, V/F, time.

V = aeration tank volume, volume.

X = biological solids concentration in reactor, mgl^{-1}.

X_R = biological solids concentration in recycle flow to reactor, mgl^{-1}.

X_w = excess biological solids (sludge wasted), $mg \times time^{-1}$.

Y_t = true cell yield, $mgmg^{-1}$.

α = recycle flow ratio, F_R/F.

θ_c = mean cell residence time; it is the reciprocal of the net specific growth rate, μ_n, time.

μ = specific growth rate, $time^{-1}$.

μ^* = critical specific growth rate, $time^{-1}$.

μ_{max} = the maximum specific growth rate of which a biomass is capable under specified conditions of temperature and pH in the absence of any restriction on growth rate by limiting nutrient concentration, e.g., oxygen, nitrogen or carbon source, $time^{-1}$.

μ_n = the net specific growth rate, $time^{-1}$.

μ_n^* = critical net specific growth rate, $time^{-1}$.

INTRODUCTION

The need for innovative and cost-effective technology for ameliorating the spread of pollution due to hazardous and toxic wastes is one which is global in nature. As industrialization and economic development increase, so does the need for base chemicals which are required in a variety of manufacturing applications; unfortunately, increased chemical production and utilization also yield increased quantities of waste products and an assortment of by-products. The challenge, then, to the pollution control field is to meet present treatment needs while assessing industrial production trends in order to anticipate pollution control requirements for tomorrow. Failure to do so will be detrimental to the overall quality of the environment.

Although biological process technology has been available for waste treatment applications for some time, pollution control professionals often remain somewhat hesitant regarding the application of these systems for the treatment of hazardous and toxic wastes. The purpose of this paper is to review some of the recent advances which have been made regarding the activated sludge treatment of toxic organics. It should be stressed that one advantage of biological systems is that they tend to be more cost-effective than physical or chemical waste treatment technology. Additionally, since the operation of biological systems such as an activated sludge system can be modified in a manner that realizes no net production of organic sludges, i.e., that results in the total oxidation

of all toxic organic material in the influent, the need for subsequent
residue disposal, which characterizes many physical or chemical operations, is eliminated, or in any event is drastically reduced.

In this paper, a predictive modelling approach which can be utilized
for the design and operation of activated sludge systems which are
treating toxic, or inhibitory, wastewaters will be presented; this
modelling approach has been shown to be suitable for substances such as
phenol, orthochlorophenol, and 2,4-dinitrophenol. Additionally, the
utility of this technique as a means for grading the relative toxicity of
an industrial discharge to a municipal treatment plant biomass will be
discussed; this application should be of use in situations involving the
joint treatment of municipal and industrial wastewaters or as a tool for
administering industrial pretreatment programs. Regarding process
technology, the application of modified extended aeration processes and a
high-rate biological treatment process for hazardous waste treatment will
be discussed. The former process which is designed to achieve zero net
sludge production could also be modified for heavy metal removal, while
the latter system appears to be particularly suited for treating concentrated toxic waste streams.

KINETICS OF BIODEGRADATION OF TOXIC ORGANICS

Current techniques for devising design and operational strategies for
biological waste treatment systems are based on the work of Monod [1] and
Novick and Szilard [2]. These workers showed that the use of an engineering control such as flow rate could control the growth rate, μ, of a
microbial culture in a biological reactor. Furthermore, if the relationship between specific growth rate, μ, and exogenous substrate, or waste,
concentration, S is quantified, then, by controlling reactor growth rate,
one can also control the waste concentration in the reactor effluent.

Kinetically, the key consideration for the biological treatment of
hazardous organics is to recognize that, although microbial populations
are capable of metabolizing and are acclimated to a toxic or hazardous
organic, these organisms will nevertheless adhere to substrate inhibition
kinetics; this has been shown to hold true for a number of different
toxic organics [3,4,5,6]. It should be stressed that for noninhibitory
or nontoxic "conventional" wastes, e.g., domestic waste, brewery waste,
or sugar refinery waste, it is appropriate to utilize the familiar Monod
equation for relating μ to S; whereas the Haldane equation which accounts
for substrate inhibition is better suited for describing the growth
kinetics of acclimated microbial populations metabolizing a toxic
organic. A graphical comparison of the two contrasting kinetic expressions, the Monod (equation (1)) and the Haldane (equation (2)), is given
in Figure 1.

The inhibitory growth curve depicted in Figure 1 illustrates a key
point regarding the biological treatment of toxic wastes. This concerns
the peak, or critical, growth rate, μ^*, which is quantified by the
inhibition function. It has been shown that biological reactors which
are treating an inhibitory organic and are operated too close to the
critical growth rate, undergo rapid effluent deterioration, i.e., reactor
failure [3,4,7]; this was demonstrated analytically for chemostats which
are treating an inhibitory substrate by Spicer [8] who showed that these
reactors fail once the dilution rate, D, equals μ^* (D = μ for a
chemostat). Thus, for biological reactors which are treating hazardous
organics, and are characterized by inhibition kinetics, it is crucial to
quantify the μ^* of the biomass which is degrading the waste and then

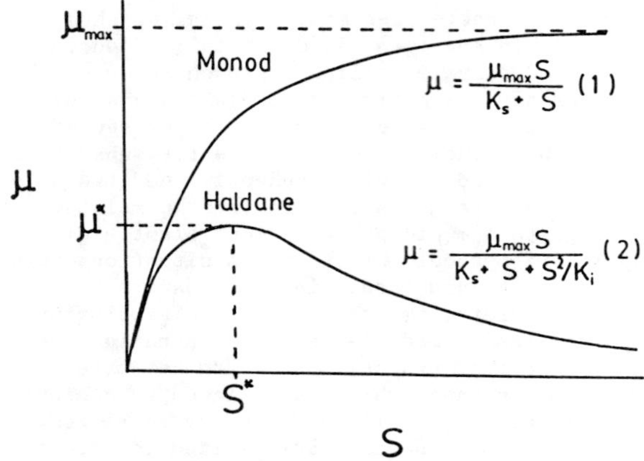

$$\mu = \frac{\mu_{max} S}{K_s + S} \quad (1)$$

$$\mu = \frac{\mu_{max} S}{K_s + S + S^2/K_i} \quad (2)$$

Figure 1. Comparison of Monod (Noninhibitory) and Haldane
(Inhibitory) growth rate expressions.

identify the operational location of this parameter, in order to avoid a
sudden deterioration in process performance, by determining appropriate
design and operational strategies which will prevent μ^* violations;
additional details of the application of this concept for toxic waste
treatment will be discussed in subsequent sections of this manuscript.

APPLICATIONS FOR INDUSTRIAL PRETREATMENT PROGRAMS

One of the main purposes of industrial pretreatment programs is to
prevent the admission of toxic waste materials that will either interfere
with the treatment processes in a municipal waste treatment plant or pass
through these plants untreated and be dispersed into the environment;
pass-through of toxic materials contributes to the effluent toxicity of a
municipal treatment plant and has a detrimental effect on the environment. In order to formulate policy for determining the types and
quantities of hazardous substances that can be satisfactorily treated by
a municipal plant's biotreatment reactor, it is important for managers to
have a predictive tool that will allow them to make admission decisions
regarding toxic materials. Since most admission decisions will be
determined primarily by the response that is given by the biological
system, then it is appropriate to screen potential inputs via a biokinetic testing protocol.

Given the fact that the kinetics of biodegradation of toxic organics
can be readily assessed [3,4], it is feasible to recommend that the
biokinetic growth response of the municipal treatment plant biomass to an
industrial discharge can be utilized as a quantitative means for
"grading" the waste in question; i.e., one can assess the growth kinetics
of the plant biomass on a particular waste and compare this response to
that which would be obtained using a standardized synthetic domestic
waste mixture. With this approach, one would be able to obtain a
comparative index of the biodegradability of various industrial inputs;
it should be noted that by doing repetitive evaluations, it should be
feasible to gauge the relative acclimatization of the plant biomass to a
given waste. That is, as the plant biomass acclimates to the industrial
input, one would expect an increase in the ability of the biomass to

treat the industrial waste. Additionally, it should be stressed that, before an industrial waste is admitted to a municipal system, some assessment of the potential for pass-through should be made. This can be done by applying some form of the ΔCOD test [9], i.e., one can assess that fraction of the waste that is readily amenable to biodegradation and analyze the residual which results from this test for generic toxicity in order to assess the potential contribution to the municipal plant's effluent toxicity. If this procedure indicates the potential for significant pass-through of toxicity, then only a small portion of the discharge should be admitted to the system. Whereas, if the test indicates that little or no toxicity will pass through, then admission would be permitted and the kinetic results would be utilized to establish operational guidelines at the plant.

Finally, the biokinetic evaluation data can also be utilized to assist with developing user charges for industrial contributors. That is, this information can be employed in concert with a pricing algorithm to insure that wastes which are more difficult to treat biologically are charged at appropriately higher rates. This will help to insure that parity is maintained in assigning user charges by accounting, on a biokinetic basis, for the level of difficulty that is associated with treating a particular industrial waste.

ACTIVATED SLUDGE PROCESS TECHNOLOGY FOR TREATING HAZARDOUS ORGANICS

The initial key consideration for optimizing the activated sludge treatment of toxic wastes is to quantify the relationship between the biomass growth rate, μ, and the waste, or substrate, concentration, S. It needs to be emphasized that appropriate methodologies must be employed in order to evaluate the growth kinetics of the acclimated biomass; details regarding the utilization of various procedures for evaluating inhibition kinetics have been presented elsewhere [3,4,10,11]. It needs to be emphasized that failure to account for the inhibitory effect of a toxic or hazardous waste on a given biomass will result in a gross overestimation of the reactor's effective operating range; reasons for this are illustrated and discussed elsewhere [12,13]. Consequently, it is mandatory for the technologist, operator or designer, to be cognizant of this consideration and to make provisions for evaluating the magnitude, range and variation of the biokinetic growth constants.

Equations which can be utilized to outline design and operational guidelines for activated sludge systems which are treating toxic wastes are derived by writing mass balances for biomass and substrate, X and S, respectively, around the reactor which is depicted in the process schematic in Figure 2. Steady-state conditions are assumed, the Haldane equation, Equation (2) is utilized to relate μ to S, and predictive equations for X and S are obtained by simultaneously solving the two mass balance equations; the detailed derivation and the predictive equations are given elsewhere [12,13].

It should be noted that the activated sludge process schematic shown in Figure 2 differs from most in that it emphasizes provision of both a controlled recycle sludge concentration and recycle flow rate via the use of a return sludge dosing tank or some other means to obtain the desired recycle sludge concentration. This operational strategy has been shown to be effective for providing stability for activated sludge systems which are treating both nontoxic and toxic wastes [7,9]; for toxics, it is imperative to provide the highest possible level of operational stability by engineering procedures. The reason for this is the fact that

426

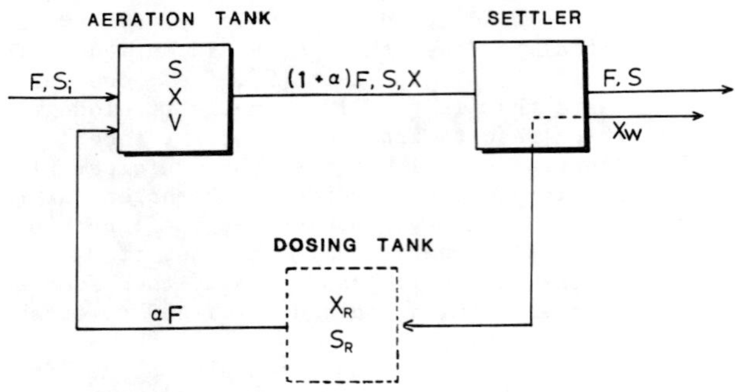

Figure 2.   Schematic for activated sludge system using recycle
sludge concentration, X_R, as a selectable engineering
control parameter for control of specific growth rate.

violation of the peak growth rate, μ^*, (see Figure 1) will result in
severe effluent deterioration [3,4,7]; these violations can be avoided by
maintaining stable substrate removal rates via the maintenance of a
steady concentration of recycle cells and a constant recycle flow rate.

Figure 3 presents a set of dilute-out curves which illustrate the
pitfalls that can be encountered if one does not account for the effect
of substrate inhibition in the design and operation of activated sludge
facilities which are handling toxic wastes. A comparison of process
performance predictions which were made for phenol using the afore-
mentioned predictive equations [12,13] indicates that the Monod-based
equations, which do not account for inhibition, provide a significant
overestimate of the effective operating range of the reactor as compared

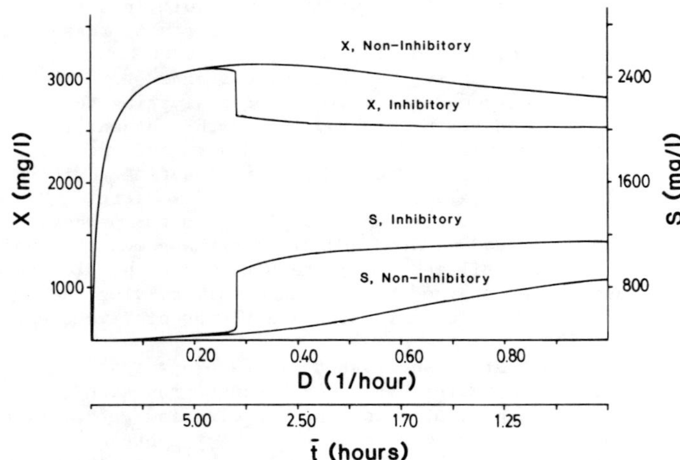

Figure 3.   Dilute-out curves for inhibitory (Haldane-based)
and noninhibitory (Monod-based) models for activated
sludge system treating phenol. Details are given in
reference [13].

to the prediction of the Haldane-based equations (inhibitory). It also
is evident in Figure 3 that the inhibition model predicts a critical
point whose violation produces a rapid decrease in effluent quality which
results from only slight decreases in hydraulic detention time, $\bar{t}$; this
failure point, which has been predicted using batch methods and verified
in continuous flow pilot systems for various toxic organics [3,4,7]
occurs when the reactor growth rate equals the peak or critical growth
rate, μ^*, which is designated by the inhibition function (see Figure 1).
The operational location of μ^* can be quantified using a critical point
analysis [14].

A critical point analysis is simply a method which allows one to
readily identify the point at which a reactor will realize a μ^*
violation, that is, that combination of detention time, $\bar{t}$,influent waste
strength, S_i, recycle sludge concentration, X_R, and recycle flow ratio,
α, that will cause the reactor growth rate to attain μ^*. Details and
examples of the derivation of critical point equations are given
elsewhere [14]. Critical point equations can be used to construct
critical point curves which can be employed as design and operational
tools for avoiding μ^* violations. Consider Figure 4. The plot

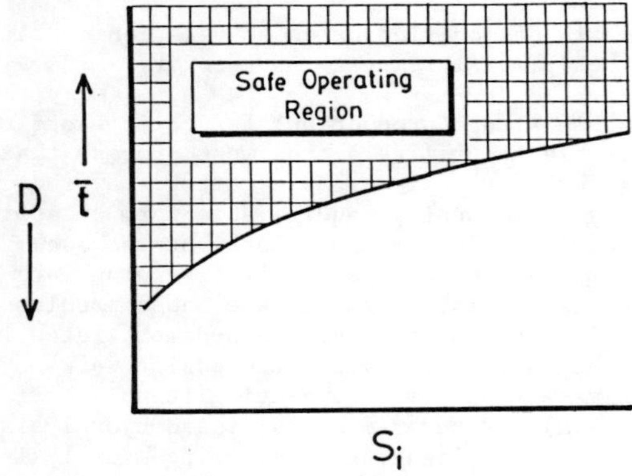

Figure 4.   Critical point curve for activated sludge reactor
            treating an inhibitory waste. The curve is defined
            by substitution of appropriate values of the biokinetic
            and engineering constants into critical point equations.

of the solid line which gives the value of $\bar{t}$ (for constant values of α
and X_R) at which the reactor attains μ^* at different S_i values can be
generated using a critical point equation. The cross-hatched area above
the curve comprises a safe operating region where the reactor is not apt
to experience sudden effluent deterioration and failure; conversely,
operating conditions which are on or below the curve risk operational
failure due to violation of the μ^* limit. Consequently, operation well
above the curve is required in order to prevent reactor failure; other
aspects and details of critical point analysis have been given in other
reports [14,15].

MODIFIED EXTENDED AERATION SYSTEMS

The application of extended aeration systems for the biological
treatment of toxic wastes is attractive for a number of reasons; a

process schematic is given in Figure 5. One goal of these systems is to

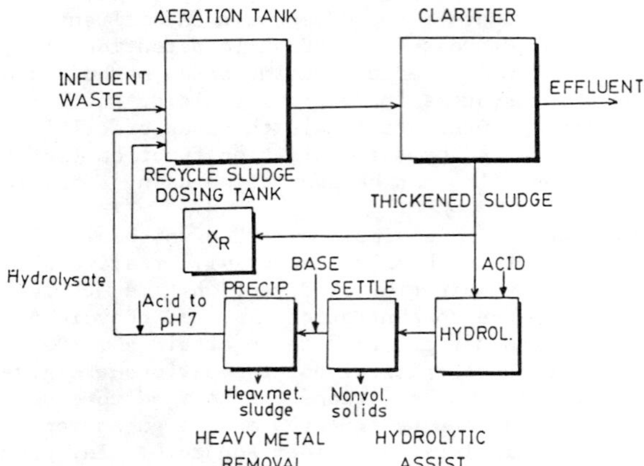

Figure 5.   Schematic of modified extended aeration system augmented
for heavy metal removal and recovery.

realize zero net organic sludge production; for toxic waste treatment, it
could be highly desirable to utilize a treatment process that realizes
the oxidation of all influent toxic organic matter. By recognizing that
biological systems may occasionally require some form of assistance to
enhance the auto-digestion of cells and to prevent the accumulation of
nonvolatile, inert material in the system, it is appropriate to recommend
the use of the hydrolytic assist to ameliorate these problems; the use of
this technique on pilot scale systems has been demonstrated in other
reports [9,16,17]. Basically, the hydrolytic assist works by acidifying
the sludge to approximately pH 1 and subjecting it to autoclaving
conditions (121°C, 1 ATM) for several hours; these hydrolysis conditions
may be subject to further optimization. Following hydrolysis, the
inorganic material is separated via settling and the organic liquor is
conveyed back to the aeration tank for treatment.

Figure 5 also depicts a heavy metal removal step. This was shown
because it has been demonstrated that this system can be modified for
heavy metal removal and recovery [18]; it should be noted that activated
sludge systems are capable of removing significant quantities of heavy
metals from waste streams [19]. Thus, if a given toxic organic waste
stream has significant quantities of heavy metals, it would be feasible
to utilize a modified extended aeration process for treatment and then
follow the hydrolysis step with a metal recovery procedure. Work in our
laboratories indicated that the process was feasible for synthetic wastes
containing very high concentrations of cadmium (Cd^{++}) [18] and has valid
potential for field application.

Finally, the most attractive feature regarding the application of
modified extended aeration systems is perhaps the fact that these systems
have the potential to be extremely cost-effective when compared to
alternative systems that utilize conventional methods of sludge disposal;
this point was made in a cost comparison analysis which indicated that,
for a 37,800 m^3/d (10 MGD) municipal treatment plant, an annual savings
of $1,000,000 (U.S.) in operation costs could be realized if the plant
utilized the modified extended aeration system in lieu of conventional

sludge disposal technologies [20]. Thus, given the expense that characterizes the disposal of residues from hazardous and toxic waste treatment facilities, it would seem highly desirable to utilize a treatment system that minimizes residue production, especially if it will be cost-effective over conventional technologies.

HIGH-RATE BIOLOGICAL TREATMENT PROCESS

The basis for a high rate process lies in the concept of selector technology, i.e., the use of one compartment of a biological reactor system to select for a particular microbial population in order to meet specific treatment goals. There are many examples of the application of this concept in biological waste treatment. For example, long mean cell residence times (MCRT's or θ_c's > 5 days) are specified to select for nitrifying populations in order to remove ammonia from wastewaters [21]. Also, the Bardenpho process advocates the application of an anoxic zone to achieve denitrification [22] while the A/O process employs an anaerobic zone to select for cells that can assimilate and store phosphorus [23]. The use of a variety of environmental conditions which are subject to control in a sequencing batch reactor is recommended for implementing feast/famine conditions by Chiesa, et al. [24] in order to cultivate populations which exhibit good settling characteristics. Likewise, it appears that one can realize the high-rate treatment of toxic organics by employing one reactor as a selector to grow high-rate organisms.

The basis for suggesting the high-rate technology is found in previously-reported work on the high-rate biodegradation of phenol [25]. Briefly, the results of initial studies showed that, by maintaining cultures in continuous flow reactors at high steady state phenol concentrations, it was possible to select organisms which were capable of substrate utilization rates that were almost ten times higher than those of cells which were cultured at low phenol concentrations [25]. These results suggested the feasibility of utilizing this concept as part of a continuous flow system. Subsequent work which reported the results of a pilot scale study of a high rate system showed that the high rate cells could be utilized in a continuous flow treatment process [26]. In addition to providing relatively high treatment rates, the system was also capable of delivering effluents which had lower phenol concentrations than conventional systems [26], which seems to indicate that a high rate process would be desirable both because of increased treatment rates and enhanced process performance. Thus, this system would have application for treating the relatively concentrated hazardous and toxic wastes that emanate from process industries.

A potential configuration for the field application of a high-rate process for treating phenolic wastes is shown in Figure 6; equations for design and operation for this system have been derived and given elsewhere [26]. The operational strategy for this system would be to grow the high rate cells in Stage 1 and use Stage 2 to bring effluent concentrations to acceptable levels. It is recommended that some provision be made to insure that a constant flux of recycle cells is conveyed to both reactors in order to maintain stable removal rates; a constant X_R dosing tank or other technique which realizes constant recycle mass flow rates is suggested. Since it is desirable to reduce target organic concentrations to the lowest possible levels in Stage 2, it is appropriate to operate this stage in an extended aeration mode since these systems have been shown to exhibit the highest percentage removal for toxic organic pollutants [27].

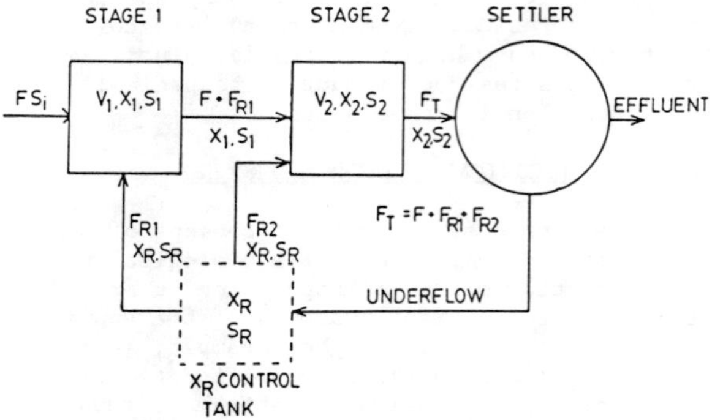

Figure 6.   Schematic of high-rate biological process for
toxic waste treatment.

Although high-rate biological systems offer a potentially innovative
method for treating toxic organics, it should be emphasized that additional developmental work is needed before these systems can be utilized
in full-scale treatment situations. However, it should be stressed that
the application of this process technology will probably not be contin-
gent on the design and/or construction of new facilities. Rather, it
appears that current activated sludge facilities could be converted to a
high-rate process with relatively little capital outlay.

SUMMARY

The utilization of biological processes for treating hazardous
organic wastes can be a feasible and cost-effective alternative to
physical or chemical treatment systems. By recognizing and quantifying
the key kinetic features that characterize the biodegradation of toxic
wastes, engineers and operators can develop guidelines which can be
utilized to enhance the design and operation of activated sludge facilities which must treat toxic organic wastes.

In this paper, the differences that exists between the biodegradation
kinetics of toxic and nontoxic wastes were discussed as well as the
implications of these differences for the design and operation of
biological facilities that are treating hazardous organics. Additionally, the use of a biokinetic approach for determining policies for
admitting toxic organic wastes into municipal treatment systems was
outlined; by basing admission policies on biokinetic data which are
obtained using municipal plant biomass as seed, it is feasible to suggest
that pollution control professionals will be able to formulate strategies
for pretreatment programs that have a relatively sound technological
basis.

With regard to process technology, methodologies for delineating
design and operational guidelines for activated sludge facilities for
toxic waste treatment were outlined. Additionally, the application of
modified extended aeration processes were discussed; these systems
feature a sludge minimization scheme which has the potential for realizing significant cost-savings in comparison to conventional technology.
Finally, the use of high-rate biological systems for toxic waste treatment was discussed. This technology has the potential of realizing
significantly enhanced biodegradation rates for toxic organics in

treatment systems and it appears that the high-rate systems may be capable of producing lower effluent concentrations of toxic organics than conventional process technology can achieve.

REFERENCES

1. Monod, J., The growth of bacterial cultures, _Annual Rev. of Microbiol._, 1949, _3_, 371.

2. Novick, A. and Szilard, J., Experiments with the chemostat on the spontaneous mutation of bacteria, _Proceedings, Nat. Acad. of Science_, _USA_, 1950, _36_, 708.

3. Rozich, A. F., Gaudy, A. F., Jr. and D'Adamo, P. C., Selection of growth rate model for activated sludges treating phenol, _Water Res._, 1985, _19_, 481.

4. Gaudy, A. F., Jr., Lowe, W. L., Rozich, A. F., and Colvin, R. J., Practical methodology for predicting critical operating range of biological systems treating inhibitory substrates, _Presented_, 59th Annual Water Pollution Control Federation Conference, Los Angeles, CA, 1986.

5. Klecka, G. M. and Maier, W. J., Kinetics of microbial growth on pentachlorophenol, _Appl. Environ. Microbiol._, 1985, _49_, 46.

6. Papanastasiou, A. C. and Maier, W. J., Kinetics of the biodegradation of 2,4-dichlorophenoxyacetate in the presence of glucose, _Biotechnology and Bioengineering_, 1982, _24_, 2001.

7. Rozich, A. F. and Gaudy, A. F., Jr., Response of phenol acclimated sludge to quantitative shock loadings, _J. Water Poll. Control Fed._, 1985, _57_, 795.

8. Spicer, C. C., The theory of the bacterial constant growth apparatus, _Biometrics_, 1955, _6_, 225.

9. Gaudy, A. F., Jr. and Gaudy, E. T., _Microbiology for Environmental Scientists and Engineers_, McGraw-Hill, New York, NY, 1980.

10. D'Adamo, P. C., Rozich, A. F. and Gaudy, A. F., Jr., Analysis of growth data with inhibitory carbon sources, _Biotechnol. Bioeng._, 1984, _26_, 397.

11. Castens, D. J. and Rozich, A. F., Analysis of batch nitrification using substrate inhibition kinetics, _Biotechnol. Bioeng._, 1986, _28_, 461.

12. Rozich, A. F., Gaudy, A. F., Jr. and D'Adamo, P. C., Predictive model for the activated sludge treatment of phenolic wastes, _Water Res._, 1983, _17_, 1453.

13. Gaudy, A. F., Jr. and Rozich, A. F., Design and operational model for activated sludge treating inhibitory carbon sources, _Civ. Eng. Design Pract. Eng._, 1983, _2_, 55.

14. Rozich, A. F. and Gaudy, A. F., Jr., Critical point analysis for toxic waste treatment, *J. Environ. Eng. Div., Proc. Am. Soc. Civ. Eng.*, 1984, **110**, 562.

15. Rozich, A. F. and Gaudy, A. F., Jr., Process technology for the biological treatment of toxic organic wastes, *Presented*, Third International Symposium on Industrial and Hazardous Waste, Alexandria, Egypt, 1985.

16. Gaudy, A. F., Jr., Yang, P. Y. and Obayashi, A. W., Studies on the total oxidation of activated sludge with and without hydrolytic pretreatment, *J. Water Poll. Control Fed.*, 1971, **43**, 50.

17. Yang, P. Y. and Gaudy, A. F., Jr., Control of biological solids concentration in extended aeration, *J. Water Poll. Control Fed.*, 1974, **46** 543.

18. Lowe, W. L. and Gaudy, A. F., Jr., Removal of cadmium at high and low dosages by an extended aeration process, *Proceedings*, 40th Industrial Waste Conference, Purdue University, Ann Arbor Press, Ann Arbor, MI, 1986, pp. 431-442.

19. Elenbogen, G., Sawyer, B., Lue-Hing, C., Rao, K. and Zeng, D., Studies of uptake of heavy metals by activated sludge, *Proceedings*, 40th Industrial Waste Conference, Purdue University, Ann Arbor Press, Ann Arbor, MI, 1986, pp. 493-505.

20. Rozich, A. F. and Gaudy, A. F., Jr., Modified extended aeration plant for minimizing sludge production, *Proceedings*, 40th Industrial Waste Conference, Purdue University, Ann Arbor Press, Ann Arbor, MI, 1986, pp. 775-784.

21. U. S. Environmental Protection Agency, Process design manual for nitrogen control, U.S.E.P.A., Office of Technology Transfer, Washington, D.C., 1975.

22. Burdick, C. R., Reflins, R. R. and Stensel, H.D., Advanced biological treatment to achieve nutrient removal, *J. Water Poll. Control Fed.*, 1982, **54**, 1078.

23. Tracy, K. D. and Flamino, A., Kinetics of biological phosphorus removal, *Presented*, 58th Annual Water Pollution Control Federation Conference, Kansas City, MO, October, 1985.

24. Chiesa, S., Irvine, R. and Manning, J., Feast/famine growth environments and activated sludge population selection, *Biotechnology and Bioengineering*, 1985, **27**, 562.

25. Colvin, R. J. and Rozich, A. F., Phenol growth kinetics of heterogeneous populations in a two-stage continuous culture system, *J. Water Poll. Control Fed.*, 1986, **58**, 326.

26. Rozich, A. F., Colvin,. R. J. and Gaudy, A. F., Jr., High-rate biological process for treatment of phenolic wastes, *Proceedings*, International Conference on Innovative Biological Treatment of Toxic Wastewaters, Arlington, VA, June, 1986, (in press).

27. Gaudy, A. F., Jr., Kincannon, D. F. and Manickam, T. S., Treatment compatibility of municipal waste and biological hazardous industrial compounds, NTIS PB83-105536, Volume I, PB83-105544, Volume II (Appendix), U. S. Department of Commerce, 1982.

PRETREATMENT OF SEWAGE SLUDGE
AND ITS SUBSEQUENT ANAEROBIC DIGESTION

R.J. Martin
Dept. of Civil Engineering,
University of Birmingham,
UK

A.M. Bruce
Water Research Centre,
Stevenage,
UK

and

S.B.O. Jallow
Dept. of Water Resources,
Banjul,
The Gambia

ABSTRACT

Attempts were made to increase the anaerobic digestibility of sewage sludge to increase digester gas yields and improve dewaterability by means of mechanical, thermal and chemical pretreatment procedures. Two types of sewage sludge were used: an activated sludge and a mixed (primary plus humus) sludge from different treatment plants. Pretreatment procedures led to increased gas production and increased reduction in volatile solids. The effects of pretreatment were greater for the activated sludge than for the mixed sludge. In some cases, sludge dewaterability was improved.

INTRODUCTION

A fundamental part of sewage treatment is the concentration of contaminants as a wet sludge. Estimates of costs associated with the handling and disposal of sewage sludge vary, of course, but a useful rule of thumb is that approximately 40-50% of the total cost of construction and operation of a sewage treatment plant is incurred in the treatment and disposal of sludge[1].

Anaerobic digestion is widely used, particularly at large works, to achieve partial stabilization of primary and secondary sludges. These sludges have solids contents of around 20,000-60,000 mg/l (2-6%), about

70% of which are volatile. Digestion reduces the volatile content to about 50% and the total solids to about 70% of the original values. The remaining organic solids are relatively stable. In addition, digestion largely destroys pathogenic micro-organisms if any are present; this is of importance if digested sludge is to be spread on agricultural land.

Anaerobic digestion is a two-stage bacterial fermentation process taking place in the absence of oxygen. In the first stage, organic matter is broken down to short chain fatty acids; in the second stage, the acids are converted to carbon dioxide and methane.

Pretreatment of sludge has largely been concerned with conditioning the sludge to enhance thickening and dewaterability. Mechanical dewatering of sewage sludges usually necessitates chemical conditioning to ensure improved sludge dewaterability and the prevention of blinding of the filter medium by small sludge particles[2]. Thermal conditioning, otherwise known as heat treatment, has been used to break down the colloidal structure of the sludge thereby facilitating separation of solids and liquor. Operational problems relating to odour, corrosion and high maintenance costs have been significant. Thus, chemical conditioning is very much more frequently encountered than thermal conditioning.

The objective of the research reported here was to employ sludge pretreatment procedures to increase the anaerobic digestibility of the sludge. Increasing the digestibility could mean an increase in gas production, a greater stabilization of the organic matter and perhaps an improved dewaterability of the final digested sludge.

Very little work has been done on sludge pretreatment for subsequent digestion (as opposed to sludge pretreatment for subsequent dewatering). Haug et al. investigated the effect of thermal pretreatment on digestibility and dewaterability of sludges[3]. Thermal pretreatment was found to improve digester gas yields and enhance the destruction of volatile solids. Sludge dewaterability was also improved.

Chemical sludge pretreatment for subsequent digestion has, essentially, been limited to the addition of lime to raise the pH of a sludge. The bacteria associated with the second stage of the digestion process are inhibited by low pH conditions. The organisms function best at around pH 7. The problem of pH sensitivity of the methane-forming bacteria is, of course, exacerbated by the fact that the first stage of the mechanism produces organic acids. A process imbalance can be corrected by the addition of lime. The addition of lime is not a pretreatment process in the true sense; it constitutes a rescue operation simply to maintain digestion.

Haug et al. reported very briefly on the use of acidic and basic conditions in the thermal pretreatment studies referred to already[3]. Acid-treated and alkali-treated sludges were more digestible for a few days, (the acid-treated sludge more than the alkali-treated sludge), but soon became less successful due to the production of inhibitory materials resulting in decreases in gas production.

In addition to chemical and thermal pretreatment prior to digestion, it was felt that mechanical pretreatment should be investigated. Sludge can be significantly heterogeneous; for large agglomerates of solids, it might be difficult for the anaerobic bacteria to reach the organic matter in the centres of these agglomerates. By mechanically breaking down such agglomerates, it is possible that digestibility could be enhanced.

Digestion is traditionally an expensive process and accounts for about 20% of the total capital cost of a new sewage treatment works. A 1981 survey concluded that not enough was being done to exploit the activities of the organisms in anaerobic sludge digestion[4]. The rationale of the objective of the research reported here is therefore self-evident.

EXPERIMENTAL PROCEDURE

Two types of sewage sludge were used in the studies: an activated sludge from Letchworth treatment works and a mixed sludge (primary plus humus) from Hatfield treatment works. These two towns, each of population around 40,000, are in the commuter belt north of London. The two plants are conventional. For both plants about 90% of the raw sewage comes from domestic waste and about 10% comes from trade waste.

At Hatfield, the primary and humus sludges are co-settled to give a sludge consisting of approx. 72% primary and 28% humus sludge. After collection, the sludge was sieved through a 5 mm sieve and concentrated to between 3.5 and 4.5% volatile solids and stored at 4°C for subsequent tests and for feeding the seed digesters.

After collection from Letchworth, the activated sludge was thickened by centrifugation to about 3.5% volatile solids and stored at 4°C as before.

The seed digesters consisted of 20 litre bottles each containing 15 litres of continuously stirred sludge in a 35°C heated water bath. Operation was semi-continuous with a litre of digested sludge removed daily and replaced by a litre of undigested sludge. The seed digesters were fed only on the Hatfield mixed sludge.

Small batch digesters were constructed. Bottles of 500 ml capacity were placed in a 35°C heated water bath. Each bottle was connected, via tubing, to the bottom of a tall glass cylinder standing in an unheated bath of water containing 0.5% hydrochloric acid and coloured with methyl orange. The acid was added to prevent loss of carbon dioxide from the digesters through dissolution in the water, and also to prevent bacterial growth in the cylinders. The colouring facilitated reading the scales on the multiple cylindrical gas collectors.

In all the tests performed, a sample of sludge, pretreated or otherwise, was mixed with freshly collected seed sludge in the correct ratio and placed in the 500 ml batch digester. After allowing about 10 minutes, the cylinder was filled to the top with the coloured, acidified, unheated water by means of a pump. As digester gas was produced in the multiple digesters, water was displaced from the multiple columns which were calibrated for volume of digester gas produced. All the batch tests were carried out in duplicate.

Preliminary experiments were carried out to ascertain the best combination of undigested and seed sludges by monitoring daily rates of gas production for the digestion of variously proportioned mixtures over

30 days. As the undigested sludge proportion was increased above 25% by volume, the initial rate of digestion decreased; this is a clear indication of overloading the digestion process. It was decided to operate all the batch tests using a combination of 25% undigested sludge and 75% seed sludge. This combination gave 90% of the total gas produced within 9 days. These experiments clearly showed that sludge digesters in practice could be operated at lower retention times without any significant reduction in total gas yield but with a significant saving in the capital cost associated with a smaller digestion plant of lower retention time. Conventional retention times for sludge digesters are usually around 20 to 30 days. Even with a combination of 62.5% undigested sludge and 37.5% seed sludge, 90% of the total gas was produced within 16 days.

Mechanical pretreatment involved homogenizing sludges for varying lengths of time using different homogenizers. A Moulinex 530 blender, a Silverson L2R heavy duty stirrer and a Braun 53030 MSK homogenizer were used. The sludges were then mixed with seed sludge and digested.

Thermal pretreatment consisted of heating undigested sludge samples at different temperatures for varying periods. The temperatures were 70, 121 and 135°C. Heating at 70°C was done in a water bath with reflux condensers to minimize loss of volatile matter. Heating at 121 and 135°C was performed in autoclaves. After heating, the sludge samples were cooled to room temperature before mixing with seed sludge and digesting.

Chemical pretreatment involved treating undigested sludge samples with hydrochloric acid to achieve a pH of 1.3, or sodium hydroxide to achieve a pH of 12.5. The sludges were then heated as before, cooled to room temperature and neutralized before mixing with seed sludge and digesting.

RESULTS AND DISCUSSION

A full record of the results has been produced by Jallow[5]; of necessity, the results presented in this paper have had to be more than somewhat abbreviated.

Mechanical pretreatment did have, in general, a positive effect on the subsequent digestion of sludge samples, but this effect was slight. Results are presented in Tables 1-4. The results include volatile solids (suspended _and_ dissolved) both before (bd) and after (ad) digestion; reductions in volatile solids are expressed as percentages. Total digester gas production is reported as litres of gas produced per kilogram of volatile solids in the undigested sludge proportion. (The gas produced by the seed sludge has been subtracted.) Enhancement of gas production resulting from pretreatment is expressed as a percentage.

The three types of homogenizer work by different actions and it is therefore unwise to compare the results achieved with different pieces of equipment. The Moulinex blender is a conventional high-speed rotating cutting blade; the Silverson apparatus is much the same but with fittings to control the sizes of the particulate matter desired. The Braun homogenizer is significantly different and achieves its disintegrating action by oscillation of the sample at high speed with glass beads; the application of this homogenizer to break down sludge flocs has been reported by Pike et al.[6].

TABLE 1
Use of Moulinex blender on activated sludge

Homogenizing time (mins)	VS bd (% wt.)	VS ad (% wt.)	VS reduction (%)	Gas production (1/kg VS added)	Gas increase (%)
0	3.40	2.01	40.9	269.1	0
2	3.40	1.96	42.4	278.9	3.64
4	3.40	1.97	42.1	289.2	7.47
6	3.40	1.97	42.1	263.2	-2.19
8	3.40	1.96	42.4	242.4	-9.92
10	3.40	1.97	42.1	238.1	-11.52

TABLE 2
Use of Silverson stirrer on mixed sludge

Homogenizing time (mins)	VS bd (% wt.)	VS ad (% wt.)	VS reduction (%)	Gas production (1/kg VS added)	Gas increase (%)
0	3.57	1.76	50.7	459.3	0
10	3.57	1.72	51.8	468.5	2.00
20	3.57	1.64	54.1	474.8	3.37
40	3.57	1.64	54.1	463.8	0.98

TABLE 3
Use of Silverson stirrer on activated sludge

Homogenizing time (mins)	VS bd (% wt.)	VS ad (% wt.)	VS reduction (%)	Gas production (1/kg VS added)	Gas increase (%)
0	3.17	2.02	36.3	309.4	0
10	3.17	1.88	40.7	340.4	10.02
20	3.17	1.90	40.1	320.7	3.65
40	3.17	1.88	40.7	334.3	8.05

TABLE 4
Use of Braun homogenizer on activated sludge

Analysis	Non-pretreated	Homogenized
Homogenizing time (mins)	0	5
VS bd (% wt.)	3.30	3.17
VS ad (% wt.)	1.95	1.35
VS reduction (%)	40.9	57.4
Gas production (1/kg VS added)	273.3	284.1
Gas increase (%)	0	3.95
CST bd (s)	283	>6000
CST ad (s)	309	371

The results do indicate that the effect of pretreatment was greater for the activated sludge samples than for the mixed sludge samples. Tables 2 and 3 arise from results obtained with the same homogenizer; a superior enhancement of gas production, in percentage terms, was observed for the activated sludge samples. In absolute terms of course, the mixed sludge samples gave considerably greater gas yields and volatile solids reductions. This is due to the fact that the mixed sludge contains more biodegradable matter via the primary sludge component.

Table 1 is interesting in that the longer times of homogenizing gave less digester gas than did the sludge sample that had not received any mechanical pretreatment; it will be observed that the percentage VS reductions are virtually identical for blending times of 2 to 10 minutes. Since the VS bd was determined before pretreatment, the results would indicate that some volatile solids were lost during the blending period, the amount lost increasing with the duration of blending.

The volatilization of material during mechanical pretreatment would be a bigger factor for the mixed sludge than for the activated sludge and this could be a reason why mechanical pretreatment had less effect on mixed sludge digestion than on activated sludge digestion. A further reason could simply be that the greater digestibility of the mixed sludge minimizes any impact tht pretreatment might make.

Thermal pretreatment results are presented in Tables 5-7. In general, thermal pretreatment did have a beneficial effect on subsequent digestion. Table 5 clearly shows that pretreatment at 70°C for 30 minutes led to improvements in VS removal and gas production; increasing the period of heating did not further these improvements. As in the case of mechanical pretreatment, loss of volatile matter during thermal pretreatment is indicated.

Tables 6 and 7 show, as before, that the effect of pretreatment was greater for the digestion of activated sludge. As before, the enhanced inherent digestibility of the mixed sludge together with its greater susceptibility to loss of volatile matter are thought to be significant. For the activated sludge samples, increasing the heating temperature was obviously successful; the same may not be stated for the mixed sludge samples.

The complexity and variability of sewage sludges may be noted from even casual observations of Tables 1-7; differences in gas production for sludge samples that had not been pretreated can be readily observed. With such variable source material, individual numerical results should not be over-interpreted.

Thermal pretreatment assists in the breakdown of those organic compounds which are only degraded with difficulty, or perhaps not degraded at all, during normal anaerobic digestion. The resulting breakdown products are more amenable to anaerobic biodegradation and conversion to digester gas. Furthermore, thermal pretreatment helps to rupture the cell walls of dead bacteria thereby releasing cytoplasmic material which is readily digestible.

The fact that thermal conditioning prior to sludge dewatering has proved to be troublesome has been commented on by Haug[7]. In a largely qualitative and predictive paper, Haug opined that most of the

TABLE 5
Effect of preheating on activated sludge digestion

Heating condition	VS bd (% wt.)	VS ad (% wt.)	VS reduction (%)	Gas production (1/kg VS added)	Gas increase (%)	CST bd (s)	CST ad (s)
No heating	3.01	1.70	43.5	299.5	0	282	309
70°C; 30 mins	3.00	1.62	46.0	322.3	7.61	332	269
70°C; 60 mins	3.30	1.78	46.1	310.1	3.54	665	340
70°C; 120 mins	3.22	1.74	46.0	307.5	2.67	775	327
70°C; 240 mins	2.43	1.27	47.7	308.9	3.14	723	325

TABLE 6
Effect of preheating on activated sludge digestion

Heating condition	VS bd (% wt.)	VS ad (% wt.)	VS reduction (%)	Gas production (1/kg VS added)	Gas increase (%)	CST bd (s)	CST ad (s)
No heating	3.04	1.52	50.0	300.1	0	247	258
60 mins; 70°C	3.13	1.52	51.4	320.9	6.93	1274	196
60 mins; 121°C	3.55	1.60	54.9	346.8	15.56	3308	393
60 mins; 135°C	3.81	1.62	57.5	385.2	28.36	3166	374

TABLE 7
Effect of preheating on mixed sludge digestion

Heating condition	VS bd (% wt.)	VS ad (% wt.)	VS reduction (%)	Gas production (1/kg VS added)	Gas increase (%)	CST bd (s)	CST ad (s)
No heating	3.63	1.86	48.8	596.8	0	283	232
60 mins; 70°C	3.55	1.85	47.9	612.3	2.60	331	229
60 mins; 121°C	4.50	1.96	56.4	593.0	−0.64	220	196
60 mins; 135°C	4.14	1.93	53.4	561.8	−5.86	129	230

operational problems associated with thermal conditioning could be overcome by having a thermal pretreatment-anaerobic digestion system in which both solids and liquor produced in the first phase would be fed into the digester. He suggested that the residual heat from the heating phase would be enough to maintain mesophilic, or even thermophilic, digestion. It was suggested that by tranferring both solids and liquor to the digester without exposure to the atmosphere, problems relating to odour and the heavy organic load of the liquor could be eliminated.

Results for combined thermal and chemical pretreatment are presented in Tables 8-11. Samples of sludge that were pH modified but not heated were left in the acidic or alkaline condition for 60 minutes so that a comparison could be made with those samples which were heated for 60 minutes. The chemical pretreatment helps to break down complex organic matter by hydrolysis; acidic conditions can hydrolyse fats, proteins and carbohydrates whereas alkaline conditions will, in general, only hydrolyse fats and proteins. The hydrolysis would be expected to be enhanced by increase in temperature.

The results clearly show that acidic and alkaline pretreatment conditions helped to boost gas production in subsequent digestion of activated sludge. A study of Tables 6, 8 and 9 confirms the expectations outlined already.

As before, pretreatment did not have the impact on mixed sludge digestion that it obviously had on activated sludge digestion. A study of Tables 7, 10 and 11 reveals no clear trends.

Throughout the experimental studies, assessments of sludge dewaterability were carried out with a capillary suction time (CST) apparatus. The CST test is one of simple convenience; CST data are not absolute as they are dependent on solids contents. The results simply give an idea as to the effect of pretreatment on dewaterability. Values are tabulated as CST bd (before digestion) and CST ad (after digestion). CST bd values were determined on the sludge samples (pretreated or not) before mixing with the seed sludge; CST ad values were determined on a blend of seed and feed sludges. This provides another reason why interpretation of CST values should be qualitative only.

CST ad values were, in most cases, lower than CST bd values. Tables 5 and 6 show that CST bd values increased with heating for the activated sludge samples; Table 7 shows that, in essence, the opposite is true for mixed sludge samples. Alkaline pretreatment tended to promote a gelatinous quality and CST bd values were therefore very high; this effect was enhanced with increase in alkaline pretreatment temperature. Acid pretreatment, in general, produced more dewaterable sludges.

There is much less variation in the CST ad values; it is clear that digestion essentially masks whatever has happened prior to digestion.

Throughout the studies, analysis of the digester gas was performed. Pretreatment procedures did not have much effect on the gas composition. In general, the mixed sludge yielded approx. 70% CH_4; the activated sludge yielded approx. 78% CH_4.

TABLE 8

Effect of acid preheating on activated sludge digestion

Pretreatment condition	VS bd (% wt.)	VS ad (% wt.)	VS reduction (%)	Gas production (l/kg VS added)	Gas increase (%)	CST bd (s)	CST ad (s)
No pretreatment	3.04	1.52	50.0	300.1	0	247	258
Acid; no heating	3.00	1.45	51.7	372.7	24.19	871	243
Acid; 70°C; 60 mins	3.04	1.43	53.0	426.1	41.99	490	235
Acid; 121°C; 60 mins	3.11	1.36	56.3	462.1	53.98	251	210
Acid; 135°C; 60 mins	3.00	1.20	60.0	487.4	62.41	140	211

TABLE 9

Effect of alkali preheating on activated sludge digestion

Pretreatment condition	VS bd (% wt.)	VS ad (% wt.)	VS reduction (%)	Gas production (l/kg VS added)	Gas increase (%)	CST bd (s)	CST ad (s)
No pretreatment	3.04	1.52	50.0	300.1	0	247	258
Alkali; no heating	3.02	1.48	51.0	351.3	17.06	2239	293
Alkali; 70°C;60 mins	3.09	1.40	54.7	397.0	32.29	3104	307
Alkali; 121°C;60 mins	3.14	1.36	56.7	437.3	45.72	3580	198
Alkali; 135°C;60 mins	3.47	1.44	58.5	449.3	49.72	2554	180

TABLE 10

Effect of acid preheating on mixed sludge digestion

Pretreatment condition	VS bd (% wt.)	VS ad (% wt.)	VS reduction (%)	Gas production (l/kg VS added)	Gas increase (%)	CST bd (s)	CST ad (s)
No pretreatment	3.63	1.86	48.8	596.8	0	283	232
Acid; no heating	3.62	1.87	48.3	572.4	-4.09	97	117
Acid; 70°C; 60 mins	3.54	1.90	46.3	584.8	-2.01	32	140
Acid; 121°C; 60 mins	4.12	1.89	54.1	597.9	0.18	26	120
Acid; 135°C; 60 mins	3.78	1.88	50.3	642.7	7.69	24	165

TABLE 11

Effect of alkali preheating on mixed sludge digestion

Pretreatment condition	VS bd (% wt.)	VS ad (% wt.)	VS reduction (%)	Gas production (l/kg VS added)	Gas increase (%)	CST bd (s)	CST ad (s)
No pretreatment	3.63	1.86	48.8	596.8	0	283	232
Alkali; no heating	3.56	1.78	50.0	596.0	-0.13	419	241
Alkali; 70°C; 60 mins	3.49	1.78	49.0	625.1	4.74	659	203
Alkali; 121°C; 60 mins	4.01	1.86	53.6	658.1	10.27	1162	203
Alkali; 135°C; 60 mins	3.85	1.83	52.5	594.4	-0.40	1651	244

CONCLUSIONS

With such a variable source material as wastewater sludge, it is prudent to state that the conclusions should only be related to the experiments reported here. It is for this reason that a direct comparison of these results and those of Haug[3] is not made. Nevertheless, certain trends did emerge in this study and they are as follows.

1. Mechanical, thermal and chemical pretreatment of sludge led to increased gas production and increased reduction in volatile solids in subsequent anaerobic digestion of the sludge.
2. The most effective procedure was combined acid and thermal treatment, followed by combined alkali and thermal treatment. Thermal treatment alone was next, and mechanical treatment was the least effective.
3. The effects of pretreatment were greater for the activated sludge than for the mixed sludge (primary plus humus).
4. Whilst there was considerable variation in the dewaterabilities of pretreated undigested sludge samples, there was much less variation for the digested sludges (pretreated or not).

REFERENCES

1. Dick, R.I., Gravity thickening of sewage sludges, _Effl. Wat. Treat. J._, 1972, _12_, 597-605.

2. Martin, R.J., Henry, R. and Karatosun, H., Chemical conditioning and mechanical dewatering of sewage sludge in an industrial region, _Management of Industrial and Hazardous Wastes_, ed. A. Hamza, United Nations ESCWA, Baghdad, 1985, 329-350.

3. Haug, R.T., Stuckey, D.C., Gossett, J.M. and McCarty, P.L., Effect of thermal pretreatment on digestibility and dewaterability of organic sludges, _J. Wat. Pollut. Control Fed._, 1978, _50_, 73-85.

4. Brade, C.E. and Noone, G.P., Anaerobic sludge digestion - need it be expensive?, _Effl. Wat. Treat. J._, 1981, _21_, 391-406.

5. Jallow, S.B.O., Studies on the effects of mechanical, thermal and chemical pretreatments on the anaerobic digestibility of sewage sludges, MSc thesis in Water Resources Technology, Department of Civil Engineering, University of Birmingham, September 1985.

6. Pike, E.B., Carrington, E.G. and Ashburner, P.A., An evaluation of procedures for enumerating bacteria in activated sludge, _J. Appl. Bact._, 1972, _35_, 309-321.

7. Haug, R.T., Sludge processing to optimize digestibility and energy production, _J. Wat. Pollut. Control Fed._, 1977, _49_, 1713-1721.

EFFECTS OF TEMPERATURE ON METHANE FERMENTATION OF INDUSTRIAL SOLVENTS

J. Diaper and E. Terzis
School of Environmental Science,
University of Bradford,
Bradford, BD7 1DP,
UK

ABSTRACT

Wastewaters from industrial chemical operations will inevitably contain toxic industrial solvents at relatively high concentrations (>1000 p.p.m.). Under adverse conditions the volatilization of these solvents can result in toxic and explosive atmospheres. Anaerobic treatment of many of these solvents is possible, in which the organic material is converted, 90% to volatile substances (carbon dioxide and methane gas) and 10% to new cells (solids).

In order to evaluate the optimum conditions for the methane fermentation of propan-2-ol and acetone the kinetic parameters of the Monod kinetic model, maximum growth yield, endogenous decay coefficient, maximum specific growth rate and half-velocity constant, were determined at the temperatures 25°C, 30°C, 35°C and 40°C .

In the propan-2-ol system, molecular hydrogen, in addition to its role in the production of methane itself and the growth of methanogenic bacteria, also regulates the conversion of propan-2-ol to acetone. The regulatory role of molecular hydrogen will be discussed, and also its use as a control parameter of the whole process.

INTRODUCTION

Organic solvents are part of the problem of industrial wastewater disposal. Wastewaters from pharmaceutical, cosmetic, textile, plastic, rubber, and other industries will inevitably contain solvents, and often in high concentrations. The total consumption of solvents in the U.K. in 1973 was 790,000 tons [1]. That amount, minus any volatilization and recovery, found its way to the wastewater recovery system. Their economic value makes them attractive for recovery. It was estimated that in the U.K. in 1973 about one half of all solvents were potentially recoverable but only 10%-20% were actually recovered [1]. The presence of solvents is undesirable in an industrial effluent. Direct disposal to a river or to the sewage system is prohibited because of their effects in living organisms [2], on the sewage reclamation system itself, and also because of their flammability. The discharge of these solvents to watercourses or to the sewage system not only can create toxic conditions within the waters themselves but can also create toxic or explosive atmospheres in closed and confined spaces such as sewers and culverts. The Water Research Centre used the Threshold Limit Values (TLV) published by the Health and Safety Executive [3] together with volatility data available in the literature to calculate the aqueous concentrations (CTLV) of compounds in equilibrium with the relevant TLVs [4], some of which are quoted in Table 1, together with the concentrations of those compounds which would be toxic to aquatic organisms [5].

TABLE 1
Threshold Limit Values with the relevant aqueous concentrations and aquatic toxicity values

	TLV (ppm)	CTLV (mg/l)	AQUATIC TOXICITY (ppm)
Ethanol	1000	8550	250
Acetone	1000	1030	14250
Propan-2-ol	400	2920	1100
Propan-1-ol	200	1890	500
Toluene	100	2	1180
Butan-1-ol	50	456	1000
Pyridine	5	32	1350

Thus, a concentration of 0.3% (w/v) in the case of propan-2-ol is hazardous not only in the aqueous phase, but also this concentration could result in a toxic atmosphere in a confined space since the CTLV is 2920 mg/l. Two strategies are apparent by which the effects of propan-2-ol can be reduced, firstly by a suitable biological oxidation process, and secondly by converting the propan-2-ol to a less hazardous material, in this case, acetone, again biologically, through a dehydrogenation reaction. The two major microbiological processes conventionally used for the biological elimination of pollution are aerobic and anaerobic treatment. This paper addresses itself to the use of anaerobic processes.

The advantages of the process make it a suitable alternative to the aerobic treatment processes and it is increasingly being applied to industrial wastewater treatment [6].

Figure 1 illustrates some of the reactions taking place in an anaerobic environment. The heavy arrows indicate methane fermentations by individual species or syntrophic associations. The remaining reactions do not generate methane and they are catalyzed by fermentative bacteria. Part of the general fermentation process according to Figure 1 involves the two solvents propan-2-ol and acetone. In addition, hydrogen is an important product of the fermentation.

The regulatory role of molecular hydrogen is being investigated, to provide a method of reducing the toxicity of a wastewater containing propan-2-ol, as also its use as a control parameter of the whole anaerobic process.

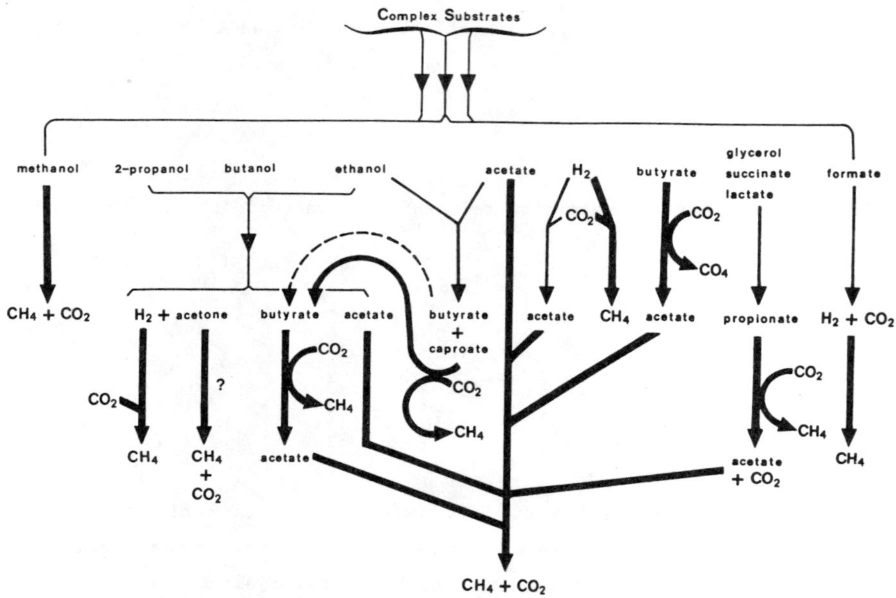

Figure 1.   Interrelationships between methanogenic and fermentative bacteria [7].

To investigate the applicability of methane fermentation to the reduction of the concentration of toxic solvents in wastewaters, the Monod kinetic model was employed. The Monod model has been presented elsewhere [8] and it will not presented here. Knowledge of the kinetic coefficients maximum specific growth (μm), half-velocity coefficient (Ks), yield (Y), and endogenous decay (Kd) permits the prediction of the the steady state concentrations of biomass (solids) and waste effluent for any solids retention time and concentration of inflowing wastewater. The expressions derived linking the flow characteristics of the reactor and kinetic equations can provide a complete quantitative description of the anaerobic degradation of propan-2-ol and acetone, and it also offers useful design information for such a treatment system. In addition, the thermodynamics of the propan-2-ol - acetone equilibrium need to be investigated in order to assess the viability of the conversion as a toxicity reducing strategy.

THE THERMODYNAMICS OF THE PROPAN-2-OL, ACETONE HYDROGEN REACTION

A large body of consistent free energy (ΔG^o) and enthalpy (ΔH^o) data is available in the literature [9,10,11] for many organic substances. Burton [12] uses standard thermodynamic data at 25°C for propan-2-ol, acetone and hydrogen for the chemical equation:

$$CH_3CHOHCH_3(aq) \rightleftharpoons CH_3COCH_3(aq) + H_2(g)$$

to determine the enthalpy change for the reduction of Nicotinamide Adenine Dinucleotide. The values of ΔG^o and ΔH^o for the above reaction are quoted as being +24.4 kJ/mole and +71.8 kJ/mole respectively. The process at first sight does not look promising because of the positive sign of ΔG^o. However, taking the value of ΔG^o and using the relationship; $\Delta G^o = -R.T.\ln K$, where K is the equilibrium constant for the above reaction, then the concentration of hydrogen at equilibrium for a 10:1 ratio of propan-2-ol to acetone is 500 ppm. Theoretically as long as the hydrogen concentration remains below 500 ppm the conversion of propan-2-ol to acetone can take place. Above this concentration the conversion is inhibited. However, in the methane fermentation process, hydrogen is utilized in the reduction of carbon dioxide to methane.

$$4H_2 + CO_2 \longrightarrow CH_4 + 2H_2O$$

Thus, the concentration of hydrogen is effectively reduced and this resulting low concentration of hydrogen allows the conversion to continue almost to completion. Therefore, the concentration of hydrogen needs monitoring, to test the above interpretation.

EXPERIMENTAL PROCEDURES AND RESULTS

In order to evaluate the kinetic constants and their behaviour with temperature, four identical, completely mixed, 5 litre bench-scale anaerobic units were operated at four temperatures, 25°C, 30°C, 35°C, and 40°C, under controlled laboratory conditions. Each digester could operated to ±1°C. The generated gas was collected using a inverted measuring cylinder, by displacement of acidified water to prevent the carbon dioxide dissolving, and the gas volume was corrected to S.T.P. The anaerobic systems were seeded with actively digesting sewage sludge to provide an active microbial population, purged of oxygen with nitrogen gas and left without feeding until no gas was produced. Then, they were fed with a synthetic wastewater containing inorganic nutrient salts solution and 1 g/l propan-2-ol.

When a stable acclimatized microbial population was established the steady state programme was started. The digesters were fed daily, and thus the operation was semi-continuous. The flowrate was set at 0.250 l/day, and thus, a hydraulic retention time of 20 days. Two synthetic wastewaters were prepared, one based on propan-2-ol and the other on acetone with variable concentrations (2.5, 3, 4, 4.5 and 5 g/l) and the inorganic nutrient salts solution.

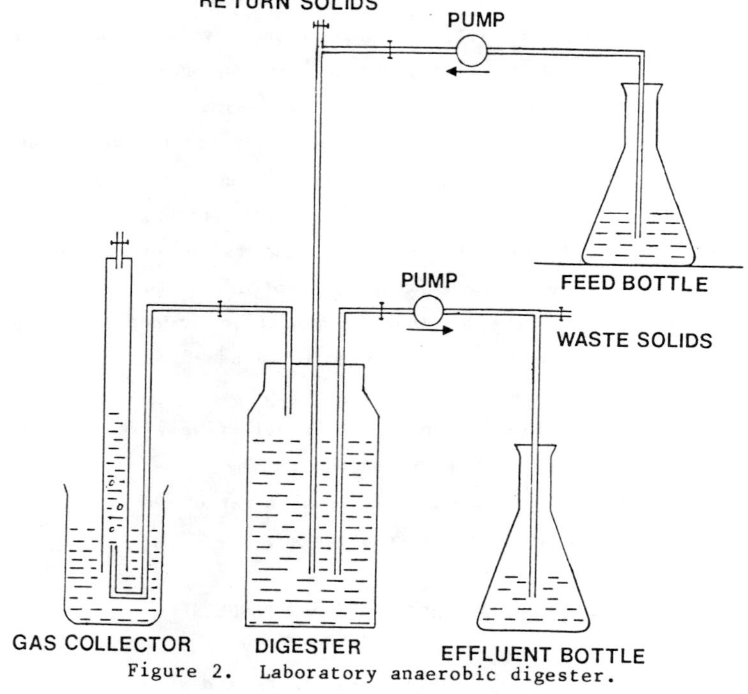

Figure 2.   Laboratory anaerobic digester.

The criteria used to assess the steady state conditions were volatile solids concentration, effluent C.O.D., gas production and gas composition. It was possible to determine the volume of the solids wasted in the first 3-5 days, after a change from one feed concentration to another, by trial and error. Steady state was assumed when the above four parameters had become constant, i.e. with variation of less than 5%. This was achieved in the next 5-7 days. The measurement of these parameters was carried on for 7 days and the mean values were used to determine the kinetic coefficients.

Tables 2 and 3 summarize the values of Ks, μm, Kd, and Y, for propan-2-ol and acetone respectively, at 25°C, 30°C, 35°C and 40°C.

TABLE 2

Kinetic coefficients of propan-2-ol for substrate utilization and biological growth

TEMPERATURE °C	Ks (mgCOD/l)	μm (day-1)	Kd (day-1)	Y (mgVSS/mgCOD)
25	670	0.047	0.013	0.066
30	1470	0.065	0.014	0.059
35	1050	0.095	0.016	0.061
40	4000	0.125	0.017	0.074

TABLE 3

Kinetic coefficients of acetone for substrate utilization and biological growth

TEMPERATURE °C	Ks (mgCOD/l)	μm (day-1)	Kd (day-1)	Y (mgVSS/mgCOD)
25	1330	0.046	0.004	0.038
30	2000	0.071	0.005	0.040
35	2380	0.111	0.006	0.043
40	2860	0.140	0.007	0.051

The Arrhenius equation is used to describe the effect of increasing temperature on the rate-limiting reaction. The Arrhenius law is applicable in biological systems within a limited temperature range, because enzymes present in these systems are inactivated at high temperatures, due to the denaturation of protein [13]. Arrhenius plot is illustrated in Figure 3 for μm for propan-2-ol. Similar plots were also produced for Ks and Kd and the kinetic coefficients of acetone. It was apparent that temperature influences the kinetic coefficients.

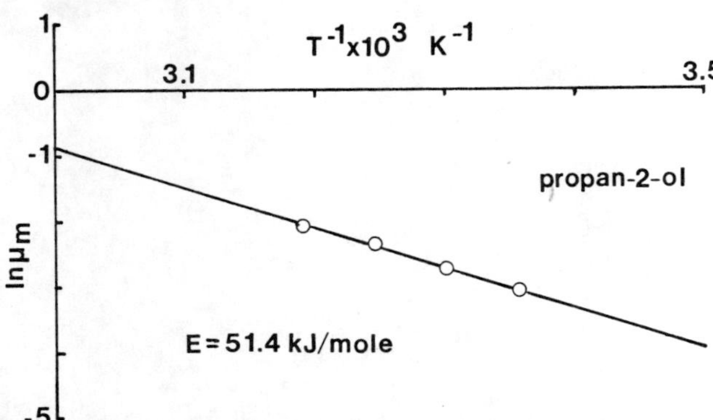

Figure 3. Dependency of μm on temperature in methane fermentation of propan-2-ol

DISCUSSION AND CONCLUSIONS

1. Maximum specific growth rate.

The effect of temperature on methane fermentation is best illustrated considering the temperature variation of the maximum specific growth rate. The rate of a chemical reaction doubles with every ten degrees Celsius increase in temperature. The rate of methane fermentation as reflected in μm, also approximately doubles. The maximum specific growth rate seems to be a very strong function of temperature, with activation energies for propan-2-ol and acetone of +51.4 kJ/mole and +58.9 kJ/mole respectively. The significance of this is that more stable and fast fermentation occurs at the higher temperatures compared with the lower temperatures. It is possible to operate an anaerobic system at a lower temperature with the same efficiency as one at higher temperature by increasing the solids retention time (SRT), e.g. by doubling the solids retention time in the system at 25°C, it should operate at a rate equivalent to one at 35°C with a shorter SRT. Conventional anaerobic digestion of mixed materials has an operational temperature in the 30-35°C region [14] one may conclude that this temperature is a result of averaging the optimum temperature of a number of different processes, each of which has a slightly different temperature optimum. For propan-2-ol and acetone this optimum has not been reached by 40°C.

2. Yield.

Temperature is expected to influence the yield coefficient but not to a great extent, provided that no significant change took part in the metabolism of the bacterial population. That seems to be the case for acetone and propan-2-ol as Figure 4 shows.

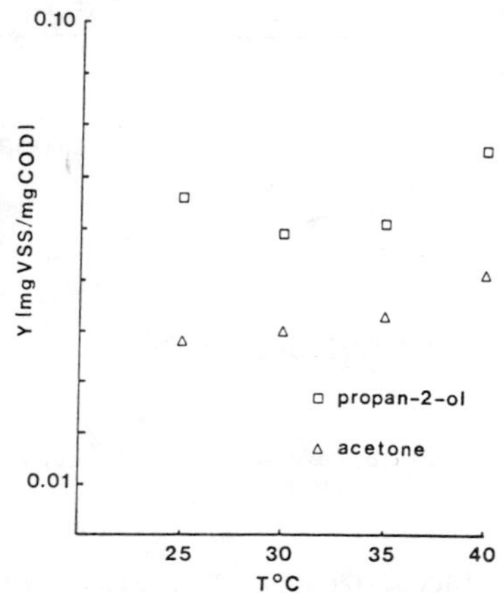

Figure 4.   The yield coefficient as a function of temperature

However, the absolute values of the yield for propan-2-ol and for acetone differ so markedly, that it may be that different bacterial populations were present in each case, especially with their different values for Ks and different values for the energy of activation for the respective Kd values.

3. Endogenous decay.

The coefficient of endogenous decay, is however a function of temperature. The activation energy for Kd was determined to be +14.6 kJ/mole for propan-2-ol and +28.9 kJ/mole for acetone.

4. Half-velocity coefficient.

There is no agreement in the reported values of Ks as to effect of temperature, apart from the point that very few values have been published on this [15]: the information gained from the half velocity coefficient cannot be considered to be anymore than that which can be gleaned from the analytical geometry of the hyperbolic curve generated by the definition of Ks - it is the numerical value of S when μ is equal μm/2. Ks might be considered as some quantitative expression of the sensitivity of μ (or of the population, expressed through the observed μ) to the concentration of the limiting substrate. In that case, it is reasonoble to assume that Ks may decrease or increase with increasing temperature depending on the specific reaction. For both propan-2-ol and acetone Ks increases with increasing temperature, with activation energies +77.6 kJ/mole and +38.4 kJ/mole respectively.

5. The effect of Hydrogen

For a complex substrate (e.g. cellulose) approximately 70% of the methane produced originates from acetate and 30% from carbon dioxide reduction [16]. In the case of propan-2-ol, stoichiometry suggests that 89% originates from acetone and 11% from carbon dioxide reduction. Large amounts of hydrogen are produced in anaerobic systems, and since hydrogen is rarely detectable, it seems that there is a rapid utilization of the hydrogen produced. The discovery of syntrophic associations, between hydrogen producing and hydrogen consuming (e.g. methanogens) bacteria, for the degradation of ethanol and volatile fatty acids [17,18,19,20], and the theory of interspecies hydrogen transfer show that this is the case.

It is known that high hydrogen concentration inhibits the growth of the hydrogen producing bacteria [17]. This was confirmed by an unexpected digester failure experienced during this study as a result of high hydrogen concentration. The feeding substrate was propan-2-ol and the efficiency of the system approached zero. Gas was not produced for no apparent reason. The pH was 7.0 and the alkalinity 2,300 mg/l measured as $CaCO_3$. All the nutrients were present in sufficient quantities. The gas composition was 23% carbon dioxide, 31% methane and 46% hydrogen, instead of 80% methane 20% carbon dioxide and traces of hydrogen, which was the normal composition of the gas. The gas was flushed out and the digester didn't recover as long as it was fed with propan-2-ol, it recovered only when it was fed with acetone. From that point the digester recovered almost immediately. The next two days the gas composition was almost 100% methane and then returned to normal. It is confirmed that the high hydrogen concentration inhibits the hydrogen-producing bacteria. The tolerant hydrogen concentration was higher than that predicted fron thermodynamic considerations. This suggests that the concentration of dissolved hydrogen is important and probably the concentration inside the syntrophic consortia, as the steady state gaseous concentration of hydrogen was always in the range of 650-1150 ppm depending on the loading.

From the above results the upper limit on the concentration of hydrogen in the evolved gases, to arrest the conversion of propan-2-ol to acetone is between 1150 ppm and 500,000 ppm. Under these conditions propan-2-ol is converted to acetone, but the acetone is itself converted at a faster rate to methane and carbon dioxide, so keeping the concentration of acetone down, allowing a higher concentration of hydrogen than the 500 ppm predicted for a 10:1 ratio of propan-2-ol to acetone in the digester mixed liquor. Therefore notwithstanding the apparent thermodynamic barrier, it is possible to convert propan-2-ol to acetone by biological dehydrogenation. In doing so the toxicity and the flammability hazard of the wastewater is reduced. There are similarities between propan-2-ol fermentation and the fermentation of ethanol [19] and other hydrogen producing fermentations. The use of butan-2-ol and methyl ethyl ketone, a pair of solvents similar to propan-2-ol and acetone, in methane fermentation process should be investigated.

It has been confirmed if hydrogen is monitored during the anaerobic process it can indicate forthcoming unbalanced conditions. It is necessary to more sharply define the cut-off conditions for the hydrogen concentration and cross check these with thermodynamic data.

REFERENCES

1. Withaff, H.A. and Reuben, B.G., _Industrial Organic Chemicals in Prespective, Part 2: Technology, Formulation and Use_, New York, John Wiley and Sons, 1980, pp280-99.

2. Browing, E., *Toxicity of Industrial Solvents*, Elsevier Publishing Company, London, 1965.

3. Health and Safety Excecutive., Threshold Limit Values for 1984. Guidance Note EH40/84. London, HMSO, 1984.

4. Toogood, S.J. and Hobson, J.A., The determination of safe limits for the discharge of volatile materials to sewers. Technical Report 142, Water Research Centre, Stevenage Laboratory, June 1980.

5. *Hazardous Chemicals Data Book*, Noyes Data Corporation, New Jersey, 1980.

6. Lettinga, G., Van Velsen L., De Leeuw, W. and Hobma S.W., The application of anaerobic digestion to industrial pollution control. In *Anaerobic Digestion*, ed. D.A. Stafford, B.I. Wheatley and D.E. Hughes, Applied Science Publishers, London, 1980, pp167-86.

7. Pine, J.M., The methane fermentations. In Anaerobic Biological Treatment Processes, ed. Gould F.R., American Chemical Society, *Advan. Chem. Ser.*, 1971, *105*, pp1-10.

8. Metcalf & Eddy, Inc., *Wastewater Engineering: Treatment Disposal Reuse*, 2nd edition, McGraw Hill, New York, 1975, pp411-35.

9. Cox, J.D., *Thermochemistry of Organic and Organometallic Compounds*, Academic Press, London, 1970.

10. Rossini, F.D., Wagman, D.D., Evans, W.H., Levine, S. and Jaffe, I., *Selected Values of Chemical Thermodynamic Properties*, National Bureau of Standards Circular no. 500, Washington, 1952.

11. Stull, D.R., Westrum, E.G. and Sinke, G.C., *The Chemical Thermodynamics of Organic Compounds*, John Wiley and Sons, New York, 1969.

12. Burton, K., The enthalpy change for the reduction of Nicotinamide-Adenine Dinucleotide. *Bioch. J.*, 1974, *143*, 365-68.

13. Laidler, K., *Physical Chemistry with Biological Applications*, Addison Wesley, 1980, pp386-446.

14. Sahm, H., Anaerobic Wastewater Treatment. *Advances in Biochemical Engineering/Biotechnology*, 1984, *29*, pp83-115.

15. Henze, M. and Harremoes, P., Anaerobic Treatment of Wastewater in Fixed Film Reactors-A Literature Review. In *Anaerobic Treatment of Wastewater in Fixed Film Reactors*, ed. M. Henze, Pergamon Press, Exeter, 1983, pp1-101.

16. Jeris, J.S. and McCarty, P.L., The biochemistry of methane fermentation using Cl4 traces. *J. of Water Pollut. Control Fed.*, 1965, *37*, 178-92.

17. Bryant, M.P., Wolin, E.A., Wolin M.J. and Wolfe, R.S., *Arch. Mikrobiol.*, 1967, *59*, 20-31.

18. Boone, D.R. and Bryant, M.P., Propionate degrading bacterium, Syntrophobacter wolinii sp. nov. gen. nov., from methanogenic ecosystems. *Appl. Environ. Microbiol.*, 1980, *40*, 626-32.

19. McInerney, M.J., Bryant, M.P. and Pfenning, N., Anaerobic bacterium that degrades fatty acids in syntrophic association with methanogens. *Arch. Microbiol.*, 1979, *122*, 129-35.

20. McInerney, M.J., Bryant, M.P., Hespell, R.B. and Costerton, J.W., Syntrophomonas wolfei gen. sp. nov., an anaerobic, syntrophic, fatty acid-oxidizing bacterium. *Appl. Environ. Microbiol.*, 1981, *41*, 1029-39.

BIOLOGICAL TREATMENT OF NITROGENOUS WASTEWATER - A NITRIFICATION STUDY

S.K. Gupta, S.A. Hashsham and S.G. Joshi
Centre for Environmental Science and Engineering
Indian Institute of Technology
Powai, Bombay 400 076
INDIA

ABSTRACT

A single-stage biological treatability study by activated sludge process was performed with synthetic wastewater containing about 190 mg/l of NH_4^+-N and 205 mg/l of urea. It was found that the wastewater could be processed at all solid retention times (SRT's). It was observed that biological nitrification did not affect the organic carbon removal capacity of the biological reactors. The yield coefficient and decay coefficient for heterotrophic bacteria were 0.23 (COD basis) and 0.024 day^{-1} (COD basis) respectively. The yield coefficient and decay coefficient for nitrifiers were 0.045 and 0.023 day^{-1}.

INTRODUCTION

The removal of nitrogen content of fertiliser industry wastewater discharged into the aquatic environment has been assuming an increasing degree of significance in recent years. The consequences of indiscriminate discharge of wastewater into receiving water bodies have been well documented [1]. The quantum of wastewater generated is appreciable and contains about 650 mg/l of NH_4^+-N and 300 mg/l of urea.

Several researchers [2,3,4] have reported the success of nitrification of industrial wastewater while working with high concentration of ammonium nitrogen in a suspended growth system. This has drawn our attention to biological nitrification by activated sludge process on highly nitrogenous wastewater. This paper presents the significant findings and conclusion emerging from a study concerning the biological treatability of synthetic wastewater containing about 190 mg/l of NH_4^+-N and 205 mg/l of urea.

<u>Objectives</u>

The primary objectives of this study were to:

1. evaluate the treatment of synthetic wastewater by activated sludge on a bench scale when glucose and sodium bicarbonate were used as organic and inorganic carbon sources respectively;

2. determine the relationship between basic biological kinetic parameters including solid retention time, substrate removal rate and biomass concentration;

3. assess the basic performance characteristics in terms of effluent quality and sludge property;

4. determine the effect of SRT on nitrification;

5. estimate the contribution of air stripping, if any, as ammonia removal mechanism in the biological waste treatment system.

<u>EXPERIMENTAL DESIGN AND PROCEDURES</u>

Biological Kinetics

The biological kinetic model used for nitrification study is based on substrate and cell mass balances, and on the growth rates of microorganisms and concentrations of rate limiting substances. The following equations are applicable to suspended growth completely mix activated sludge reactors with sludge wasting from the mixed liquor. Net growth rate for heterotrophic bacteria in a mixed culture, θ_c^{-1}, per day:

$$\frac{1}{\theta_c} = Y_b\, q_b - k_d \qquad \ldots (1)$$

COD removal rate, q_b, per day:

$$q_b = \frac{S_o - S}{X_b\, \theta} \qquad \ldots (2)$$

Net growth rate for nitrifying bacteria, θ_c^{-1}, per day:

$$\frac{1}{\theta_c} = Y_N\, q_N - k_{dn} \qquad \ldots (3)$$

Ammonium nitrogen removsl rate, q_N, per day:

$$q_N = \frac{N_o - N}{X_N\, \theta} \qquad \ldots (4)$$

Unit nitrification rate, r_N, per day:

$$r_N = q_N\, f \qquad \ldots (5)$$

Fraction of nitrifiers, f, is given by:

$$f = \frac{1}{(S_o/N_o)\,(Y_b/Y_N) + 1} \qquad \ldots (6)$$

Hydraulic retention time, θ, in days;

$$\theta = \frac{V}{Q} \qquad \qquad \ldots (7)$$

Sludge washing rate, Q_w, in l/day:

$$Q_w = \frac{\dfrac{V}{\theta_c}(X) - Q X_e}{X} \qquad \qquad \ldots (8)$$

in which θ_c = solid retention time; Y_b = yield coefficient for heterotrophs; Y_N = yield coefficient for nitrifiers; k_d = decay coefficient for heterotrophs, day^{-1}; k_{dn} = decay coefficient for nitrifiers, day^{-1}; S_o = influent COD, mg/l; S = effluent COD, mg/l; N_o = influent TKN, mg/l; N = effluent TKN, mg/l; X_b = concentration of heterotrophic bacteria, mg/l; X_N = concentration of nitrifiers, mg/l; V = reactor volume, in litres; Q = influent flow rate, l/day; X = total MLVSS concentration, mg/l; X_e = suspended solids in the effluent, mg/l.

During the entire experiment run θ_c, Q, V, S_o and N_o were held as independent variables and S, N, and X were measured as dependent variables.

All measurements, except as noted, were done according to Standard Methods [5]. The effluent COD samples from the biological reactor were filtered through Whatman GF/c filter paper to remove suspended solids. Urea measurement was done according to Watt and Chrip Method [6].

Synthetic Wastewater Characterization

Synthetic waste was prepared by taking appropriate quantities of ammonium chloride and urea. Glucose and sodium carbonate were added as carbon sources. Phosphorous was added as potassium dihydrogen phosphate. The average characteristics of wastewater are given in Table 1.

TABLE 1

Average characteristics of synthetic wastewater

Item	Concentration, mg/l
Glucose	450
NH_4Cl	190
Urea	205
$NaHCO_3$	2,500
KH_2PO_4	100
$MgCl_2 \cdot 6H_2O$	75

Experimental Set-up

The biological nitrification unit consisted of an aeration chamber of
5 litres of liquid volume with a built-in settling unit of 1 litre separated
by an adjustable baffle. Compressed air was supplied in a controlled manner
with the help of two diffusers and a mechanical mixer to keep the dissolved
oxygen concentration in the range of 2.5 to 4.0 mg/l. The overflow from
the settling tank was sent to an effluent collection reservoir. The schematic of the bench scale activated sludge process is shown in Figure 1.

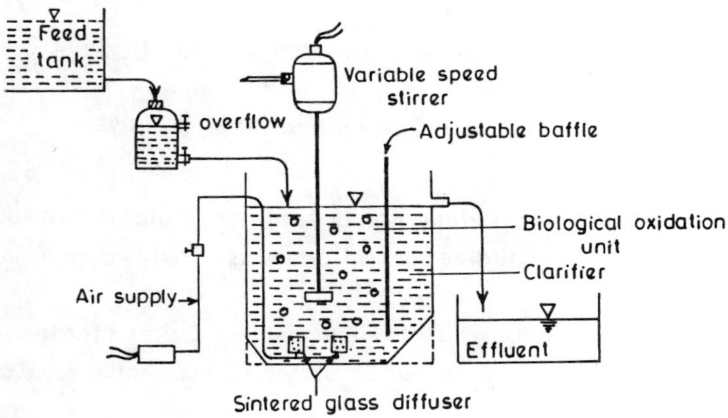

Fig. 1. Schematic of bench scale activated sludge system

The synthetic waste was fed to the system from a storage tank placed
30 cm above the reactor. Two such units were operated in the laboratory at
ambient temperature (25 to 37°C).

Experimental Design

The strategy used to study the biological treatability of synthetic
wastewater is presented in Figure 2. Solid retention times were selected
on the basis of literature review recommendation for nitrification in domestic and industrial wastewater. The sludge wasting rate of about 5 per cent
per day gave an SRT of 20 days. This value was considered as a point from
which the parametric studies were performed to determine a range of suitable
solid retention times.

It may be noted here that one of the objectives of the biological
oxidation study was to determine the functional relationship between the
parameters given in equations (1) and (3). The experiment was also aimed

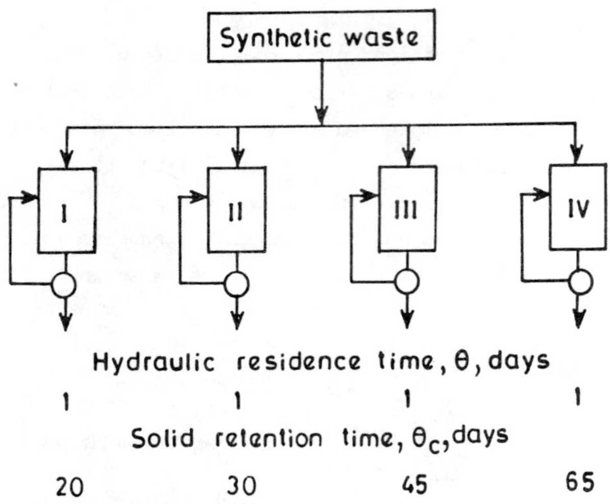

Fig. 2. Experimental design for biological oxidation studies on nitrogenous wastewater

at determining the upper limit of ammonium and urea nitrogen concentration in the wastewater which would not affect nitrification significantly.

RESULTS AND DISCUSSION

Two bench scale completely stirred reactors were operated successfully for about two-and-a half months after the acclimation period upto 190 mg/l of NH_4^+-N and 205 mg/l of urea without any difficulty. However, problems regarding the stability of the process above this concentration were observed. The performance data were collected when the unit reached a steady state (HRT basis) in terms of nitrogen removal and mixed liquor suspended solids. In all, the performance of the process was evaluated at four different SRT's, viz. 20 days, 30 days, 45 days, and 65 days. The steady state operation period for the four SRT's varie from 3 to 5.5 weeks. Table 2 gives the average values for the analysis performed on the influent, effluent and mixed liquor during this study.

It can be observed from the table that the COD and TKN removal efficiency increased with the decrease in organic and TKN loadings respectively. It is also evident that the COD removal rates and nitrification rates employed were not excessive. The long term stability of the reactor during

TABLE 2

Average steady state performance data for biological oxidation of synthetic wastewater

| Operating data[a] | | | | | REACTOR NUMBER | | | | | | | |
	I	II	III	IV	I Influent	I Effluent	II Influent	II Effluent	III Influent	III Effluent	IV Influent	IV Effluent
Steady state period (d)	21	21	27	39	–	–	–	–	–	–	–	–
SRT (days)	20	30	45	65	–	–	–	–	–	–	–	–
HRT (days)	1.07	1.07	1.09	1.09	–	–	–	–	–	–	–	–
D.O.	2.5	2.5	2.5	2.5	–	–	–	–	–	–	–	–
COD/TKN ratio	1.58	1.58	1.65	1.65	–	–	–	–	–	–	–	–
MLSS	1648	2216	2968	3531	–	–	–	–	–	–	–	–
MLVSS	1353	1839	2309	2708	–	–	–	–	–	–	–	–
F/M—COD basis	0.334	0.246	0.202	0.172	–	–	–	–	–	–	–	–
F/M—TKN basis	0.211	0.155	0.122	0.104	–	–	–	–	–	–	–	–
COD removal rate (d^{-1})	0.313	0.235	0.194	0.169	–	–	–	–	–	–	–	–
Nitrification rate (d-1)	1.66	1.15	1.02	0.87	–	–	–	–	–	–	–	–
SVI (ml/g MLSS)	83	78	94	86	–	–	–	–	–	–	–	–
Effluent SS	25.5	27	24	22.5	–	–	–	–	–	–	–	–
pH	7-7.8	7-7.8	7-7.8	7-7.8	7.9-8.2	–	7.9-8.2	–	7.9-8.2	–	7.9-8.2	–
COD	–	–	–	–	452.0	48.4	452.0	39.6	466.0	29.7	466.0	22.0
NH_4^+-N	–	–	–	–	190.0	12.6	190.0	7.24	188.34	3.59	189.9	2.92
Urea	–	–	–	–	207.5	N.D.	207.5	N.D.	203.3	N.D.	203.9	N.D.
NO_3^--N	–	–	–	–	–	225.7	–	230.5	–	232.7	–	238.2
NO_2^--N	–	–	–	–	–	3.30	–	3.77	–	3.15	–	2.79
Alkalinity	–	–	–	–	1621.4	101.4	1621.4	113.6	1698.5	72.0	1713.3	72.2
TKN	–	–	–	–	285.4	30.5	285.4	22.5	282.45	14.13	282.45	11.25

a – All units in mg/l except as noted,   N.D. – Not Detectable

steady state operation lends further support to this observation.

The concentration of heterotrophic and nitrifying bacteria in the mixed culture of biomass was obtained by the procedure given by Gupta [4], in order to determine the kinetic coefficients. The net growth rate, θ_c^{-1}, data are plotted against COD removal rate, q_b, according to equation (1) by the least square method as shown in Figure 3. This figure shows that the data correlate well with the biological kinetic model chosen for this study. The data gave biological yield coefficient, Y_b, of 0.237 (COD basis) and a decay coefficient, k_d, of 0.024 day^{-1}.

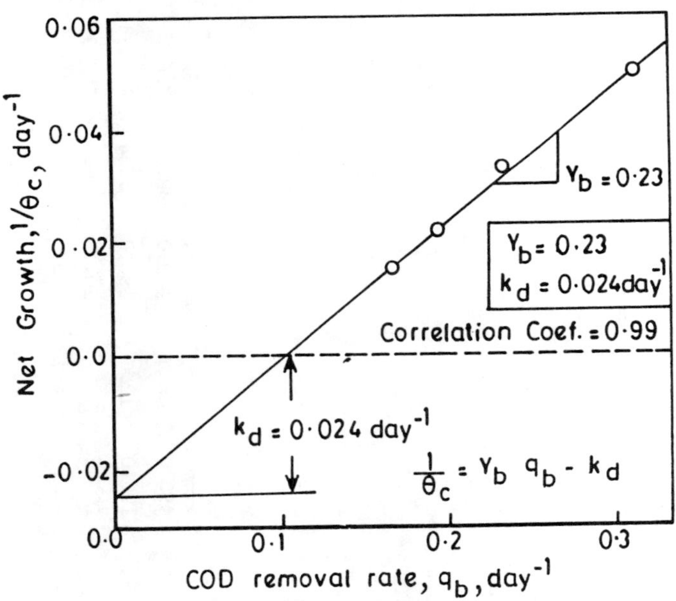

Fig. 3. Relationship between COD removal rate and net growth to give Y_b and k_d

Figure 4 shows the inverse of COD removal rate, q_b^{-1}, vs inverse of effluent COD, s^{-1}, which is plotted to find the values of K_s, the half velocity constant and k, the maximum substrate utilization rate per unit mass of heterotrophic bacteria. From the figure the values of K_s and k were found as 78.3 mg/l and 0.75 day^{-1} respectively.

The data in Table 2 were also plotted to estimate kinetic coefficient for nitrifying bacteria. The yield coefficient, Y_N, and decay coefficient,

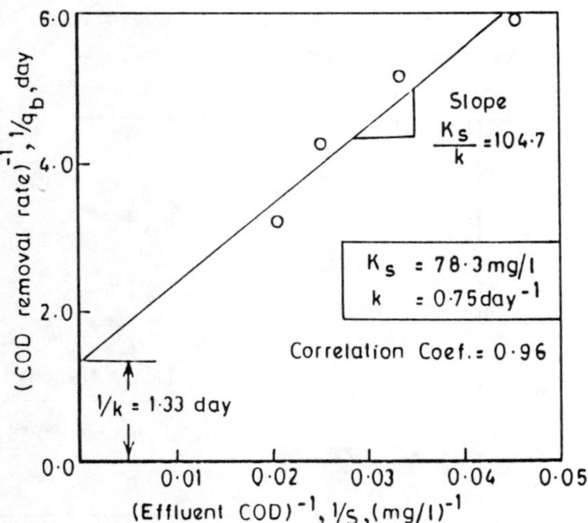

Fig. 4. Relationship between effluent COD and
COD removal rate to give K_S and k

k_{dn}, values were obtained as 0.045 and 0.023 day^{-1} respectively as shown in
Figure 5. The values of half velocity constant, K_N, and maximum substrate

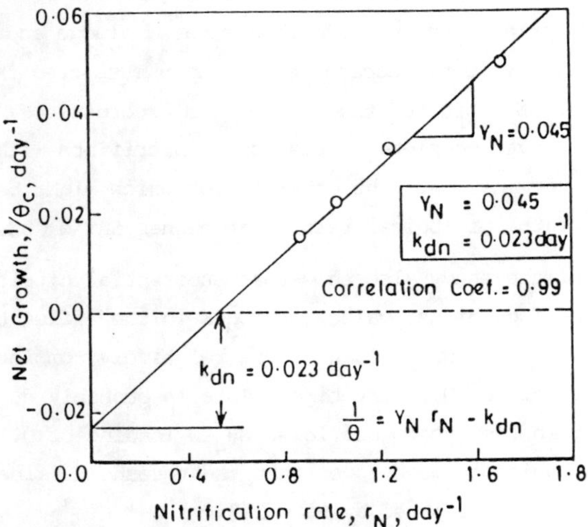

Fig. 5. Relationship between nitrification rate
and net growth to give Y_N and k_{dn}

utilization rate per unit mass of nitrifiers, k_n, were found to be 3.84 mg/1 and 2.0 day^{-1} respectively as given in Figure 6.

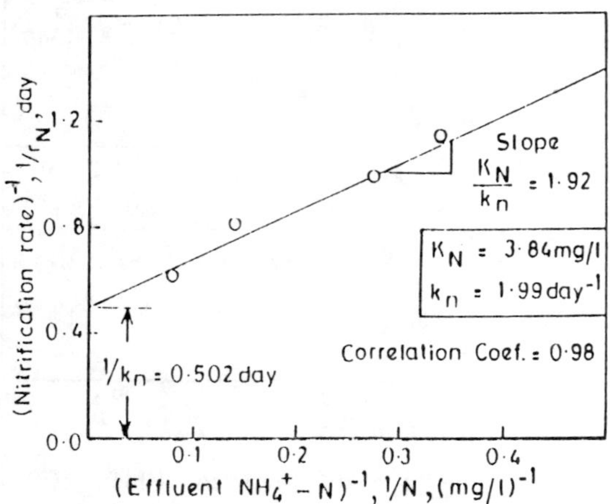

Fig. 6. Relationship between effluent ammonia nitrogen and nitrification rate to give K_N and k_n

The values of the kinetic coefficients thus observed show a lower value of Y_b and Y_N than normally encountered in activated sludge and nitrification process (pure culture values) respectively. It is suspected that the high concentration of NH_4^+-N inhibited the growth of microorganisms. Siddiqi et al.[7] have observed complete supression of nitrifiers with 0.1 M NH_4Cl in five day BOD determination. The other factor which might be responsible is the operation of the biological reactor at higher SRT's [8].

The results of the study also show that substantial nitrification was achieved as evidenced by higher values of NO_3^--N in the effluent averaging between 225–238 mg/l. The nitrification was relatively complete and there was no accumulation of NO_2^--N at any time. This is probably due to sufficient alkalinity present in the system and lower NH_4^+-N loading of 0.1–0.2 kg total NH_4^+-N/kg MLVSS–day which is much lower than the threshold value of 0.45–0.55 kg NH_4^+-N/kg MLVSS–day as reported by Ford [9].

A nitrogen balance was performed at different SRT's as shown in Table 3. The values shown in the table are the averages of 4 readings taken on different days. The effluent TKN accounted for 4 to 10.7 per cent, and NO_2^--N accounted for 0.8 to 1.5 per cent only, wherezs NO_3^--N accounted for maximum of 74.9–80.3 per cent of the influent TKN. About 8.7–19.3 per cent of the influent TKN remained unaccounted. This unaccounted portion of influent TKN may be due to assimilation of nitrogen into the cell, possible denitrification, air stripping, and analytical errors.

TABLE 3

Nitrogen balance at different solid retention time

Reactor SRT	Nitrogen balance, as a percentage			
	NO_2^--N	NO_3^--N	Effluent TKN	Unaccounted
(1)	(2)	(3)	(4)	(5)
20	1.5	79.1	10.7	8.71
30	0.9	77.7	7.9	13.5
45	0.8	74.9	5.0	19.3
65	0.8	80.3	4.0	14.9

The effect of SRT on ammonia nitrogen concentration was also investigated. The results show that the ammonia nitrogen concentration in the effluent declines with the increase in SRT above 20 days.

The pH of the effluent was always in the range of 7.0–7.8 and it never fell below 7.0. This is because the system was adequately buffered by the addition of sodium hydrogen carbonate. The alkalinity consumption calculated ranged from 5.42–5.85 mg/l as $CaCO_3$ per mg NH_3-N oxidized. Low consumption of alkalinity than theoretically required (7.14 mg/l as $CaCO_3$ per mg NH_3-N oxidized) may be due to the loss of TKN through cell synthesis, incidental denitrification and ammonia stripping.

The sludge settling characteristics were found to be reasonably good and the SVI value averaged 85 ml/g.SS. Under normal conditions, the effluent suspended solids were always below 30 mg/l. However, because of difficulties in adjusting the baffle and the stirrer speed in the aeration unit, occasional sludge accumulations at the bottom of the settling unit were observed. If these sludge accumulations were left unattended, it resulted in the production of a large block of floating sludge in the clarifier, which in turn gave rise to high suspended solids in the effluent.

CONCLUSIONS

A single stage activated sludge nitrification system was successfully used for the removal of ammonia nitrogen and urea from a synthetic wastewater containing about 190 mg/l of NH_4^+-N and 205 mg/l of urea.

The main parameters of concern in a biological treatability study are solid retention time and hydraulic residence time. In evaluating the treatability of a new process wastewater at high SRT's (greater than 10 days), it is essential that equilibrium be attained prior to investigation period. The drawback in the available data base is that the attainment of steady state has been defined on the basis of hydraulic residence time rather than solid retention time.

The amenability of nitrogenous wastewater treatment by activated sludge process must include provisions for the addition of (i) organic carbon for the growth of heterotrophic microorganisms, and (ii) alkalinity for preventing the pH to fall below 7.0. This may be achieved by adding a suitable wastewater having high BOD and $NaHCO_3$ respectively. About 1600 mg/l of alkalinity (as $CaCO_3$) is required to provide sufficient buffering capacity in the reactor. The COD/TKN ratio of about 1.5 is appropriate from the point of view of process stability and nitrification.

The wastewater was treatable at HRT of 1 day with solid retention time varying from 20-65 days. It was possible to remove TKN between 89-96 %, NH_4^+-N between 93-98 % and urea hundred per cent for nitrogen loading of 0.1-0.2 kg NH_4^+-N/kg MLVSS-day. The major NH_4^+-N removal mechanism was the oxidation of NH_4^+-N to NO_3^--N in the range 75-80 per cent. The COD removal was found to average between 89-95 per cent for the organic loadings of 0.16-0.31 kg COD/kg MLVSS-day.

Recommended design criteria for the treatment of nitrogenous wastewater by activated sludge process are solid retention time of 20-65 days, hydraulic residence time of 1 day, nitrification rates of 0.87-1.66 day^{-1}, COD removal rates of 0.17-0.31 day^{-1}, influent alkalinity of about 1600 mg/l as $CaCO_3$, and pH between 7.0-7.8.

It is recommended that additional biological treatability study should be conducted with higher concentrations of NH_4^+-N and urea. This can probably be achieved by providing a higher acclimation period. It is also suggested that a study be carried out for a longer period under steady state

465

conditions (based on SRT) to evaluate the kinetic coefficients more rigorously.

REFERENCES

1. Culp, R.L., Werner, M.G. and Culp, L.G., Handbook of Advance Wastewater Treatment, 2nd Edn., Von Nostrand Reinhold Co., New York, 1978, pp.298-408.

2. Luthy, R.G. and Tallon, J.T., Biological Treatment of a Coal Gassification Process Wastewater, Water Res., 14, 1980, 1269-1282.

3. Bridle, T.R., Climenhage, D.C. and Stelzig, A., Operation of a Full-scale Nitrification-Denitrification Industrial Waste Treatment Plant, J. Wat. Pollut. Control Fed., 5, 1979, 127-139.

4. Gupta, S.K., Nitrogenous Wastewater Treatment by Activated Algae, J. environ. Engng. Div., 111, 1985, 61-77.

5. Standard Methods for the Examination of Water and Wastewater, 14th Edn., American Public Health Association, Washington, D.C., 1975.

6. Watt, G.W. and Chrip, J.D., Spectrophotometric Determination of Urea, Anal. Chem., 26, 1954, 452-456.

7. Siddiqi, R.H., Speece, R.E., Engelbrecht, R.S. and Schmidt, J.W., Elimination of Nitrification in the BOD Determination with 0.10 M Ammonia Nitrogen, J. Wat. Pollut. Control Fed., 39, 1967, 579-589.

8. Schroeder, E.D., Water and Wastewater Treatment, McGraw-Hill Inc., New York, 1977.

9. Ford, D.L., Comprehensive Analysis of Nitrification of Chemical Processsing Wastewaters, J. Wat. Pollut. Control Fed., 52, 1980, 2726-2745.

A STUDY OF BIODEGRADATION USING PURE SPECIES KINETICS TO MODEL THE PERFORMANCE OF MIXED POPULATIONS

G. Lewandowski, B. Baltzis and C. Peter Varuntanya
Department of Chemical Engineering, Chemistry
and Environmental Science
New Jersey Institute of Technology
Newark, New Jersey 07102 (USA)

ABSTRACT

In an effort to gain a more fundamental understanding of the performance of mixed microbial cultures in the biodegradation of toxic organic chemicals, studies have been conducted using three phenol degrading species isolated from a municipal treatment plant. The rate of phenol degradation was investigated for each of the three pure phenol degrading species, and various combinations of the three species. A simple competitive model was used to predict the behavior of the mixed cultures by using the pure culture Monod rate constants. The model fit the growth data for total biomass very well, although (as with the pure culture experiments) the fit of the phenol degradation data was less accurate.

INTRODUCTION

Biological degradation of toxic organic compounds generally involves a highly heterogeneous biomass, which is treated as a single functional population for purposes of approximating kinetic parameters. As has been (very properly) argued [1] "any formal model even if it `works' presents just a more or less reasonable approximation to real processes and phenomena".

The rate of phenol biodegradation has been reported to be as simple as first or second order, or to follow Monod or Haldane (substrate inhibition) kinetics [2-7]. Any attempt to compare or generalize the results of these studies is very difficult, since they exhibit a wide spectrum of different conditions, including the composition of the culture (types of microorganisms present) which often is not even reported.

An important improvement to the current state of kinetic modelling would take into account the individual contributions of the species which make up a mixed population. Clearly, this will be difficult to achieve. However, unless the system stops being treated as a single functional population, it seems that little progress toward improved models can be made. The benefits to be derived from such improvements can be particularly significant when hazardous wastes are involved.

The present study reports the first results of a systematic effort to understand the biodegradation of phenol by a mixed culture (obtained from the secondary treatment units at the Passaic Valley Sewerage Commissioners (PVSC) plant in Newark, New Jersey). The following areas will be stressed:

- characterization (identification) of the species present in the mixed culture.

- determination of the ability of each of the isolated species to degrade phenol as a sole carbon source.

- determination of the rate of phenol degradation by mixed cultures consisting of combinations of two of the pure phenol degraders.

- comparison of the mixed culture data with a model assuming pure and simple competition [8] between the two pure species.

ISOLATION PROCEDURES

Two liters of PVSC mixed liquor were aerated at about 1.0 liter/min in a clear plastic batch reactor. Phenol was added to the reactor to bring the concentration to 100 ppm. When the phenol concentration fell below 1 ppm (the accuracy of the GC analysis using direct aqueous injection), phenol was again added to bring it back up to 100 ppm. After the third exposure, 10 ml of mixed liquor were diluted and inoculated on nutrient agar plates. After incubating for 24 hours, representative bacterial colonies showing morphological differences were streaked onto fresh nutrient agar plates and incubated once again. Successive species isolations were needed in order to assure that the isolated cultures were purified.

Once purified, the individual species were maintained in nutrient broth, and identified using a modification of standard taxonomic techniques employing diagnostic tubes.

Figure 1 shows the predominant genera in PVSC mixed liquor. Three categories of microbial organisms are indicated: bacteria, yeast and fungi, and protozoa. They are shown in two groups: those which were isolated from the fresh mixed liquor, and those which were isolated after phenol acclimation. Of these, only the phenol-acclimated bacteria were of interest in this research. The population density of the bacteria decreased on phenol exposure (as did the population diversity), and gram negative bacteria predominated.

Table 1 shows the results of three investigators in the

FIGURE 1

PREDOMINANT MICROBIAL GENERA IN PVSC MIXED LIQUOR

	After Phenol-acclimation[*]
Fresh	
10^9 bacteria/cm^3	10^8 bacteria/cm^3

$\underline{\text{gram positive}}$
gram negative   1.0   0.3

```
Gram positive rods (Bacillus) ───────────────────────>
Gram positive cocci (Micrococcus,Staphylococcus) ───>
Pseudomonas ─────────────────────────────────────────>
Acinetobacter ───────────────────────────────────────>
Enterobacter ────────────────────────────────────────>
Providencia ─────────────────────────────────────────>
Pasturella.
Alcaligenes.
```

10^6 yeast cells/cm^3 10^6 yeast cells/cm^3

```
Candida ─────────────────────────────────────────────>
Cryptococcus ────────────────────────────────────────>
Trichosporon ────────────────────────────────────────>
Debaromyces ─────────────────────────────────────────>
Saccharomyces.

Penicillium ─────────────────────────────────────────>
Aspergillus ─────────────────────────────────────────>
Streptomyces ────────────────────────────────────────>
Trichophyton.
Geotrichum.
Rhodotorula.
```

10^5protozoa/cm^3 10^5 protozoa/cm^3

```
Epistylis ───────────────────────────────────────────>
Opercularia ─────────────────────────────────────────>
Peranema ────────────────────────────────────────────>
Colpidium ───────────────────────────────────────────>
Stylonichia.
Carchesium.
Paramecium.
Podophyra.
```

*   100 ppm for 10 days

same laboratory, using various batches of the PVSC mixed liquor. Some differences in dominant species are noted, but in general the results are very similar. This paper will concentrate on the third group of results.

TABLE 1
DOMINANT BACTERIAL SPECIES IN PHENOL-ACCLIMATED MIXED LIQUOR

Investigator I

Achromobacter sp.
Acinetobacter lwoffii
Bacillus cereus
Enterobacter agglomerans
Escherichia coli
Micrococcus sp.
Moraxella sp.
Pseudomonas cepacia
Pseudomonas fluorescens
Pseudomonas sp.
Group 2K-1 Pseudomonas-like
Serratia marcescens
Staphylococcus sp.

Investigator II

Acinetobacter anitratus
Acinetobacter lwoffii
Alcaligenes faecalis
Bacillus cereus
Enterobacter agglomerans
Micrococcus sp.
Providencia stuarti
Pseudomonas aeruginosa
Pseudomonas cepacia
Pseudomonas fluorescens
Group 2K-1 Pseudomonas-like

Investigator III (present work)

Acinetobacter lwoffii
Aeromonas hydrophilia
Bacillus cereus
Enterobacter cloacae
Escherichia coli
Klebsiella pneumoniae
Pseudomonas cepacia
Pseudomonas fluorescens
Pseudomonas putida
Pseudomonas sp.
Serratia liquefaciens

EXPERIMENTS WITH PURE PHENOL-DEGRADING SPECIES

Each isolated bacterial colony was tested for its ability to degrade phenol over a two-week period. Only three of the species listed for Investigator III (Table 1) consumed phenol in pure culture:_Klebsiella_ pneumoniae,_Serratia_ liquefaciens, and _Pseudomonas_ putida.

Inocula from the three pure stock cultures were transferred to three successive subcultures utilizing phenol as the sole source of carbon. That is, the pure cultures used to determine the Monod kinetic rate parameters were exposed to phenol in two prior subcultures.

The cells were grown in 250 ml nephelometric shaker flasks, with the initial phenol concentrations ranging from 10 to 200 mg/liter. The cultivation time was 10-14 hours. There were no baffles, and no aeration other than that transferred by shaking. The course of growth in each flask was assessed optically as percent transmittance at 540 nm, and converted to biomass concentration (see Figure 2). The initial slope of the optical density plot on semi-log paper is the specific growth rate (μ) and is representative of the exponential growth phase. Figure 3 shows the results from eight batch reactors using different initial phenol concentrations.

These results were then used with the Monod model to determine the maximum specific growth rate and saturation constant for each of the three phenol-degrading species (Table 2).

TABLE 2 – Monod Parameters for Phenol Degrading Species

	μ_{max} (hr^{-1})	K_s (mg/liter)
Klebsiella pneumoniae	0.467	35.5
Serratia liquefaciens	0.409	43.6
Pseudomonas putida	0.158	15.3

Figure 4 shows the data points for phenol degradation for one of the batch experiments (along with the theoretical Monod curve), and Figure 5 shows the relationship between biomass growth and phenol utilization (with the slope [Y] equal to the yield coefficient).

MIXED-CULTURE EXPERIMENTS AND KINETICS

The equations below describe a simple competitive model for two species degrading the same substrate. This model was then used to calculate the theoretical rates of growth and phenol degradation for the mixed cultures by incorporating the previously determined pure species Monod parameters. That is, all of the parameters below were defined by the pure culture experiments.

$$\frac{dC_1}{dt} = \frac{\mu_{m1}\, S}{K_{s1} + S}\, C_1$$

$$\frac{dC_2}{dt} = \frac{\mu_{m2}\, S}{K_{s2} + S}\, C_2$$

$$\frac{dS}{dt} = -\frac{1}{Y_1}\frac{\mu_{m1}\, S}{K_{s1} + S}\, C_1 - \frac{1}{Y_2}\frac{\mu_{m2}\, S}{K_{s2} + S}\, C_2$$

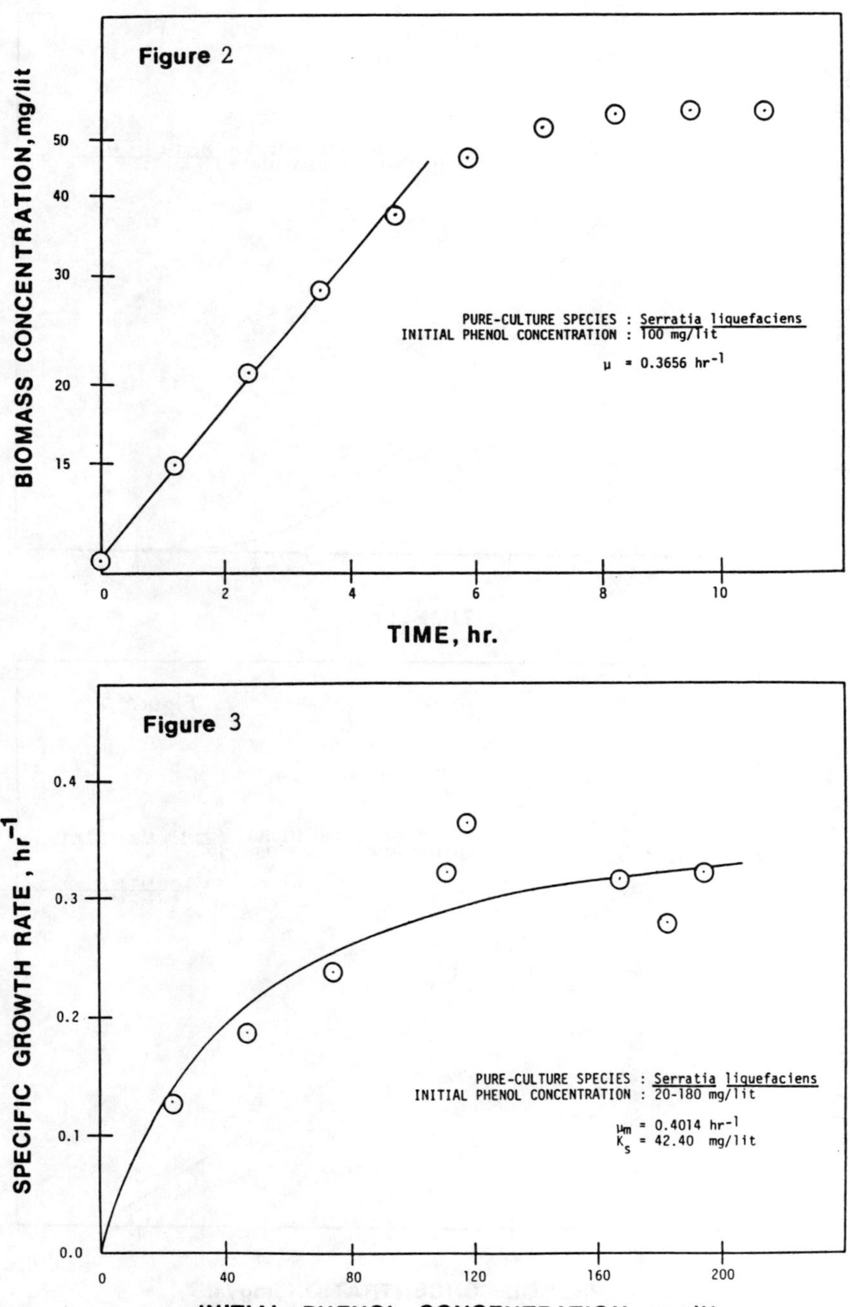

Figure 2
BIOMASS CONCENTRATION, mg/lit
50
40
30
20
15
TIME, hr.
0   2   4   6   8   10
PURE-CULTURE SPECIES : Serratia liquefaciens
INITIAL PHENOL CONCENTRATION : 100 mg/lit
μ = 0.3656 hr^{-1}

Figure 3
SPECIFIC GROWTH RATE , hr^{-1}
0.4
0.3
0.2
0.1
0.0
INITIAL PHENOL CONCENTRATION, mg/lit
0   40   80   120   160   200
PURE-CULTURE SPECIES : Serratia liquefaciens
INITIAL PHENOL CONCENTRATION : 20-180 mg/lit
μ_m = 0.4014 hr^{-1}
K_s = 42.40 mg/lit

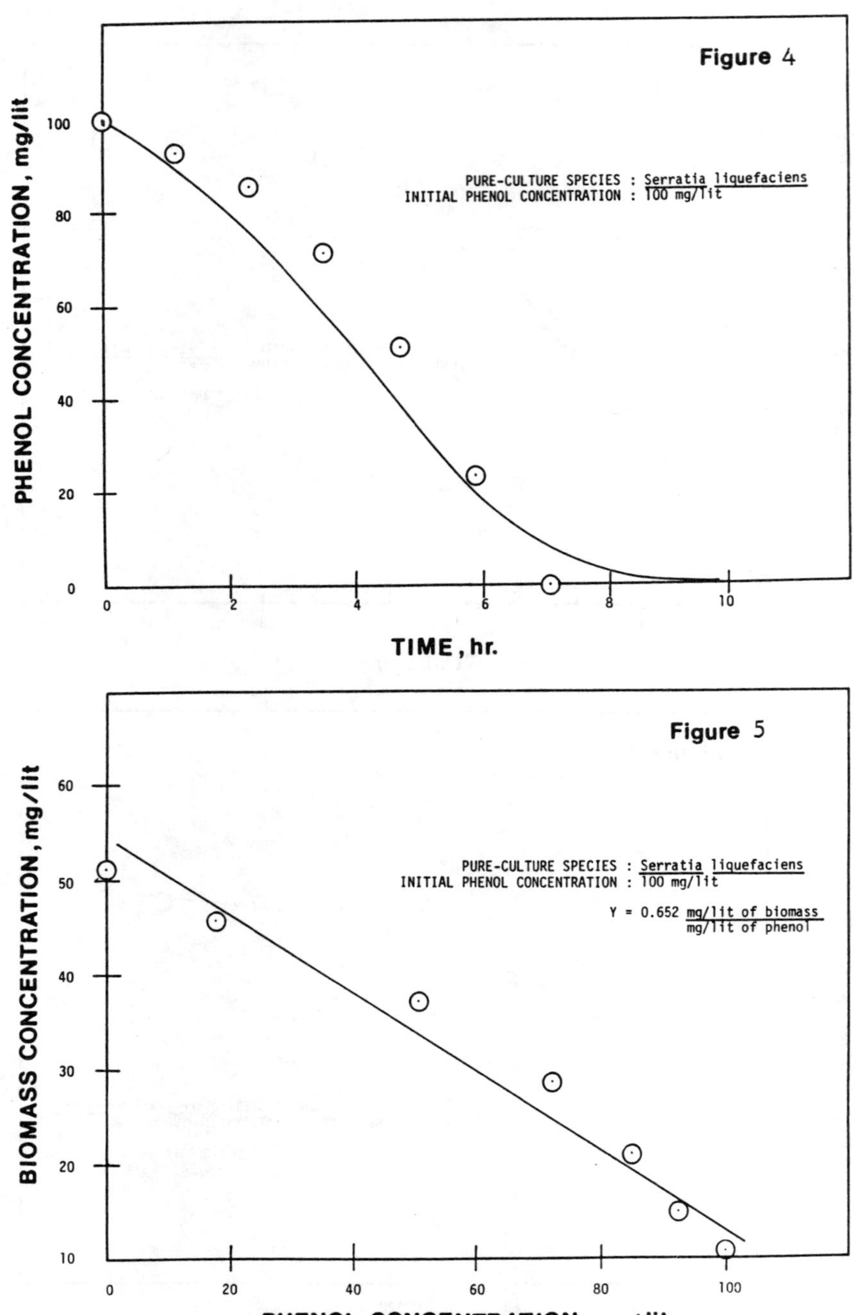
Figure 4
PHENOL CONCENTRATION, mg/lit
100
80
60
40
20
0
0 2 4 6 8 10
TIME, hr.
PURE-CULTURE SPECIES : Serratia liquefaciens
INITIAL PHENOL CONCENTRATION : 100 mg/lit
Figure 5
BIOMASS CONCENTRATION, mg/lit
60
50
40
30
20
10
0 20 40 60 80 100
PHENOL CONCENTRATION, mg/lit
PURE-CULTURE SPECIES : Serratia liquefaciens
INITIAL PHENOL CONCENTRATION : 100 mg/lit
Y = 0.652 mg/lit of biomass
mg/lit of phenol

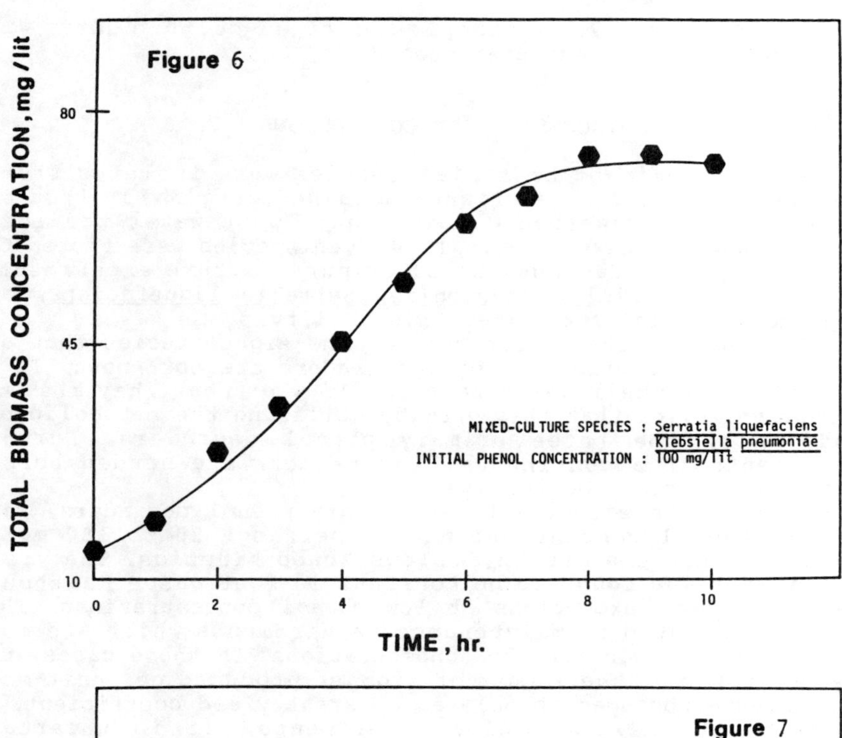

Figure 6
TOTAL BIOMASS CONCENTRATION , mg / lit
80
45
10
MIXED-CULTURE SPECIES : Serratia liquefaciens
Klebsiella pneumoniae
INITIAL PHENOL CONCENTRATION : 100 mg/lit
0 2 4 6 8 10
TIME , hr.

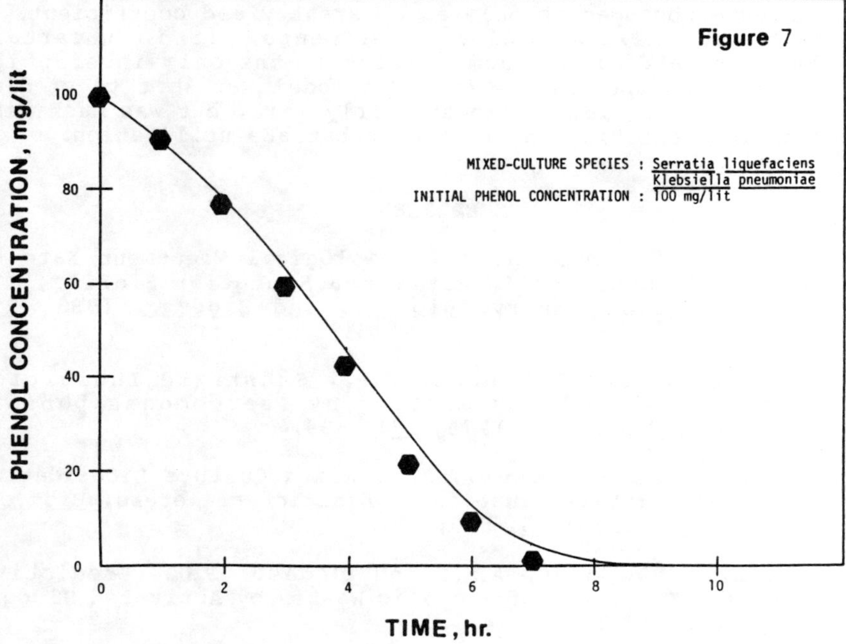

Figure 7
PHENOL CONCENTRATION , mg/lit
100
80
60
40
20
0
MIXED-CULTURE SPECIES : Serratia liquefaciens
Klebsiella pneumoniae
INITIAL PHENOL CONCENTRATION : 100 mg/lit
0 2 4 6 8 10
TIME , hr.

Figures 6 and 7 are examples of mixed-culture data, with the theoretical curves superimposed.

DISCUSSION AND CONCLUSIONS

Eleven dominant bacterial species were isolated from a phenol-acclimated mixed liquor obtained originally from the Passaic Valley Sewerage Commissioners wastewater treatment plant in Newark, New Jersey. All eleven species were tested for their ability to degrade phenol in pure culture experiments. Only three (<u>Klebsiella</u> <u>pneumoniae</u>, <u>Serratia</u> <u>liquefaciens</u>, and <u>Pseudomonas</u> <u>putida</u>) exhibited this ability.

Reasons for the persistence of the eight species that did not appear to be primary phenol degraders are not known. It is possible that their death rate is slow, or that they are in a dormant phase, or that they grow by utilizing the metabolic by-products of the three primary phenol degraders. Further experiments in a continuous flow reactor are needed before these questions can be resolved.

Experiments with the three primary phenol degraders, with initial phenol concentrations in the range 20 to 180 mg/l, indicated that the growth follows Monod kinetics. The yield coefficient was found to be constant in most cases, although there were some exceptions at low phenol concentrations. This may be attributed to maintenance requirements which are more pronounced at low substrate concentrations. In those cases, the average ratio of the amount of biomass produced per unit mass of substrate consumed is only an apparent yield coefficient.

With the mixed species experiments, it is uncertain whether pure and simple competition is the only interaction between the two species present. The model was able to predict the rate of total biomass growth fairly well, but was much less accurate in predicting the rate of substrate utilization.

REFERENCES

1. Vavilin, V.A., Dependence of Biological Treatment Rate on Species Composition in Activated Sludge or Biofilm, II: From Models To Theory, <u>Biotech.</u> <u>and</u> <u>Bioeng.</u>, 1983, <u>25</u>, 1539.

2. Hill, G.A., and Robinson, C.W., Substrate Inhibition Kinetics: Phenol Degradation by Pseudomonas putida, <u>Biotech.</u> <u>and</u> <u>Bioeng.</u>, 1975, <u>17</u>, 1599.

3. Pawlowsky, U., and Howell, J.A., Mixed Culture Biooxidation of Phenol, I: Determination of Kinetic Parameters, <u>Biotech.</u> <u>and</u> <u>Bioeng.</u>, 1973, <u>15</u>, 889.

4. Rozich, A.F., Gaudy, A.F., and D'Adamo, P.D., Predictive Model for Treatment of Phenolic Wastes by Activated Sludge, <u>Water</u> <u>Res.</u>, 1983, <u>17</u>, 1453.

5. Beltrame, P., Beltrame, P.L., Carniti, P., and Pitea, D., Kinetics of Phenol Degradation by Activated Sludge in a Continuous Stirred Reactor, *JWPCF*, 1980, *52*, 126.

6. Luthy, P.G., Treatment of Coal Coking and Coal Gasification Wastewaters, *JWPCF*, 1981, *53*, 325.

7. Sokol, W., and Howell, J.A., Kinetics of Phenol Oxidation by Washed Cells, *Biotech. and Bioeng.*, 1981, *23*, 2039.

8. Fredrickson, A.G., and Tsuchiya, H.M., Microbial Kinetics and Dynamics. In *Chemical Reactor Theory, A Review*, ed. L. Lapidus, and N.R. Amundson, Prentice-Hall, Englewood Cliffs, NJ, 1977, pp. 405-483.

THE TREATMENT OF INDUSTRIAL WASTE WATER IN INTEGRATED ACTIVATED SLUDGE/POWDERED ACTIVATED CARBON SYSTEMS

A.B. van Luin, L.V.M. Teurlinckx
Institute for Inland Water Management and Waste Water Treatment
Rijkswaterstaat, D.B.W./RIZA
P.O. Box 17, 8200 AA Lelystad, The Netherlands

ABSTRACT

A comparitive study has been carried out between the conventional activated sludge process (AS), the activated sludge process supported by powdered activated carbon (Powdered Activated Carbon Treatment Process, PACT) and the PACT-process in combination with wet-air regeneration of activated carbon containing surplus sludge. The latter process is commercially marketed under the name "Waste Water Reclamation System (WRS)". During this comparative study the effects of shock-loading were studied by addition of aniline and o-cresol to the influent.
Attention was also paid to the biological decomposition of the adsorbed components and the biological activity of the sewage sludge. The investigation was carried out using waste water of a tank-accomodation storage company.
The supplementation of activated carbon to the activated sludge process might be suitable for the treatment of industrial effluents with strongly changing compositions and the presence of hazardous substances. A more stable purification process is obtained with activated carbon together with a more extensive removal of toxic substances.

INTRODUCTION

Practice has shown that the performance of conventional biological treatment systems is influenced by the presence of specific toxic organic substances. Addition of powdered activated carbon (PAC) to the aeration basins of activated sludge facilities is one of the developed processes to improve the removal of toxic and harmful organic impurities (1,2,4).

This process is known as the Powdered Activated Carbon Treatment Process (PACT).
In this process biological degradation and adsorption on carbon are combined. The retention time of poorly biodegradable organic compounds adsorbed to the activated carbon will increase from the hydraulic to the sludge retention time. This will enlarge the chance of adaptation of the biomass to and biodegradation of the adsorbed compounds. An additional advantage of the use of powder activated carbon is the fact that adsorbed biodegradable organic substances will be desorbed and biologically degraded. In the case of the use of granulated carbon adsorption columns as a pre-treatment both biological and non-biological organic compounds will be adsorbed thus decreasing the adsorption capacity (4,5). In the PACT-process, due to the biodegradation of the desorbed biodegradable organic compounds a more effective use of the carbon is achieved. This phenomenon is known as bioregeneration of the carbon.
Based on the PACT-process Zimpro Inc. Wisconsin U.S.A. has developed and commercially marketed the so-called Waste Water Reclamation System (WRS). In the WRS-process PACT is used in combination with wet-air regeneration of the waste sludge containing powdered activated carbon (6).
Regenerated powdered activated carbon is reused in the aeration basin with a certain supply for carbon losses by oxidation.
The WRS-process is claimed to have a better performance, a more effective removal of toxic substances and lower running costs than the PACT-process. Since little information was available on the effectiveness of both processes with respect to the treatment of specific waste waters from multipurpose plants, tank storage companies, and harbour reception facilities,investigations have been carried out to get more information on the performance of these processes. To investigate the performance of the conventional activated sludge process (AS), the PACT-process and the WRS-process, a study was carried out by the Dutch Institute for Inland Water Management and Waste Water Treatment (D.B.W./RIZA) in cooperation with Tebodin Consulting Engineers, Zimpro Inc. USA, Paktank, Afvalstoffen Terminal Moerdijk (ATM) and Norit.

<u>Set-up of the study</u>

Two phases can be distinguished:
phase 1: comparative study of the performance of the AS-, PACT- and
         WRS-process.

phase 2: physical and biological aspects of the AS- and PACT process

The composition of the waste water is listed in table 1.

TABLE 1
Waste water composition

		average	range
COD	(mg/l)	1250	500-3000
BOD	(mg/l)	500	200-1500
Kj-N	(mg/l)	80	10- 140
COD/BOD		2.5	2- 3

The process conditions are listed in table 2.

TABLE 2
Process conditions phase 1 and 2

		phase 1			phase 2	
		AS	PACT	WRS	AS	PACT
influent	l/h	40	40	1.5	20	20
aeration tank volume	l	2000	2000	75	50	50
biomass	g/l	2-6	2-6	2-6	2-6	2-6
powdered activated carbon	g/l	–	8	8	–	8
sludge retention time (SRT)	d	25	25	25	40	40
hydraulic retention time (HRT)	d	2	2	2	2.5	2.5
coagulant	mg/l	–	–	–	–	20

The activated carbon content of 8 g/l (Hydrodarco H) is based on process conditions proposed by Zimpro Inc. The wet-air regeneration unit of the WRS-process contains a 4 l reactor in which batches of 2 l activated sludge/activated carbon are regenerated at 230° C and 90 bar during 1 hour.

Phase 1

Industrial waste water of a tank storage company is treated in three parallel connected pilot plants. A flocculation/flotation unit is used as pre-treatment (fig. 1).
At steady state conditions shock-loading was carried out with aniline (1000 mg/l) and o-cresol (600 mg/l) during 2 days.

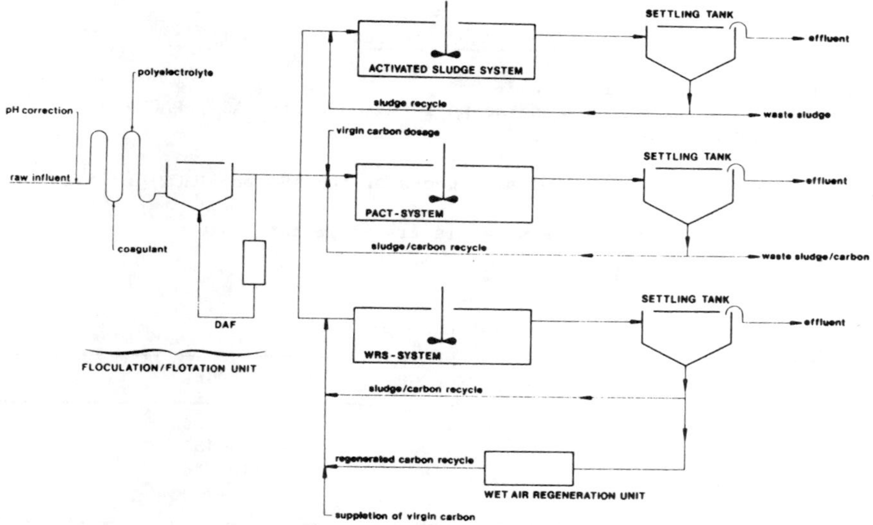

Figure 1: Flow sheet pilot plant

Phase 2

Physical and biological aspects of the AS-and PACT-processes were studied at laboratory scale. The conditions were kept simular to those employed in phase 1 (table 1). A higher SRT (40 days) was chosen because the main aim here was to investigate the biodegradation of carbon adsorbed organic compounds (bioregeneration) and the changes in the adsorption capacity of the carbon during use.
At steady state conditions aniline and 1,2,4,-trichlorobenzene (TCB) were successively spiked separately for a week. The amount of aniline and TCB in the influent was 250 and 8 mg/l respectively. The latter being the solubility limit of TCB in the treated waste water. After an intervening steady state period, both additions were repeated to investigate the adaptation of the sludge to aniline and TCB. In the second period of spiking TCB was used first because accumulation was expected.

RESULTS AND DISCUSSION PHASE 1

Physico-chemical pre-treatment by flocculation/flotation

The effect of the pre-treatment of the waste water in the flocculation/flotation unit (FFU) on the amount of some specific organic compounds was also studied. The results indicated that the pre-treatment partially removed the volatile aromatics and, to a considerable extent, the volatile organic chlorine compounds, largely caused by stripping. Since the FFU pre-treatment hardly removed the present oil and the waste water contained only a small amount of suspended solids, it was decided to omit this pre-treatment during the rest of the study.

Steady state conditions

The removal of specific organic compounds is shown in table 3. The values in this table represent the values of 7-14 measurements.

TABLE 3
Removal of specific compounds

	influent		effluent		
---	---	---	AS	PACT	WRS
acrylates/acetates	mg/l	5- 120	b.d.	b.d.	b.d.
alcohols/ethers	mg/l	5- 50	b.d.	b.d.	b.d.
t-butanol	mg/l	50-1500	1-4	1-2	1-7
t-methyl-butyl ether	mg/l	108	16	6	5
aromatics	ug/l	50-1500	1-4	1-2	1-7
dichloroethane	μg/l	39560	620	220	266
1,1,1,-trichloromethane	μg/l	424	3	2	8
trichloroethane	μg/l	1810	54	13	18
trichloromethane	μg/l	5260	340	170	240
tetrachloromethane	μg/l	4	1	n.d.	n.d.
EOCl	μg/l	132	45	9	14

b.d. means below detection level
n.d. means not determined

Under steady conditions only a small difference in the COD and BOD levels in the effluent of the PACT and WRS-processes in comparison with the AS-process is observed. Addition of carbon has little effect with respect to the removal of biodegradable substances.

Shock-loading

The results of shock-loading with o-cresol are shown in table 4.

TABLE 4

o-Cresol content, COD and BOD levels in the in-and effluent
during shock-loading experiments

| | influent | | | | | effluent | | | | | |
| | o-cresol mg/l | COD mg/l | | o-cresol mg/l | | | COD mg/l | | | BOD mg/l | |
			AS	PACT	WRS	AS	PACT	WRS	AS	PACT	WRS
day 1	530	3990	55	b.d.	b.d.	425	47	50	149	78	8
day 2	560	5270	200	b.d.	b.d.	760	50	94	358	10	33
day 3	645	5755	210	b.d.	b.d.	1170	69	240	560	20	159
day 4	13	5860	170	b.d.	b.d.	1710	160	455	780	69	159
day 7	–	4240	–	b.d.	–	1670	640	1020	–	195	105
day 9	–	4540	–	–	–	1330	940	1410	56	45	80
day 10	–	4720	–	–	–	725	685	1010	36	40	15

b.d. means below detection level

In contrast to the AS-process, no o-cresol was found in the effluents of the PACT-and-WRS-processes, indicating that o-cresol is largely adsorbed on the powder activated carbon. Aniline was not measured. The higher concentrations in the COD-and BOD-levels in the effluents caused by shock-loading with o-cresol are particularly noticeable in all three processes. The peak in the BOD of the effluent found in the AS-process during spiking did not occur in the other two processes.

Wet-air regeneration

The powdered activated carbon of the WRS-process was analysed after wet air regeneration in order to determine the efficiency of this process for the removal of biomass and the adsorbed organics and to study the changes in the properties of the carbon. Apart from biomass, the carbon was analysed for aniline, o-cresol and TCB. The mean and values for the carbon before and after regeneration are listed in table 5.

TABLE 5

Removal of biomass and chemicals adsorbed on powdered activated carbon
by wet-air regeneration (WAR)

Parameter		Before WAR	After WAR
Biomass	(g/l)	6.8	0.5
Aniline	(mg/g dry matter)	3.1	0.1
o-Cresol	(mg/g dry matter)	0.8	0.1
1,2,4,-TCB	(mg/g dry matter	1.5	0.1

The biomass content of the thickened sludge was 6.8 g/l before
regeneration and only 0.5 g/l after, which means that the biomass removal
was effective (6). As a result of the wet-air regeneration about 10% of
the adsorption capacity of the used carbon was lost based on its micro-
and macropores. When regenerated carbon is reused in the WRS-process,
about 5-10% of fresh carbon must be added to eliminate losses mainly
caused by oxidation.

RESULTS AND DISCUSSION PHASE 2

Performance of the processes during spiking experiments

Spiking with aniline in a concentration of 250 mg/l had no
detrimental effect on the treatment during the test periods in either the
AS- or PACT-process. Whether based on the reduction in COD or BOD, the
treatment efficiency was well above 90%.
Spiking with 1,2,4,-trichlorobenzene (TCB) in a concentration of 8 mg/l
for a week led to a smaller COD-reduction in the AS-process and in the
PACT-process, but this was more noticeable in the former. On the other
hand, the decreases in the BOD-reduction were only minor in the
PACT-process but more pronounced in the AS-process. All these changes were
noted during both TCB test periods (fig. 2).

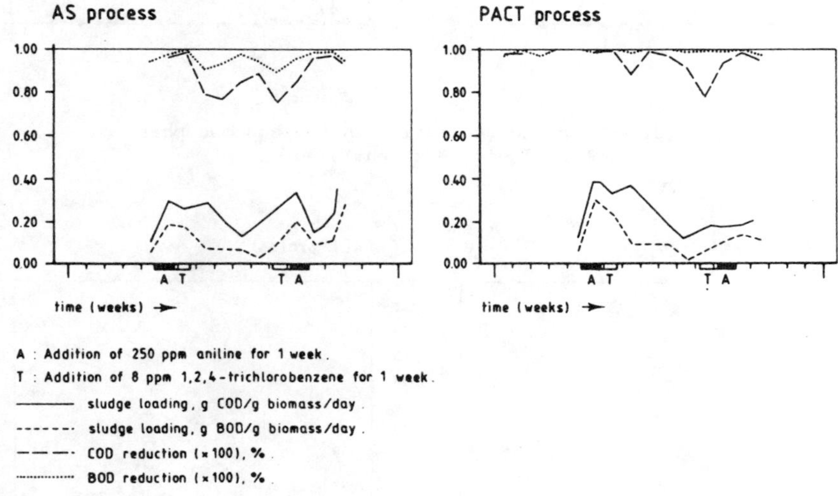

Figure 2. Purification yield during spiking periods with aniline and
1,2,4,-TCB

It can be concluded that the purification efficiency was more stable throughout the whole experimental period, but especially during the spiking period, in the PACT-process than in the AS-process.
The higher COD-based treatment efficiency found in the PACT might be explained by the additional adsorption capacity introduced with the powdered carbon. The high purification efficiency based on the BOD reduction may be explained by the higher substrate respiration or by the extra adsorption capacity coming from the carbon, or by combination of these two.

Behaviour of aniline

The figures below show the aniline content of the aqueous phase of the sludge (figure 3) and the amount of aniline adsorbed on the sludge (figure 4).

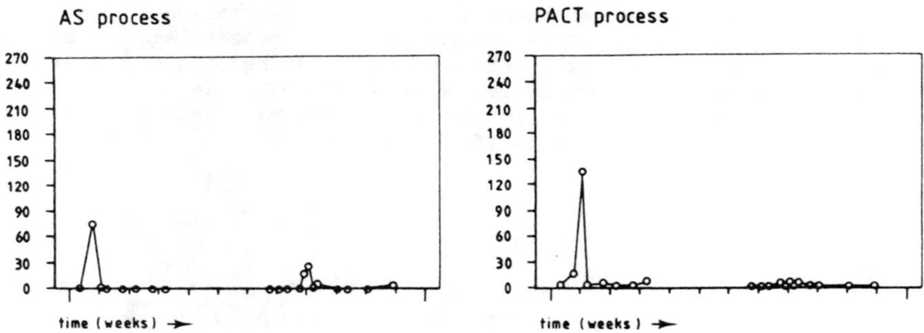

Figure 3. Aniline concentration in the aqueous phase
of the sludge (effluent), mg/l

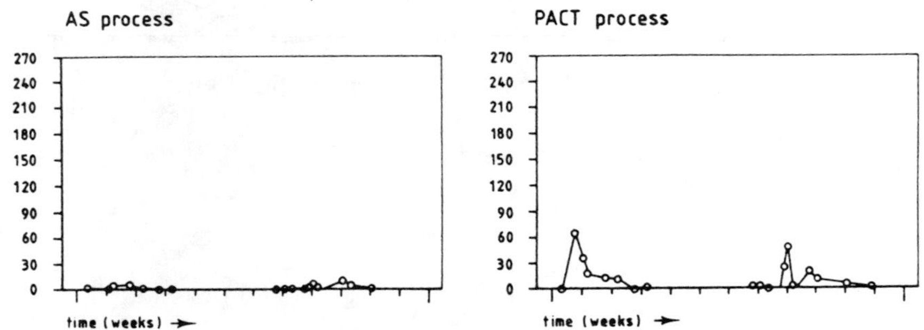

Figure 4. Amount of aniline adsorbed on the sludge, mg/l

The graphs for the AS-process show that the aniline concentration rose in the aeration tank 2-3 days after the start of the aniline addition. Aniline was adsorbed on the sludge only to a very small extent.

Subsequently, the aniline concentration of the effluent fell below the
detection limit of 1 mg/l, even though aniline was still being added to
the influent. This indicates that after a 2-3 days adaptation period,
aniline is efficiently removed.
The same picture emerged in the second aniline dosing period. In this case
the maximum aniline concentration of the effluent was reached after 2 days
and was found to be considerably lower. This means that the sludge
adapted faster to aniline than in the first aniline spiking period.

The graphs for the PACT-process show the same picture for the aniline
concentration, except for the fact that a considerable portion of the
aniline in the aeration tank was adsorbed on the sludge (either on biomass
or on the carbon). The adsorbed aniline showed a gradual decomposition
over a period of 2 weeks.
Since hardly any aniline was adsorbed on the AS-sludge, the aniline on the
PACT-sludge will to a considerable degree have been adsorbed on the
activated carbon. This part of the aniline is not directly available as a
microbial substrate or nutrient for the activated sludge and is not
decomposed until it has been gradually desorbed.
In the second aniline dosing period the sludge was already adapted to this
compound. As a result, the aniline was biodegraded to a large extent, the
rest of it being adsorbed.
These data indicate that the activated carbon used in the PACT-process is
regenerated by the desorption and decomposition of the adsorbed aniline;
an effect described as "bio-regeneration" (3,5)

Behaviour of 1,2,4,-trichlorobenzene

The results of the measurements during the periods when
1,2,4,-trichlorobenzene (TCB) was introduced into the influent are
illustrated in figure 5 and 6.

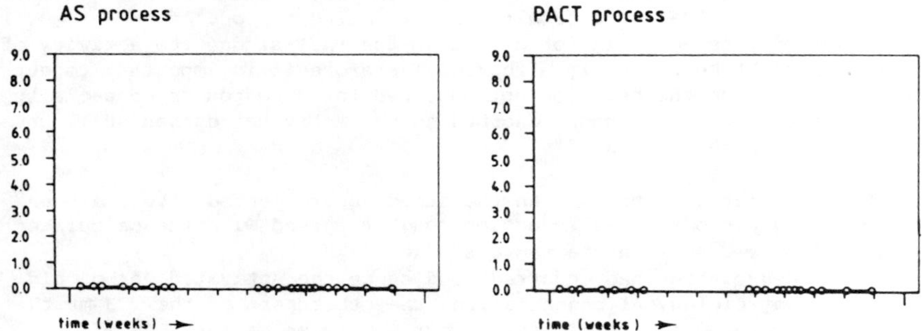

Figure 5. 1,2,4,-trichlorobenzene concentration in
the aqueous phase of the sludge (effluent), mg/l

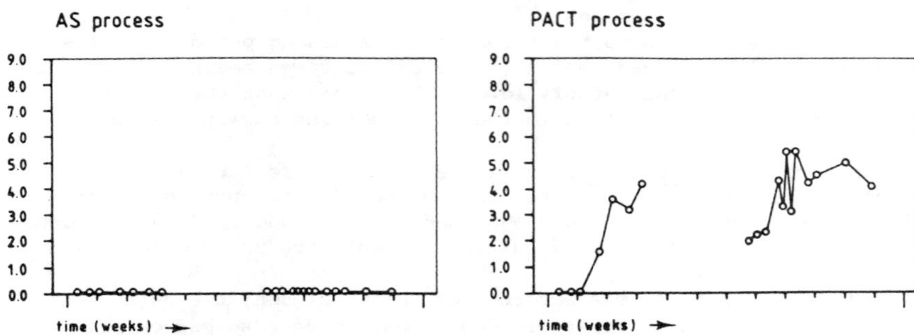

Figure 6. Amount of 1,2,4,-trichlorobenzene adsorbed
on the sludge, mg/l

No TCB was detected in the aeration tank of the AS-process in the entire dosing period, suggesting that this compound is not adsorbed on the sludge and it will probably be stripped out of the aqueous phase quickly. In the PACT-process the load of TCB introduced into the influent was 17.5 mg/l of aeration tank volume. When addition of TCB stopped, its concentration in the aeration tank was 4.3 mg/l, i.e. about 25% of the load introduced. No TCB was found in the aqueous phase of the sludge, e.g. in the effluent. If we also take in account the fact that no adsorption was detected in the case of the AS-sludge, we can conclude that TCB found in the PACT-sludge was adsorbed on the carbon.

The same pattern was observed in the second TCB dosing period, but the TCB concentration in the aeration tank was higher now, because of TCB residues remaining in the PACT-sludge from the first dosing period. As a consequence of this accumulation of TCB in the PACT-sludge the activity of the biomass might be adversely affected. Therefore it is important to get more information on the behaviour of adsorbed toxic compounds, especially from the fact if the compounds adsorbed to the activated carbon still have toxic effects on the activated sludge.

The TCB concentration decreased in the steady state period, i.e. between the two dosing periods. This reduction roughly agreed with the calculated amount of TCB removed with the waste sludge.
Since no biodegradation had occurred, and hence the activated carbon did not undergo any biological regeneration, the other part of the TCB must have been evaporated off (stripping effect). This means that TCB will almost fully evaporate during the AS-process.

CONCLUSIONS

- The PACT- or WRS-processes have a small effect on the BOD and COD reduction under steady state conditions.

- Especially by shock-loading, the presence of activated carbon has a stabilizing effect on the purification degree of the biological treatment process.

- The use of powdered carbon has a desirable effect on the elimination of organic micro-pollutants, and especially those collectively called extractable organic chlorine compounds (EOCl).

- By wet-air regeneration, the carbon is freed almost completely from the biomass, and the organic compounds adsorbed on it are fully or partly oxidized.

- Volatile organic compounds are evaporated to a large extent in the AS-process and partially adsorbed in the PACT-process.

- At present the PACT-process appears to be suitable for the treatment of industrial waste water containing organic pollutants that are hardly or slowly biodegradable.

RECOMMENDATIONS

- It is advisable to carry out further investigations to the type and content of the powdered carbon to be used in the PACT-process in order to optimize the removal of specific organics and to decrease the costs.

- Because not or partly degradable organic compounds can accumulate on the carbon it is necessary to get more information on the behaviour of adsorbed toxic organics to the biological part of the sludge.

- Though volatile compounds are adsorbed partially to the activated carbon and stripping is a main effect in the removal of harmful volatile organics, investigations should be done to increase this adsorption.

REFERENCES

1. Sublette. Kerry L., Snider Eric H. and Sylvester N.D., A review of the mechanism of powder activated carbon enhancement of activated sludge treatment, Wat Res., 16 (1982), 1075-1082.

2. Grieves, C.G., Crane, L.W., Venardos D.G., IJling, W., Powdered carbon enhancement versus granular carbon adsorption for oil refinery BATEA waste water treatment. 51st Annual Conference Water Pollution Control Federation Anaheim, California. (October 3, 1978).

3. Li, Alan Y.L., Di Giano, Francis A., Availability of sorbed substrate for microbial degradation on granular activated carbon. Journal WPCF, Volume 55, numer 4 (1983).

4. Weber, Jr., Walther J., Corfis, Nora H., Jones, Bruse E., Removal of priority pollutants in integrated activated sludge-activated carbon treatment systems. Journal WPCF, Volume 55, number 4. (1983).

5. Schultz, J.R. and Keinath, T.M., Organic removal mechanism in biophysical treatment systems. Water Sci. Techn. Vol. 17, pp 1043-1054 (1984).

6. Randall, Tripton L., Wet Oxidation of PACT-process carbon loaded with toxic compounds, Proc. 38th Annual Purdue Ind. Waste Conf. West Lafayette, Indiana, May 1983.

Chapter 5

CHEMICAL TREATMENT METHODS

CATALYTIC COMBUSTION - THE SEARCH FOR EFFICIENT COMMERCIAL
TECHNIQUES FOR THE DESTRUCTION OF POLYCHLORINATED BIPHENYLS

S.T. Kolaczkowski, B.D. Crittenden, U. Ullah and N. Sanders
School of Chemical Engineering
University of Bath
Claverton Down
Bath BA2 7AY
U.K.

ABSTRACT

Much criticism has been directed recently at thermal incineration as
an environmentally satisfactory means of destroying polychlorinated biphenyls (PCBs).   An overview is presented of alternative thermal techniques that are aimed at completing the destruction, to high efficiencies,
of wastes which contain high concentrations of PCBs.   Although the trend is
generally towards higher temperatures, research at Bath University is
directed towards using catalysts to complete the combustion process at
relatively low temperatures.   Prior to using PCBs themselves, combustion
trials have been carried out on 1,2,4-trichlorobenzene (TCB), which is a
constituent of askarels, flowing through a tubular reactor packed with
catalyst.   The preliminary results show that high conversions of TCB can
be obtained at temperatures greater than 750 °C.   The results of these
experiments will provide valuable guidelines on the experimental difficulties that are likely to be encountered when selecting suitable catalysts
for the combustion of PCBs.

INTRODUCTION

PCBs, (see Figure 1), are synthetic organic chemicals that possess
excellent dielectric and fire-resistant properties and, as a result, they
were widely used throughout the world in electrical equipment and other
specialised applications.   In the 1960's, awareness of the environmental,
health and safety risks of PCBs increased and most countries now have a
commitment to phase out their use completely.   It is expected that this
will occur over the next twenty years or so, the options being:

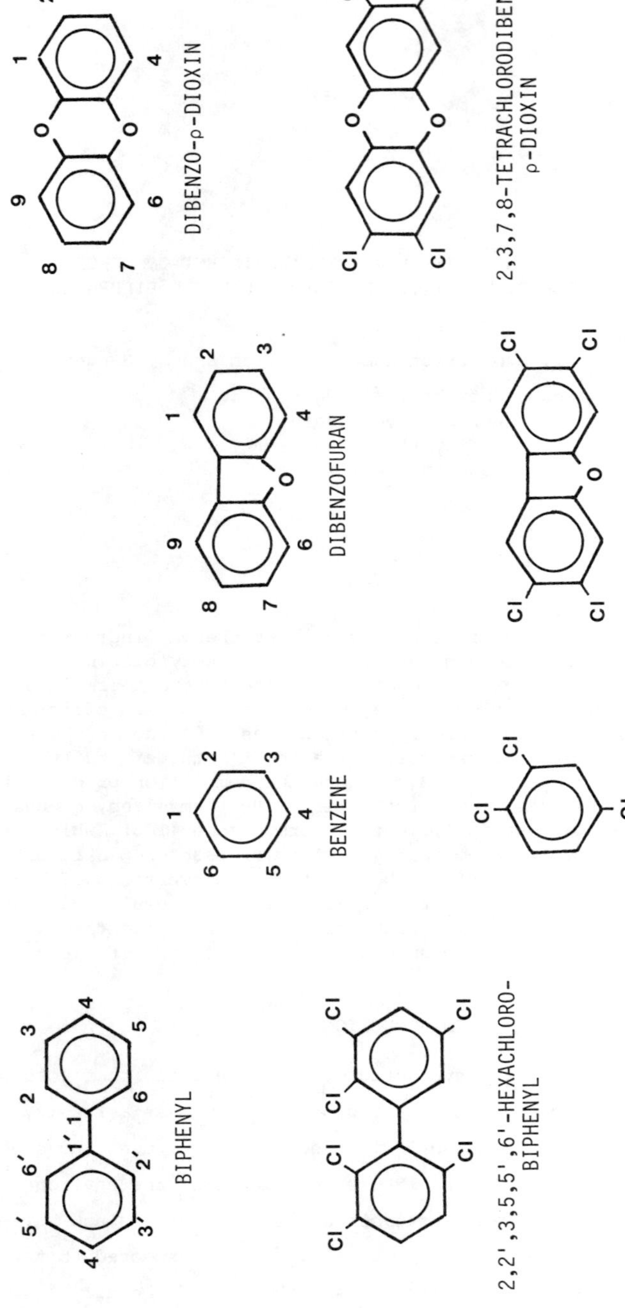

Figure 1.   Relevant chemical structures.

- to scrap existing equipment and replace with hardware
  which contains environmentally acceptable fluids; for
  example, silicones, esters, high molecular weight hydrocarbons and non-flammable halogenated hydrocarbons.

- to drain existing equipment, flush with solvent and
  retrofill with environmentally acceptable fluids.

Retrofilling is practicable for transformers, but not for capacitors.
Unfortunately, the retrofilled fluid does not remain completely free from
PCBs because some, typically 5%, would have remained trapped in the transformer windings and insulation to leach out subsequently with time into
the retrofilled fluid. After one retrofill the concentration of PCBs in
the replacement fluid may rise as high as 1.5% over two years. Thus,
chemical or physical processes have been developed for destroying or
removing the PCBs so that the parent oil can continue to be used without
the need for an expensive second retrofill.

WASTE DISPOSAL

Askarels, the fluids used in PCB filled electrical equipment usually
consist of a mixture of PCBs, typically up to 70%, together with other
chlorinated organic compounds such as 1,2,4-trichlorobenzene (Figure 1).
Therefore, the wastes arising from phasing out the use of PCBs include
scrapped hardware, liquids which contain PCB concentrations ranging from
parts per million to over 70%, contaminated soil and water, sludges, *etc*.
It is important in any waste disposal process that virtually complete conversion of PCBs to environmentally acceptable products is obtained. It
is especially important that the more toxic compounds such as PCDFs (polychlorinated dibenzofurans) and PCDDs (polychlorinated dibenzodioxins) are
not formed in the destruction process (Figure 1).

Methods for treating PCB wastes fall into thermal, chemical, physical
and biological classifications, but with considerable overlap. There is
a number of excellent reviews of these processes [1-5]. Most of the
chemical, physical and biological methods are being developed for cleaning
up contaminated oils, soils, waters, *etc*. In the U.S.A. particularly,
research has been directed at developing chemical processes for the decontamination of mineral oil transformers. Most processes depend on the
removal of chlorine atoms from PCB molecules by the powerful action of

alkali metals or their compounds. Mobile units have been developed to detoxify transformer oils on site, thereby eliminating the need to transport PCB bearing materials. However, the upper concentration that chemical processes can detoxify PCBs is typically of the order of 1%. Even if technically viable, it is unlikely that a chemical process for the destruction of askarel fluids would be an economic proposition. Therefore attention is being focussed on the development of alternative high temperature processes to thermal incineration.

INCINERATION

This is the method allowed by the U.S. Environmental Protection Agency for fluids which contain more than 500 ppm PCBs, for high and low voltage PCB capacitors and for certain small PCB capacitors [6]. Alternative disposal methods may be used only if they meet with the approval of the regional EPA administration. The U.S. Regulations [7] for the incineration of $\geq$ 500 ppm PCB fluids require:

- a combustion efficiency[*] of at least 99.9%, and either
    (i) maintenance of the introduced liquids for a two second dwell time at 1200 °C (±100 °C) and 3% excess oxygen in the stack gas, or
    (ii) maintenance of the introduced liquids for a $1\frac{1}{2}$ second dwell time at 1600 °C (±100 °C) and 2% excess oxygen in the stack gas
- acceptable analytical and control specifications

A destruction efficiency[†] in excess of 99.9999% is expected of U.S. incinerators, although this figure is not definitive. Instead, approval to burn PCBs is given only if the incinerator is capable of achieving acceptable emission levels. For comparison, the U.K. Code of Practice [8] states that PCB incinerators should:

---

[*]Combustion efficiency $= \dfrac{p}{p+q} \times 100$, where p = concentration of CO_2 in the flue gas

q = concentration of CO in the flue gas

[†]Destruction efficiency $= \dfrac{a-b}{a} \times 100$, where a = mass of PCB waste fed into process

b = mass of undestroyed PCB waste leaving process

- be fitted with gas scrubbing devices to remove hydrogen chloride, a product of combustion.
- be capable of operating at sustained temperatures greater than 1100 °C.
- have a mean residence time of at least 2 seconds.
- have a minimum excess oxygen content of 3%.

Despite this, opinion towards PCB incineration continues to be adverse [9] and attention is being focussed on improving destruction efficiencies, mainly by raising combustion temperatures.

Tests performed in cement kilns operating between 1450 and 2000 °C have shown that high destruction efficiencies can be obtained [10], but this method has met with considerable public opposition in North America [2,3]. When pure PCBs are burnt in pure oxygen, rather than in air, destruction efficiencies in excess of 99.99999% can be achieved without the need for auxiliary fuel [11]. Without the diluting effect of nitrogen, furnace temperatures of 1500 °C can be achieved. However, despite the fuel savings, the cost of providing pure oxygen would render this an expensive process to operate.

Even higher temperatures can be achieved in fluid wall reactors which typically consist of a porous carbon core surrounded by carbon electrical heaters which take the core temperature up to around 2200 °C [12]. As waste falls through the reactor, it is exposed to intense thermal radiation and is subsequently turned into a glass-like slag. The highest temperatures, up to 50000 °C, have been achieved by plasma arc pyrolysis, giving destruction efficiencies above 99.9999999% [13-15]. Within microseconds, organic compounds are atomised into constituent elements. The plasma, a gaseous cloud of charged particles, is generated in an arc by electrodes on each side of a reactor vessel.

CATALYTIC COMBUSTION

Research at the University of Bath is aimed at completing the destruction process at temperatures below 1000 °C by using catalysts. A suitable catalyst will lower the activation energy for the heterogeneous reaction to proceed. Consequently, higher oxidation rates can be achieved at temperatures and fuel concentrations much lower than those required for homogeneous reactions. Also, because of the large thermal inertia of a catalyst bed, catalytic combustors can sustain stable combustion at lower fuel concentrations than those encountered in conventional combustors.

A recent review of developments in catalytic combustion [18] focusses on combustion applications to generate heat and power, in order to attain low NO_x emissions and to maintain stable combustion for lean fuel/air mixtures.

Catalytic combustion has also been applied to the destruction of chlorinated hydrocarbons in the B.F. Goodrich Catoxid process [19]. A chlorinated by-product stream from the manufacture of vinyl chloride monomer is fed into a fluidised bed reactor which operates at temperatures below 540 °C. The feedstream may contain dichloroethylene, trichloroethane, carbon tetrachloride, chloroform and dichlorobutene. The identity of the catalyst used was not disclosed, as this was proprietary information.

In a project funded by the U.S. Department of Energy, researchers at Rockwell International conducted a catalytic combustion trial of poly-chlorinated biphenyls (PCBs) in a fluidised bed reactor [20]. The oxidation catalyst used was 20% Cr_2O_3 on alumina and high destruction efficiencies (more than 99.9999%) were reported in the pilot plant test.

Because of the proprietary and competitive nature of catalyst technology, relatively little information has been published on the kinetics of catalytic oxidation of chlorinated wastes. In a recent study [21], catalytic oxidation kinetics were investigated for di-, tri- and per-chloroethylene in a recycle reactor with a commercially available chromia on alumina catalyst over a temperature range of 350-550 °C.

A catalytic reactor to combust chlorinated waste would consist of a catalyst bed through which a premixed chlorinated waste/air mixture with auxiliary fuel (if required) would be fed. Although the feed would be preheated, in the first part of the bed the reaction is likely to be kinetically controlled and the overall process is controlled by the intrinsic surface kinetics. As the bed temperature increases, the overall rate of reaction may become limited by mass transfer effects. As bulk gas temperatures increase further, homogeneous gas phase reactions may occur simultaneously with the heterogeneous catalytic reactions as the combustion goes to completion. A schematic of these stages is illustrated in Figure 2.

In this paper, early results of the first phase of performing catalytic screening trials are reported. Prior to using PCBs, it was decided to conduct combustion trials with 1,2,4-trichlorobenzene (TCB), which is a

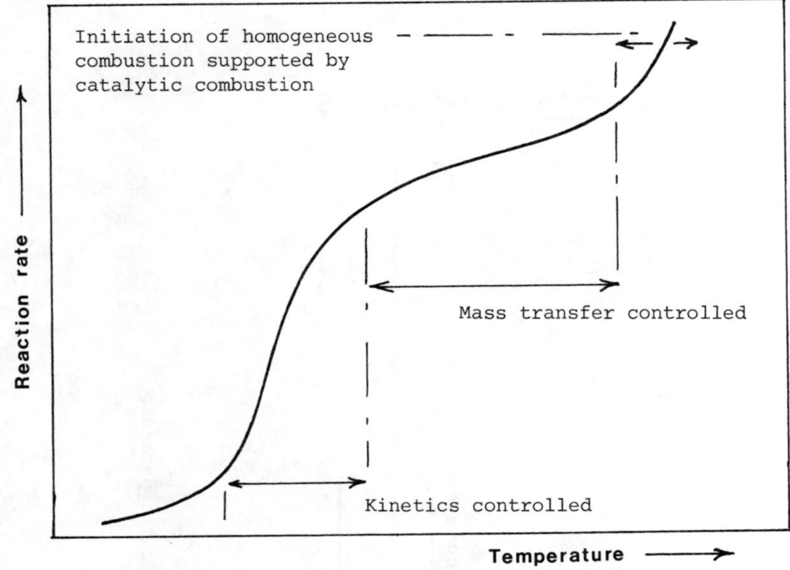

Figure 2. Overall rate of reaction *versus* temperature
relationship for a catalytic combustor.

constituent of askarel fluids.   The relative performance of platinum on
alumina and chromium oxide on alumina have been evaluated in a packed bed
reactor under comparative mass transfer resistance conditions. It is
planned to screen a number of other catalysts prior to excluding external
and internal mass transfer effects and evaluating intrinsic kinetic data.

EXPERIMENTAL

A flow diagram of the experimental assembly is illustrated in Figure 3.
Catalytic combustion trials were performed in a 12.7 mm diameter tubular
reactor which contained a 47.5 mm bed of catalyst.   The TCB fluid was
stored in a 250 ml graduated glass reservoir that was maintained at 30 °C.
The TCB was pumped into a preheater where it was vaporised and preheated
prior to being mixed with a preheated air stream and fed into the catalytic
reactor.   The reactor assembly was inserted into an electrically heated
furnace tube, the temperature of which was controlled by a proportional,
integral and derivative (PID) controller.   The packed bed reactor was
operated in the temperature range 500-900 °C, and a pressure of 1.01 bar.
The catalyst bed furnace temperature was measured with a thermocouple to
within an estimated error of ±1 °C.   The flow of air was monitored by a

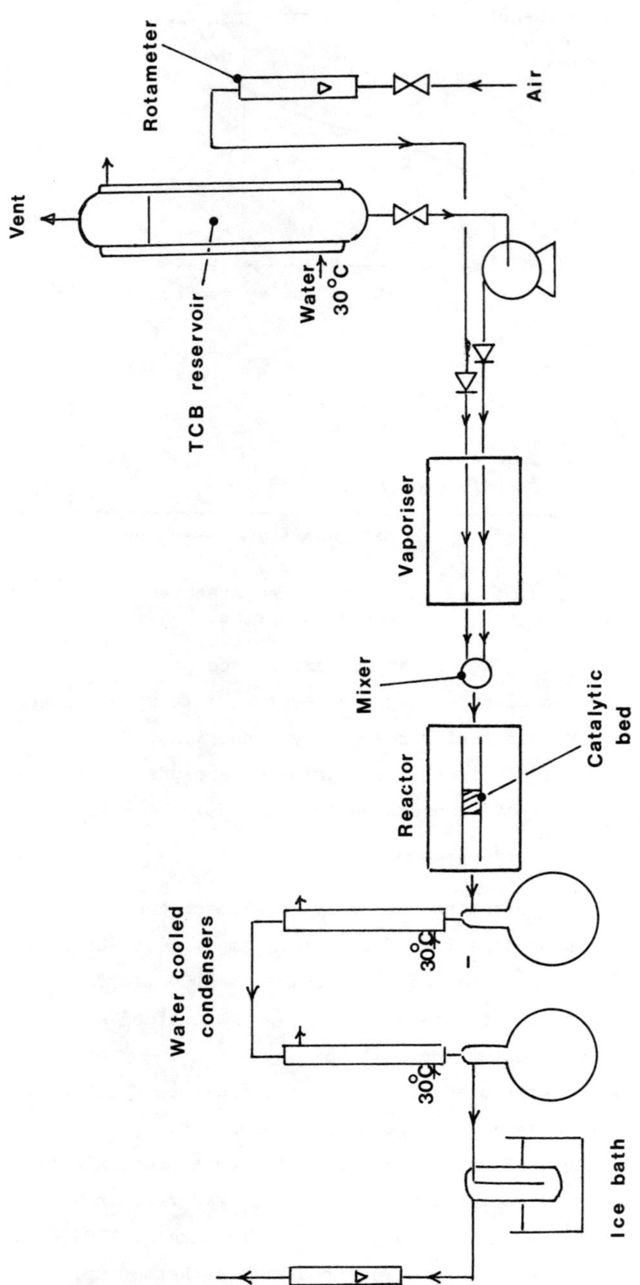

Figure 3. Flow diagram of experimental apparatus.

calibrated rotameter.   The flow rate of TCB was regulated with a peri-
staltic pump whose performance was regularly checked and calibrated.

The gaseous products from the reactor were quenched to 30 °C in a water
cooled condenser, where unreacted TCB and other liquid products were
collected.   The gaseous stream was then passed through a condenser maintained at 0 °C to remove water vapour and remaining condensible material
prior to being analysed on an infra-red spectrophotometer (I.R.).   The
liquid condensate was analysed on a gas-liquid chromatograph equipped with
an electron-capture detector and on the infra-red spectrophotometer.

RESULTS AND DISCUSSION

The experimental conditions over which the comparative performance of
the two catalysts was explored are given in Table 1.   In this initial stage
of catalyst screening, no attempt was made to exclude the effects of interparticle mass transfer on the rate of reaction.   In the first phase of the
trial this was not considered important, since the catalysts were compared
under similar interparticle mass transfer resistance conditions.   Combustion experiments were conducted both for a stoichiometric air to TCB
ratio and for a 25% excess air ratio.   The fractional conversions of TCB
achieved as a function of temperature are shown in Figures 4(a) and 4(b).

TABLE 1
Experimental conditions for comparative performance trials

Catalyst	20% wt Cr_2O_3/Al_2O_3, 0.3% Pt/Al_2O_3
Pellet size, mm	3.175 cylindrical
Bed ϕ, mm	12.7
Bed depth, mm	47.5
Temperature, °C	500, 600, 700, 800
Press, bar	1.01
$\frac{W}{F_T}$, g.s mol^{-1}	$4.1 \cdot 10^3$
TCB : oxygen, molar ratio	1 : 6, 1 : 7.5
Excess air %	0, 25

It can be seen that as the temperature approached 800 °C, fractional
conversion of TCB approached a value of 1.   The Cr_2O_3/Al_2O_3 catalyst exhibited higher levels of activity in the 550-700 °C temperature range.   The
shape of these curves resembles the transition between kinetic and mass
transfer regions anticipated in catalytic combustion (see Figure 2).

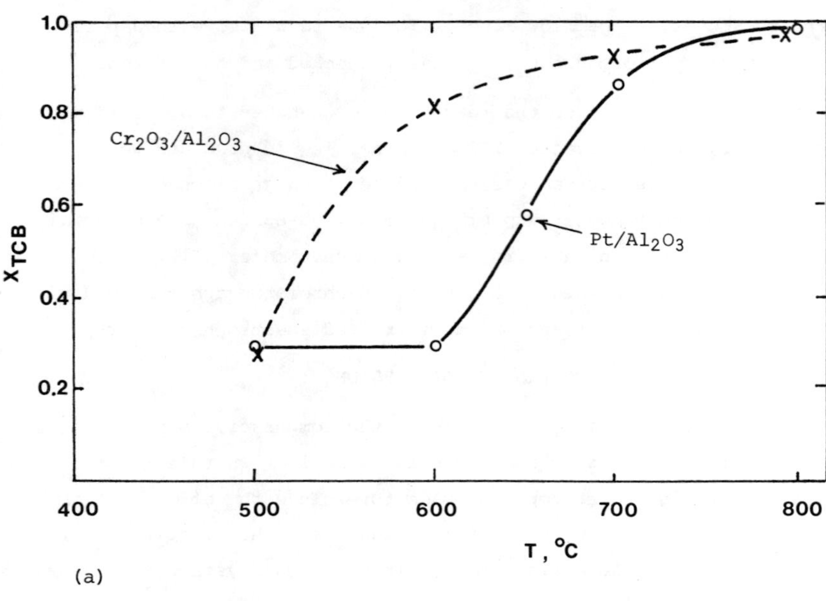

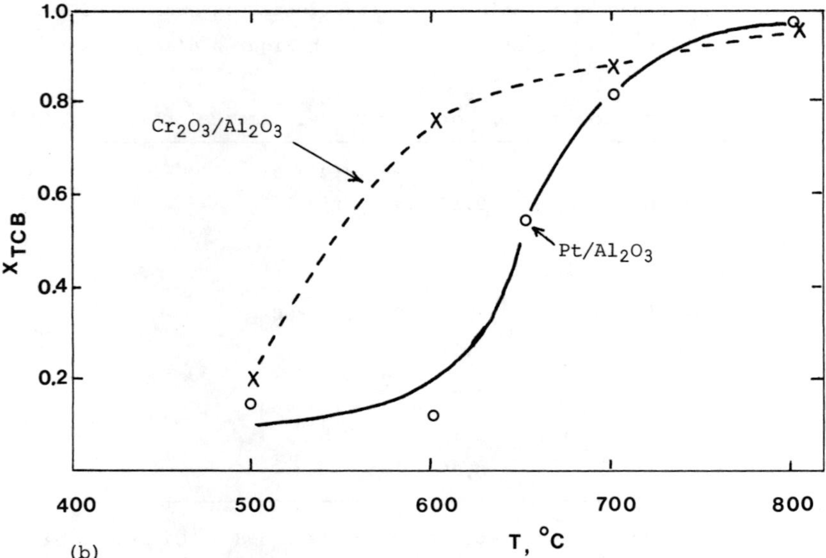

Figure 4. Fractional conversion of TCB, X_{TCB} as a function of temperature, T. (a) Stoichiometric TCB : air ratio; (b) 25% excess air.

At the higher temperatures ($\approx$ 750 °C), I.R. analysis of the gaseous stream clearly identified the existence of CO_2 and HCl in the combustion products (see Figure 5).   In the condensed material from the gaseous stream, besides TCB, a number of other compounds have been produced which still remain to be identified.

RELEVANCE TO THE CATALYTIC COMBUSTION OF PCBs

In extending the work to investigate the catalytic combustion of PCBs, due regard should be given to the following stoichiometric relationships which are associated with the homogeneous combustion of chlorinated hydrocarbons: for y greater than z:

$$C_xH_yCl_z + (x + \frac{y-z}{4})O_2 \rightarrow xCO_2 + zHCl + (\frac{y-z}{2})H_2O \qquad (1)$$

for y less than z:

$$C_xH_yCl_2 + xO_2 \rightarrow xCO_2 + yHCl + (\frac{z-y}{2})Cl_2 \qquad (2)$$

Deacon equilibrium reaction:

$$H_2O + Cl_2 \rightleftharpoons 2HCl + \tfrac{1}{2}O_2 \qquad (3)$$

The stoichiometry of equation (2) should be avoided, since chlorine is toxic, corrosive and difficult to remove from the other products of combustion.,   Free chlorine can be avoided by using auxiliary fuel to:

(i)   ensure that overall y is greater than z.   This is a requirement for PCBs containing 54% chlorine at which y = z (*e.g.*, Aroclor 1254)

(ii)   raise the calorific value of the waste feed, thereby raising the combustion temperature and driving the Deacon equilibrium to the right of equation (3).

In addition, it is difficult to stabilise a flame using air when the calorific value of the fuel is less than about 17000 kJ kg^{-1} [16].   The lower heating value of Aroclor 1254 is about 15000 kJ kg^{-1}.

In the absence of a catalyst and air, PCBs are thermally stable up to 550 °C, but decompose rapidly at 690 °C.   Oxidation in air takes place from about 350 °C upwards and leads to the formation of PCDFs by intramolecular cyclisation [17].   The amount of PCDFs formed depends on the PCB isomer (there are 209 possible isomers, but in commercial products only some 100 individual isomers have been identified), the temperature,

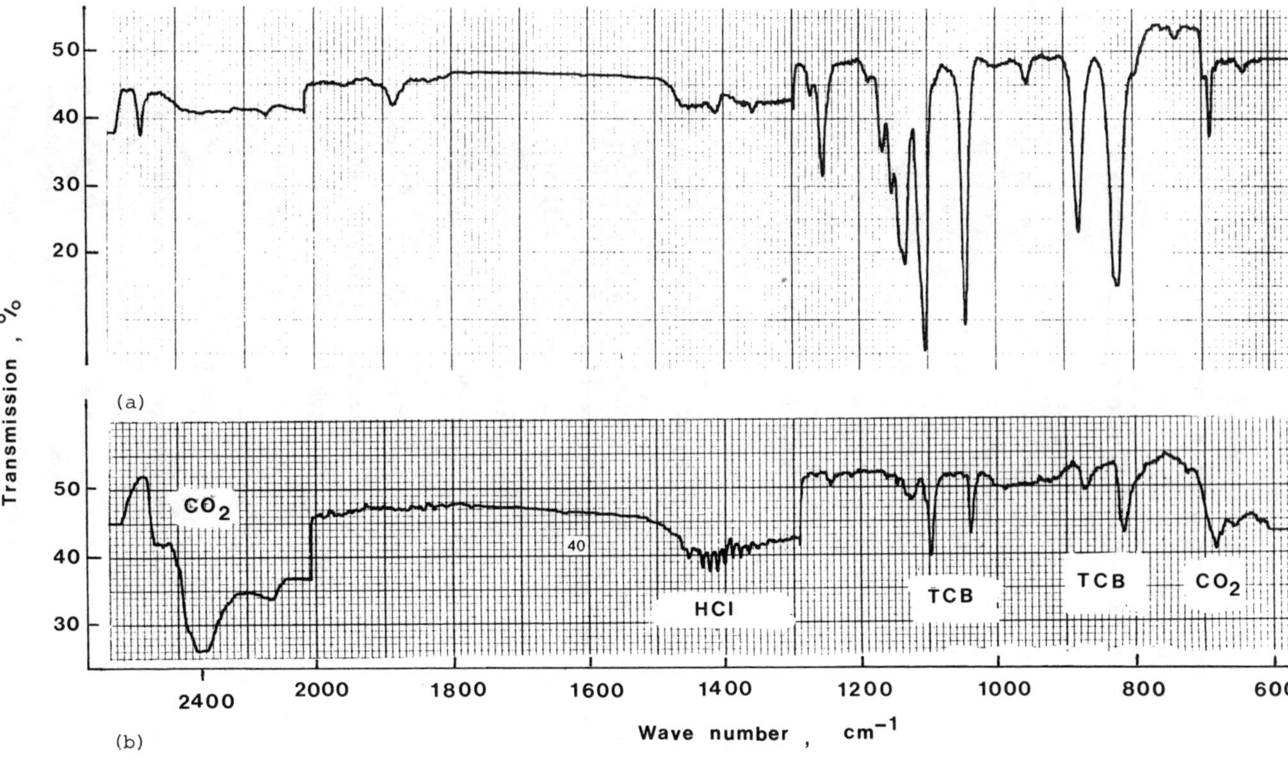

Figure 5.   (a) I.R. spectrum of pure TCB fluid;   (b) I.R. spectrum of gaseous product of the reaction at: 750 °C, 25% excess air, Pt/Al$_2$O$_3$.

and the oxygen excess.   Laboratory yields of PCDFs have ranged from 0.1% to several % in low temperature oxidation trials [17].

Care must be taken in selecting excess air requirements for combustion. Excess air lowers the combustion temperature by dilution of the product gases and also inhibits the formation of HCl by shifting the Deacon equilibrium to the left.   Nonetheless, excess air must be used to provide oxygen concentrations which are sufficient to complete the combustion process.

CONCLUSION

Preliminary results indicate that high conversions of TCB can be achieved at temperatures greater than 750 °C in a catalytic combustor. The Cr_2O_3/Al_2O_3 catalyst (*versus* Pt/Al_2O_3) exhibited a higher level of activity in the 550-700 °C temperature range.

Since the stoichiometric relationships of likely catalytic PCB combustion reactions are unknown, careful attention will need to be given to ensure that undesirable by-products such as soot, PCDFs and PCDDs are not formed.

ACKNOWLEDGEMENT

The authors are grateful to the SERC for supporting this research.

REFERENCES

1. Berry, R.I., New ways to destroy PCBs. *Chemical Engineering*, 1981, **88**, (10th August), pp.37-41.

2. Childs K., PCB treatment and destruction technologies.   Proceedings of Ontario Industrial Waste Conference, 1982, **29**, pp.121-148.

3. Craddock, J.H., Polychlorinated biphenyls (PCB's) disposal and treatment technologies - an update. *Proc. Environ. Symp.*, 1982, pp.161-188.

4. Akerman, D.G., *et al.*, Destruction and disposal of PCBs by thermal and non-thermal methods, 1983, Noyes Data Corporation, Park Ridge.

5. Webber, I., Pilgrim, D.B. and Thompson, M.A., The safe disposal of polychlorinated biphenyls, *IEEE Trans. Ind. App.*, 1984, **1A-20**(1), pp.159-166.

6. Kokoszka, L. and Flood, J., A guide to EPB-approved PCB disposal methods. *Chemical Engineering* 1985, **92** (8th July), pp.41-43.

7. Environmental Protection Agency, Polychlorinated biphenyls (PCBs) manufacturing, processing, distribution in commerce and use prohibitions. Federal Register, 1979, 40 CFR Part 761, **44**(106), pp.31514-31568.

8. Department of the Environment, Waste Management Paper No.6, 1976, Polychlorinated biphenyl (PCB) waste, HMSO, London.

9. Anon, PCB pollution lands Britain in Court. *New Scientist*, 1987, **113** (29th January), p.27.

10. McDonald, L.D., Skinner, D.J., Hopton, F.J. and Thomas, G.H., Burning waste chlorinated hydrocarbons in a cement kiln, 1978, U.S. NTIS PB Report 280118.

11. Komamiya, K. and Morisaka, S., Incineration of polychlorinated biphenyls with oxygen burner. *Environ. Sci. and Tech.*, 1978, 12(10), pp.1205-1208.

12. Arthurs, J., Management strategies for disposal of difficult industrial wastes, *Chemistry and Industry*, 1986, pp.410-422.

13. Barton , T.G. and Arsenault, G.P., Ultimate disposal of poly-chlorinated biphenyls.   In Detoxication of Hazardous Waste (edited by Exner, J.R.), 1982, Ann Arbor Science, Ann Arbor, pp.185-199.

14. Hiraoka, K., Aoyama, K., Nakamura, T., Mochizuki, S., Mitsumori, K. and Matsunaga, K., Decomposition of PCBs in the radio-frequency glow discharge plasmas of oxygen, hydrogen and water vapour. *Canad. J. Chem.*, 1982, 60(22), pp.2876-2882.

15. Anon, *Environ. Sci. and Tech.*, 1985, 19(2), p.105.

16. Senkan, S.M., Combustion characteristics of chlorinated hydrocarbons. In detoxication of Hazardous Wastes (edited by Exner, J.R.), 1982, Ann Arbor Science, Ann Arbor, pp.61-92.

17. Eggleton, A.E.J., Chemistry, origin and fate of dioxins and PCBs, 1986, paper presented to The Royal Society, Discussion Meeting, 16th September, 1986.

18. Prasad, R., Kennedy, L.A. and Ruckenstein, E., Catalytic combustion. *Catal. Rev-Sci. Eng.*, 1984, 26(1), pp.1-58.

19. Benson, J.S., Catoxid for chlorinated by-products. *Hydrocarbon Processing*, 1979, 58(10), pp.107-108.

20. Johnson, A.J., Meyer, F.G., Hunter, D.I. and Lombardi, E.F., Incinceration of polychlorinated biphenyl using a fluidised bed incinerator, 1981, Rockwell International (18th September), RFP-3271.

21. Manning, M.P., Fluid bed catalytic oxidation: an underdeveloped hazardous waste disposal technology. *Hazardous Wastes*, 1(1), pp.41-65.

HYDROTHERMAL DECOMPOSITION OF POLYCHLORINATED BIPHENYLS

N. Yamasaki
Research Laboratory of Hydrothermal Chemistry,
Faculty of Science, Kochi University,
Akebono-cho, Kochi 780, Japan

ABSTRACT

Decomposition of chlorinated organic substances by dechlorination were studied with a microautoclave. PCBs,being most stable, were decomposed completely in the presence of a methanol and sodium hydroxide solution under hydrothermal conditions of 300-320°C and 180 kg/cm^2 pressure. Similar dechlorinating decomposition processes using 4-chlorobiphenyl, chlorobenzene, and BHC were tested. A possibility was suggested for the industrial treatment of chlorinated organic substances with a continuous pipeline system. The decomposed mixture was easily treated with activated bacteria.

INTRODUCTION

Environmental contamination by PCBs, BHCs and trichloroethylene, stable and fatphillic materials(easily invaded animal organs) is a widespreaded scial problem.

Some method[1-3] have been investigated for the decomposition of PCBs, including open systems using radiant ray or oxygen flame of temperature above 2000°C.

The author has confirmed that the chlorine atoms of chlorinated organic molecules are substituted by hydroxyl groups under alkaline hydrothermal conditions. A closed system with an autoclave for the hydrothermal dechlorination of chlorinated organic compounds is described in this report. The treatment proposed here is safe and simple method, which is convenient for the large scale treatment of waste sludge containing hazardous materials as PCBs, BHCs, trichloroethylene, and so on. The released chlorine from chlorinated compounds forms a chloride ion or sodium chloride, and the dechlorinated products(being mainly hydroxide) are safely burned or are treated easily by activated sludge process.

The dechlorinating velocity is very large under alkaline hydrothermal conditions and so a pipeline system autoclave may be suitable rather than a large volume butch type one for the treatment of slurry materials of various waste form as soil, incineration ash, sludge, etc.

EXPERIMENTAL

Apparatus

Figure 1 shows the microautoclave(a) and the induction heater(b) used in this study. The stirring ball in the reaction chamber can be rolled up and down with the rocking motion of the heater. This heating method is convenient to select the conditions in using of pipeline system for having rapid temperature increasing(100°C/min).

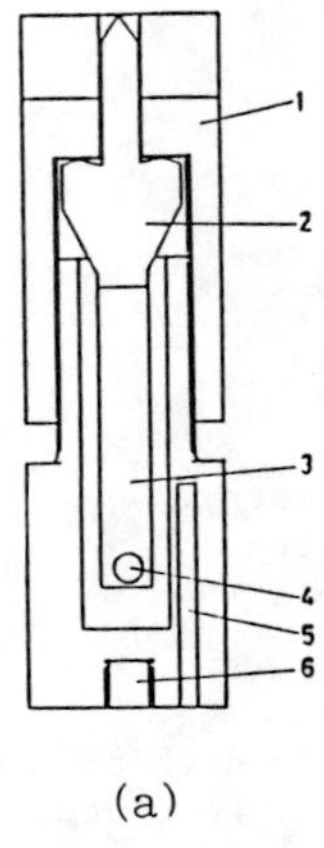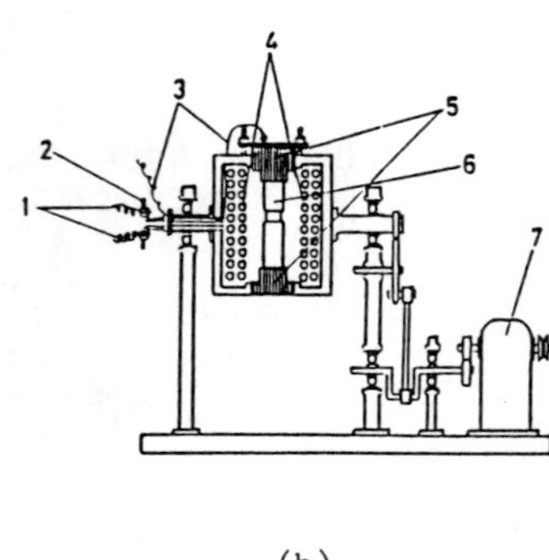

(a)   (b)

Fig.1 (a) Microautoclave: 1.lid, 2.cone packing, 3.reaction chamber, 4.ball for stirring, 5.thermocouple
well, 6.socket of steel rod for taking
out the autoclave.
(b) Induction heater: 1.lead wire for alterating current,
2.cooling water, 3.lead wire of thermocou-
ple, 4.copper tube(induction coil), 5.
steelblades of ferrosilicon, 6.microauto-
clave, 7.reduction gear.

Test sample

PCB used in this study has four chlorine atoms in one PCB molecule and
is called KC-400(produced by Kanegafuchi Chemical Co. Japan). Chlorobenzene, hexachlorocyclohexane(BHC), 4-chlorobiphenyl, trichloroethane were
selected for examination of dechlorinating procedure under alkaline hydrothermal conditions.

Procedure

The starting chlorinated organic compounds(0.5-15g) and 5-30mL of sol-
vent(a sodium hydroxide solution and dispersing agents) were poured into
the microautoclave. The total volume of the mixture in the autoclave was
adjusted to half the volume of the reaction chamber. The mixture in the
autoclave was heated with the induction heater, and then cooled quickly
with fan. The product in the autoclave was run off with water and hexane.
The solid product was collected by filtration. The aqueous and organic layer were separated in a separating funnel. The aqueous solution was washed
with hexane, and the organic layer was washed with 0.25M sodium hydroxide
solution repeatedly. All aqueous solutions were mixed and decolored with
1g of activated charcoal. The amount of chloride ion in the aqueous solution was determined with Mohr's titrimetric method. On the other hand, all
hexane solution were mixed, dried with 20g of anhydrous calcium chloride,
and then evaporated to 10 mL. The amount of undecomposed chlorinated compounds in the haxane solution was determined using gas chromatography and
mass spectrometry.

Measurement of pressure

The pressure in the reaction chamber was measured with a specially made

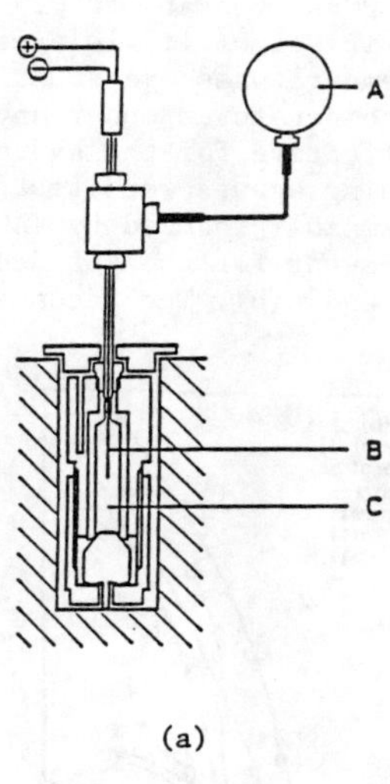

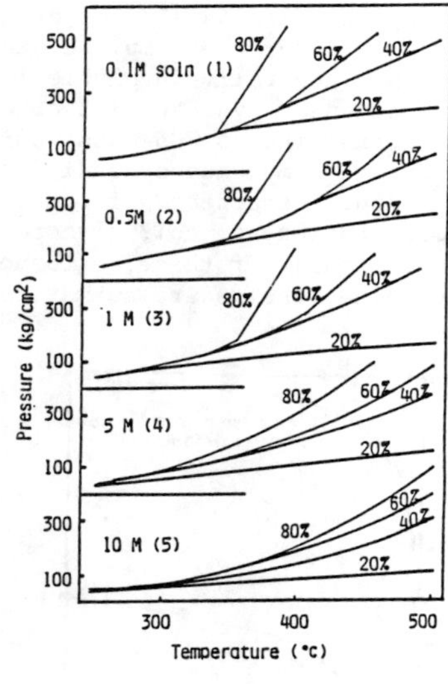

(a)

(b)

Fig.2   (a) Microautoclave equipped pressure gauge: A.pressure gauge,
                 B.thermocouple, C.sample chamber.
            (b) P-T curves of NaOH solutions(0.1M–10M): %;degree of filling.

autoclave equipped with a pressure gauge(Fig.2(a)). Figure 2(b) shows that
the pressure decreased as the concentration of sodium hydroxide solution
increased. In this study, the pressure was under 200 kg/cm² at 300°C.

RESULTS AND DISCUSSION

Dechlorinating conditions
Figure 3 shows the effects of the concentration of the sodium hydroxide

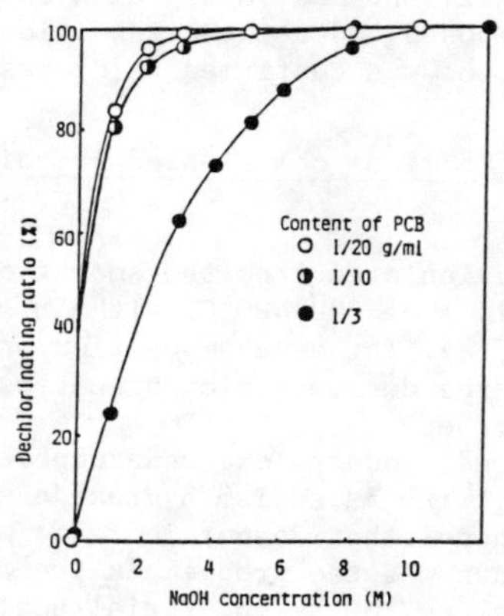

Fig.3   The effect of the concentration
of NaOH solution on dechlorinating
ratio.

solution on the dechlorinating decomposition of PCBs in the absence of dispersing agents such as alcohol or glycol. A large amount of sodium hydroxide was needed for the complete dechlorination of PCBs even at 350°C.

PCBs are generally insoluble in water, but their solubility is expected to increase under hydrothermal conditions. High temperatures are found to be more conductive to PCB decomposition than low temperatures under any conditions. A homogeneous condition may also be effective for the hydrolysis reaction. The effects of the following dispersing agents were tested: methanol, ethylene glycol, glycerol, phenol, and Emazol(produced by KAO ATRAS Co. Japan). Of these, methanol was the best agent for the PCB decomposition. The results are summarized in Fig. 4(a) and 4(b). The recommended

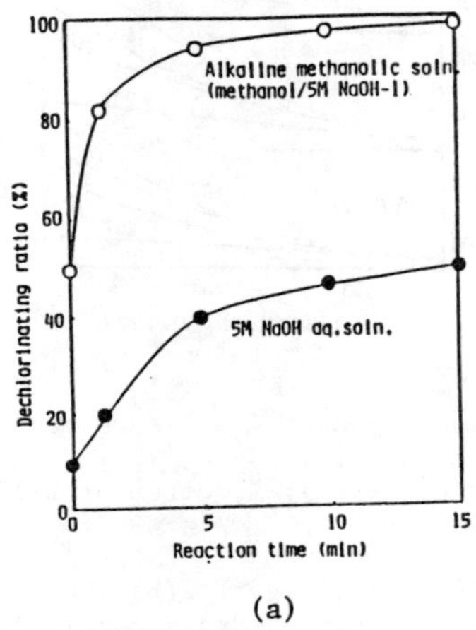

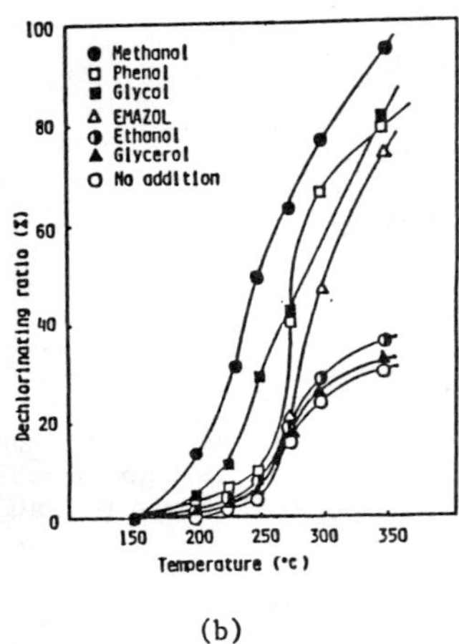

Fig.4 (a) Comparison of dechlorinating ratio between alkaline methanolic and NaOH solution only,
(b) Effects of dispersing agent on dechlorinating ratio.

conditions for the dechlorination of PCBs with the microautoclave are: temperature, 300-320°C; PCBs, 1g; methanol, 12.5mL; and 5M sodium hydroxide solution; 12.5mL. PCBs were completely decomposed under these conditions within a few minutes. The decomposition of PCBs was confirmed using gas chromatography and mass spectrometry.

Comparison of dechlorinating procedure among various chlorinated organic compounds

1) Chlorobenzene

The variations of dechlorinating ratio calculated from the amount of undecomposed chlorobenzene under the conditions of 300-400°C, with 5M sodium hydroxide solution, are shown in Fig. 5(a). These decomposition profoles are similar to those of PCBs and then the decomposition processes of these aromatic chlorides may resemble each other.

Figure 5(b) shows the amount of decomposed products and undecomposed chlorobenzene under the conditions of 350°C, with 5M sodium hydroxide solution. This figure shows that phenol and phenylether do not increase with the decrease of chlorobenzene. This phenomenon was the problem in the production of phenol by hydrolysis of chlorobenzene in the industrial scale. Phenylether is formed as the following equation;

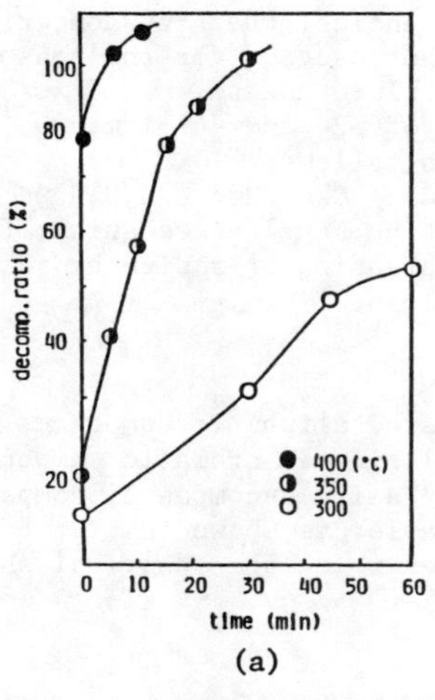
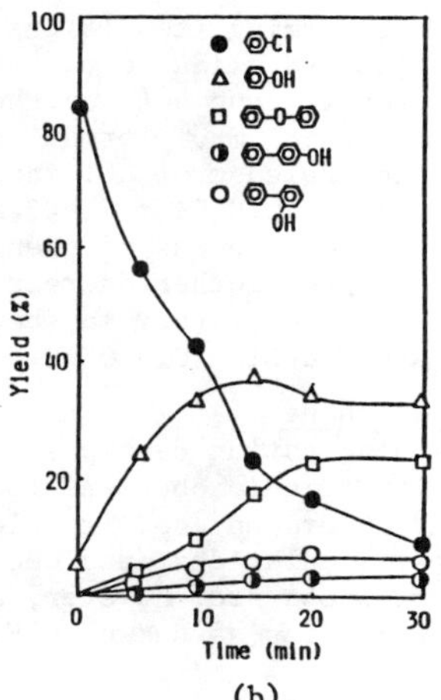

Fig.5   (a) Dechlorinating ratio calculated from the amount of undecom-
           posed chlorobenzene,
        (b) Yields variation of decomposed products.

TABLE 1
Yields of reaction products at 350°C.

Time(min)	NaOH (concentration)	◎-OH (%)	◎-0-◎ (%)	◎◎OH (%)	◎-◎-OH (%)
0	2 (M)	17.9	—	—	—
	3.5	15.2	—	—	—
	5	8.39	—	—	—
5	2 (M)	45.4	8.03	1.60	0.10
	3.5	54.7	4.59	2.81	0.62
	5	64.6	2.42	4.48	1.38
10	2 (M)	40.2	11.3	3.93	0.69
	3.5	45.4	11.3	4.68	1.04
	5	56.1	9.18	5.33	1.83

$$C_6H_5ONa + C_6H_5Cl = C_6H_5OC_6H_5 + NaCl.$$

Table 1 shows reaction products at 350°C. The yields of phenol and bi-
phenyl-2- and -4-ol increase with the increasing of alkaline concentration.
These results support the benzine mechanism[4-5]. That is,

If the rate determing step is the formation of benzine, the high concentration of sodium hydroxide solution accerelates the reaction for the reason why the hydroxyl group acts on the proton from chlorobenzene.

Another mechanism of formation of biphenyls of -2- and -4-ol may be explained by the release of HCl inter molecules as follows,

$$2C_6H_5Cl \rightarrow C_6H_5C_6H_4Cl \rightarrow C_6H_5C_6H_4OH \quad or \quad C_6H_5Cl + C_6H_5OH \rightarrow C_6H_5C_6H_4OH$$

The formation process of phenylether may not be benzine mechanism, for the amount of phenylether decreases with the increasing of sodium hydroxide concentration. Phenylether may be formed as follows,

$$C_6H_5ONa + C_6H_5Cl = C_6H_5OC_6H_5 + NaCl$$

2) Trichloroethane

The dechlorinating decomposition of chlorinated aliphatic compounds was examined with trichlorobenzene for comparing chlorinated aromatic compounds as PCBs and chlorobenzene. Trichlorobenzene was easily decomposed, compared with aromatic chlorides, and decomposition behavier is shown in Fig. 6(a). The solid products are, however, amorphous carbon which are spherical shape of μm order radious as shown in Fig. 6(b).

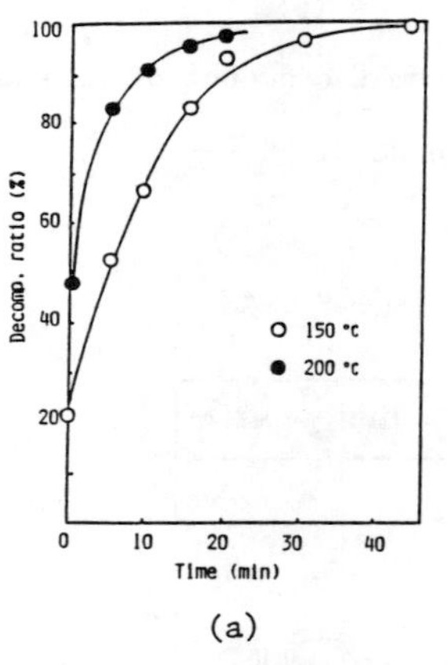

(a)

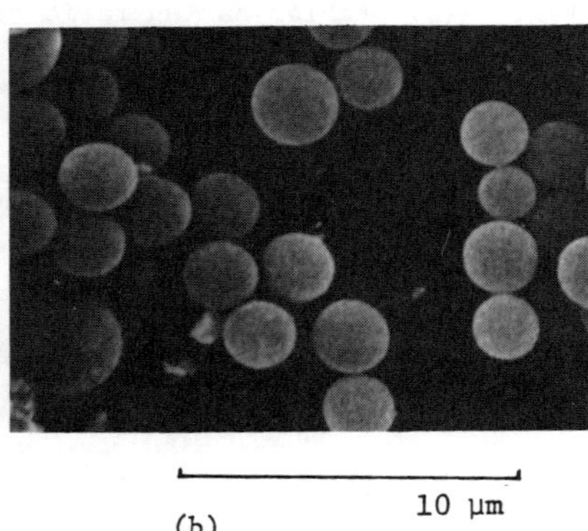

(b)

Fig.6　(a) Decomposition ratio of trichloroethane,
　　　　(b) SEM photograph of decomposition product of trichloroethane.

3) 1,2,3,4,5,6-hexachlorocyclohexane(BHC)

Figure 7(a) shows the decomposition ratio against temperature using γ-BHC. The initial dechlorination of BHC was easily occurred above 150°C. However, trichlorobenzene was formed at around of 50% dechlorination. Oily products were formed between 200 to 300°C, and their gas chromatograms are shown in Fig. 6(b). Four gas chromatograms in this figure are those of product from four kinds of BHC isomers. 1,2,3-trichlorobenzene, one of the products, was most easily formed on every cases. The amount of 1,2,3 type was least, and in the case from α-BHC nondetected. The amorphous carbon was formed under the conditions above 350°C.

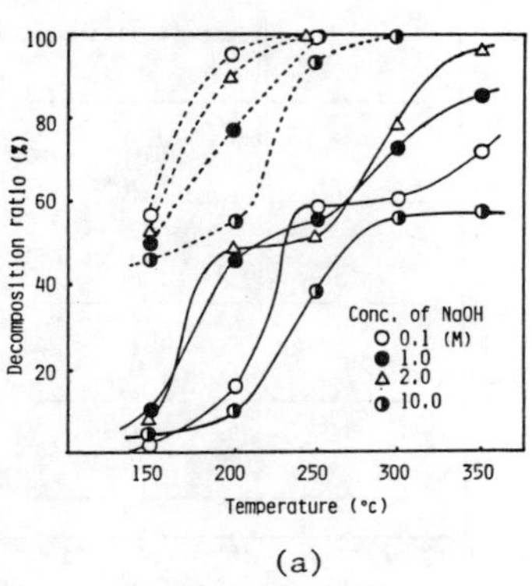

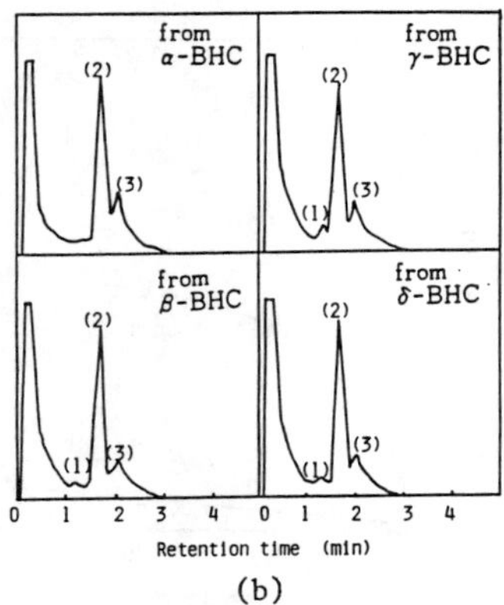

Fig.7   (a) Decomposition ratio of BHC(broken line;calculated from un-
              decomposed BHC, solid line;from dechlorinating ratio),
        (b) Gas chromatograms of decomposed products of BHC isomers.

4) 4-chlorobiphenyl

Decomposition ratio of 4-chlorobiphenyl against temperature is shown in
Fig. 8(a) and against the concentration of alkaline solutions in Fig. 8(b).

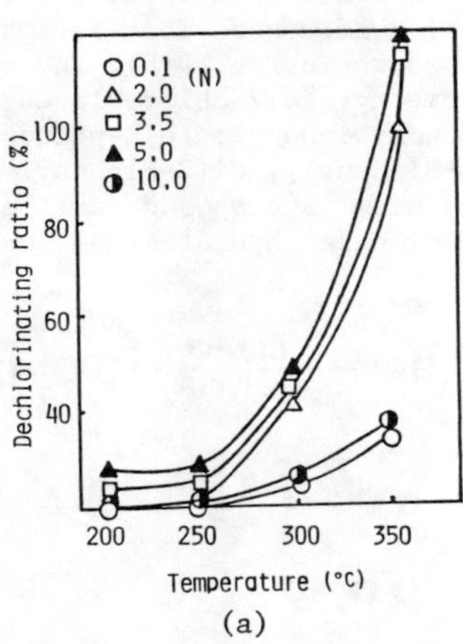

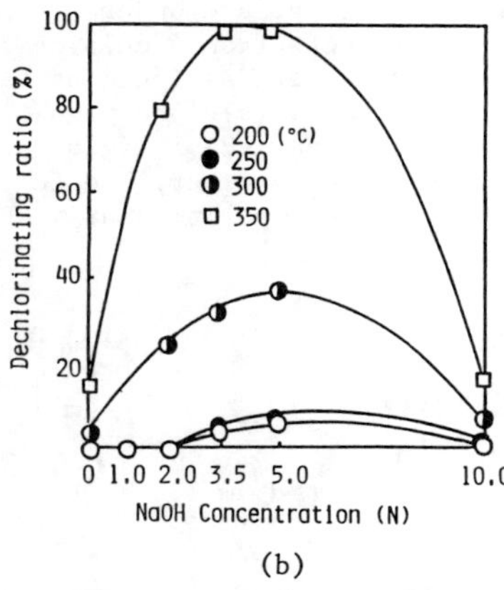

Fig. 8   (a) Dechlorinating ratio of 4-chlorobiphenyl against temperature
         (b) Dechlorinating ratio of 4-chlorobiphenyl against NaOH con-
             centration.

The reaction occurs at 250-350°C and 5M sodium hydroxide solution acts most
effectively. The decrease of decomposition ratio at above 5M may be explained as the high alkaline concentrate solution has a low activity of
hydroxyl ion for its high viscosity.

The comparison between PCBs and 4-chlorobiphenyl is shown in Fig. 8(a).
The decomposition ratio of 4-chlorobiphenyl is lower than that of PCBs, and

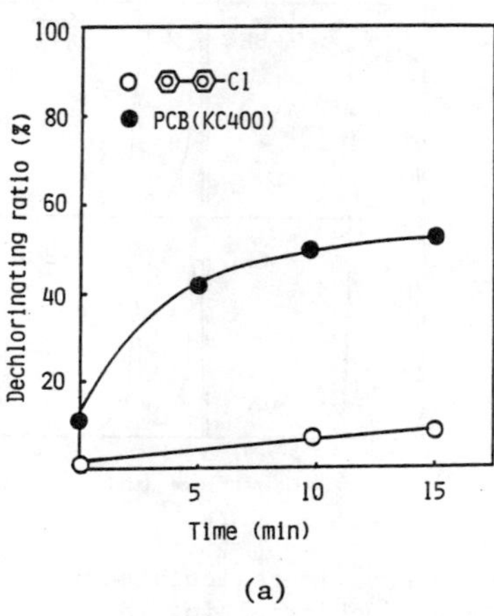

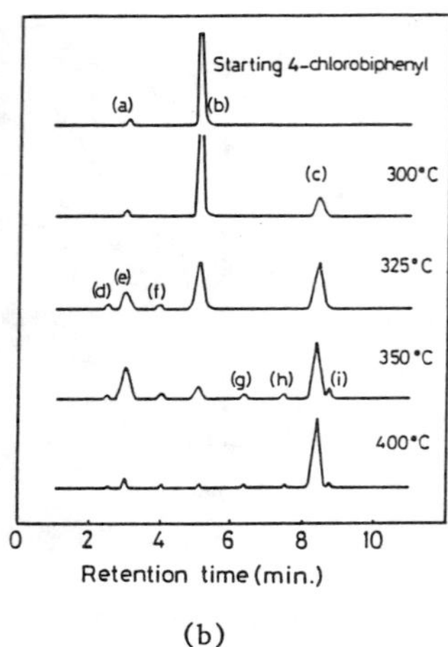

Fig.9   (a) Comparison of dechlorinating ratio between 4-chlorobiphenyl
         and PCBs,
        (b) Gas chromatograms of decomposition products of 4-chlorobi-
            phenyl under various temperature conditions.

as can be seen from this figure, the molecules containing fewer chlorine
number are more stable under the hydrothermal conditions. The gas chromatograms of products from 4-chlorobiphenyl are shown in Fig. 9(b). The peaks
in this figure were assigned by mass spectrometry: (a)2-chlorobiphenyl; (b)
4-chlorobiphenyl; (c)biphenyl-4-ol; (d)4-phenoxybiphenyl; (e)2-phenoxybi-
phenyl; (f)biphenyl-2-ol; (g,h, and i) di(4-biphenyl), di(2-biphenyl), 2-
biphenyl-4-biphenylether. The following mechanism is proposed for the dechlorinating decomposition of 4-chlorobiphenyl under hydrothermal conditions:

The main reaction at low temperature was suggested to be stage I in the
reaction process. Reaction II -X, as well as reaction I, were presumed to
occur at 325-400°C. The decomposition mechanism of 4-chlorobiphenyl is useful model for the dechlorination of PCBs, though the reaction of PCBs may

be more complex than that of 4-chlorobiphenyl.

These main reactions were the release of hydrogen chloride from inner or inter molecules. And so, the formation of chlorinated dibenzofuran is presumed as intermediate. However, these were not detected in the decomposition products of 4-chlorobiphenyl.

In the case of PCBs, the analysis of products was not performed because the starting PCBs were mixtures of many isomers.

POSSIBILITY OF INDUSTRIAL TREATMENT

The continuious-pipeline capillary system seems to be the most suitable for the treatment of PCBs on the industrial scale. PCBs were decomposed within a few minutes above 300°C, and a long pipeline may be needless. The author assembled a "mini" test plant, as shown in Fig. 9.

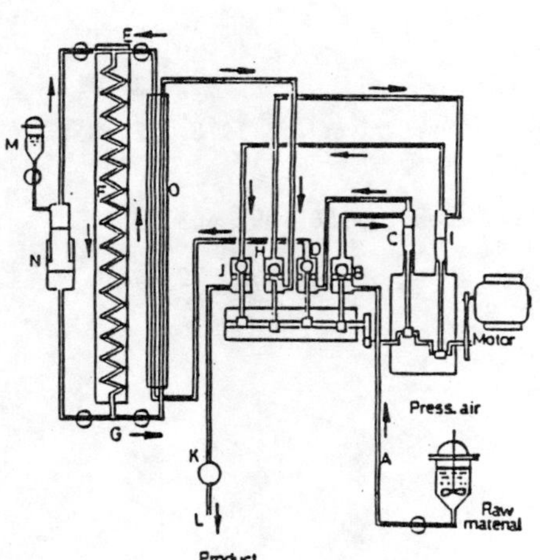

Fig.10 Flow diagram of continuous pipeline system for industrial treatment。
N and M are the injection system for PCB treatment only, and the raw material tank is filled with an alkaline methanolic solution only. In the treatment of industrial waste solutions containing PCBs, N and M are not necessary and raw material tank is filled with waste solution; (A)inlet pipe, (B)ball valve; (C) injection pump; (D)ball valve; (E) reaction column inlet; (F)reaction column; (H)ball valve; (I)back pressure pump; (J)ball valve; (K) pressure control valve; (L)outlet; (O)heat exchanger.

This pipeline system has four characteristics; (a)low power cost by recovery of discharge pressure, (b)safety system for use of small diameter pipe (20 mm inner diameter) as reactor, (c)treatability of slurry or muddy materials as sludge, (d)application to a large scale treatment. The Zimmermann process, using similar equipment, has already been applied to waste disposal process. However, PCBs can not be decomposed in this system, because PCB is not oxidized under wet combustion conditions. And so, the treatment of PCBs may be more easily for the system of oxygen free.

About 400mL/min of PCBs solution(containing 20mL of PCBs) was able to decomposed successively by the test plant.

Biological treatment of decomposed solution under hydrothermal conditions was studied. A PCBs solution dechlorinated by the proposed method was diluted with water to 2350 ppm of the biochemical oxygen demand(BOD) value. The solution was treated with a mixture of Enzobac species(produced by American Enzyme Co.). The BOD value was decreased by about 98% after 9 days. These results support the view that the highly toxic materials such as chlorinated dibenzofuran do not exist. Furthermore, the waste solutions containing dechlorinated PCBs can be easily treated with activated sludge without hazard.

512

CONCLUSIONS

1.Chlorinated organic substances such as PCBs, BHCs, trichloroethane, etc.
were easily decomposed under alkaline hydrothermal conditions.
2.Decomposed conditions were mild(-350°C, -200kg/cm^2) compared with other
method.
3.Decomposition processes are mainly hydrolysis and products are phenolic
and alcolic compounds, and these derivatives such as ether.
4.Decomposition products were easily treated with activated sludge without
hazard.
5.Theclosed and continuous pipeline system for industrial treatment was
developed,and safety, low energy cost and application of large scale slurry
material were confirmed.

REFERENCES

1. Nasu, H., *Sangyo Kogai*, 1973, 11(3), 221

2.Imamura, A.,*Kagaku*, 1973, 28(7), 581

3. Komamiya, K. and Morisaki, S., *Environ. Sci. Technol.*, 1978, 12(10),
   1205

4. Lüttringhaus, A.t. and Ambros, D., *Chem. Ber.*, 1956, 89, 463

5. Bottini, A.T. and Roberts, J.D., *J. Am. Chem. Soc.*, 1957, 79, 4458

CATALYTIC OXIDATION OF CHLORINATED HAZARDOUS WASTES

by

Howard L. Greene

Prasad Subbanna

The University of Akron
Akron, Ohio 44325

1. ABSTRACT

Products from the combustion of Aroclor 1254 (a mixture of
polychlorinated biphenyls) were passed through a 22.86 cm long
section of 0.64 cm I.D. monolithic α-alumina support, which had
been previously treated with one or more catalytic agents.
Analysis of the results showed that the overall PCB destruction
efficiency (amount destroyed per gram vaporized) dropped in the
order: CuO, Co_3O_4, $CuCr_2O_4$, Pt-Pd and Cr_2O_3. The PCB destruction
rates (amount destroyed per gram of catalytic agent) decreased in
the order: Pt-Pd, Co_3O_4, Cr_2O_3, $CuCr_2O_4$, CuO; Pt-Pd destroyed the
most PCBs. Activity and selectivity for deep oxidation were in
accordance with Sabatier's volcano plot, decreasing as the heat of
chemisorption for O_2 increased. The metallic oxide catalysts
caused the concentrations of the more toxic, higher chlorine-
content PCBs to decline by a greater amount than the noble metal
catalysts. Results obtained by mathematically modeling reactor
conversions were found to be in close agreement with the
experimental data.

2. INTRODUCTION

2.1 Polychlorinated_Biphenyls

Polychlorinated biphenyls (PCBs) are a class of chlorinated
hydrocarbons, derivatives of the compound biphenyl, with one to ten
of the hydrogen atoms replaced by chlorine atoms (1,2). They were
introduced commercially in 1929, and soon found a multitude of
applications, principally as dielectric fluids in transformers and

capacitors (14); over 340,000 tons are estimated to be still in use (10). Manufactured by Monsanto under the tradename "Aroclor," these non-volatile, and water-insoluble fluids were later found to be potential chronic toxicants. Studies have linked PCB exposure to liver, pancreatic and melanoma skin cancer (13,18).

By the time manufacture of these chemically- and thermally-stable compounds was halted in 1977, over one million tons had been produced; most still exists today, either in storage, transformer service, or as a contaminant in the environment (15,21). Because thermal destruction requires temperatures of at least $1200^{o}C$ and residence times of over one second, municipal incinerators merely vaporize and release PCBs into the atmosphere (16), or result in the formation of even more toxic polychlorinated dibenzofurans (PCDFs) and polychlorinated dibenzo-p-dioxins (PCDDs) (6,7,22). Hence, there is a compelling need for lower temperature processes, capable of effectively treating incinerator flue gas or other effluents which may be contaminated with PCB vapors.

Previous work by Lombardi, et al. (20) has shown some success in destroying PCB vapors by two-stage contact with $Cr_2O_3-Al_2O_3$ particles in fluidized-bed reactors at temperatures between 595 and $695^{o}C$. However, the relatively complex system suffers from a high pressure drop, and requires pumps, cyclones and multiple stages for good conversion. In addition, no fundamental kinetic data (which could be applied to the design of simpler reactor geometries) are reported.

Chemical treatment methods (Franklin Institute, SunOhio, Goodyear), which use a reagent containing sodium or calcium to strip chlorine from the PCBs, are portable but involve dangerously exothermic reactions (9). Other methods (activated carbon adsorption, microwave plasma, chlorinolysis, dehydro-chlorination, wet-air oxidation, biological destruction) of PCB disposal suffer from various defects; none has been shown to be economically viable for large-scale use (1,2).

2.2 Scope

The scope of this study was primarily to obtain relative activity for each catalyst, utilizing a low pressure drop monolith support treated with noble metals or transition metal oxides, in PCB destruction at known temperatures and reactor residence times. Mathematical modeling, to quantify the results, was of secondary importance.

3. EXPERIMENTAL

3.1 Apparatus

A schematic of the reactor/furnace system is shown in Figure 1. Approximately 1 gram samples of Aroclor 1254, a mixture of chlorinated biphenyl isomers with an average of five chlorine atoms per molecule, 54% chlorine by weight and a molecular weight of 325, were obtained from the Biology department at the University of Akron. These were placed in the sample vaporizer (a shallow cup attached to a long quartz tube) and vaporized at controlled temperatures in the lower tube furnace; a stream of dry, pre-heated primary air from a Linde compressed air cylinder carried the PCB vapors upwards, through the reactor tube in the separately heated/controlled catalytic reaction chamber (upper tube furnace). Condensible organics and water were collected in a train of dry ice/acetone traps attached to the reactor, and residual PCBs

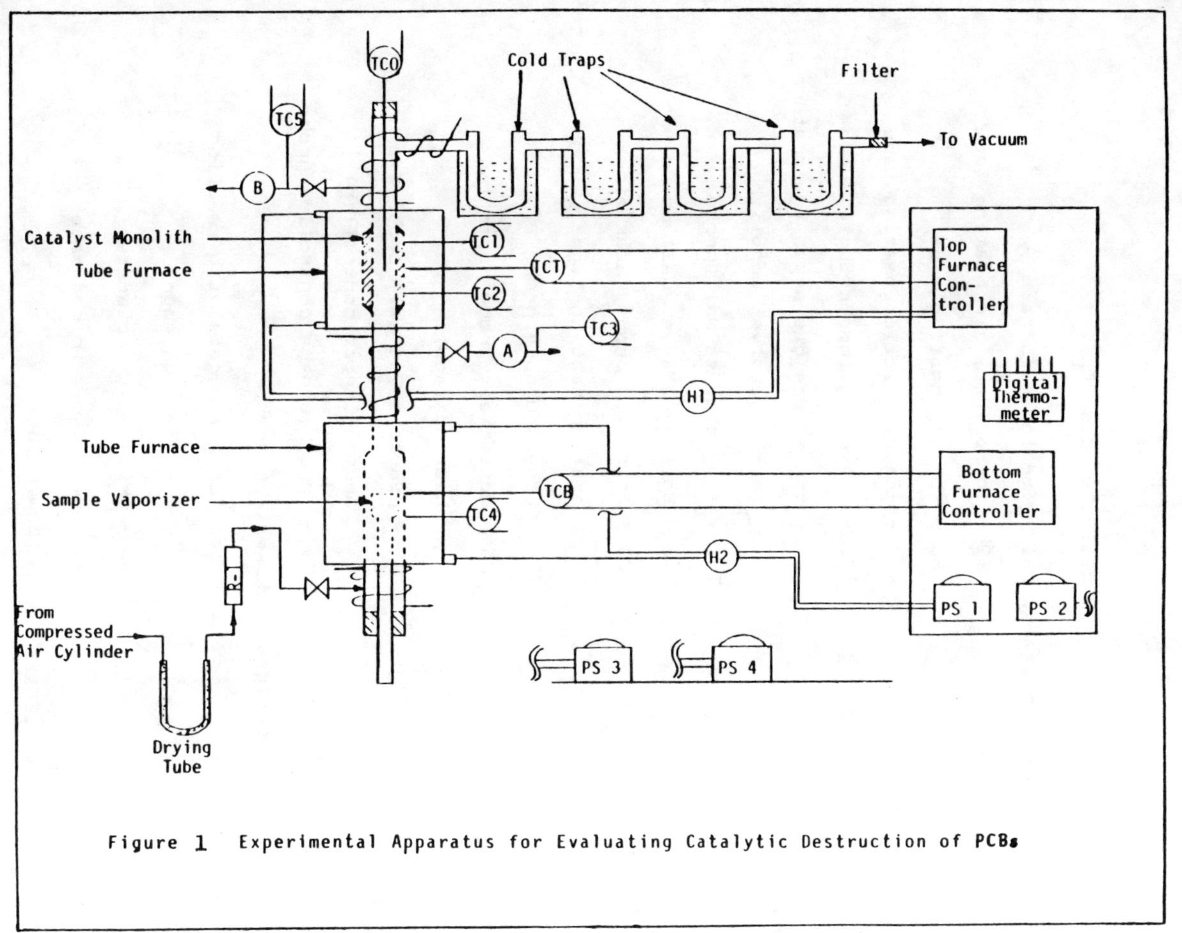

Figure 1   Experimental Apparatus for Evaluating Catalytic Destruction of PCBs

<u>Notation For Figure 1</u>

<u>Power Rheostats:</u> PS1 — for lower furnace

PS2 — for primary air heating tape

PS3 — for top heating tape

PS4 — for mid-section heating tape

<u>Thermocouples:</u> TC0 — traveling thermocouple, used

to measure gas phase temperature

in the catalyst monolith

TC1 — Top catalyst wall probe

TC2 — Bottom catalyst wall probe

TC3 — Inlet air probe (near mid-

section heating tape)

TC4 — Lower furnace probe

TC5 — Outlet air probe (near top

heating tape)

TCT — Mid-catalyst wall probe (for

upper furnace controller)

TCB — Lower furnace controller probe

<u>Tube furnaces:</u> Lower furnace

Upper furnace — catalytic reaction

chamber

<u>Rotameters:</u> R-1 — primary air rotameter

R-2 — secondary air rotameter

<u>Sample Ports:</u> A — Lower port

B — Upper port

estimated using a Florisil 'Sep-Pak' cartridge and electron-capture gas chromatography (26). Vapors emerging from the cellulose filter at the exit of the traps were vented to the hood where the experimental system was located. Gas samples were taken periodically, above and below the catalyst tube, from sampling ports A and B, using syringes with 15.24 cm long needles in order to reach the center of the gas stream.

The lower furnace, manufactured by Lindberg, was controlled by an ERC Co. West Model 10 on/off controller. The heating rate was manipulated by means of a Superior Electric Co. powerstat. The catalytic reaction chamber, a Hevi-Duty Electric Co. Tube 70 furnace, was controlled by a West Gardsman II proportional controller. The tube furnaces were capable of maximum temperatures of $1010\,^{\circ}C$, but the limit on the upper furnace controller was $600\,^{\circ}C$. Temperatures during a run were measured with chromel-alumel thermocouples connected to an Analog Devices Type K scanning digital thermometer. Thermolyne Co. heating tapes connected to Superior Electric Co. powerstats and insulated with fiberglass, prevented condensation on exposed surfaces of the quartz tube.

3.2 Catalyst Preparation

The catalyst support used in all experiments was a ceramic α-alumina monolith, 15.24 cm long by 1.59 cm O.D. and 0.64 cm I.D., except the pyrex blank (which was a 15.24 cm by 0.64 cm I.D. tube). Monoliths were provided by the Norton Co., with the following specifications:

Surface Area, m^2/g 0.483

Apparent Porosity, % (by weight) 31.94

Water Absorption, % (by weight) 14.61

```
Apparent Specific Gravity  ........................  3.21
                       3
Particle Density, g/cm     .........................  2.19
                       3                                      *
Mercury Pore Volume, cm /g  ........................  0.229
                                                              *
Median Pore Diameter, μ  ...........................  1.45
                                          *
                                               Unglazed Tube
```

The outer surface of the tubes was glazed, to prevent the
diffusion (and consequent loss) of reactants and/or products, from
the gas stream to the surrounding furnace. The hollow ceramic
tubes are also characterized by excellent thermal shock resistance,
high strength, light weight, chemical inertness and reasonably high
melting temperatures (23).

The method used to impregnate the various catalysts on the
substrates was to dissolve a known weight of metal ions (amount
decided by the loading desired), usually in the form of the nitrate
salt, in water and/or dilute HCl. This solution was used to
physically saturate the support, care being taken to ensure
uniformity in distribution. Water and other volatile matter were
driven off by heating the treated monoliths at $110\,^{\circ}C$ in a drying
oven for 24 hours. Asbestos baffles were placed at various
intervals along the outer length of the monolith, to minimize the
effect of free convection currents. Carbon steel adaptors were
placed on the ends of the monoliths, which were then calcined in
the upper tube furnace at $600\,^{\circ}C$ with air flow, for a period of 24
hours. The resultant catalyst tubes were subsequently used
directly in the oxidation studies.

In the case of the $CuCr_2O_4$ catalyst, an unglazed tube was
used, and a solution of the nitrates of copper and chromium (with

excess copper) was impregnated on the support. The monolith was oven-dried and the nitrates decomposed at $200\,^{o}C$; it was then fired to $600\,^{o}C$ for 6 hours, when CuO and $CuCr_2O_4$ were formed. The monolith was leached in 3M HCl, thoroughly washed and subsequently fired to $800\,^{o}C$, to leave $CuCr_2O_4$ as the only remaining species.

3.3 Procedures

The basic run procedure was to meter air at a flowrate of 500 ml/minute (STP) through the catalyst tube/monolith section. Product traps (250 ml acetone in 600 ml pyrex beaker with dry ice to fill, and pre-weighed pyrex U-tube inside) were attached to the top of the reactor; the U-tubes were connected by Tygon tubing and a cellulose filter fitted after the last U-tube. The system was allowed to attain thermal equilibrium.

At time zero, the sample holder, containing a pre-weighed amount of Aroclor 1254, was inserted into the lower furnace, and a timer started up. The PCB vaporizer was maintained at $300\,^{o}C$ for 12 minutes, elevated to $400\,^{o}C$ at the rate of $10\,^{o}C$/minute, and held at this temperature for 5 minutes. Temperature data were recorded for all thermocouples, using the digital scanner. Vapor samples were taken at designated times with syringes inserted through sampling ports A and B.

At the end of the run (27 minutes), the sample holder was removed, cooled and re-weighed. The U-tubes in the beakers were disconnected, removed, dried and re-weighed; power to the system was disconnected, and the air flow stopped.

The gas samples removed periodically from ports A and B were analyzed for H_2, O_2, N_2, CO, CH_4 and CO_2 (water-free basis), using a Varian 90P-3 gas chromatograph (GC) equipped with a thermal conductivity detector maintained at 225 ^{o}C. The column was 2.44 m long by 0.32 cm I.D. stainless steel and packed with 100/120 mesh Carbosieve-S. The samples were injected with the column temperature at 27 ^{o}C; this was raised to 190 ^{o}C at the rate of 20 ^{o}C/minute, once the O$_2$ peak had been observed on the associated Leeds and Northrup Speedomax W strip-chart recorder. A CSI-Model 38 digital integrator was used to find the peak areas.

To estimate the residual PCBs in the condensate in the U-tubes, 20 ml of pesticide grade hexane was injected into each tube. Solutions from all U-tubes were thoroughly mixed in a beaker, where PCBs deposited on the filter were also dissolved. One ml of the mixed solution was injected into a Florisil 'Sep-Pak' cartridge (Waters Associates) which had been pre-eluted with 10 ml of hexane. The eluate was collected in a 100 ml volumetric flask, and PCBs deposited in the cartridge were flushed into the flask by repeated injections of hexane; interfering organics and unnecessary compounds remained behind (19,28). The volume in the flask was brought up to the mark with hexane. Five-μl samples were injected into a Bendix-2200 GC equipped with a Tritium foil electron-capture detector maintained at 200 ^{o}C. The glass column (1.83 m x 0.64 cm I.D., packed with 3% SE-30 on 100/120 mesh Gas-Chrom Q), was operated in the isothermal mode at 192 ^{o}C. Peak areas were determined using a Columbia CSI-Model 38 digital integrator, and the concentration of PCBs in the solution injected was computed from them (26). The product of the mean PCB concentration and the solution volume (in appropriate units) gave the weight of PCB in the condensate. A typical chromatogram of Aroclor 1254 is shown in

Figure 2; the weights of each PCB peak before and after catalysis was calculated, as were the peak and overall destruction efficiencies (PCB destroyed/PCB charged). The average rate of PCB destruction (amount destroyed/gram of catalytic agent) was found.

4. RESULTS_AND_DISCUSSION

4.1 Homogeneous_vs._Heterogeneous_Reaction

Table I lists the results for the pyrex tube and the five catalytic systems studied (runs for which the material balances were above 95%). It may be seen that the homogeneous reaction was negligible (< 0.2% PCB conversion) at $500\,^{\circ}C$ and 4-9% at $575-600\,^{\circ}C$, which were at least one order of magnitude less than the reaction with the catalysts. Strictly speaking, the runs with the active catalysts should be interpreted with respect to the pyrex tube, which may be considered to be the blank for the system. It should be noted that the amount of Aroclor 1254 available for reaction was maintained approximately constant for each run. The reactor tube residence times were 0.33 s at $500\,^{\circ}C$ and 0.29 s at $600\,^{\circ}C$. Exotherms based on temperature rise during Aroclor 1254 oxidation, given in Table II ($Pt-Pd > CuCr_2O_4$, $Co_3O_4 > CuO > Cr_2O_3 >$ pyrex), did generally confirm catalytic activity, but gave no clues as to the oxidized species or their relative importance. Moreover, the reactor was not adiabatic and the accuracy of the digital scanner was another limiting factor.

4.2 Gas_Phase_Analysis

The concentrations (by volume) of O_2, N_2, CO and CO_2 (no CH_4 or H_2 was ever detected), before and after reaction in the monolith, were monitored by means of the gas samples. It should be noted that excess O_2 was available, even during the period of

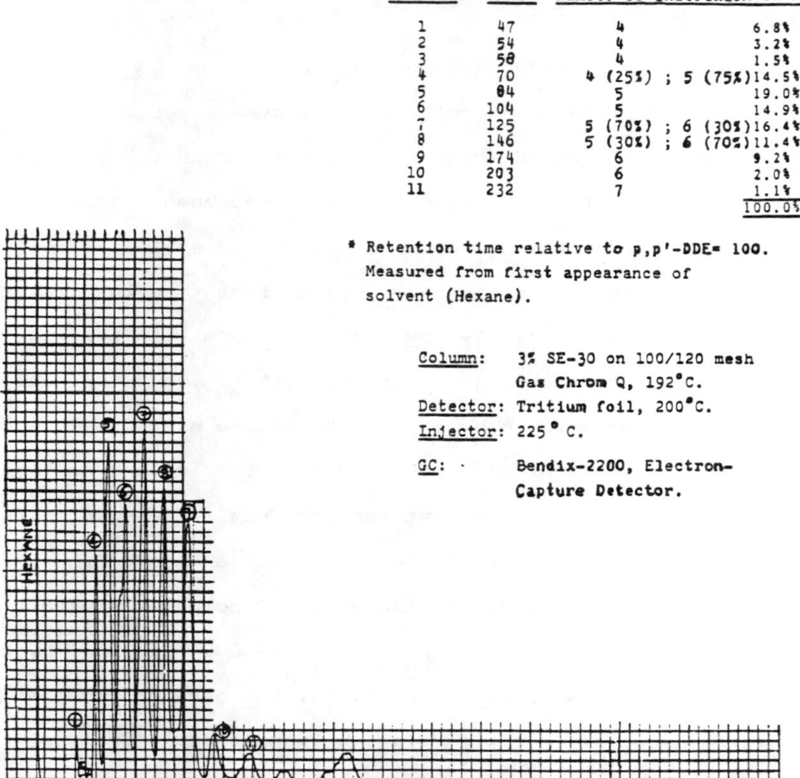

Peak No.	RRT*	Number of Chlorines	Mean Wt.
1	47	4	6.8%
2	54	4	3.2%
3	58	4	1.5%
4	70	4 (25%) ; 5 (75%)	14.5%
5	84	5	19.0%
6	104	5	14.9%
7	125	5 (70%) ; 6 (30%)	16.4%
8	146	5 (30%) ; 6 (70%)	11.4%
9	174	6	9.2%
10	203	6	2.0%
11	232	7	1.1%
			100.0%

* Retention time relative to p,p'-DDE= 100.
Measured from first appearance of
solvent (Hexane).

Column: 3% SE-30 on 100/120 mesh
          Gas Chrom Q, 192°C.
Detector: Tritium foil, 200°C.
Injector: 225° C.

GC: Bendix-2200, Electron-
          Capture Detector.

Figure 2   Typical chromatogram of Aroclor 1254 (Bendix-2200 GC)

Table I Summary of Results

Catalyst, Temperature, Run No.			Initial A-1254 in holder, g	A-1254 vaporized, g	A-1254 vaporized / Initial A-1254	Condensate, g	Condensate / A-1254 vaporized	Mean PCB concn., mg/l	Relative Std.Devn., %	PCB in the condensate, g	Overall PCB Destrn.Eff., %
Pyrex,	500°C,	1	0.9110	0.8540	0.937	0.853	0.999	8.52	5.9	0.852	0.2
Pyrex,	500°C,	2	0.9467	0.9109	0.962	0.910	0.999	9.10	3.0	0.910	0.1
Pyrex,	575'C,	1	0.8532	0.8521	0.999	0.832	0.976	8.16	4.2	0.816	4.2
Pyrex,	600°C,	1	0.9522	0.9144	0.960	0.889	0.972	8.63	6.1	0.863	5.6
Pyrex,	600°C,	2	0.7122	0.6844	0.961	0.654	0.956	6.24	5.3	0.624	8.8
.05% Pt-Pd,	500°C,	2	0.9597	0.9336	0.973	0.755	0.809	2.02	9.0	0.202	78.4
.05% Pt-Pd,	575°C,	5	1.0570	1.0003	0.946	0.544	0.544	2.00	12.5	0.200	80.0
.05% Pt-Pd,	600°C,	1	0.9840	0.9600	0.976	0.460	0.479	1.21	12.3	0.121	87.4
4% Cr$_2$O$_3$,	500°C,	1	1.0075	0.9929	0.985	0.827	0.833	5.88	12.5	0.588	40.9
4% Cr$_2$O$_3$,	500°C,	4	0.8870	0.8804	0.993	0.674	0.765	5.36	4.4	0.536	39.1
4% Cr$_2$O$_3$,	575°C,	1	0.8715	0.8581	0.985	0.466	0.543	2.48	9.3	0.248	71.1
4% Cr$_2$O$_3$,	600°C,	1	0.9380	0.9180	0.979	0.433	0.472	3.37	5.3	0.337	67.3
4% Cr$_2$O$_3$,	600°C,	2	0.9007	0.8950	0.994	0.410	0.458	2.58	7.5	0.258	71.2
6% Co$_3$O$_4$,	500°C,	3	0.9355	0.8800	0.941	0.558	0.634	1.72	6.9	0.172	80.4
6% Co$_3$O$_4$,	500°C,	4	0.9692	0.9196	0.949	0.644	0.700	1.87	19.5	0.187	79.7
6% Co$_3$O$_4$,	575°C,	2	0.7739	0.7321	0.946	0.501	0.685	1.42	9.0	0.142	80.5
6% Co$_3$O$_4$,	575°C,	3	0.7002	0.6988	0.998	0.381	0.544	1.36	13.7	0.136	80.6
6% Co$_3$O$_4$,	600°C,	1	0.7847	0.7571	0.965	0.427	0.564	0.91	2.6	0.091	88.0
6% Co$_3$O$_4$,	600°C,	2	0.6800	0.6522	0.959	0.311	0.477	1.01	6.5	0.101	84.5
6% Co$_3$O$_4$,	600°C,	4	0.7300	0.7060	0.967	0.363	0.514	1.04	20.0	0.104	85.3
10% CuO,	500°C,	2	0.7963	0.7566	0.950	0.621	0.821	0.42	13.9	0.042	94.4
10% CuO,	500°C,	3	0.7630	0.6693	0.877	0.552	0.824	0.67	9.7	0.067	90.4
10% CuO,	575°C,	1	0.6371	0.5981	0.939	0.418	0.699	0.56	15.2	0.056	90.6
10% CuO,	600°C,	2	0.7117	0.6630	0.931	0.338	0.510	0.13	15.9	0.013	98.1
10% CuO,	600°C,	3	0.8227	0.7528	0.915	0.490	0.650	0.26	22.9	0.026	96.6
10% CuO,	600°C,	4	0.5424	0.4909	0.905	0.377	0.768	0.13	16.2	0.013	97.3
10% CuCr$_2$O$_4$,	500°C,	1	0.5984	0.5603	0.936	0.487	0.869	1.67	8.4	0.167	70.4
10% CuCr$_2$O$_4$,	500°C,	2	0.5641	0.4947	0.877	0.429	0.867	1.68	4.5	0.168	66.1
10% CuCr$_2$O$_4$,	575°C,	1	0.6380	0.5783	0.906	0.455	0.787	0.98	7.7	0.098	83.1
10% CuCr$_2$O$_4$,	600°C,	1	0.8488	0.7828	0.922	0.536	0.685	0.69	17.2	0.069	91.2
10% CuCr$_2$O$_4$,	600°C,	2	0.8248	0.7801	0.946	0.499	0.639	1.40	12.5	0.140	82.1
10% CuCr$_2$O$_4$,	600°C,	3	0.6861	0.5252	0.766	0.329	0.626	0.85	14.3	0.085	83.8

Table II Vapor Phase Analysis of Runs

Catalyst, Temperature, Run No.			Maximum Exotherm, °C	Weight of PCB destroyed, g	PCB Destruction Rate		Weight of CO emitted, g	Weight of CO$_2$ emitted, g	Combustion Efficiency, %	Carbon Conversion, %
					$\frac{g\ PCB}{catalyst\cdot s}$	$\frac{g\ PCB}{cc\ tube\cdot s}$				
Pyrex,	500°C,	1	0	0.002	–	7.31×10^{-8}	0.000	0.019	100.0	1.4
Pyrex,	500°C,	2	0	0.001	–	3.65×10^{-8}	0.000	0.009	100.0	0.6
Pyrex,	575°C,	1	0	0.036	–	1.32×10^{-6}	0.000	0.032	100.0	2.4
Pyrex,	600°C,	1	1	0.051	–	1.86×10^{-6}	0.000	0.119	100.0	8.3
Pyrex,	600°C,	2	1	0.060	–	2.19×10^{-6}	0.000	0.016	100.0	1.5
.05% Pt-Pd,	500°C,	2	5	0.732	1.18×10^{-2}	2.67×10^{-5}	0.102	0.401	79.8	38.3
.05% Pt-Pd,	575°C,	5	5	0.800	1.29×10^{-2}	2.92×10^{-5}	0.197	0.724	78.6	65.9
.05% Pt-Pd,	600°C,	1	6	0.839	1.35×10^{-2}	3.07×10^{-5}	0.212	0.806	79.2	75.6
4% Cr$_2$O$_3$,	500°C,	1	2	0.405	1.06×10^{-4}	1.48×10^{-5}	0.061	0.492	89.0	31.7
4% Cr$_2$O$_3$,	500°C,	4	3	0.344	9.02×10^{-5}	1.26×10^{-5}	0.099	0.526	84.2	49.3
4% Cr$_2$O$_3$,	575°C,	1	3	0.610	1.60×10^{-4}	2.23×10^{-5}	0.186	0.737	79.8	76.4
4% Cr$_2$O$_3$,	600°C,	1	3	0.581	1.52×10^{-4}	2.12×10^{-5}	0.143	0.738	83.8	66.9
4% Cr$_2$O$_3$,	600°C,	2	3	0.637	1.67×10^{-4}	2.33×10^{-5}	0.086	0.338	79.6	33.7
6% Co$_3$O$_4$,	500°C,	3	3	0.708	2.92×10^{-4}	2.59×10^{-5}	0.000	0.691	100.0	50.1
6% Co$_3$O$_4$,	500°C,	4	1	0.733	3.02×10^{-4}	2.68×10^{-5}	0.000	0.591	100.0	40.9
6% Co$_3$O$_4$,	575°C,	2	2	0.590	2.43×10^{-4}	2.16×10^{-5}	0.037	0.767	95.4	71.9
6% Co$_3$O$_4$,	575°C,	3	7	0.563	2.32×10^{-4}	2.06×10^{-5}	0.000	0.761	100.0	69.4
6% Co$_3$O$_4$,	600°C,	1	5	0.666	2.75×10^{-4}	2.43×10^{-5}	0.025	0.884	97.3	77.7
6% Co$_3$O$_4$,	600°C,	2	4	0.551	2.27×10^{-4}	2.01×10^{-5}	0.012	0.750	98.4	75.1
6% Co$_3$O$_4$,	600°C,	4	9	0.602	2.48×10^{-4}	2.20×10^{-5}	0.033	0.946	96.6	90.2
10% CuO,	500°C,	2	5	0.715	1.05×10^{-4}	2.61×10^{-5}	0.000	0.743	100.0	62.6
10% CuO,	500°C,	3	2	0.602	8.82×10^{-5}	2.20×10^{-5}	0.000	0.507	100.0	48.3
10% CuO,	575°C,	1	6	0.542	7.94×10^{-5}	1.98×10^{-5}	0.000	0.676	100.0	72.0
10% CuO,	600°C,	2	2	0.650	9.52×10^{-5}	2.38×10^{-5}	0.000	0.861	100.0	82.7
10% CuO,	600°C,	3	12	0.727	1.06×10^{-4}	2.66×10^{-5}	0.000	0.694	100.0	58.8
10% CuO,	600°C,	4	2	0.478	7.00×10^{-5}	1.75×10^{-5}	0.000	0.653	100.0	84.8
10% CuCr$_2$O$_4$,	500°C,	1	1	0.393	7.13×10^{-5}	1.44×10^{-5}	0.026	0.329	92.7	42.1
10% CuCr$_2$O$_4$,	500°C,	2	3	0.327	5.93×10^{-5}	1.19×10^{-5}	0.079	0.346	81.4	60.6
10% CuCr$_2$O$_4$,	575°C,	1	4	0.480	8.71×10^{-5}	1.75×10^{-5}	0.114	0.437	79.3	67.9
10% CuCr$_2$O$_4$,	600°C,	1	7	0.714	1.30×10^{-4}	2.61×10^{-5}	0.123	0.831	87.0	82.5
10% CuCr$_2$O$_4$,	600°C,	2	5	0.640	1.16×10^{-4}	2.34×10^{-5}	0.206	0.702	77.3	83.8
10% CuCr$_2$O$_4$,	600°C,	3	4	0.440	7.98×10^{-5}	1.61×10^{-5}	0.140	0.513	78.5	89.1

maximum combustion ($\sim$19 minutes). Profiles at 575 and 600 $^\circ$C were similar, and the inlet concentrations remained constant during runs.

Assuming applicability of the ideal gas law, the amounts of CO and CO_2 emitted during each run were calculated (areas under the curves). From these weights, the Combustion Efficiency [weight of CO_2 x 100 % / weight of (CO and CO_2)] and the Carbon Conversion [carbon emitted x 100 % / initial carbon in Aroclor 1254 vaporized] were computed; these are given in Table II. However, a poor catalyst could cause modest CO emissions that could be readily oxidized to CO_2 in a later step, distorting the significance of the catalyst's Combustion Efficiency.

The point data of exit CO and CO_2 concentrations, are useful in comparing catalysts at each temperature. At 500 and 575 $^\circ$C, Cr_2O_3, Pt-Pd and $CuCr_2O_4$ (in descending order) showed CO emissions; CO_2 emissions were headed by Co_3O_4 and Cr_2O_3, closely followed by CuO, Pt-Pd and $CuCr_2O_4$. At 600 $^\circ$C, CO emissions dropped (in order) from Cr_2O_3, Pt-Pd, $CuCr_2O_4$ to Co_3O_4 (CuO did not emit any CO); the order for CO_2 emissions was: $CuCr_2O_4$, then Co_3O_4 and Cr_2O_3, followed by CuO and Pt-Pd. It may be concluded that the efficiency in oxidizing CO to CO_2 dropped in the order: CuO > Co_3O_4 > $CuCr_2O_4$ > Cr_2O_3. A possible explanation is similar to that provided by Il'Chenko and Golodets for ammonia oxidation (17): catalyst selectivity for deep oxidation products (CO_2) is favored by lower heats of chemisorption for O_2. Since p-type catalysts have the lowest heats of chemisorption for O_2, they should be the best for deep oxidation; in this study, the metal oxide catalysts selected (all of the p-type) had heats of chemisorption which dropped in the order: Cr_2O_3 > $CuCr_2O_4$ > Co_3O_4 > CuO, which is

exactly opposite to the order of oxidation efficiency and in line
with the above hypothesis.

4.3 Condensate_Composition

Conforming to the scope of this research work, the
condensate was analyzed to determine the total amount of residual
PCBs (Table I). However, separate runs were carried out, with the
CuO catalyst, to determine the amounts of water (using the Karl-
Fischer apparatus) and acidic compounds such as HCl (using NaOH in
the U-tubes, and titrations) formed during Aroclor 1254 oxidation.
Results showed that water formed approximately 32% of the
condensate at 500 $^\circ$C and 45% at 600 $^\circ$C; acidic compounds made up 15%
of the condensate at 500 $^\circ$C and 45% at 600 $^\circ$C. Since CuO displayed
the highest overall PCB destruction efficiency, these are probably
the maximum amounts in the present system. Poor elemental material
balances on chlorine and carbon suggested the presence of a number
of chlorinated compounds in the condensate, as well as organics
(aromatics/aliphatics) and/or acids.

4.4 Liquid_Phase_Analysis

Table III lists the mean PCB weight for each of the 11
calibrated peaks of Aroclor 1254, before and after catalytic
oxidation (selected runs were used for each catalyst). As shown by
the typical Aroclor 1254 chromatogram (Figure 2), it is generally
true that the higher molecular weight PCBs are present in the peaks
with higher residence times; these have been documented to be more
toxic and more stable than those with lower molecular weights (13).
A study of Table III shows that the overall effect of the
homogeneous reaction was to increase the proportion of the higher

TABLE III VAPOR PHASE ANALYSIS OF SELECTED RUNS (CONTINUED).

Catalyst, Temperature, Run No.	Peak No. (RRT)	1 (47)	2 (54)	3 (58)	4 (70)	5 (84)	6 (104)	7 (125)	8 (146)	9 (174)	10 (203)	11 (232)
10% CuO, 600°C, 2	Initial Weight, g.	0.0484	0.0226	0.0109	0.1030	0.1350	0.1061	0.1171	0.0812	0.0656	0.0140	0.0078
	Final Weight, g.	0.0004	0.0000	0.0002	0.0009	0.0044	0.0029	0.0038	0.0002	0.0001	0.0001	0.0000
	Rel. Std. Dev., %	15.2	0.0	8.4	14.1	7.7	5.9	13.8	73.8	77.9	22.6	187.3
	Change, %	-99.17	-100.0	-98.17	-99.13	-96.74	-97.27	-96.75	-99.75	-99.85	-99.29	-99.83
10% CuCr$_2$O$_4$, 500°C, 1	Initial Weight, g.	0.0407	0.0190	0.0092	0.0866	0.1135	0.0892	0.0984	0.0682	0.0551	0.0118	0.0066
	Final Weight, g.	0.0110	0.0000	0.0033	0.0295	0.0374	0.0300	0.0335	0.0134	0.0080	0.0009	0.0000
	Rel. Std. Dev., %	2.6	0.0	4.2	4.4	3.7	3.5	3.9	20.9	27.4	82.5	217.2
	Change, %	-72.97	-100.0	-64.13	-65.94	-67.05	-66.37	-65.96	-80.35	-85.48	-92.37	-99.79
10% CuCr$_2$O$_4$, 575°C, 1	Initial Weight, g.	0.0434	0.0203	0.0098	0.0923	0.1210	0.0951	0.1049	0.0728	0.0588	0.0126	0.0070
	Final Weight, g.	0.0054	0.0000	0.0020	0.0159	0.0238	0.0194	0.0222	0.0071	0.0001	0.0001	0.0000
	Rel. Std. Dev., %	6.6	0.0	4.5	4.2	4.2	5.3	7.1	39.0	87.5	158.3	0.0
	Change, %	-87.56	-100.0	-79.59	-82.77	-80.33	-79.60	-78.84	-90.25	-96.43	-99.21	-100.0
10% CuCr$_2$O$_4$, 600°C, 1	Initial Weight, g.	0.0577	0.0270	0.0130	0.1229	0.1610	0.1266	0.1396	0.0968	0.0782	0.0168	0.0093
	Final Weight, g.	0.0015	0.0000	0.0014	0.0088	0.0202	0.0157	0.0175	0.0027	0.0008	0.0003	0.0001
	Rel. Std. Dev., %	27.8	0.0	11.9	14.0	3.9	5.4	7.7	22.5	48.4	23.1	68.2
	Change, %	-97.40	-100.0	-89.23	-92.84	-87.45	-87.60	-87.46	-97.21	-98.98	-98.21	-98.92

TABLE III LIQUID PHASE ANALYSIS OF SELECTED RUNS.

Catalyst, Temperature, Run No.	Peak No. (RRT)	1 (47)	2 (54)	3 (58)	4 (70)	5 (84)	6 (104)	7 (125)	8 (146)	9 (174)	10 (203)	11 (232)
Pyrex, 500°C, 2	Initial Weight, g.	0.0644	0.0300	0.0145	0.1370	0.1796	0.1412	0.1557	0.1080	0.0872	0.0187	0.0104
	Final Weight, g.	0.0646	0.0000	0.0144	0.1305	0.1697	0.1320	0.1449	0.1143	0.1006	0.0243	0.0147
	Rel. Std. Dev., %	1.2	0.0	0.9	0.7	0.8	0.6	1.6	1.6	1.9	3.7	15.2
	Change, %	0.32	-100.0	-0.69	-4.74	-5.51	-6.51	-6.94	5.83	15.37	29.95	41.35
Pyrex, 575°C, 1	Initial Weight, g.	0.0580	0.0271	0.0131	0.1235	0.1617	0.1272	0.1403	0.0973	0.0786	0.0168	0.0094
	Final Weight, g.	0.0582	0.0000	0.0128	0.1168	0.1516	0.1171	0.1311	0.1026	0.0912	0.0219	0.0127
	Rel. Std. Dev., %	1.3	0.0	0.6	0.6	0.8	0.6	1.4	1.2	1.6	2.1	15.5
	Change, %	0.34	-100.0	-2.29	-5.43	-6.30	-7.94	-6.58	5.45	16.43	30.36	35.11
Pyrex, 600°C, 1	Initial Weight, g.	0.0647	0.0303	0.0146	0.1378	0.1806	0.1420	0.1566	0.1086	0.0877	0.0188	0.0104
	Final Weight, g.	0.0601	0.0000	0.0133	0.1230	0.1591	0.1241	0.1356	0.1110	0.0975	0.0239	0.0155
	Rel. Std. Dev., %	5.9	0.0	4.5	4.0	3.0	2.5	1.9	5.2	7.0	11.0	29.0
	Change, %	-7.11	-100.0	-8.90	-10.74	-11.90	-12.60	-13.41	2.21	11.17	27.13	49.04
0.05% Pt-Pd, 500°C, 2	Initial Weight, g.	0.0652	0.0305	0.0147	0.1389	0.1820	0.1431	0.1578	0.1094	0.0884	0.0189	0.0105
	Final Weight, g.	0.0103	0.0000	0.0028	0.0237	0.0372	0.0302	0.0329	0.0360	0.0212	0.0064	0.0012
	Rel. Std. Dev., %	2.0	0.0	1.0	0.9	1.1	0.3	0.7	1.3	2.8	2.4	38.0
	Change, %	-84.20	-100.0	-80.95	-82.94	-79.56	-78.90	-79.15	-67.09	-76.02	-66.14	-88.57
0.05% Pt-Pd, 575°C, 5	Initial Weight, g.	0.0719	0.0336	0.0162	0.1530	0.2005	0.1576	0.1738	0.1205	0.0974	0.0209	0.0016
	Final Weight, g.	0.0101	0.0000	0.0027	0.0230	0.0371	0.0291	0.0320	0.0356	0.0211	0.0071	0.0022
	Rel. Std. Dev., %	7.2	0.0	3.8	3.2	1.6	1.0	0.6	2.9	3.9	6.9	9.6
	Change, %	-85.95	-100.0	-83.33	-84.97	-81.50	-81.54	-81.59	-70.46	-78.34	-66.03	-81.03
0.05% Pt-Pd, 600°C, 1	Initial Weight, g.	0.0669	0.0313	0.0151	0.1424	0.1867	0.1467	0.1618	0.1122	0.0906	0.0194	0.0108
	Final Weight, g.	0.0057	0.0000	0.0016	0.0138	0.0224	0.0180	0.0196	0.0231	0.0125	0.0034	0.0007
	Rel. Std. Dev., %	5.0	0.0	1.7	1.2	1.5	0.9	0.8	2.9	1.6	4.4	3.2
	Change, %	-91.48	-100.0	-89.40	-90.31	-88.00	-87.73	-87.89	-79.41	-86.20	-82.47	-93.52
4% Cr$_2$O$_3$, 500°C, 4	Initial Weight, g.	0.0603	0.0282	0.0136	0.1282	0.1684	0.1323	0.1459	0.1012	0.0817	0.0175	0.0097
	Final Weight, g.	0.0466	0.0000	0.0110	0.0848	0.1155	0.1011	0.1247	0.0000	0.0462	0.0099	0.0069
	Rel. Std. Dev., %	3.0	0.0	2.0	2.3	1.3	1.3	5.9	0.0	7.0	11.4	14.8
	Change, %	-22.72	-100.0	-19.12	-33.96	-31.37	-23.58	-14.58	-100.0	-43.45	-43.43	-28.87

TABLE XII LIQUID PHASE ANALYSIS OF SELECTED RUNS (CONTINUED).

Catalyst, Temperature, Run No.	Peak No. (RRT)	1 (47)	2 (54)	3 (58)	4 (70)	5 (84)	6 (104)	7 (125)	8 (146)	9 (174)	10 (203)	11 (232)
4% Cr_2O_3, 575°C, 1	Initial Weight, g.	0.0592	0.0277	0.0134	0.1261	0.1653	0.1300	0.1433	0.0994	0.0803	0.0172	0.0096
	Final Weight, g.	0.0222	0.0000	0.0051	0.0399	0.0578	0.0452	0.0564	0.0000	0.0179	0.0023	0.0013
	Rel. Std. Dev., %	23.1	0.0	5.2	3.5	4.3	3.2	8.2	0.0	32.7	75.9	70.9
	Change, %	-62.50	-100.0	-61.94	-68.36	-65.03	-65.23	-60.64	-100.0	-77.71	-86.63	-86.46
4% Cr_2O_3, 600°C, 2	Initial Weight, g.	0.0612	0.0286	0.0138	0.1304	0.1709	0.1343	0.1481	0.1027	0.0830	0.0178	0.0099
	Final Weight, g.	0.0079	0.0000	0.0019	0.0321	0.0338	0.0416	0.0422	0.0936	0.0000	0.0039	0.0011
	Rel. Std. Dev., %	17.2	0.0	3.4	2.8	2.4	1.7	0.6	2.4	0.0	6.1	51.2
	Change, %	-98.71	-100.0	-86.23	-75.38	-80.22	-69.02	-71.51	-8.86	-100.0	-78.09	-88.89
6% Co_3O_4, 500°C, 4	Initial Weight, g.	0.0659	0.0308	0.0149	0.1403	0.1838	0.1445	0.1594	0.1105	0.0893	0.0191	0.0106
	Final Weight, g.	0.0047	0.0000	0.0063	0.0378	0.0622	0.0457	0.0141	0.0000	0.0133	0.0025	0.0004
	Rel. Std. Dev., %	141.9	0.0	50.2	17.0	18.1	15.4	142.4	0.0	10.3	31.2	88.7
	Change, %	-92.87	-100.0	-57.72	-73.06	-66.16	-68.37	-91.15	-100.0	-85.11	-86.91	-96.23
6% Co_3O_4, 575°C, 2	Initial Weight, g.	0.0526	0.0246	0.0119	0.1120	0.1468	0.1154	0.1273	0.0883	0.0713	0.0153	0.0085
	Final Weight, g.	0.0082	0.0000	0.0028	0.0196	0.0355	0.0272	0.0374	0.0000	0.0085	0.0028	0.0000
	Rel. Std. Dev., %	5.9	0.0	5.2	7.2	4.4	5.1	9.1	0.0	20.6	38.0	163.5
	Change, %	-84.41	-100.0	-76.47	-82.50	-75.82	-76.43	-70.62	-100.0	-88.08	-81.70	-99.92
6% Co_3O_4, 600°C, 1	Initial Weight, g.	0.0533	0.0249	0.0120	0.1136	0.1488	0.1170	0.1291	0.0895	0.0723	0.0155	0.0086
	Final Weight, g.	0.0043	0.0000	0.0015	0.0110	0.0213	0.0169	0.0267	0.0000	0.0060	0.0031	0.0001
	Rel. Std. Dev., %	7.8	0.0	6.8	4.3	5.3	1.7	4.3	0.0	8.3	10.8	165.2
	Change, %	-91.93	-100.0	-87.50	-90.32	-85.69	-85.56	-79.32	-100.0	-91.70	-80.00	-98.84
10% CuO, 500°C, 3	Initial Weight, g.	0.0519	0.0242	0.0117	0.1104	0.1447	0.1138	0.1255	0.0870	0.0703	0.0151	0.0084
	Final Weight, g.	0.0023	0.0000	0.0013	0.0066	0.0207	0.0160	0.0187	0.0007	0.0003	0.0003	0.0001
	Rel. Std. Dev., %	89.7	0.0	22.2	19.5	3.9	8.6	12.1	122.1	180.7	18.6	86.9
	Change, %	-95.57	-100.0	-88.89	-94.02	-85.69	-85.94	-85.10	-99.20	-99.57	-98.01	-98.81
10% CuO, 575°C, 1	Initial Weight, g.	0.0433	0.0202	0.0098	0.0922	0.1209	0.0950	0.1048	0.0727	0.0587	0.0126	0.0070
	Final Weight, g.	0.0013	0.0000	0.0010	0.0062	0.0171	0.0133	0.0142	0.0016	0.0040	0.0008	0.0001
	Rel. Std. Dev., %	13.2	0.0	8.9	14.4	5.8	5.9	9.1	42.6	71.6	36.0	77.1
	Change, %	-97.00	-100.0	-89.80	-93.28	-85.86	-86.00	-86.45	-97.80	-93.19	-93.65	-98.57

532

molecular weight PCBs, which is totally undesirable. In sharp contrast, the catalysts caused reductions in all peaks; the metallic oxide semi-conductors generally resulted in decreases of greater amounts (than the noble metals) for the high molecular weight PCB peaks. For the more active catalysts (CuO and Co_3O_4), the reductions were independent of temperature, suggesting diffusion control of reaction rates.

Based solely on the overall PCB destruction efficiency, the activity series for this work was: 10% CuO > 6% Co_3O_4 > 10% $CuCr_2O_4$ > 0.05% Pt-0.05% Pd > 4% Cr_2O_3; this is in broad agreement with results reported for similar systems in the literature (5,8). The probable explanation for the order is that activity is favored (along with selectivity for deep oxidation) for oxidation reactions catalyzed by p-type semi-conductors with low heats of O_2 chemisorption. Figure 3 shows that catalyst activity (as overall PCB destruction efficiency) at each temperature dropped as the heat of chemisorption increased (CuO < Co_3O_4 < $CuCr_2O_4$ < Cr_2O_3). The graph, akin to the volcano plot predicted by Sabatier's principle, suggests that the catalysts chosen were appropriate. In addition, Cr_2O_3, CuO and Co_3O_4 probably have maxima in activity because of their d-shell configurations (3,5,11,12).

4.5 <u>Catalyst Evaluation</u>

Although the CuO catalyst had the highest overall PCB destruction efficiency, there are many other criteria involved in selecting a catalyst for PCB oxidation. The catalyst that destroyed the greatest amount (absolute) of PCB was Pt-Pd; its destruction efficiency may not have been the highest because of its low loading. In addition, Pt-Pd had by far the highest PCB

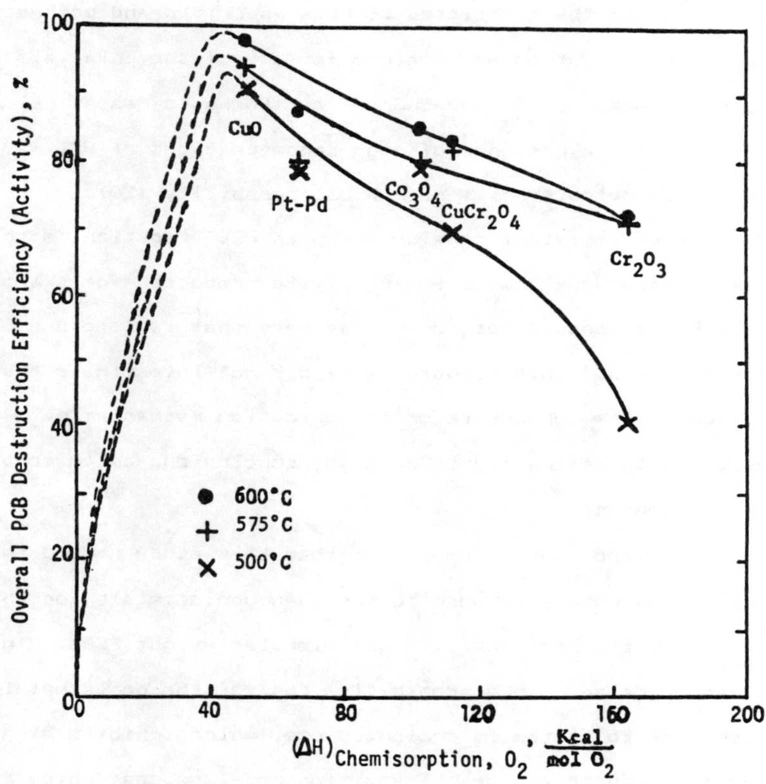

Figure 3   Plot of Activity vs. Heat of Chemisorption for O_2

destruction rate (Table II). However, absolute PCB destruction
(and PCB destruction rate) would be important in commercial
operation.

In large-scale use, significant amounts of HCl would be
formed because of the quantities of PCBs vaporized and combusted.
Although no deactivation was noticed for any of the catalysts in
the present system, Cr_2O_3 and $CuCr_2O_4$ are known to resist HCl
attack, even in the presence of high concentrations of HCl (20,27).
Also, Cr_2O_3 is noted for its mechanical durability (20).

Another important consideration in PCB oxidation is that
potentially hazardous combustion/catalysis products (for example,
PCDDs and PCDFs) should not form in amounts that could cause new
toxicity concerns. This research work did not investigate the
composition of the condensate or the oxidation mechanism(s)
involved, and therefore cannot make any conclusions as to complex
oxidation products.

It is important to emphasize that this study showed that
the metal oxide catalysts were better than noble metals for the
destruction of the more toxic, high molecular weight PCBs. This is
significant because of the prohibitive cost of the noble metals and
their tendency to sinter on prolonged use, which inhibits high
loadings for industrial use. Transition metal oxides (which are
capable of withstanding high temperatures) are thus also attractive
for the cost reductions possible with their commercialization.

The values of apparent surface reaction rate constant
(ηk_s), mass transfer coefficient (k_g) and the concentration profile
in the tube, obtained from the reactor model (26) are given in
Table IV; the closeness of the values of observed and calculated

Table IV   Results from Reactor Model

Catalyst, Temperature, Run No.	Conversion	Concentration, C^* ($C^* = C/C_i$), Z^* ($Z^* = Z/L$)										$(k_g)_{Z^*=1}$ (cm/s)	Observed C^*_{Out}	Calculated C^*_{Out}	Calculated k_s (cm/s)
		0.1	0.2	0.3	0.4	0.5	0.6	0.7	0.8	0.9	1.0				
Pyrex, 500°C, 2	0.000	1.00	1.00	1.00	1.00	1.00	1.00	1.00	1.00	1.00	1.00	0.00	1.000	1.000	0.00
Pyrex, 575°C, 1	0.042	0.98	0.98	0.97	0.96	0.95	0.94	0.93	0.92	0.92	0.91	3.86	0.958	0.954	2.88×10^{-6}
Pyrex, 600°C, 1	0.056	0.99	0.98	0.97	0.97	0.96	0.95	0.95	0.94	0.94	0.93	3.68	0.944	0.940	4.16×10^{-6}
.05% Pt-Pd, 500°C, 2	0.784	0.61	0.51	0.43	0.37	0.32	0.27	0.23	0.20	0.17	0.15	2.28	0.216	0.218	1.35×10^{-4}
.05% Pt-Pd, 575°C, 5	0.800	0.60	0.49	0.48	0.35	0.30	0.26	0.22	0.18	0.16	0.13	2.61	0.200	0.198	1.59×10^{-4}
.05% Pt-Pd, 600 C, 1	0.874	0.50	0.40	0.32	0.26	0.21	0.17	0.14	0.12	0.09	0.08	2.66	0.126	0.130	2.35×10^{-4}
4% Cr_2O_3, 500°C, 4	0.391	0.87	0.82	0.78	0.74	0.70	0.67	0.64	0.60	0.57	0.55	2.42	0.609	0.606	3.37×10^{-5}
4% Cr_2O_3, 575°C, 1	0.711	0.70	0.60	0.53	0.47	0.41	0.36	0.32	0.28	0.25	0.22	2.67	0.288	0.292	1.08×10^{-4}
4% Cr_2O_3, 600°C, 2	0.712	0.70	0.60	0.53	0.47	0.41	0.36	0.32	0.28	0.25	0.22	2.79	0.288	0.291	1.11×10^{-4}
6% Co_3O_4, 500°C, 4	0.797	0.59	0.49	0.41	0.35	0.30	0.26	0.22	0.19	0.16	0.13	2.27	0.203	0.203	1.45×10^{-4}
6% Co_3O_4, 575°C, 2	0.805	0.60	0.49	0.42	0.35	0.30	0.25	0.22	0.18	0.16	0.13	2.61	0.195	0.196	1.60×10^{-4}
6% Co_3O_4, 600°C, 1	0.880	0.49	0.39	0.31	0.25	0.21	0.17	0.14	0.11	0.09	0.07	2.66	0.120	0.125	2.43×10^{-4}
10% CuO, 500°C, 3	0.900	0.43	0.33	0.26	0.21	0.17	0.13	0.11	0.08	0.07	0.05	2.17	0.100	0.105	2.63×10^{-4}
10% CuO, 575°C, 1	0.906	0.44	0.35	0.28	0.23	0.20	0.15	0.13	0.10	0.09	0.07	2.48	0.094	0.094	2.90×10^{-4}
10% CuO, 600°C, 2	0.981	0.28	0.18	0.11	0.10	0.09	0.08	0.07	0.05	0.03	0.01	1.01	0.020	0.020	1.56×10^{-3}
10% $CuCr_2O_4$, 500°C, 1	0.703	0.69	0.60	0.53	0.47	0.41	0.36	0.32	0.29	0.25	0.22	2.32	0.297	0.299	9.74×10^{-5}
10% $CuCr_2O_4$, 575°C, 1	0.831	0.57	0.46	0.38	0.32	0.27	0.23	0.19	0.16	0.13	0.11	2.59	0.170	0.173	1.80×10^{-4}
10% $CuCr_2O_4$, 600°C, 1	0.912	0.42	0.32	0.25	0.20	0.16	0.12	0.10	0.08	0.06	0.05	2.60	0.088	0.092	3.13×10^{-4}

outlet concentrations ($\pm$5%) lends credibility to the model.
Details of the procedures used in the calculations may be found in
Subbanna's MS thesis (26). The values of the actual rate constant
(k_s), Thiele parameter (ϕ), and catalyst effectiveness factor (η),
computed from the model results, are given in Table V. Internal
diffusional limitations, suspected by the temperature-independent
behavior of PCB destruction for the active catalysts, were
confirmed by the values of η and their variation with conversion
(Figure 4). Although k_s did vary with temperature (Figure 5), the
trend was generally not the same as that predicted for diffusion-
limited reactions in the literature (25). No definite conclusions
can be drawn because of the inaccuracies and the lack of data.
Activation energies calculated from the results are given in Table
V.

The model results, which show that diffusion limits depend
on the catalyst activity, are in agreement with those reported for
similar systems. Prasad and Kennedy reported, for the oxidation of
propane in binary transition metal oxide catalytic combustors, that
the monoliths are mass transfer-limited at high conversions,
depending on the fuel oxidized and on catalyst properties (23,24).
Boldyreva, et al. reported the oxidation of fuels on Pd-promoted
V_2O_5 and Cr_2O_3 to proceed by a homogeneous-heterogeneous mechanism,
and to be diffusion-limited (4).

Table V Calculations from model results

Catalyst, Temperature, Run No.	Conversion	(k_s) Model (cm/s)	ϕ	η	(k_s) Actual (cm/s)	Activation Energy (Kcal/mol)	$(k_g)_{Z^*=1}$	Sherwood Number
Pyrex, 500°C, 2	0.000	0.00	0.0	0.00	-	-	0.00	0.00
Pyrex, 575°C, 1	0.042	2.88×10^{-6}	0.6	0.90	3.22×10^{-6}	-	3.86	6.30
Pyrex, 600°C, 1	0.056	4.16×10^{-6}	-	-	-	-	3.68	5.74
.05% Pt-Pd, 500°C, 2	0.784	1.35×10^{-4}	30.1	0.03	4.08×10^{-3}		2.28	4.27
.05% Pt-Pd, 575°C, 5	0.800	1.59×10^{-4}	31.3	0.03	4.93×10^{-3}	10.3	2.61	4.27
.05% Pt-Pd, 600°C, 1	0.874	2.35×10^{-4}	44.5	0.02	1.05×10^{-2}		2.66	4.16
4% Cr_2O_3, 500°C, 4	0.391	3.37×10^{-5}	7.5	0.13	2.53×10^{-4}	38.3	2.42	4.53
4% Cr_2O_3, 575°C, 1	0.711	1.08×10^{-4}	21.2	0.05	2.29×10^{-3}	1.3	2.66	4.35
4% Cr_2O_3, 600°C, 2	0.712	1.11×10^{-4}	21.0	0.05	2.34×10^{-3}		2.79	4.35
6% Co_3O_4, 500°C, 4	0.797	1.45×10^{-4}	32.3	0.03	4.68×10^{-3}		2.27	4.25
6% Co_3O_4, 575°C, 2	0.805	1.60×10^{-4}	31.6	0.03	5.06×10^{-3}	9.0	2.61	4.26
6% Co_3O_4, 600°C, 1	0.880	2.43×10^{-4}	46.0	0.02	1.12×10^{-2}		2.66	4.15
10% CuO, 500°C, 3	0.900	2.63×10^{-4}	58.6	0.02	1.54×10^{-2}		2.17	4.06
10% CuO, 575°C, 1	0.906	2.90×10^{-4}	57.1	0.02	1.66×10^{-2}	34.2	2.48	4.06
10% CuO, 600°C, 2	0.981	1.56×10^{-3}	295.0	0.003	4.60×10^{-1}		1.01	1.57
10% $CuCr_2O_4$, 500°C, 1	0.703	9.74×10^{-5}	21.7	0.05	2.11×10^{-3}		2.32	4.34
10% $CuCr_2O_4$, 575°C, 1	0.831	1.80×10^{-4}	35.4	0.03	6.37×10^{-3}	26.5	2.59	4.23
10% $CuCr_2O_4$, 600°C, 1	0.912	3.13×10^{-4}	59.4	0.02	1.85×10^{-2}		2.60	4.05

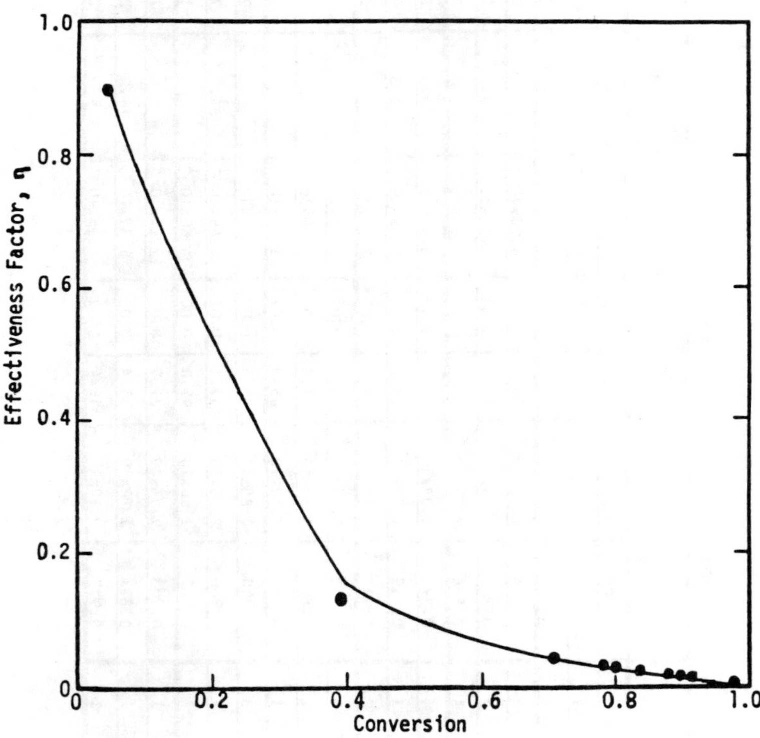

Figure 4   Plot of Effectiveness Factor vs. Conversion

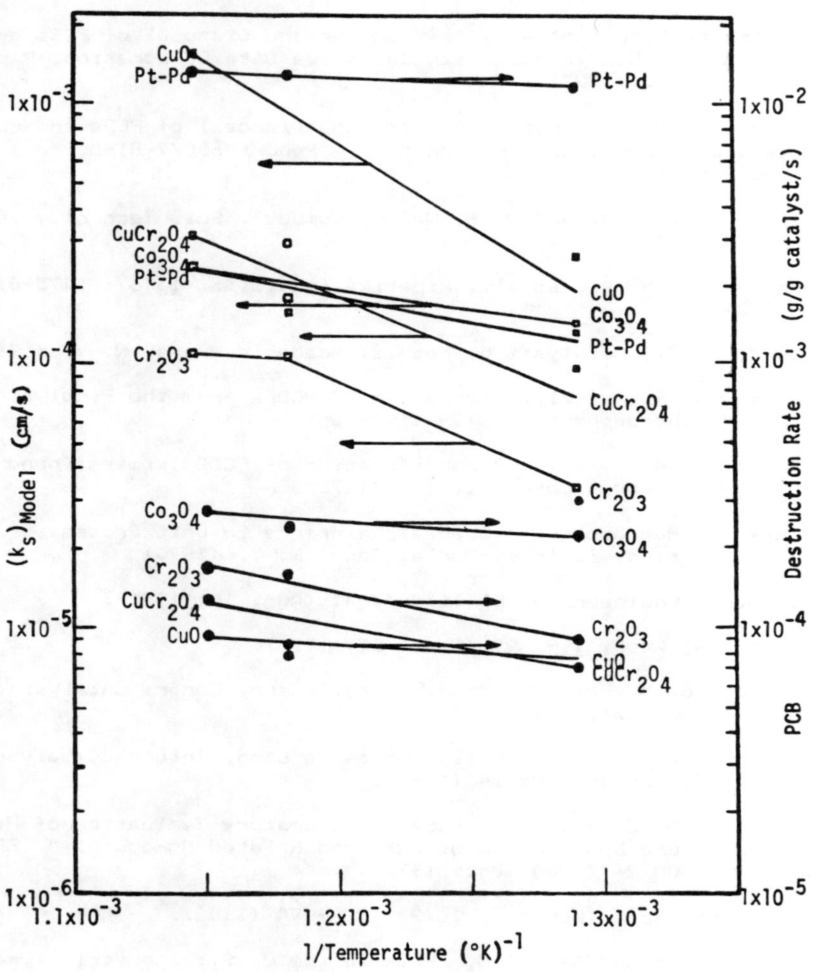

Figure 5   Plots of Reaction Rate Constant and PCB Destruction Rate vs. 1/Temperature

<u>LITERATURE CITED</u>

(1) Ackerman, D. G., et al., Destruction and Disposal of PCBs by Thermal and Non-Thermal Methods, Noyes Data Corporation, Park Ridge, N. J. (1983).

(2) Ackerman, D. G., "Guidelines for the Disposal of PCBs and PCB Items by Thermal Destruction," EPA Report 600/2-81-002 (Feb. 1981).

(3) Blazowski, W. S. and D. E. Walsh, Combust. Sci. Technol., 10: 233-44 (1975).

(4) Boldyreva, N. A., et al., Kinetika i Kataliz, 13(6): 1475-81 (1972).

(5) Bond, G. C., Catalysis by Metals, Academic Press, N. Y. (1962).

(6) Buser, H. R., et al., "Formation of PCDFs from the Pyrolysis of PCBs," Chemosphere, 1: 109 (1978).

(7) Buser, H. R., et al., "Identification of PCDD Isomers found in Fly Ash," Chemosphere, 2: 165 (1978).

(8) Catalyst Handbook with Special Reference to Unit Processes in Manufacturing, Springer-Verlag Inc., N. Y. (1970)

(9) Chemical Engineering, 88(16): 37-41 (Aug. 10, 1981).

(10) Chemical Week, 119: 29 (Sept. 22, 1976).

(11) Dixon, G. M., et al., Proc. Third Intern. Congr. Catalysis, 2, 815 (Amsterdam, 1964)

(12) Dowden, D. A. and D. Wells, Aetes 2e Cong. Intern. Catalyse, 2, 1489, Technip, Paris (1961).

(13) Duvall, D. S. and W. A. Rubey, "Laboratory Evaluation of High-Temperature Destruction of PCBs and Related Compounds," EPA Report 600/2-77-228 (Dec. 1977).

(14) Environ. Sci. Technol., 16(2): 98A-99A (1982).

(15) EPA Report 560/6-76-005, "PCBs in the U. S.: Industrial Use and Environmental Distribution" (1976).

(16) Farrell, J. B. and B. V. Salotto, "Effect of Incineration on Metals, Pesticides and PCBs in Sewage Sludge," National Symposium on the Ultimate Disposal of Wastewaters and Their Residuals, Raleigh, N. C., 186-98 (1973).

(17) Il'Chenko, N. I. and G. I. Golodets, J. Catal., $\underline{39}$: 73-86 (1975).

(18) Kimbrough, R. D. and R. E. Linder, J. Natl. Cancer Inst., $\underline{33}$: 547 (1947).

(19) Lerman, S. I., et al., American Laboratories, $\underline{14}$(2): 176-81 (1982).

(20) Lombardi, E. F., et al., "Incineration of PCBs using a Fluidized-Bed Incinerator," Rockwell International Corporation, Golden, CO, Rocky Flats Plant (Sept. 1981).

(21) Nisbet, I. C. and A. F. Sarofin, Environ. Hlth. Perspect., $\underline{1}$: 21 (1972).

(22) Olie, K., et al., Chemosphere, $\underline{8}$: 455 (1977).

(23) Prasad, R., et al., Combust. Sci. Technol., $\underline{22}$: 271-80 (1980).

(24) Prasad, R., et al., Catal. Rev. - Sci. and Eng., $\underline{26}$(1): 1-58 (1984).

(25) Satterfield, C. N., Mass Transfer in Heterogeneous Catalysis, MIT Press, Cambridge, MA (1970).

(26) Subbanna, P., MS Thesis, University of Akron, in progress.

(27) Vlasenko, V. M., et al., Kataliz i Katalizatory, $\underline{18}$: 33-7 (1980).

(28) Waters Associates, "Sep-Pak Cartridges, A New Dimension in Sample Cleanup" (1981).

DECHLORINATION OF PCB'S, DIOXINES AND DIFURANES IN ORGANIC LIQUIDS

P.F. van den Oosterkamp, L.J.M.J. Blomen,
H.J. ten Doesschate, A.S. Laghate, R. Schaaf
Kinetics Technology International
Zoetermeer, The Netherlands

ABSTRACT

Many organic liquids, like spent lubrication oils, contain various concentrations of chlorinated hydrocarbons, including hazardous substances like PCB's and especially dioxines and difuranes. A chemical process has been developed in which hydrodechlorination under certain conditions with commercially available catalysts has proven to be very successful in breaking up all chlorine containing hydrocarbons. Feedstreams containing large quantities of chlorine and including various amounts of PCB's, dioxines and difuranes, were found to contain undetectably low limits of these compounds after the mild treatment. The organic feed itself is almost not affected, and the present process offers for the first time a selective destruction method for the chlorinated hydrocarbons. Several plant designs have been made, and the process is very economical. The kinetics of the hydrodechlorination reactions have also been determined for several substances, including 2,3,7,8-TCDD, the supposedly most dangerous compound of chlorinated hydrocarbons. The presentation will show the kinetic results and some design information.

INTRODUCTION

Since about seven years, KTI is occupied with some environmental product lines. One of these product lines is concerned with the rerefining of waste lube oil streams. This process, the KTI RELUBE process, is presently being operated at a number of sites worldwide. One of the features of the RELUBE process is the removal of sulfur- and chlorinated compounds. Further research on this specific aspect has resulted in a second process. This process, the KTI CHLOROFF process, is quite similar to the RELUBE process and incorporates a catalytic hydrodechlorination step in which chlorinated hydrocarbons can be effectively removed. In Fig. 1, the development of the two processes is shown.

KTI's **RELUBE** Process for waste lube oil treatment by a catalytic hydrogenation process:

°	Pilot Plant FRG	(1972)
°	Plant Greece	(1982)
°	Plant Tunesia	(1983)
°	Plant FRG	(1983)
°	Plant USA	(1985)

KTI's **CHLOROFF** Process, catalytic hydrodechlorination:

°	Semi pilot plant in The Netherlands	(1984)
°	Plant FRG	(1985)
°	Several designs	

Figure 1. Development of KTI's **RELUBE**- and **CHLOROFF** process.

Extensive testing of the **CHLOROFF** process has been carried out on laboratory scale, some results of which will be presented in this paper. First, however, a brief outline of chlorinated waste streams will be presented, mainly focussing on PCB's, TCDD's and TCDF's.

a. PCB's:
PCB's are being synthesised by reaction of bifenyl (C_6H_5 - C_6H_5) with chlorine. Replacement of one or more hydrogen atoms by chlorine takes place and theoretically, 209 different isomers can be synthesised in this way. Fig. 2 gives the general basic structure for PCB's:

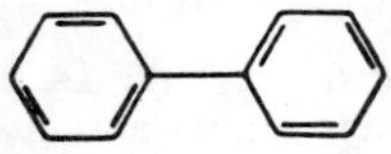

Figure 2. General chemical structure of PCB's.

b. PCDD's, PCDF's:
Polychlorodibenzodioxines (PCDD's) and polychlorodibenzofuranes (PCDF's) consist of two aromatic rings, bridged by 1 or 2 oxygen atoms. Totally, 75 PCDD isomers and 135 PCDF isomers can be found. Most toxic compounds are reviewed in fig. 3:

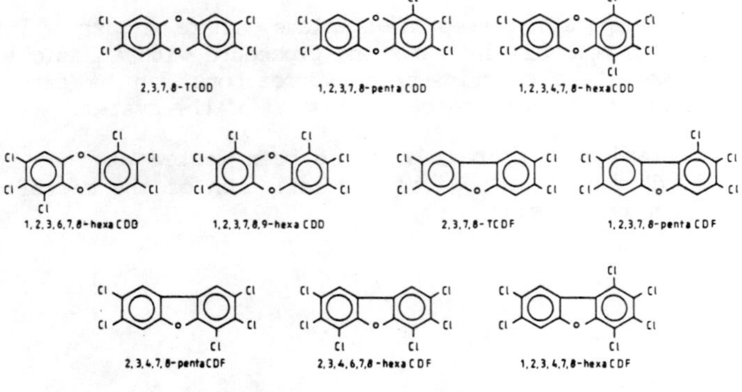

Figure 3. The most toxic PCDD's and PCDF's.

METHOD AND MATERIALS

The principle of the CHLOROFF process is the catalytic hydrogenation whereby chlorine is removed from the chlorinated hydrocarbon compound. The exothermal substitution reaction is as follows:

$$RCl_n + xH_2 \longrightarrow RH_xCl_{n-x} + xHCl$$

ΔH for this substitution reaction lies between -40 and -90 kJ/kmol.

Reaction takes place at a temperature between 250 and 400°C over a heterogeneous catalyst. As an illustration, in Fig. 4 the CHLOROFF process for the hydrodechlorination of a lube oil, containing chlorinated hydrocarbons, is schematically drawn:

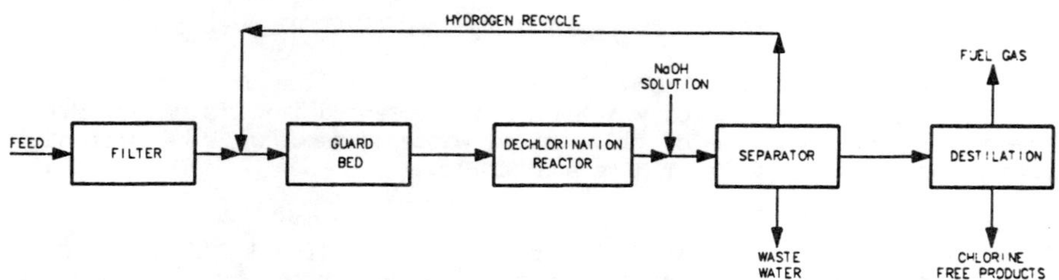

Figure 4. Schematic description of waste lube oil treatment by the CHLOROFF process.

Lube oil is first freed from particulates and is then heated to a temperature of 250°C to 400°C before being converted over a heterogeneous catalyst at a pressure between 30 to 80 bars. Under these reaction conditions not only chlorinated hydrocarbons are converted. Other heterocompounds behave in a similar manner: sulfur is converted to hydrogensulfide, nitrogen to ammonia and oxygen to water.

Besides hydrodechlorination, some cracking reactions take place, whereby light compounds are produced. Reaction conditions determine the extent to which these cracking reactions occur.

After reaction, a separation step takes place, where excess hydrogen is recycled and the HCl removal via a washing procedure with a basic solution is established. Resultant chlorine-free hydrocarbons can be fractionated or be burnt in order to supply the energy demand of the system.

Conversion of contaminated hydrocarbons has been tested at the Technical University of Eindhoven. Fig. 5 gives a description of the microflowreactor system used for the experiments.

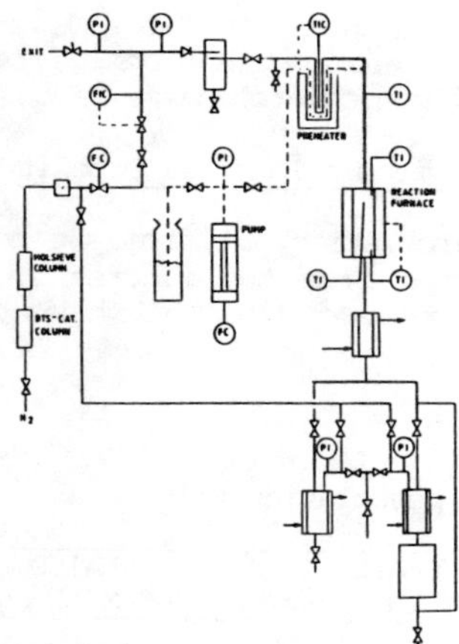

Figure 5. The high pressure microflowreactor system for testing of the CHLOROFF process.

The contaminated stream is introduced into the system with a high pressure pump, mixed with hydrogen, preheated and fed into the microflowreactor. Reaction products are cooled and separated in a liquid- and gasphase, respectively. Analyses of reaction products were performed offline and were carried out by several laboratories in both the Netherlands and Germany. PCB analysis was performed by Technical University of Eindhoven using High Pressure Liquid Chromatography and GC-MS (Gas Chromatograph - Mass Spectrometer). Most of the PCDD and PCDF analyses have been performed by the Toxicological Department of Amsterdam University.

Following contaminated streams have been processed:
1. Gasoil with different levels of chlorobenzene and PCB.
2. Percolation oil from landfill deposit site.
3. Waste stream from pesticide production site.
4. Lube oil, contaminated with PCDD's and PCDF's.
5. Gasoil, spiked with PCDD's and PCDF's.

RESULTS AND DISCUSSION

Conversion of Gasoil Contaminated with Chlorobenzene and PCB

In table 1, results of the conversion experiments with a contaminated gasoil are presented.

TABLE 1

Conversion experiments with contaminated gasoil

COMPONENT	UNTREATED SAMPLE mg/kg	TREATED SAMPLE mg/kg	DETECTION LIMIT mg/kg
PCB	530	nd	0.1
Chlorobenzene	0	-	-
PCB	530	nd	0.1
Chlorobenzene	4600	nd	10

As can be concluded from the figures in table 1, concentrations of PCB around 500 ppm are reduced to less than 0.1 ppm, a reduction of more than 99.9%. The same conversion of monochlorobenzene is observed.

Conversion of Percolation Oil from a Landfill Deposit Site

In table 2, conversion of percolation oil from a landfill deposit site is presented. As can be concluded from the figures in table 2, conversion factors for both dibenzofuranes and dibenzodioxines are mostly more than 99%.

TABLE 2

Conversion of a contaminated percolation oil

COMPONENT	UNTREATED SAMPLE $\mu g/kg$	TREATED SAMPLE $\mu g/kg$	CONVERSION %
2,3,7,8,-tetra-CDD	26	0.1	99.6
Sum penta-CDD	196	7.5	96.2
Sum hepta-CDD	3840	3.6	99.9
Octa-CDD	7610	4.1	99.9
2,3,7,8-Tetra-CDF	nd	nd	-
Sum tetra-CDF	298	1.5	99.5
Sum hexa-CDF	1650	1.3	99.9
Octa-CDF	4980	1.6	99.9

Conversion of Waste Stream from Pesticide Production Site

In tables 3a and 3b, the conversion of a contaminated waste stream from a pesticide production site is given.

TABLE 3a

Dechlorination of residual mixture (pesticide production) in the
CHLOROFF reactor
- Dibenzodioxines and -furanes -

COMPONENT	UNTREATED SAMPLE $\mu g/kg$	TREATED SAMPLE $\mu g/kg$	DETECTION LIMIT $\mu g/kg$
2,3,7,8-tetra-CDD	19	nd	0.02
Sum pentra-CDD	7	nd	0.05
1,2,3,4,6,7,9-hepta-CDD	20	nd	0.1
1,2,3,4,6,7,7-hepta-CDD	23	nd	0.1
Octa-CDD	64	nd	0.2
2,3,7,8-tetra-CDF	0.1	nd	0.02
Octa-CDF	12	nd	0.2

TABLE 3b

Dechlorination of residual mixture (pesticide production) in the
CHLOROFF reactor
- Chloroaromatics and -alicyclics -

COMPONENT	UNTREATED SAMPLE μg/kg	TREATED SAMPLE μg/kg	DETECTION LIMIT μg/kg
2,3,4-trichlorophenol	0.1	nd	0.1
2,4,5-trichlorophenol	10.9	nd	0.1
3,4,5-trichlorophenol	0.1	nd	0.1
2,3,4,5-tetrachlorophenol	0.2	nd	0.1
Pentachlorophenol (PCP)	0.1	nd	0.1
1,3-dichlorobenzene 1,4-dichlorobenzene	10.3	nd	0.1
1,2,3-trichlorobenzene	72.0	nd	0.1
1,2,4-trichlorobenzene	297	nd	0.1
1,2,3,4-tetrachlorobenzene	76.9	0.1	0.1
1,2,3,5-tetrachlorobenzene 1,2,4,5-tetrachlorobenzene	53.6	0.1	0.1
Pentachlorobenzene	2.0	nd	0.1
Hexachlorobenzene (HCB)	0.5	nd	0.1
alpha-HCH	0.1	nd	0.1

As can be concluded from the data in table 3a, PCDD and PCDF concentrations were reduced below detection limits.

Table 3b shows that chlorinated aromatic compounds like chlorophenols and chlorobenzenes can be removed effectively. Aliphatic ring compounds like hexachlorohexane behave in a similar manner. Exact conversion factors could not be established due to the relatively high detection limits of the method of analysis. However, conversion factors are in general more than 98%. Only conversion of 1,2-dichlorobenzene (table 3b) shows a relatively low value of 92.2%.

Conversion of Lube Oil, Contaminated with PCDD's and PCDF's

Conversion of lube oil, spiked with some PCDD and PCDF compounds in the ppb concentration range and PCB in the ppm concentration range is presented in table 4:

TABLE 4

Conversion of contaminated lube oil

COMPONENT	UNTREATED SAMPLE μg/kg	TREATED SAMPLE μg/kg	DETECTION LIMIT μg/kg
1,2,4-Tri-CDD	4.5	nd	0.2
1,2,3,4-tetra-CDD	19.5	nd	0.2
Sum penta-CDD	nd	nd	0.2
Sum hexa-CDD	5.2	nd	0.2
Sum hepta-CDD	17.3	nd	0.2
Octa-CDD	64.1	nd	0.2
Octa-CDF	40.4	nd	0.2
PCB	15000	nd	100

As can be concluded from table 5, conversion factors for the PCDD and PCDF compounds are at least larger than 95.5%. For the elimination of PCB's a conversion factor of 93.3% can be calculated. In a second experimental series, we used the specialised know-how of the Toxicological Chemical Laboratory of Amsterdam University and could use very sensitive analysis techniques. In this experiment, a contaminated lube oil fraction was processed and the very sensitive analysis techniques were employed.

TABLE 5
Elimination of chlorinated hydrocarbons from lube oil samples

COMPONENT	UNTREATED LUBE OIL (ppt)	HYDROTREATED LUBE OIL (ppt)	ELIMINATION (%)
Sum tetra-CDD	39622	9	99.98
Sum hexa-CDD	17937	488	97.28
Sum hexa-CDF	3605	467	87.05
Sum hepta-CDD	61112	394	99.36
Sum hepta-CDF	5880	162	97.24
Octa-CDD	103418	168	99.84
Octa-CDF	3597	33	99.08
TOTAL	235171	1721	99.26

Typical detection limits were now around 20 ppt and more exact conversion data could be calculated. Elimination of PCDD and PCDF compounds was achieved with an average conversion factor of about 99.2%.

Conversion of a Contaminated Gasoil

For kinetic reasons, experiments have been performed with gasoil, spiked with dibenzodioxine compounds and samples have been processed in the microflowreactor. In table 6, results of a conversion experiment with gasoil, contaminated with OCDD are illustrated.

TABLE 6
Conversion of gasoil, contaminated with OCDD

COMPONENT	UNTREATED GASOIL (ppt)	HYDROTREATED GASOIL (ppt) T = 225°C	HYDROTREATED GASOIL (ppt) T = 250°C
Sum tetra-CDD	-	7227	< 80
Sum penta-CDD	-	5468	< 80
Sum hexa-CDD	-	3351	< 70
Sum hepta-CDD	-	1129	< 70
Octa-CDD	550000	298	199

Besides the elimination of OCDD, which increases from 68% to 99% (based on the sum of concentrations of dioxines) with the small rise in reaction temperature, we observe a redistribution of dioxines with a lower amount of chlorine per molecule. In fact, at the higher temperature, levels of these smaller molecule compounds strongly decrease.

In a second experiment, a gasoil spiked with 2,3,7,8-TCDD, the most notorious dioxine compound, at a concentration of 80000 ppt was processed at different reaction temperatures. See table 7.

TABLE 7
Conversion of gasoil, contaminated with 2,3,7,8-TCDD

COMPONENT	UNTREATED GASOIL (ppt)	HYDROTREATED GASOIL (ppt)			
		200°C	225°C	250°C	275°C
2,3,7,8-TCDD	80000	6107	1152	56	40
OCDD	–	75	58	26	20

Also, we observe here some redistribution of chlorine and an increase in conversion with temperature. Note that at or above 250°C dioxine levels are very low indeed.

As an illustration, fig. 6 shows the conversion of TCDD as a function of the reaction temperature. We observe an almost complete ($\sim$99.9%) elimination of TCDD at a temperature above 250°C.

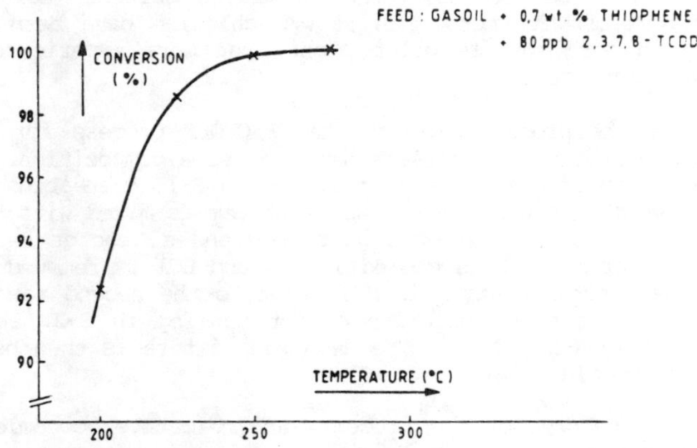

Figure 6. Conversion of 2,3,7,8-TCDD as function of reaction temperature.

DESCRIPTION OF THE CHLOROFF PROCESS

Based upon the results of the pilot plant study and the experiences of the very similar RELUBE process, which is being operated already on a world-wide scale, we developed the CHLOROFF process. The Process Flow Diagram of the CHLOROFF process is given in figure 7:

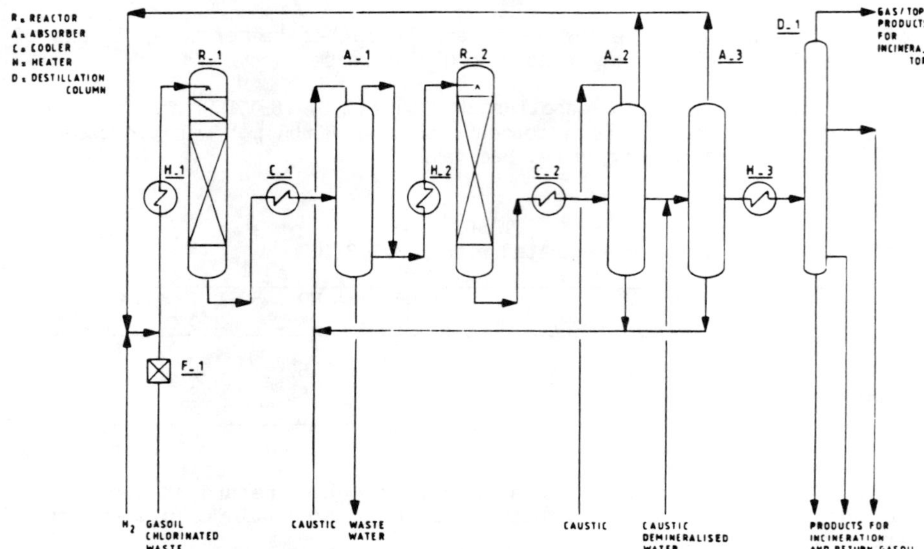

Figure 7: Process Flow Diagram of the CHLOROFF Process.

For the time being, smooth treatment of chlorinated hydrocarbons up to 5% w/w chlorine can be guaranteed. When a waste stream with a higher content of chlorine has to be treated, dilution with a suitable solvent will be done. Waste streams with more than 5% w/w chlorine have been treated satisfactorily in our labscale pilot plant, but more experience has to be gained on a larger scale.

Figure 8 gives the process flow of the CHLOROFF process for large quantities of hydrocarbons more or less with the same composition. The chlorinated waste stream from storage is filtered in F-1. Preheating takes place in H-1 after which the chlorinated waste stream is mixed with hydrogen and then sent to R-1, consisting of a guard bed and a reactor section. After reaction, the sour mixture is cooled in C-1 and HCl is removed in A-1. The mixture is then, after heating in H-2, sent to the second reactor R-2 for removal of the last bound chlorine. After cooling in C-2, absorption of HCl takes place in A-2 and A-3. The degassed mixture is then heated in H-3 and distilled in column D-1.

Based upon the features of the process, an indicative economic evaluation is presented in table 8.

Table 8:
Indicative economic evaluation of the CHLOROFF process.

Basics:	° 10,000 t/yr of waste stream, chlorine content less than 5% w/w
	° operating capacity 7800 h/yr
Main	° Hydrogen: 5 - 7 kg H_2/ton feed
variable $\longrightarrow$	about Dfl 20/ton feed
costs:	° Catalyst: about Dfl 40/ton feed
Fixed	° Personnel (manager + 2 operators): 400 kDfl/yr
costs:	° Capital cost (depreciation + interest): 1350 kDfl/yr
	° Maintenance: 3% of investment or 270 kDfl/yr
$\longrightarrow$	Total fixed costs are 2020 kDfl/yr or Dfl 202/ton feed.

As can be concluded from table 9, total costs (OPEX + fixed-) for the CHLOROFF process amount to about 262 Dfl/ton. It should be noted that operating costs can be considerable higher for lower capacity plants, because capital costs are relatively high in that case. Credit of the processed product can be as high as Dfl 100 - 1000/ton.

CONCLUSIONS

Based on the experimental results of the hydrodechlorination experiments, some conclusions can be made, although we realise that some experimental data are not yet fully known.

However, based upon the information sofar, we can state that we developed a catalytic process which is able to eliminate chlorinated hydrocarbons of a wide range of origin with a remarkably high efficiency.

Preliminary characteristic features of the process are:

- Reaction temperature   250 - 300°C
- Reaction pressure   50 - 60 bar

The CHLOROFF process has lots of characteristics of the KTI RELUBE process with which already a lot of experience has been gained on a worldwide scale. The CHLOROFF process is a sophisticated application of the RELUBE process and is therefore not an essentially new process design, but a different application of an existing technique.

The conversion of chlorinated hydrocarbons is going to be done in a closed system - unlike the incineration of chlorinated hydrocarbons. No flue gases, which form a central problem in the incineration process, will be present.

We therefore are of the opinion that the CHLOROFF process can contribute significantly to a solution for treating chlorinated waste stream in a safe and acceptable way from an environmental point of view.

CATALYTIC HYDROGENATION OF POLYCHLORINATED BIPHENYLS

by

R. W. Johnson, K. J. Youtsey,
L. Hilfman, T. Kalnes
UOP Inc.
Box 5017
Des Plaines, Illinois   60017-5017
USA

ABSTRACT

Polychlorinated biphenyls represent one member of a growing list of materials that are recognized as harmful to the environment. A process employing catalytic hydrogenation has been expanded from current industrial applications to the conversion of hazardous materials such as PCB's. The process catalytically reacts hydrogen with the chlorine moiety of the PCB compounds to form biphenyl and hydrochloric acid. A key element of this process is that a high quality fuel oil or specialty chemical product can be produced. The detoxified oil is recovered as a liquid product which can be recycled or sold. This process eliminates the possibility of releasing products of incomplete combustion into the atmosphere.

INTRODUCTION

Catalytic hydrogenation of polychlorinated biphenyls (PCB's) provides an environmentally sound, commercially viable alternative to high temperature incineration or chemical treatment. Through the UOP Decontamination process, selective hydrogenation can be utilized to transform regulated hazardous substances, such as PCB's, into useable hydrocarbon products.

This process is an outgrowth of petroleum refining and petrochemical technology which UOP has commercialized throughout the world during the last 70 years [1]. Typical applications for UOP's hydrogenation technology range from the desulfurization/demetallization of residual oils to the production of specialty chemicals. Table 1 summarizes some of UOP's existing hydrogenation processes.

TABLE 1
UOP HYDROGENATION TECHNOLOGY

<u>Petroleum Refining</u>   <u>UOP* Process</u>

Production of low aromatic solvents   AH Unibon*
Desulfurization of residual oils   RCD Unibon*
Production of LPG   LPG Unibon*
Hydrogenation of transportation fuels   HD Unibon*

<u>Chemical Industry</u>

Conversion of benzene to cyclohexane   HB Unibon*
Conversion of butadiene to butene   Huels SHP

---

*   UOP, AH Unibon, RCD Unibon, LPG Unibon, HB Unibon and HD Unibon
     are registered service marks of UOP Inc.

UOP's wide range of catalytic hydrogenation technology is now
extended into hazardous waste management through the UOP Decontamination process. Some examples of the potential applications of this
process are provided in Table 2.

TABLE 2
UOP DECONTAMINATION PROCESS
POTENTIAL APPLICATIONS

<u>Pollutant Chemical</u>   <u>Transformed Hydrocarbon</u>

Trichloroethylene   Ethane
Tetraethyl Lead   Ethane
Cresol   Toluene
PCB   Biphenyl

<u>Polluted Stream</u>   <u>Treated Product(s)</u>

Still Bottoms from Solvent Recycle   Solvent/Fuel
Used Motor Oil   Lube Base Stock
Spent Solvent   LPG/Fuel
Coal Tar   Fuel Oil
Transformer Oil   PCB-free Mineral Oil

This paper describes the UOP Decontamination process and presents
pilot plant test results confirming the application of catalytic hydrogenation to three different PCB contaminated waste oils.

PROCESS DESCRIPTION

Catalytic hydrogenation of PCB-contaminated organic waste streams is carried out at moderate temperatures and pressures. Proprietary UOP catalysts are utilized to effect rapid reaction of hydrogen with the pollutant molecule(s). Reactor design configuration is based on waste oil composition, desired end products, regulatory specifications and overall processing flexibility requirements. Within the scope of the UOP Decontamination process, additional unit operations such as phase separation, extraction, adsorption, evaporation and distillation are integrated with catalytic hydrogenation to meet necessary pre- and post-treatment requirements.

A schematic diagram of the hydrogenation section of this process is shown in Figure 1. The waste oil is heated and combined with a gas stream that is rich in hydrogen. This mixture is charged to a reactor containing proprietary catalyst(s) where pollutants such as PCB's are converted. Fresh hydrogen gas is charged to the inlet of the reactor to balance consumption.

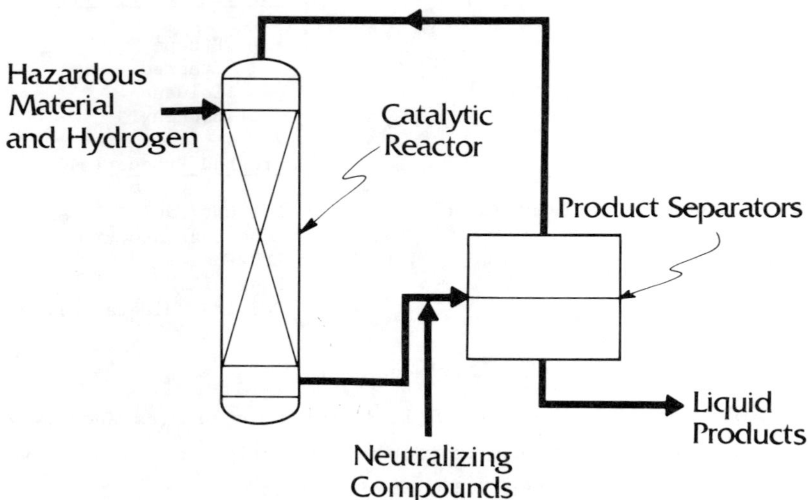

FIGURE 1. Hydrogenation section of the UOP Decontamination process.

The effluent from the catalytic reactor is cooled and combined with an aqueous solution of a basic compound. The stream is then transferred to a product separation system. The gas phase is recycled to the reactor inlet to provide a large excess of hydrogen gas in the catalytic reaction zone. The aqueous fraction is recovered as a solution of non-hazardous salts. The treated organic phase is generally suitable for re-use as a fuel oil. The organic product may also be converted into a slate of higher value products depending upon the nature of the charge stock. Products are collected in receiver tanks and certified prior to removal from the treatment site. Post-treatment required to produce specific products varies with the charge stock and site-specific economic considerations.

PROCESS ADVANTAGES

Environmental and Safety

The UOP Decontamination process inherently produces no acutely toxic by-products. Chlorinated dioxins and furans are always potential products of an oxidative process [2, 3], whereas the formation of these materials is fundamentally inconsistent with operation under the reducing conditions present in the UOP process. In addition, the opportunity to certify the conversion products before release from the treatment facility, and to reprocess the materials in the event of a plant upset, offers extra protection to the environment not possible with incineration.

Commercial hydrogenation catalysts offered by UOP minimize undesirable side reactions involving the hydrocarbon matrix, thus eliminating many of the safety concerns associated with other technologies. For example, water is known to react violently with metallic sodium, whereas no reaction occurs between water and hydrogen. The UOP process handles waste streams containing entrained water without pre-treatment. The chlorine atom is replaced by a hydrogen atom, thus preventing any by-product sludge formation. The hydrogen chloride formed by the hydroconversion of PCB's is neutralized in a uniquely engineered product separation system which minimizes the corrosivity and volatility of the product streams while emphasizing safety of operation.

Flexibility/Product Reuse

The UOP Decontamination process allows treatment of many hazardous materials, in addition to PCB's, in a cost-effective, environmentally acceptable manner. Typical organic waste streams, such as transformer oils, still bottoms, waste solvents, cutting oils and used lubricating oils all fall within the scope of this technology. The reactivity of heteroatoms, such as nitrogen, chlorine, sulfur, oxygen and metal complexes, under catalytic reducing conditions, provides a powerful tool for conversion of chemical waste streams into hydrocarbon products that have value as petrochemical charge stocks, solvents, lubricants or fuels.

Economic

In general, the added value derived from converting a hazardous waste oil to a usable hydrocarbon product provides the economic driving force when considering alternatives to incineration. Hydrogen, as the reducing agent for the UOP Decontamination process, is less expensive than sodium or sodium-based compounds and acts to upgrade, not degrade the treated product. Overall processing costs are highly competitive with other existing, environmentally acceptable technologies.

EFFECTIVE OPERATING RANGE

Hydrogenation technology can be applied to almost any organic hazardous waste stream. Catalytic reaction kinetics are favorable over a wide range of pressures, temperatures and flow rates. Operating range for the technology is primarily constrained by economic, not technical, criteria. Treatment plant cost is proportional to the amount of flexibility incorporated into its design basis. A reasonable amount of flexibility can be achieved within the following physical constraints:

Operating Pressure:	14 - 70 bar
Operating Temperature:	65 - 454°C

Typical treatment plant capacity would be ~40 m^3/d but could be easily scaled upward or downward. Viscosity of the waste stream is not critical unless it prevents the material from being pumped up to the required operating pressure. If required, a diluent oil can be utilized as a carrier solvent for the pollutant(s) or to enhance unit operations on highly viscous materials.

PILOT PLANT TEST RESULTS

Pilot plant test results support the applicability of the UOP Decontamination process to a broad class of hazardous waste types. For this discussion, polychlorinated biphenyls have been chosen as the principal focus to illustrate the capabilities of this new treatment technology.

PCB Waste Streams

The following waste streams were continuously processed to demonstrate process feasibility.

o   Shell Diala AX transformer oil adjusted to a PCB concentration of 1,400-15,000 ppm using Aroclor 1016.

o   Waste lubrication oil obtained from a contaminated oil recycling site. Sample was contaminated with 270 ppm of Aroclor 1260, 750 ppm of lead and an ash content of 0.32 wt-%.

o   Bunker fuel obtained from a large source of contaminated fuel oil. This oil was contaminated with 100 ppm of a mixture of Aroclors 1242 and 1248 and an ash content of 0.05 wt-%. Principal metal contaminants included nickel, vanadium, sodium and iron. Carbon residue was measured at 15.6 wt-% and heptane insolubles were found to be present at 8.6 wt-% level.

In all cases, the waste oils were processed without dilution in a pilot plant designed to evaluate catalytic activity, stability and selectivity. Demonstration data were obtained using a recycle gas flow configuration to eliminate scale-up assumptions.

Pilot Plant Operation

Charge stock is combined with the hydrogen-rich recycle gas stream and preheated in a section of the catalytic reactor. The hydrogen gas and oil then pass through the catalyst at elevated temperature and pressure. The reactor effluent is cooled and sent to a high pressure separator. The gas from the high pressure separator is combined with a sufficient amount of fresh hydrogen gas to maintain the system pressure. The liquid separator underflow is cooled and collected for analysis. A nitrogen purge is maintained over the liquid product receiver to recover any light gases produced.

Analytical

Both product liquid and gas streams were subjected to a variety of detailed analyses to quantify pollutant conversion, hydrogen consumption and hydrocarbon product quality. PCB conversion was monitored using the GC/ECD analysis [4]. Analyses of PCB compounds in products were more difficult than comparable analyses for the charge stocks. The charge stocks had PCB concentrations from 100-15,000 ppm and well recognized congener patterns. Chemical reduction by hydrogen (or sodium) removes chlorine atoms from the PCB molecule and, therefore, eliminates in the product the standard pattern normally seen in the detector. Confirmation of product PCB concentration was, therefore, performed using either a HP 5987 GC/MS system or a HP 5890 GC equipped with a HP 5970 MSD detector. The limit of detection established using a hexane calibration matrix was found to be below 2 ppm per congener. This detection limit is influenced by hydrocarbon matrix and may vary with waste oil type. For this reason, the conversion of PCB compounds in the bunker fuel was confirmed by independent analysis at Battelle Laboratories (Columbus, Ohio).

Reduction in metals, ash, sulfur and other undesirable compounds was monitored using a combination of ASTM, UOP and EPA derived methods.

Results and Discussion

High conversion levels of contaminants were achieved for all three of the waste streams processed. In addition to reducing product PCB concentrations to non-detectable levels, other contaminants, such as lead and sulfur were also converted. Table 3 is provided as a comparative summary of waste oil and corresponding hydrocarbon product analyses. A discussion of each demonstration follows.

TABLE 3
PILOT PLANT CHARGE STOCK AND PRODUCT ANALYSIS

	TRANSFORMER OIL		WASTE LUBE		BUNKER FUEL	
	Charge Stock	Product	Charge Stock	Product	Charge Stock	Product
ANALYSIS:						
PCB Concentration, ppm	1400	No PCB's	270	No PCB's	100	No PCB's
Aroclor Type	1016	Detected	1260	Detected	1242/1248	Detected
Ash, wt-%	<0.01	--	0.32	--	0.05	0.008
Lead, wt-ppm	--	--	750	5.5	6.2	0.03
Zinc, wt-ppm	--	--	285	1.7	5.6	--
Vanadium, wt-ppm	--	--	2.9	<0.1	52	50
Nickel, wt-ppm	--	--	3.8	2.6	49	47
Sulfur, wt-%	0.07	0.0017	0.49	0.13	1.57	1.35
Nitrogen, wt-ppm	1	--	600	250	6000	6000
Carbon Residue, wt-%	--	--	0.89	--	15.6	13.9
Heptane Insols, wt-%	--	--	0.15	--	8.6	6.0
Specific Gravity	0.886	0.878	0.887	0.875	0.996	0.990
Distillation, °C						
Method (ASTM)	D-2887	D-2887	D-1160	D-1160	D-1160	D-1160
IBP	222	220	163	122	105	88
BPC 50% Over	327	329	416	405	438	431
FBP	522	493	485	545	516	523
% Vacuum Residuum	--	--	20	6	42	38
Heating Value, Btu/lb	--	--	--	19,000	--	18,000

Transformer Oil - PCB contaminated transformer oil was processed for approximately 30 days to demonstrate the activity and stability of the process. The GC/ECD analytical method was used to provide rapid characterization of the product during the run. As shown in Figure 2, the concentration of PCB's in the product was reduced to typical limits of detection. The higher product concentration measured at 15 days on stream was caused by taking the recycle compressor out of service to simulate a possible plant upset. As shown in the figure, high PCB conversion returned as soon as the recycle compressor was restarted. This experiment demonstrates that this severe upset does not damage the ability of the plant to achieve high PCB conversion at subsequent normal conditions. (Off-specification product produced during this time could be returned to the reactor for further processing.) Measured PCB conversion for this stream was greater than 99.9%. Sulfur species present in the mineral oil were converted at 97.5%. Catalyst activity was maintained over the entire 30-day period demonstrating a very stable operation.

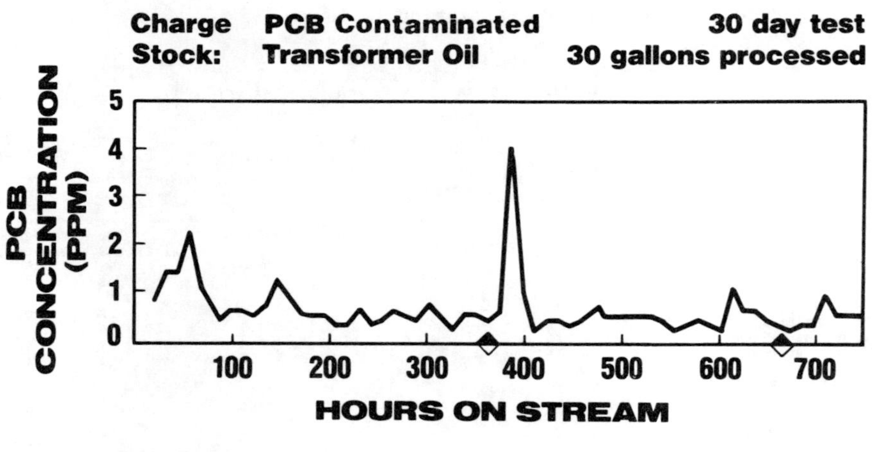

FIGURE 2. PCB Conversion

Waste Lubrication Oil - The composition of this material is typical of used automotive motor oil. The viscosity characteristics of this material (Viscosity Index = 128) make it an excellent candidate for reuse after environmentally undesirable contaminants are removed. As illustrated in Table 3, the treated product contains no detectable level of PCB's. In addition, lead content was reduced to <10 ppm. Desulfurization was ~73% with an overall ash reduction estimated at >90%. Processing retained the excellent viscosity characteristics of the starting material (Product Viscosity Index = 126).

Bunker Fuel Oil - Based on hydrocarbon type, this waste material is considered the most difficult to catalytically decontaminate. Heteroatoms such as sulfur, nitrogen and metals are tightly bound within high molecular weight, multiple ring compounds. Processing of this material resulted in a very selective conversion of the PCB contaminants present. Desulfurization was only 10% and removal of nickel and vanadium was <5%. Similarly, the boiling point distribution of the product was not significantly changed except for a reduction in the initial boiling point and slight conversion of residuum. Product PCB's were non-detectable at all levels of chlorination. If required, process conditions could be adjusted to further remove sulfur and metal contaminants.

PROCESS ECONOMICS

Pro forma economics for the UOP Decontamination process are presented in Table 4. The calculations are based on a plant size of 40 m^3/d operating 255 days per year and conservative assumptions with respect to capital and operating costs. On stream factors of greater than 90% should be easily achievable. The positive contributions to this payback calculation arise from avoidance of disposal costs and from product credits, and are based on current competitive values in the United States.

TABLE 4
PAYBACK ECONOMICS OF UOP DECONTAMINATION UNIT[a]

Waste Type	Payback (yrs)[b]
Fluorocarbon Still Bottoms	0.7
Waste Lube Contaminated with PCB's	1.3
Askarel	0.6
Highly Chlorinated Still Bottoms	1.3
Spent Chlorinated Solvent	<1
Low Chlorine Still Bottoms[c]	3.5

[a] Assumes standard plant design 40 m^3/d capacity, 70% utilization.
[b] Revenue in payback calculation based on avoidance of current disposal costs and credit for reusable product
[c] Less than 0.5 wt-% chlorine.

Commercially attractive paybacks can be associated with a wide range of organic waste streams.

STATE OF DEVELOPMENT

UOP catalytic hydrogenation processes have received wide acceptance from the petroleum and chemical industries over the past three decades. State-of-the-art processing technologies coupled with over two dozen commercially available proprietary hydrogenation catalysts provide a firm basis for the commercialization of the UOP Decontamination process. Pilot plant confirmation of the treatment of many different waste oils provides the detailed data required to scale-up plant design(s) for application to today's waste management problems.

Finally, a highly qualified engineering staff make possible the custom design of treatment plants to satisfy a wide variety of environmental and industrial needs. Plant start-up, operator training and trouble-shooting activities are all fully supported by UOP's technical service staff of trained engineers.

REFERENCES

1. Achilladelis, Basil, "History OF UOP: From Petroleum Refining to Petrochemicals." Chemistry and Industry, 1975, 6, p. 337

2. Olie, K., Vermeulen, E. E., and Hutzinger, O., "Chlorodibenzo-p-dioxins and Chlorodibenzofurans Are Trace Components of Fly Ash and Flue Gas of Some Municipal Incinerators in the Netherlands." Chemosphere, 1977, 6, 7, p. 455.

3. Weerasinghe, N. C. A., Meehan, J. L., Cross, M. L., and Harless, R. L., "The Analysis of Tetrachlorodibenzo-p-dioxins and Tetrachlorodibenzofurans in Chemical Waste and in the Emission from Its Combustion." Chlorinated Dioxins and Dibenzofurans in the Total Environment II, ed. Lawrence H. Keith, et al., Butterworth Publishers, Boston, 1985, p. 425.

4. United States Environmental Protection Agency, "The Determination of Polychlorinated Biphenyls in Transformer Fluid and Waste Oils", Method EPA-600/4-81-045.

EVALUATION OF ULTRAVIOLET LIGHT/OXIDIZING AGENT AS A MEANS FOR
THE DEGRADATION OF TOXIC ORGANIC CHEMICALS IN AQUEOUS SOLUTIONS

Patience C. Ho
Chemistry Division
Oak Ridge National Laboratory
Oak Ridge, Tennessee 37831-6201
USA

ABSTRACT

Experiments have been performed to demonstrate the synergistic
effect of ultraviolet light on the decomposition of toxic organic
polynitro and polyhydroxy benzenes in aqueous solutions by hydrogen
peroxide. These substances, which are hazardous to biota, can be
degraded through photooxidation to carbon dioxide and water (and a small
amount of nitric acid in the case of polynitrobenzenes).
The degradation of these aromatic compounds is induced by the
presence of hydroxyl radicals which are produced by the irradiation of
hydrogen peroxide with ultraviolet light in water. Compounds such as
2,4-dinitrotoluene and phenol are first hydroxylated to di- and
tri-hydroxybenzenes, followed by benzene-ring cleavage to produce
lower-molecular-weight carboxylic acids and aldehydes via photooxidation. On further photooxidation, these carboxylic acids and
aldehydes will eventually be converted to harmless carbon dioxide and
water.

INTRODUCTION

The manufacturing and handling operations in many industrial
facilities generate significant volumes of aqueous effluents
contaminated with toxic organic compounds, such as dinitrotoluenes (DNT)
and phenol, which are hazardous to biota [1]. Release of these
compounds into the surrounding environment constitutes an environmental
hazard. The important motivation for our study of degradation of these
organic compounds is to elucidate a clean and effective method to remove
these compounds from the industrial waste water.

It has been known for some time that high-energy electromagnetic
radiation, such as ultraviolet light, can induce oxidation (and degradation) of organic molecules in water. The effect is enhanced by the
addition of oxidizing agents such as hydrogen peroxide. In a previous

paper [2] we showed the synergistic effect of hydrogen peroxide and uv radiation from a medium- pressure mercury vapor lamp on the decomposition of 2,4-dinitrotoluene in aqueous solution. Dinitrotoluene in aqueous solution was degraded to CO_2, H_2O and HNO_3 via photooxidation by uv/H_2O_2. The degradation mechanism was carried out by hydroxyl radicals, which were produced by photolysis of H_2O_2 in water [3-8].

Phenolic compounds, especially phenol, which commonly exist in industrial waste and are frequently found in drinking water, not only give unpleasant tastes and odors, but also pose a health hazard to humans and aquatic life [1]. A number of treatments have been developed in order to remove them from the water stream. The well-known methods are: oxidation with Fenton's reagent [9-14] or ozone [10-22] and photo-oxidation with ozone/uv [23,24] or hydrogen peroxide/uv [25-28]. Although several research groups [25-28] have dealt with photooxidation of phenolic compounds in aqueous solutions by H_2O_2/uv, the experiments were carried out either under nitrogen [25], or by using uv light with wavelength longer than 280 nm [26-28], which cannot completely degrade these compounds to CO_2 and H_2O. The degradation mechanism of phenolic compounds in aqueous solution by H_2O_2/uv was nowhere to be found.

In this paper we compare the mechanism of photooxidation of 2,4-dinitrotoluene in water with new studies on phenolic compounds.

EXPERIMENTAL SECTION

Materials

Phenol and pyrogallol (Mallinckrodt, analytical reagent), catechol (Baker), hydroquinone (MCB, reagent), 1,2,4-benzenetriol (Aldrich, 99%), trans,trans-muconic acid (Fluke, purum, > 98 % (T)), maleic acid (Fluka, >99% (T)), and hydrogen peroxide (Mallinckrodt, 30%) were used as received without further purification. Water was deionized and double distilled in glass.

Photolysis Procedures and Analytical Methods

Photolysis procedures and analytical methods used were the same as in Ref. 2. The uv source used was either a 200-watt or a 450-watt medium-pressure mercury vapor lamp (Hanovia), with or without a Vycor filter (to filter out light < 220 nm). The medium-pressure mercury lamp has a broad spectrum from the far uv to the infrared region [29] and 28% of the emitted energy from the Hanovia lamp is in the ultraviolet region.

The aqueous organic solutions were prepared with double distilled deionized water. Hydrogen peroxide was added to the aqueous organic solution just before it was transferred into the photo-reactor. Samples were collected after the desired time of irradiation. The rates of the disappearance of phenol were monitored by high-performance liquid chromatograph (HPLC) equipped with a Varian Vari-Chrom uv-vis detector. The HPLC measurements were conducted by monitoring the absorption at 270 nm for phenol reaction mixtures. A reverse-phase C18 column (Alltech) was used for the HPLC. Reaction intermediates were characterized by HPLC, GC (Hewlett-Packard 5880 gas chromatograph) and GC/MS (Hewlett-Packard HP 5995A gas chromatograph/mass spectrometer system). Both GC and GC/MS systems were fitted with a 30-m Durabond I narrow-bond fused silica column (J.W. Scientific). The column temperature was programmed from 50°C (60°C for GC) (3-min hold) to 325°C (275°C for GC) with an increase of 10°C/min. Confirmation of the structure of intermediates whenever possible was done by comparison of the spectra with authentic compounds. Phenols, acids, and the reaction mixtures were silyated with MSTFA [N-methyl-N-(trimethylsilyl) trifluoroacetamide; Pierce] before GC and GC/MS analyses.

RESULTS AND DISCUSSION

In a previous paper [2] we have discussed the mechanism of photo-oxidation of 2,4-dinitrotoluene (DNT) in aqueous solution in the presence of hydrogen peroxide (H_2O_2) and irradiation with a medium-pressure mercury lamp. The degradation of DNT in aqueous solution to CO_2, H_2O and HNO_3 was induced by hydroxy radicals ($\cdot OH$) which were produced by photolysis of H_2O_2 in water [3-8]. The mechanism including hydrogen abstraction and addition of hydroxyl radicals can be summarized as: (a) side-chain oxidation and decarboxylation, which converts DNT to 1,3-dinitrobenzene, (b) hydroxylation of the benzene ring, which converts 1,3-dinitrobenzene to hydroxynitrobenzene derivatives, (c) benzene-ring cleavage of these hydroxynitrobenzenes, which produces lower-molecular-weight carboxylic acids and aldehydes, and (d) further photooxidation, which eventually converts the lower-molecular-weight acids and aldehydes to CO_2, H_2O, and HNO_3. From the similarity of nitrohydroxy- and hydroxy-aromatics, one might infer that the photooxidizing mechanism of the latter is similar to the former. It is known that the reaction rates of hydroxyl radicals with various organic compounds are high and approach the diffusion controlled limit [30].

Photolysis of Hydrogen Peroxide

Hydrogen peroxide (H_2O_2) is known to be photosensitive [31] and this property has often been used to induce oxidation of compounds which are not attacked by H_2O_2 in the dark. Light with wavelengths shorter than 400 nm can decompose H_2O_2 to form free hydroxyl radicals ($\cdot OH$) and hydroperoxyl radicals ($HO_2\cdot$). Photodecomposition of H_2O_2 alone is a chain reaction [31]:

$$H_2O_2 \xrightarrow{k_1} 2\cdot OH \tag{a}$$

$$\cdot OH + H_2O_2 \xrightarrow{k_2} H_2O + HO_2\cdot \tag{b}$$

$$HO_2\cdot + H_2O_2 \xrightarrow{k_3} H_2O + O_2 + \cdot OH \tag{c}$$

$$2HO_2\cdot \xrightarrow{k_4} H_2O_2 + O_2 \tag{d}$$

When the concentration of H_2O_2 is low ($1 - 4\times10^{-2}$M (moles/liter)) and the light intensity is high ($> 3\times10^{17}$ quanta/1.sec), the reactions ceases to be a chain reaction [3]. This will be further enhanced by the presence of substances which can trap hydroxyl radicals, such as phenol [25]. Under these conditions, the quantum yield for H_2O_2 decomposition is independent of intensity, concentration, and pH value of the solution [3,4]. The primary quantum yield for H_2O_2 decomposition into $(\cdot OH)$ radicals is one-half of the limiting quantum yield which is $2k_1$ [3,4]. At 253.7 nm (Hg line) and 25°C, the quantum yield was found to be ca 0.5 [4-7] and was 0.3 at 313 nm and 25°C [8].

Optimum Conditions of Photooxidation of Phenol

On the study of photooxidation of DNT in aqueous solution by uv/H_2O_2, we found there were optimum ranges of the concentration of H_2O_2 and wavelength of the irradiating light [2]. At these optimum conditions, the degradation rate of DNT was not only fast, but it could also degrade DNT all the way to CO_2, H_2O and HNO_3.

Several research groups have reported that when aqueous phenol solution was exposed to uv light without the presence of an oxidizing agent (H_2O_2 or O_3), dimers of phenolic compounds were formed [32,33]. When aqueous mixtures of phenol and H_2O_2 were exposed to uv light with wavelength greater than 280 nm, not only was the degradation rate hindered, but dimers of phenolic compounds and other complexes of stable photo products were also produced [25,28]. Only with the combination of light with wavelength shorter than 280 nm and the presence of an oxidizing agent can phenol in aqueous solution be degraded to CO_2 and H_2O.

In order to elucidate the optimal concentration of H_2O_2 on the photooxidation of phenol in aqueous solution, a 0.001 M aqueous phenol solution was mixed with various molar ratios of H_2O_2. The mixtures then were exposed to the 450-w medium-pressure Hg lamp without filter. Hydrogen peroxide was added to the aqueous solution prior to light exposure. Table 1 gives the disappearance of phenol as a function of time. The pH values and colors of each reaction mixture vs time are also given in Table 1. When the molar ratio of H_2O_2/phenol was < 52,

TABLE 1

Effect of hydrogen peroxide concentration and time of exposure on the extent of photooxidative degradation of phenol

Conc. of phenol 0.001 M; pH of phenol/H_2O_2 mixture ~ 5.80
Incident light: 450-w medium-pressure mercury vapor lamp (no filter)

(H_2O_2/phenol molar ratio)		Time of exposure (min)								
		10	20	30	45	57	60	85	94	95
6.5	pH	4.50	4.10	3.95	3.75		3.80			3.85
	Color	l.brown*	brown	d.brown***	d.brown		d.brown			golden
	% Elimination	40.7	63.9	84.9	90.1		94.8			98.2
13.0	pH	4.25	3.95	3.95	3.85		3.85	3.85		
	Color	l.brown	l.brown	l.brown	l.brown		l.yellow	yellow		
	% Elimination	45.4	66.9	80.2	94.2		97.2	99.4		
26.0	pH	4.05	3.85	3.70	3.60		3.50			
	Color	v.l.brown**	v.l.brown	v.l.brown	~colorless		colorless			
	% Elimination	50.0	74.4	91.3	97.6		100			
52.0	pH	3.85	3.50	3.30	3.25	3.55				
	Color	colorless	colorless	colorless	colorless	colorless				
	% Elimination	51.8	90.7	96.5	100	100				
130.0	pH	3.40	3.55	3.55	3.50		3.45		3.70	
	Color	colorless	v.l.brown	~colorless	~colorless		colorless		colorless	
	% Elimination	72.6	82.5	88.3	92.7		98.2		100	

*l. - light
**v.l. - very light
***d. - dark

the colorless initial mixture (pH $\sim$ 5.80), turned into light brown in
ca. 10 minutes; it then became dark brown before turning to yellow and
then fading away as the exposure continued. At the molar ratio of
H_2O_2/phenol = 52, after 10 minutes of exposure, the reaction mixture was
colorless and the HPLC spectra showed much lower amounts of the existing
intermediates, although the degradation rate was not as obviously fast
as compared to other systems. As the molar ratio increased to 130, and
after 10 minutes of exposure, a very light brown color of the solution
was noticed again. At the beginning, the degradation rate of the
H_2O_2/phenol = 130 system seemed faster than the others; however, it was
discovered that small amounts of phenol and large amounts of reaction
intermediates (revealed on the HPLC spectra) existed in the solution for
quite a long time before their disappearance. It is clear, that in the
vicinity of H_2O_2/phenol = 52, there is an optimal concentration of H_2O_2.

Mechanism of Photooxidation of Phenol

Photooxidation of Phenol. In order to study the degradation
pathway of phenol in aqueous solution with uv/H_2O_2, reaction intermediates were collected and analyzed. For better understanding of the
reaction mechanism, the reaction rate was slowed by reducing the amount
of H_2O_2 and lowering the energy of the incident light by using a 200-w
medium-pressure Hg lamp equipped with a Vycor filter (to filter out
light under 220 nm). To this end an aqueous solution of 0.0068 M phenol
and 0.044 M H_2O_2 was photolyzed. The HPLC spectra revealed that several
intermediates were formed after 10 minutes of exposure; however, maximum
amounts of intermediates were produced after 1.8 hours. After removal
of water, drying under nitrogen, and silylating with MSTFA [N-methyl-N-
(trimethylsilyl)trifluoroacetamide], the reaction mixture was analyzed
by GC/MS. The identified intermediates, besides the unreacted phenol,
are given in Table 2. In the 1.8-hour-exposed sample, catechol was
found as the major intermediate along with smaller amounts of hydroquinone, pyrogallol, 1,2,4- benzenetriol, and several lower-molecular-
weight carboxylic acids and aldehydes. Determination of glyoxylic acid,
glyoxal, and formic acid were not performed in this study as we did in
Ref. 2. Among these lower-molecular-weight intermediates, to our
surprise, there is a compound, whose MS spectra has a base ion peak at
m/e 147 and a fragment ion of m/e 233. It is very interesting to note
that the retention time and the MS spectra of this unknown compound
matches with those of malonic acid, which also has a base ion peak at
147 (a typical fragment ion, $[(CH_3)_3-Si-O-Si^+(CH_3)_2]$, of silylated
dicarboxylic acids, hydroxy acids, and terminal diols [34,35]), M-15 (M
minus methyl group) ion peak at m/e 233 and molecular ion peak at m/e
248. It is questionable, however, whether a saturated diacid can be
obtained in the process of oxidative cleavage of unsaturated compounds
(probably muconic acid derivatives, which will be discussed later in
this report) with uv/H_2O_2 in aqueous solution. We name this unknown
compound as unknown I.

The result of photooxidizing the phenol/H_2O_2 mixture to give
catechol as a major intermediate along with a small amount of hydroquinone is in agreement with Omura's work [25], who claimed that hydroxylation of phenol by photo-decomposing H_2O_2 with light at 253.7 nm in
aqueous solution under nitrogen produces the main products catechol and
hydroquinone (the weight percent ratio of catechol/hydroquinone was
$\sim$ 1.9). The result is also in agreement with our former report on
photooxidation of 3-nitrophenol in aqueous solution with uv/H_2O_2 [2].

TABLE 2
Photooxidation of phenol polyhydroxybenzenes, muconic acid and maleic acid
in aqueous solution

Starting Compound	H_2O_2/org (molar ratio)	uv Source (w)	Time Exposure (min)	Major Intermediate(s)	Other Identifiable Intermediates
phenol(a)	6.5	200/vycor	108(1.8 hrs)	b	c,d,e,i,j,k,p
catechol(b)	2.0	200/vycor	72(1.2 hrs)	d	e,f,i,k,n,p
catechol(b)	6.5	200/vycor	20	d	e,k,l
catechol(b)	6.5	200	10	d,e,l,p	f,i,j,k,o
catechol(b)	6.5	200	108(1.8 hrs)	j,p	d,i,k,m,n,o
hydroquinone(c)	6.5	200/vycor	22	j,k	e,i,l,m,n,p
pyrogallol(d)	6.5	200/vycor	20	l	h,j,k
pyrogallol(d)	26.0	450	30	l	g,i,m,n,p
1,2,4-benzene-triol(e)	6.5	450	15	o	
trans,trans-muconic acid(h)	13.0	450	20	l	k,n,p
maleic acid(i)	13.0	450	20	l	k,p

a phenol
b catechol
c hydroquinone
d pyrogallol
e 1,2,4-benzenetriol
f muconic acid derivatives [cis,cis-2,4-hexadienedioic acid or cis,cis-4-hydroxy-6-oxo-2,4-hexadienoic acid]
g 2-hydroxymaleic acid
h trans,trans-muconic acid
i maleic acid
j hydroxy-maleic acid semialdehyde [2 (or 3)-hydroxy-4-oxo-butenoic acid]
k unknown I
l (CHO)O(COOH)
m α-ketomalonic acid or mesoxalic acid [(COOH)CO(COOH)]
n 2,2-dihydroxy propanedioic acid [(COOH)C(OH)$_2$(COOH)]
o mesoxalic acid semialdehyde [(HO)$_2$C(COOH)(CHO)]
p oxalic acid

On the study of photooxidation of aqueous 3-nitrophenol/H_2O_2 mixture, we proposed that the reaction mechanism as the first step is the addition of a hydroxyl radical (which was produced by photolysis of H_2O_2) on the benzene ring to produce dihydroxynitrobenzenes (3 isomers; including 4-nitrocatechol) as major intermediates. The presence of catechol and hydroquinone in the photolyzed phenol/H_2O_2 mixture presumably implies that a similar reaction route is followed. That is

Photooxidation of Catechol and Hydroquinone. For further study on the degradation route of phenol in aqueous solution by uv/H_2O_2, several experiments have been carried out with catechol and hydroquinone at various molar ratios of H_2O_2/catechol and with different incident light sources. The results are given in Table 2. With catechol (0.0068 M mixed with 0.014 to 0.044 M H_2O_2), depending on the length of exposure at various concentrations of H_2O_2, either pyrogallol (with small amounts of 1,2,4-benzenetriol) (besides the unreacted catechol) or oxalic acid was found as the major reaction intermediate along with lower-molecular-weight carboxylic acids (besides oxalic acid) and aldehydes. Among the lower-molecular-weight intermediates were two which were not seen in the photolyzed-phenol reaction mixture. One of them has a base ion peak at m/e 147, a fragment ion of m/e 247 (presumably is the M-15 ion peak), and a molecular ion peak at m/e 262. This is an oxidation product from hydroxy maleic acid, α-ketomalonic acid [(COOH)CO(COOH)] [41]. The other intermediate, which has a base ion peak at m/e 89, and a fragment ion of m/e 147 (presumably corresponds to the M-15 ion peak), supposedly has a formula like [(CHO)O(COOH)], is probably produced through the Baeyer-Villigar oxidative rearrangement of glyoxylic acid [21,42].

$$\begin{array}{c} \text{CHO} \\ | \\ \text{COOH} \end{array} + H_2O_2 \longrightarrow \begin{array}{c} \text{CHO} \\ | \\ \text{O} \\ | \\ \text{COOH} \end{array} + H_2O \qquad\qquad (2)$$

The reaction sequence of photooxidation of catechol/H_2O_2 is parallel to the photooxidation of aqueous 4-nitrocatechol/H_2O_2 mixture [2], of aqueous catechol/H_2O_2 mixture under nitrogen [25], and of aqueous catechol/O_3 mixture [24]. This pathway is also similar to the ozonation of catechol in aqueous solution [22]. o-Quinone was neither found in any of our photolyzed catechol/H_2O_2 reaction mixtures, or in the photolyzed phenol/H_2O_2 mixture as it had been found in reaction mixtures from ozonation [15,21,22] and photo-ozonation [23,24] of phenol and catechol.

The production of pyrogallol as a major intermediate at relatively short exposure time indicated that in photooxidation of catechol in aqueous solution by uv/H_2O_2, hydroxylation of the benzene ring occurred faster than the benzene-ring cleavage. That is

Photooxidation of aqueous hydroquinone/H_2O_2 mixture was carried out under similar conditions. An aqueous solution of 0.0051 M hydroquinone and 0.033 M H_2O_2 was exposed to the 200-w Hg lamp/Vycor filter for 22 minutes. The identified intermediates besides unreacted hydroquinone in this reaction mixture are also given in Table 2. It is surprising that in 22 minutes of reaction time only very small amounts of 1,2,4-benzenetriol were produced. The major intermediates in the reaction mixture were hydroxy-maleic acid semialdehyde [2(or 3)-hydroxy-4-oxo-2-butenoic acid] [2] and unknown I. Other intermediates found in fairly large amounts were α-ketomalonic acid, maleic acid, mesoxalic acid, [(CHO)O(COOH)], and oxalic acid. The unexpectedly low yield of 1,2,4-benzenetriol and the high yields of lower-molecular-weight carboxylic acids and aldehydes presumably indicate that the degradation mechanism of hydroquinone is similar to the photooxidation of 2,4-dinitrophenol, which we reported in Ref. 2. We suggested there that photooxidation of 2,4-dinitrophenol by uv/H_2O_2 undergoes two possible reaction routes: (A) hydroxylation and benzene-ring cleavage of the phenolic intermediates, both occurring very rapidly (in other words, the formation and consumption of the formed polyhydroxybenzene intermediates counterbalanced by each other) or (B) benzene-ring cleavage occurring prior to hydroxylation. Further study proved that pathway A is the predominant route, which will be discussed later in this report.

Photooxidation of Pyrogallol and 1,2,4-Benzenetriol. For further proof that benzene-ring cleavages occurred after hydroxylation of catechol and hydroquinone, photooxidation of pyrogallol and 1,2,4-benzenetriol by uv/H_2O_2 were performed separately. Aqueous solutions of 0.0062 M pyrogallol/0.040 M H_2O_2 and 0.0050 M pyrogallol/0.13 M H_2O_2 were exposed to the 200-w/Vycor and 450-w Hg lamp without filter for about 20 and 30 minutes, respectively. Besides the large amount of unreacted pyrogallol, comparatively large amounts of [(CHO)O(COOH)] were produced as the major reaction intermediate in both cases. Not a trace of muconic acid derivatives [acid or aldehyde (the supposed products after benzene-ring cleavage)] could be detected. Other identified intermediates are given in Table 2. This finding verified that after benzene-ring cleavage, the muconic acid and its derivatives produced were further degraded rapidly through many competing reactions to smaller molecules. The fairly high yield of [(CHO)O(COOH)] may provide evidence that oxidative rearrangement of glyoxylic acid, i.e., reaction 2 was predominant. In order to confirm this assumption, trans, trans-muconic acid (commercially available) and maleic acid were photooxidized with uv/H_2O_2 later.

When an aqueous solution of 0.0032 M 1,2,4-benzenetriol and 0.021 M H_2O_2 was exposed to the 450-w Hg lamp for 15 minutes, it was very astonishing that not only did all the 1,2,4-benzenetriol decompose, but all the other intermediates, except mesoxalic acid semialdehyde [2] had also disappeared. This suggested that the degradation of 1,2,4-benzenetriol was indeed fast. This also explains why on the photolysis of aqueous hydroquinone/H_2O_2 that only very small amounts of the supposed major intermediate, i.e., 1,2,4-benzenetriol, can be found in the 22-minute-exposed reaction mixture. This also supports the hypothesis as suggested before that hydroxylation of hydroquinone and the benzene-ring cleavage both occurred rather expediently.

Photooxidation of trans,trans-Muconic Acid and Maleic Acid. Aqueous solutions of 0.0031 M trans,trans-muconic acid/0.040 M H_2O_2 and 0.0058 M maleic acid/0.075 M H_2O_2 were exposed to the 450-w Hg lamp without filter, each for 20 minutes respectively. In both reaction mixtures, large amounts of [(CHO)O(COOH)] were found as the major intermediate. This finding is in agreement with the results of photooxidation of pyrogallol. Other intermediates identified, which were in relatively small amounts, are also given in Table 2. Glyoxylic acid, glyoxal, and formic acid were again not determined.

The presence of [(CHO)O(COOH)] as the major intermediate, so far as we know, was not reported in similar studies of degradation of either trans,trans-muconic acid [22] or maleic acid [22,43,44]. Further photooxidation of acids and aldehydes whose molecular weight are lower than maleic acid, such as glyoxylic acid, glyoxal, oxalic acid, and formic acid, in aqueous solutions by uv/H_2O_2 were not performed in this study, but there is considerable information in the literature. For example, glyoxylic acid and glyoxal can be decomposed to formic acid, CO_2, and H_2O through Baeyer-Villiger oxidation by H_2O_2 [22,42]. Photooxidation of aqueous glyoxal solution with uv/O_3, produced oxalic acid and CO_2 [43]. On the other hand, oxidation of oxalic acid and formic acid in water by ozone [22] and formic acid in water by uv/H_2O_2 [45] produced CO_2 and water as the end products. It is assumed that similar degradation sequences will occur through photooxidation by uv/H_2O_2.

Based on the experiments performed in this study, photooxidation of phenol in aqueous solution by uv/H_2O_2 to CO_2 and H_2O can be summarized as follows:

$$\text{Phenol} \xrightarrow[\text{uv}]{H_2O_2} \left.\begin{array}{l}\text{catechol}\\\text{hydroquinone}\end{array}\right\} \longrightarrow \left.\begin{array}{l}\text{pyrogallol}\\\text{1,2,4-benzenetriol}\end{array}\right\} \longrightarrow$$

$$\text{muconic acid derivatives} \longrightarrow \left.\begin{array}{l}C_2 - C_4 \text{ carboxylic acids and}\\\text{aldehydes}\end{array}\right\} \longrightarrow$$

$$C_1 - C_3 \text{ carboxylic acids and aldehydes} \longrightarrow CO_2 + H_2O$$

ACKNOWLEDGMENTS

This research was sponsored by the U. S. Army Toxic and Hazardous Materials Agency under Interagency Agreement 40-1327-83, Army No. D-3-78-15. Oak Ridge National Laboratory is operated by Martin Marietta Energy Systems, Inc., for the U. S. Department of Energy under contract DE-AC05-84OR21400.

REFERENCES

1. Kaiser, K. L. E., Dixon, D. G. and Hodson, P. V., in QSAR in Environmental Toxicology, ed. K. L. E. Kaiser and D. Reidel, Dordrecht, Holland, 1984, pp. 189-206.
2. Ho, P. C., J. Environ. Sci. Technol., 1986, 20, 260-267.
3. Lea, D. E., Trans. Faraday Soc., 1949, 45, 81-84.
4. Volman, D. H. and Chen, J. C., J. Am. Chem. Soc., 1959, 81, 4141-4144.
5. Hunt, J. P. and Taube, H., J. Am. Chem. Soc., 1952, 74, 5999-6002.
6. Weeks, J. L. and Matheson, M. S., J. Am. Chem. Soc., 1956, 78, 1273-1278.
7. Baxendale, J. H. and Wilson, J. A., Trans. Faraday Soc., 1957, 53, 344-356.
8. Dainton, F. S., J. Am. Chem. Soc., 1956, 78, 1278-1279.
9. Stein, G. and Weiss, J., J. Chem. Soc., 1951, 3265-3274.
10. Brodskii, A. I. and Vysotskaya, N. A., J. Gen. Chem. USSR, 1962, 32, 2241-2245.
11. Eisenhauer, H. R., J. Water Pollut. Contr. Fed., 1964, 36, 1116-1128.
12. Shevchuk, L. G. and Vysotskaya, N. A., Zhur. Organ. Khim., 1968, 4(11), 1936-1939.

13. Jefcoate, C. R. E. and Norman, R. O. C., J. Chem. Soc. (B), 1968, 48-53.

14. Günter, K., Filby, W. G. and Eiben, K., Tetrahedron Letter, 1971, 3, 251-254.

15. Eisenhauer, H. R., J. Water Pollut. Contr. Fed., 1968, 40, 200-208.

16. Bischoff, Ch., Fortschr. Wasserchem. Ihrer Grenzgeb., 1968, 9, 121-130.

17. Bauch, H., Burchard, H. and Arsovic, H. M., Gesundh.-Ing., 1970, 91(9), 258-262.

18. Eisenhauer, H. R., J. Water Pollut. Contr. Fed., 1971, 43, 1887-1899.

19. Skarlatos, Y., Barker, R. C. and Haller, G. L., J. Phys. Chem., 1975, 79, 2587-2592.

20. Gould, J. P. and Weber, W. J., Jr., J. Water Pollut. Contr. Fed., 1976, 48, 47-60.

21. Dore, M., Langlais, B. and Legube, B., Water Res., 1978, 12, 413-425.

22. Yamamoto, Y., Nike, E., Shiokawa, H. and Kamiya, Y., J. Org. Chem., 1979, 44, 2137-2142.

23. Wako, H., Nippon Kagaku Kaishi, 1976, 10, 1530-1534.

24. Otake, T., Tone, S., Kono, K. and Nakao, K., J. Chem. Eng. Japan, 1979, 12, 289-295.

25. Omura, K. and Matsuura, T., Tetrahedron, 1968, 24, 3475-3487.

26. Massie, S. N., U. S. Patent 3,600,287, 1971.

27. Mansour, M., Parlar, H. and Korte, F., Stud. Environ. Sci., 1984, 23, 457-461.

28. Mansour, M., Bull. Environ. Contam. Toxicol., 1985, 34, 89-95.

29. Calvert, J. C. and Pitts, J. N., Jr., Photochemistry, Wiley, New York, 1966.

30. Anbar, M., Meyerstein, D. and Neta, P., J. Phys. Chem., 1966, 70, 2660-2662.

31. Symons, M. C. R., in Peroxide Reaction Mechanisms, ed. J. O. Edwards, Interscience, New York-London, 1967.

32. Joschek, H.-I. and Miller, S. I., J. Am. Chem. Soc., 1966, 88, 3273-3281.

33. Audureau, J., Filiol, C., Boule, P. and Lemaire, J., J. Chim. Phys. Phys.-Chim. Biol., 1976, 73(6), 613-620.

34. Diekman, J., Thomson, J. B. and Djerassi, C., J. Org. Chem., 1968, 33, 2271-2284.

35. Draffan, G. H., Stillwell, R. N. and McCloskey, J. A., Org. Mass Spectrom., 1968, 1, 669-685.

36. Jefcoate, C. R. E., Lindsaysmith, J. R. and Norman, R. O. C., J. Chem. Soc., 1969, 1013-1018.

37. Günter, K., Filby, W. G. and Eiben, K., Tetrahedron Letters, 1971, 3, 251-254.

38. Vysotskaya, N. A. and Shevchuk, L. G., Zhur. Organ. Khim., 1973, 9(10), 2080-2083.

39. Walling, C. and Johnson, R. A., J. Am. Chem. Soc., 1975, 97, 363-367.

40. Eberhart, M. K., Martinez, G. A., Rivera, J. I. and Fuentes-Aponte, A., J. Am. Chem. Soc., 1982, 104, 7069-7073.

41. Storesund, H. J. and Bernatek, E., Acta. Chem. Scand., 1970, 24(9), 3237-3242.

42. Bunton, C. A., in Peroxide Reaction Mechanisms, ed. J. O. Edwards, Interscience, New York-London, 1962.

43. Leitis, E., Investigation into the chemistry of the UV-ozone
    purification process, Annual Report, Westgate Research Lab, Los
    Angeles, Calif., 1979, U. S. Dept. of Commerce PB-296485.
44. Gilbert, E. Z., Naturforsch. B: Anorg. Chem. Org., 1977, 32B,
    1308-1313.
45. Ogata, Y., Tomizawa, K. and Fuji, K., Bull. Chem. Soc. Japan, 1978,
    51(9), 2628-2633.
46. Loeff, I. and Stein, G., J. Chem. Soc., 1963, 2623-2633.

THE REMOVAL OF HYDROGEN CHLORIDE FROM
HOT GASES USING CALCINED LIMESTONE

J.K. Walters and M. Daoudi
Department of Chemical Engineering,
University of Nottingham,
Nottingham NG7 2RD,
U.K.

ABSTRACT

The reaction of hydrogen chloride gas with calcined A.R. grade limestone was carried out in an automatic thermorecording balance (ATB). Initial reaction rates were used to interpret the results obtained. In the absence of both external mass transfer and product layer diffusion, the reaction was found to be first order with respect to the gas phase for concentrations ranging from 0.5 to 5 mole %. The reaction rate constant was found to obey the classical Arrhenius equation with an activation energy of 5000 cal/mole in the temperature range of 500-650°C.

INTRODUCTION

Incineration is now considered to be a good alternative disposal method for toxic and hazardous chemical waste. However, some major design features need to be improved for successful operation. Among these is concern for the removal of air pollutants such as hydrogen chloride (HCl), sulphur oxides (SO_x), nitrogen oxides (NO_x) as well as other contaminants. Hydrogen chloride gas is a most troublesome material not only because of environmental and health hazards associated with air pollution but also because of corrosion of equipment. Hydrogen chloride gas emitted from incinerators has been considered to result mainly from the destruction of toxic halogenated organic wastes, such as pesticides, various chlorinated hydrocarbons such as polyvinyl chloride (PVC) contained in solid refuse and waste polychlorinated biphenyls (PCBs) from industrial use. However HCl has also been reported even when plastic-free solid wastes have been incinerated. Hiraoka [1] also reported that 84% of the total chlorine in refuse was combustible chlorine, defined as chlorine transferred to the gas phase following incineration at 700 to 800°C. When paper and garbage were burnt 50% of the total chlorine in the flue gas was reported to result from the incineration of plastics. Table 1 shows the chloride content of domestic waste reported in a recent British survey (1986) [2]. Such experimental findings have led to increasingly tougher measures from environmentalists to limit prescribed concentrations of HCl in flue gases. Consequently industry is having to consider ways of reducing these levels to acceptable values.

TABLE 1

Soluble and insoluble chloride contents of (air dried) domestic waste. The Chemical Engineer, July/August 1986, p 4.

Category	% of total waste	% soluble chloride	% insoluble chloride	soluble chloride % of total waste	insoluble chloride % of total waste
Paper & card	30.55	0.25	0.115	0.046	0.035
Plastic film	6.5	0.26	0.385	0.017	0.025
Dense plastic	3.1	0.165	5.145	0.005	0.159
Textiles	5.85	0.095	0.24	0.006	0.01
Combustibles	3.4	0.18	1.48	0.006	0.05
Non combustibles	8.75	0.04	0.035	0.003	0.003
Putrescible	20.6	0.65	0.11	0.134	0.023
Metal	5.75	-	-	-	-
Batteries	0.25	-	-	-	-
Unclassified	15.25	0.365	0.075	0.056	0.011
TOTAL	100	-	-	0.273	0.32

METHODS OF HYDROGEN CHLORIDE REMOVAL

The processes that can be considered for HCl gas removal from flue gases may be broadly classified into "wet" and "dry" processes. The wet methods employ a slurry or solution of some suitable absorbent, whereas the dry systems use dry particles of absorbent in such gas-solid contacting devices as fluidized beds, packed beds or entrainment reactors. Kyte and Bettelheim [3] studied in a laboratory experiment the possibility of removing selectively HCl gas in a $SO_2/HCl/N_2$ simulated flue gas. Their experimental findings confirmed that the controlled neutralisation of HCl gas in contact with lime slurry could be used as a pre-scrubbing stage for a possible modification of the regenerative Wellman-Lord FGD wet process installed on a typical coal fired 2000 MW power station. Unlike wet processes, dry processes have the advantages of avoiding scaling and plugging of slurry pipelines, and corrosion and effluent disposal problems which would result in high maintenance costs and low plant availabilities. Furthermore dry processes do not require cooling of the hot gases for scrubbing to take place. However both types of process use common scrubbing agents, and limestone and its derivatives as absorbents for flue scrubbing agents, and limestone and its derivatives as absorbents for flue gas decontamination have been proved to be very efficient and their low cost favours greatly their use in such processes.

The purpose of this research investigation is to elucidate experimentally the kinetics of the removal of hydrogen chloride from simulated flue gases using calcined limestone.

APPARATUS

The conversion of calcium oxide (CaO) to calcium chloride ($CaCl_2$) produces an increase in weight which is directly proportional to the overall fractional conversion of the solid reactant. Thus a convenient method for following the progress of the reaction is to use a thermogravimetric balance which continuously measures and records the change in weight of the sample under specified reaction conditions. A Stanton automatic thermobalance (ATB) has been used for this purpose, and a schematic representation is shown in Figure 1. Further details on the (ATB) may be found in the Stanton instruction manual.

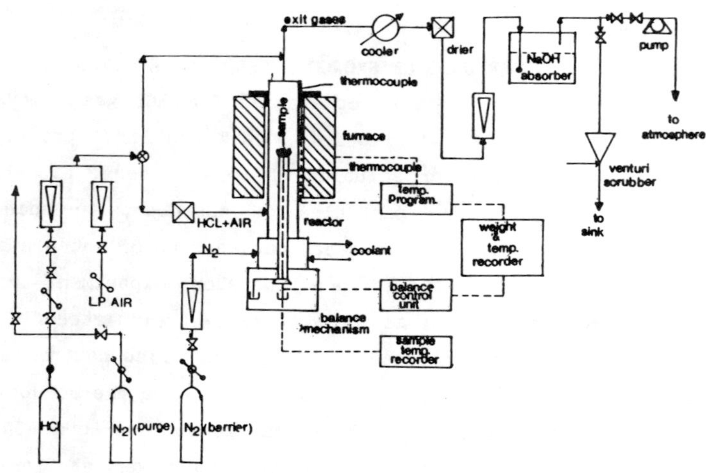

Figure 1. Apparatus

The balance was modified so that a small flow of corrosive gas (HCl) could be passed safely over the solid sample. This was achieved by adding a mullite reaction chamber which fitted over the sample and ceramic support rod and inside the furnace. HCl gas from a regulated cylinder and low pressure (LP) air were monitored through calibrated rotameters equipped with needle valves for flow control, premixed and allowed to enter via a side inlet of the reaction tube. Dried nitrogen also from a regulated cylinder was metered, controlled and allowed to enter the system just above the balance mechanism, thus serving the dual purpose of protecting the balance from corrosion (as a barrier gas) and diluting the HCl/air mixture. The resulting gas mixture HCl-air-N_2 flowed upwards past a ceramic sample pan containing the solid reactant with which the gas reacted. The sample temperature was monitored by a 13% Rh-Rh/Pt thermocouple fitted into the ceramic support rod and placed just below the sample pan. The thermocouple output was fed to a servoscribe recorder which continuously recorded the sample temperature during reaction. Furnace wall temperature was measured with another 13% Rh-Rh/Pt thermocouple and used with a controller and temperature programmer to control the furnace. Figure 1 gives a schematic representation of the whole apparatus.

MATERIALS AND PRETREATMENT

Calcium carbonate ($CaCO_3$) AR grade powder was used in all runs so as to avoid complications due to competing reactions with impurities. The powder was sieved using conventional woven-wire sieves and the batch retained for reaction studies was that consisting of grains that passed through the 350 mesh sieve (i.e. <45 μm).

OPERATING PROCEDURE

Calibration of the balance was first carried out before each run to ensure accuracy. Each sample under study was placed in a ceramic crucible and weighed on an ordinary laboratory balance and then transferred to the A.T.B. to record the starting weight. The sample was then calcined at around 750°C in the presence of nitrogen while the CO_2-N_2 mixture was being continuously removed by suction. In most cases complete calcination was achieved with the remaining $CaCO_3$ content less than 0.5%. The resulting product (CaO) was then brought to the desired temperature for the hydrochlorination to take place isothermally. While still maintaining the desired suction and inlet barrier gas flowrate the reactive gas mixture (HCl/Air) were set to the required flowrate and allowed to stabilise while being bypassed. After thermal equilibrium was achieved, the gas mixture was switched to the reactor and registered a step change

578

on the balance recorder. The correct initial time was obvious from the weight gain produced on the recorder trace. A typical thermobalance response curve is shown on Figure 2.

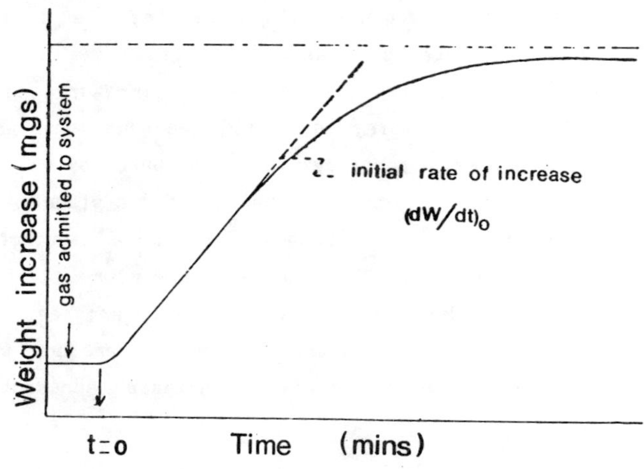

Figure 2. Weight response curve

The unreacted gases were removed at a pre-set flowrate by suction through a series of scrubbers containing NaOH solution with methyl red indicator to trace the passage of HCl gas through the system. As the reaction proceeded the weight increase of the sample reached a constant value (not necessarily complete conversion) as shown in Figure 2. At the end of the run the system was purged with high purity nitrogen before it was cooled down and the product removed for analysis. Among the methods used to analyse the end product ($CaCl_2$) were titration with $AgNO_3$ in the presence of K_2CrO_4 solution (5%) as indicator for quantitative measurements and a Jeol JSM 35 C scanning electron microscope (SEM) equipped with a microprobe connected to an energy dispersive X-ray detector (EDX) for product identification and surface topography. The reaction was carried out over the temperature range 500-650°C with hydrogen chloride concentrations in the range of 1 to 5 mole per cent. The specific surface area of the solid reactant was measured using nitrogen adsorption techniques and the BET method.

THEORETICAL CONSIDERATIONS

The reaction under study could be represented by the general stoichiometric equation

$$aG \text{ (gas)} + S \text{ (solid)} \rightarrow \text{products} \qquad (1)$$

where a is the stoichiometric coefficient for the gaseous species. In particular the reaction under study is:

$$2HCl + CaO \rightarrow H_2O + CaCl_2 \qquad (2)$$

The powder of solid reactant is visualised as consisting of a number of small particles or grains or as a "crushed" grain model as described by Szekely, Evans and Sohn [4]. The reaction occurs at the surface of each grain according to the unreacted core model of Wen [5]. Under this description and, with account for the reaction order with respect to gas phase, the intrinsic reaction rate expression can be written as follows.

$$R_G = a \, k_s \, C_s \, C_G^n \qquad (3)$$

Where R_G is the molar rate of reaction of gas G per unit surface area of solid S, k_s is the intrinsic rate constant per unit surface area, C_s is the initial molar concentration of solid S, C_G is the gas molar concentration at the surface of the solid, and n is the reaction order with respect to the gas phase.

If the analysis is based on the initial reaction rate, the initial surface area (BET surface area) can be considered as a basis for the analysis of the experimental data. Furthermore if equation (3) is used for kinetic evaluation then mass transfer resistances should be minimised. Product layer diffusion is considered unimportant in small grains and could be rendered negligible if only initial rates are to be considered. The mass transfer rate at which gas G diffuses to the solid is given by:

$$R_m = k_m \, (C_{GB} - C_G) \qquad (4)$$

where R_m is the molar rate at which gas reaches the solid per unit surface area of gas film, k_m is the mass transfer coefficient in the gas film and C_{GB} is the bulk gas molar concentration. A method of accounting for external mass transfer effects has been developed by Smith [6] for a first order reaction and is used in this study. The assumption of a first order rate was verified independently. Taking A_s as the surface area of the solid and A_m as the characteristic area for mass transfer we may equate the reaction rate to the mass transfer rate to get:

$$R_G \, A_s \; = \; R_m \, A_m \eqno(5)$$

Substitution of R_G and R_m from equation (3) and (4) and taking n = 1 then leads to an expression for C_G in terms of C_{GB} which may be put back into equation (3) to give:

$$R_G \; = \; k' \, C_{GB} \eqno(6)$$

$$\text{where} \quad \frac{1}{k'} \; = \; \frac{1}{ak_s C_s} \; + \; \frac{A_s}{A_m k_m} \eqno(7)$$

By examining equation (7) it can be seen that for a given pan geometry (A_m) and a given k_s the global reaction rate equation (6) approaches the intrinsic rate equation (3) as the ratio (A_s/k_m) approaches zero. Most correlations show that k_m increases as the gas velocity past the sample increases. Therefore the term (A_s/k_m) could be reduced either by increasing the gas flowrate or decreasing the available surface area for reaction (initial weight of sample). Mass transfer effects may also be minimised by operating the system at a lower temperature; since mass transfer involves gaseous diffusion through a boundary layer, which is not an activated process, while chemical reaction between the gas and the solid has a substantial activation energy.

RESULTS AND DISCUSSION

Experimentally the initial reaction rate was calculated from the slope of the thermobalance curve (Figure 2) since this is proportional to the conversion of solid reactant to solid product. This is expressed as follows:

581

$$(R_G)_o = \frac{-a\,(dW/dt)_o}{A_s\,(56 - 111)} \qquad (8)$$

where $(R_G)_o$ is the initial reaction rate, $(dW/dt)_o$ is the initial rate of change of weight, A_s is the total surface area of the solid reactant and 56 and 111 are the relative molecular masses of the solid reactant (CaO) and solid product ($CaCl_2$) respectively. The stoichiometry of reaction (2) was verified by X-ray analysis and from the maximum weight gain corresponded to full conversion of CaO to $CaCl_2$. Reaction (2) was the only reaction consistent with the weight change.

Measured initial rates obtained from equation (8) were used to calculate the apparent first order rate constant k' in equation (6) for constant gas phase concentration C_{GB} but various ratios of solid sample surface area to standard gas flowrate at a number of temperatures. Plots of the results obtained at 530°C, 570°C and 650°C are shown on Figure 3.

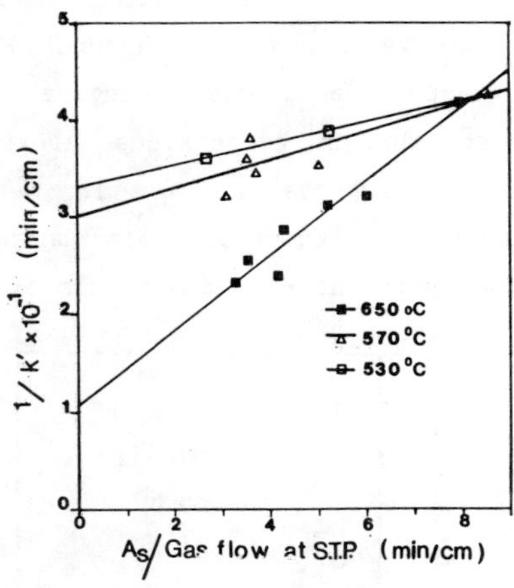

Figure 3.   External mass transfer effects

From equation (7) we see that k' approaches ak_sC_s as the ratio (A_s/k_m) approaches zero, a point where external mass transfer resistance becomes negligible. For small ratios (i.e. minimum mass transfer effects), initial rate data obtained from equation (8) were used in equation (3) to examine the reaction order with respect to the gas phase. Figure 4 shows a plot of the results obtained at 530°C and 650°C. From this plot, the reaction was found to be first order ($n = 0.975$) with respect to HCl concentration.

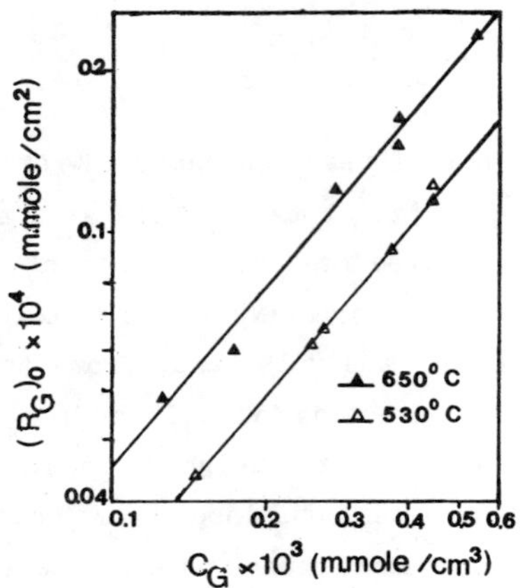

Figure 4. Reaction order with respect to gas phase

The rate constant determined from equation (6) over the temperature range was found to match the Arrhenius expression. A plot of the results obtained is shown on Figure 5. The activation energy computed from the slope of the best fit line was found to be around 5000 cal/mole which was in the range of other gas-solid activation energies. However it should be pointed out at this stage that at temperatures above 600°C mass transfer effects become increasingly important. This is shown by the use of k_s, the intercept from Figure 3, on Figure 5. This considerably increases the activation energy of the system at high temperatures.

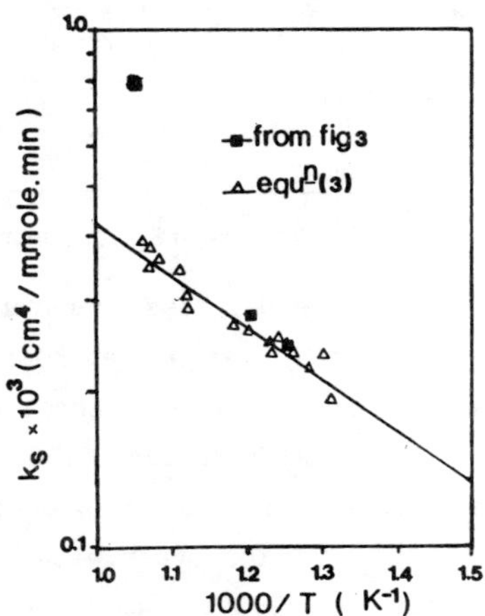

Figure 5. Arrhenius plot

REFERENCES

1. Hiraoka, M., Paper presented at 12th Fall Meeting of the Society of Chemical Engineering, Japan, Okayama, Japan, 1978.

2. Katan, L.L., Incineration of waste, _The Chemical Engineer_, July/August 1986, No. 427, p 4.

3. Kyte, W.S., & Bettelheim, T., _Environ. Prog._ Aug. 1984, _3_, No. 3, p 183.

4. Szekely, J., Evans, J.W., Sohn, H.Y., "Gas-solid reactions", Academic Press, New York, 1976.

5. Wen, C.Y., _Industrial & Engineering Chemistry_, 1968, _60_, 9, p 34.

6. Smith, J.M., "Chemical Engineering Kinetics", McGraw-Hill, New York, 1970.

THE SELECTIVE REMOVAL OF NOx FROM OFF-GASES OF TOTAL ENERGY INSTALLATIONS IN AGRICULTURAL INDUSTRY AND FROM STACK GASES OF THE PROCESS INDUSTRY

C.M. van den Bleek, A.G. Montfoort, P.J. van den Berg
Department of Chemical Engineering
Delft University of Technology
Julianalaan 136, 2628 BL Delft
The Netherlands

ABSTRACT

Based on a theory, presented earlier by the authors, it is made clear that besides NH_3 other N-containing substances are able to perform a selective conversion of NOx to N_2. So experiments were performed using, among other species, urea and even N-containing wastes. In the presence of O_2 a complete conversion of NOx to N_2 and N_2O is possible at about 100 °C. A process is proposed to deNOx the off-gases of TE-installations as used in greenhouses, based on the NOx/urea system. The CO_2 in these gases can cheaply be recovered and used to stimulate plant growing.
It is elucidated that the mechanism is actually based on the nitrosation/ nitration property of NOx, first resulting in intermediate complexes which, in the case of N-containing agents, finally dissociate into N_2 and N_2O. Extending the same mechanistic principles to species not containing nitrogen, research was done on SO_2. Then the formed complex is $NOHSO_4$, a well known intermediate in process industry. A process is proposed to recover $NOHSO_4$ from combined NOx and SO_2 stack gases which can for instance be used in caprolactam production.

INTRODUCTION.

The acid rain problem finds its origin in the NO and NO_2 content (generally written as NOx) of stack gases from nitric acid plants and power stations as well as exhaust gases from engines. Up till now an emission of 200 $\mu g/dm^3$ has been accepted as rather safe for the ecosystem. However, the recently found damage of the woods makes a further decrease of that emission level unavoidable. The problem is tackled at two fronts. First processes are improved, which is by far the most attractive way. Examples of these are nitric acid processes proposed [1] or realized [2], based upon a better NO_2 absorption [3] and the realization of lower reaction temperatures in engines [4] and power stations [5], which avoids the NOx-formation from air. The second possibility is to remove the remaining NOx from off-gases by processes of which the reduction to nitrogen has the main interest. Because all off-gases also contain oxygen, a large part of research has been directed towards the reduction of NOx selectively. Only NH_3 and its derivates have proved to be capable of converting NOx to N_2 and N_2O without first consuming all the oxygen present. The selective behaviour of NH_3 versus the non-selectivity of the other reducing agents like methane, carbon monoxide etc. was elucidated by a hypothesis proposed by

Van den Bleek and Van den Berg [6]. The key to the hypothesis was the reaction :

$$2NO + O_2 \longrightarrow 2NO_2 \tag{1}$$

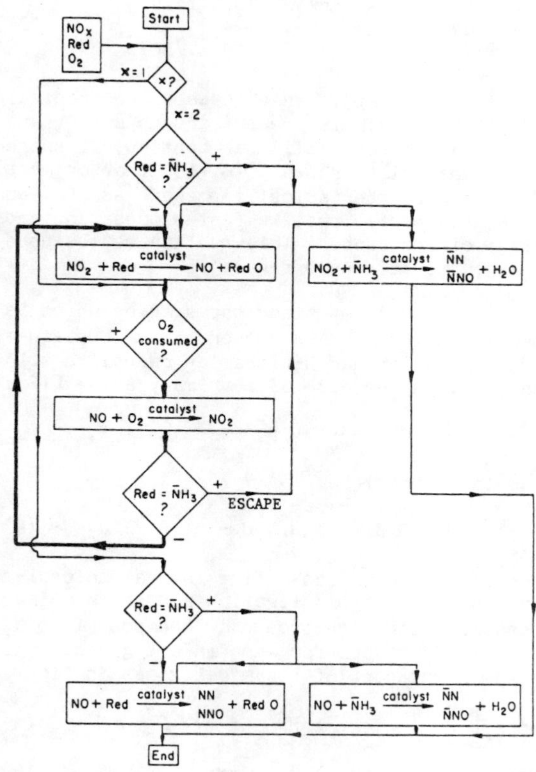

Figure 1. General scheme for the reduction of NO_x in the presence of O_2 [6]
N = nitrogen atom belonging to or originating from NO_x
N̄ = nitrogen atom belonging to or originating from NH_3
 = reductor consuming loop

This is catalyzed on nearly every surface. They showed that this reaction is responsible for the impossibility to reduce NO to N_2 and N_2O as long as oxygen is present. However, they demonstrated that in the case of NH_3 the very same reaction also provides the key to the solution of this problem by promoting a possible escape to nitrogen containing complexes which, more or less by coincidence, are converted to N_2 and N_2O. The hypothesis is represented as a model in Figure 1.

The route to escape the reductor consuming loop in fig.1 consists essentially of 'encouraging' NO_2 to take part in another reaction pattern, resulting into valuable, removable and/or harmless products, not necessarily being N_2 and N_2O. A further insight in this route will provide new possibilities to remove NOx selectively. It will be shown, that a proper choice of the reaction partner can result into N_2 and N_2O, but to more valuable products as well.

THE ESCAPE ROUTE.

The pictured escape route is in essence a nitrosation and/or nitration reaction, both well known and important in other area's of chemistry and chemical technology [7]. Nitrosation produces nitroso compounds (X-NO), while nitration results in nitro compounds (X-NO_2), where X can be carbon, sulfur, oxygen, nitrogen etc. The reactions can be very fast; in the liquid phase many diffusion controlled nitrosations and nitrations are known [8]. Also the gas phase nitrosation and nitration can be fast, often involving free radicals [9]. Nitrosation can take place by electrophilic attack of an ionic nitrosation agent. The order of reactivity for the most common species is :

$$NO^+ > ONCl > ONBr > ONSCN > ONNO_2 > ONOAc \gg ONOH$$

The nitration takes place by NO_2^+ ion attack. In the presence of oxygen these nitrosation and nitration agents can be formed from the nitrogen oxides, according to the following reactions [12] :

$$2NO + O_2 \longleftrightarrow N_2O_4 \longleftrightarrow O_2N^+...NO_2^-$$
$$N_2O_4 \longleftrightarrow 2NO_2$$
$$NO + NO_2 \longleftrightarrow N_2O_3 \longleftrightarrow ON^+...NO_2^- \tag{2}$$

N_2O_3 and N_2O_4 can dissociate, forming NO_2, which causes free radical nitrosation and nitration [8]. If no oxygen is present and thus N_2O_3 and N_2O_4 cannot be formed, the free radical process still can continu, if there are other radicals present. It was also shown [10,11], that on most catalytic surfaces, adsorbed NO is characteristically present as NO^+. So the nitrogen oxides NO and NO_2 can participate in fast nitrosation and nitration processes, under various circumstances. This quality may be used to selectively remove NOx from oxygen containing off-gas.

There are now two possibilities. In the first place the species which is nitrosated or nitrated (formerly called the reductor) may contain a nitrogen atom itself. These N-nitrosation and N-nitration reactions will lead to denitrification. For instance, in the case of a primary amine [12]:

$$R.NH_2 \longrightarrow R.NH.NO \longrightarrow R.N{:}N.OH \longrightarrow R.N_2^+ \longrightarrow R^+ + N_2 \tag{3}$$

For nitration a similar scheme would be possible (14) :

$$R.NH_2 \longrightarrow R.NH.NO_2 \longrightarrow R.N{:}N.O_2H \longrightarrow R.N_2O^+ \longrightarrow R^+ + N_2O \tag{4}$$

These reactions make the overall process **look** like a real selective reduction; de facto it is however a nitrosation/nitration and it is rather a coincidence, that the nitrosation/nitration products decompose into N_2 and N_2O. If R = H, then the reaction products for ammonia are given, just as already overall predicted by the right part of the model given in fig.1:

$$NH_3 \longrightarrow NH_2NO \longrightarrow HNNOH \longrightarrow HN_2^+ \longrightarrow H^+ + N_2 \tag{5}$$

$$NH_3 \longrightarrow NH_2NO_2 \longrightarrow HNNO_2H \longrightarrow HN_2O^+ \longrightarrow H^+ + N_2O \tag{6}$$

In fact nitramide H_2NNO_2 is a known compound, which decomposes to N_2O [15]; nitrosamide H_2NNO is an unstable compound which forms N_2 [16]. Recently [17] a mass spectroscopy study of the reaction between NH_3 and NO on vanadium oxide in the temperature range 300-400°C showed NH_2NO to be the major reaction product, while NH_2, NH and OH were also observed. HN_2 could not be found mass spectroscopically; this may also mean however, that its life time is less than 100 µs.

A second possibility may occur if the species which is nitrosated or nitrated does **not** contain a nitrogen atom itself. Without oxygen SO_2 will reduce NO_2 to NO and NO finally to N_2/N_2O at the same time producing SO_3. This fits perfectly well in the left part of the model demonstrated in Figure 1. If there is no oxygen present the reaction path will directly leave the reductor consuming loop to produce equal amounts of N_2/N_2O and SO_3. In the presence of oxygen, however, this loop cannot be left in that way before all oxygen is completely consumed by the key reaction (1), overall producing an excess of sulfur trioxide by the repeated reduction of NO_2 to NO. Because SO_2 can also be nitrosated and nitrated, the loop can be escaped as suggested, producing $NOSO_3$, $(NO)_2S_2O_7$ (and $NOHSO_4$ in the presence of water), which are solid, removable and useful products [13].

Based upon the nitrosation/nitration escape route presented, two possible kinds of processes can now be suggested to selectively remove NOx from off-gases:
- a process which uses nitrogen containing species, resulting in complexes which can dissociate in N_2 and/or N_2O.
- a process which uses SO_2 producing nitrosyl-compounds which are useful after separation.

Both possibilities were investigated rather extensively to check the escape route hypothesis. Research on the removal of NOx and SO_2 from the combined effluent gases of nitric and sulfuric acid plants was directed to the production of $NOHSO_4$ which is used in caprolactam production. On the denitrification route research was performed using supported urea and melamine, unsupported urea and even pelleted dry cow manure as possible and applicable N-containing substances. From the NOx/SO_2 experiments only the results will be presented and a process based upon these results is proposed; details on the research were published elsewhere [18]. The denitrification experiments will be presented into more detail, resulting in the concept of a process to selectively remove NOx from the off-gas of a gasmotor. Such a process may be used in the agriculture industry to utilize the carbon dioxide content of the off-gas from a Total Energy installation for plant growing purposes.

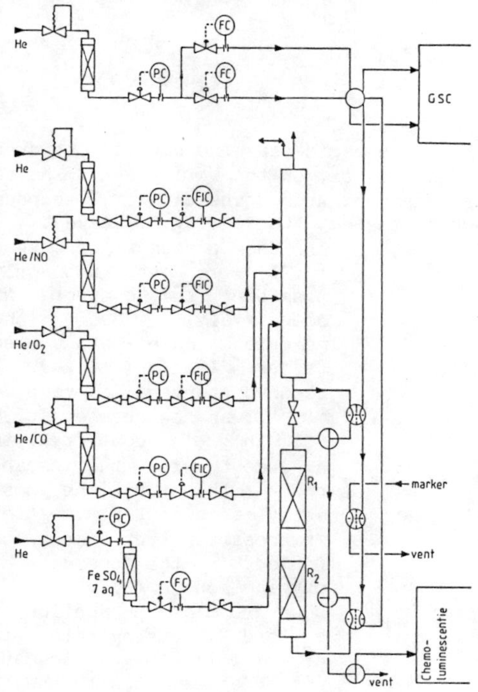

Figure 2. Experimental setup

EXPERIMENTAL SETUP.

For the experiments between NOx and the N-containing substances a 12.5 mm ID glass tubular reactor was used, consisting of two bed sections, which could be electrically heated independently (see figure 2). In the top section R_1 the N-containing component was situated, either impregnated (urea and melamine) on an alumina support (average particle diameter 0.72 mm) or as 2-5 mm particles (urea and cow manure). The bottom section R_2 contained a Pd/Al_2O_3 catalyst (typically at 210 °C) to decompose the nitrogen complexes formed.

The feed gas was made up of known flows of helium, helium/oxygen mixture (80/20 mol/mol) and helium/NO mixture (1.2 mol% NO). Helium and the helium/oxygen mixture were dried over molsieves 5A, helium/NO over supported phosphor pentoxide. Dosing of ethylene, carbon monoxide and water vapour (by passing a known flow of helium over a thermostated bed of

FeSO$_4$.7aq particles) was optional. A typical flow consisted of a mixture of 2.7 mol% O$_2$, 2500 ppm NO and balance helium; a typical space time with respect to the N-containing component was 8.2 seconds.

Feed and product gases were analyzed by gas chromatography (Perkin & Elmer Sigma 2 / Packard 433). CO, CO$_2$ and N$_2$O were separated on porapak QS, N$_2$ and O$_2$ on molsieves 5A. The NO and NO$_2$ content of both the feed and the product gas was measured by chemoluminescence (built and developed at the Delft University of Technology). The nitrogen content of urea and melamine on alumina and of the non-decomposed complexes formed, was analyzed by a Kjeldahl procedure on a Technicon Auto-analyzer II.

Urea on alumina (Ketjen Catalysts, crushed 001-1.5E alumina 0.65<d$_D$<0.80 mm) was prepared by wet impregnation in a rotating film evaporator. Batches of 10 and 20 wt% on alumina were prepared. Melamine on alumina was prepared in the same way, though only very dilute impregnation solutions could be used because of the low solubility of the melamine. Batches of 2.7 wt% on alumina were made for further use. Dried, pelleted cow manure was available, cylindrical in shape; it was crushed and particles in the range of 2-5 mm were used in the experiments without further treatment.

RESULTS AND DISCUSSION.

The Use of N-Containing Compounds.

Using only the first reactor (R1) experiments were carried out over urea (10 wt%) on alumina. In the absence of oxygen the well known S-shaped conversion vs temperature curve was measured. At a space time of 8 seconds a complete conversion of NO was reached at about 105 °C (Fig.3 ◆ curve).

In the presence of oxygen (2.7%) it was not only possible to convert NO selectively, but the reaction rate was even faster (Fig.4 ◆ curve). A hundred percent conversion of NO was now reached at 95 °C. This is completely in agreement with the escape route theory : the most active species for nitration /nitrosation can only be formed in the presence of oxygen (scheme (2)).

However, the production of N$_2$ and N$_2$O during both sets of experiments was low and not in agreement with the NO converted, as can be seen in Fig.3 and 4, (+curve) where

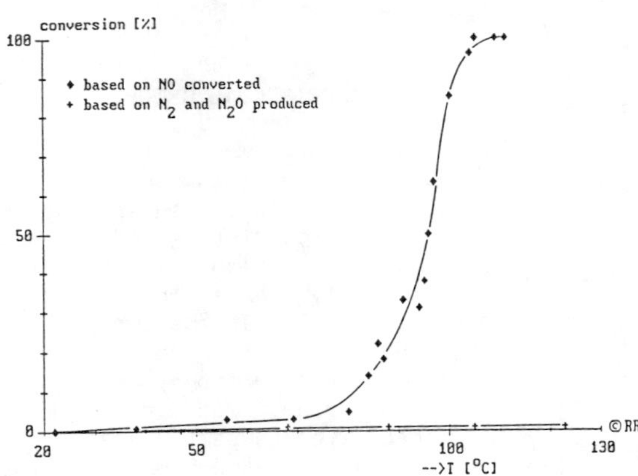

Figure 3. NO$_x$-conversion over urea without O$_2$ as a function of temperature (τ = 8 s)

the conversion based upon the amount of N$_2$ and N$_2$O produced is given. White depositions in the glass tubing after the reactor made us suppose that the nitration/nitrosation complexes were formed, but were only partly decomposed to N$_2$ and N$_2$O.

So the experiments were repeated with a second reactor (R2) installed, filled with inert alumina and operated at temperatures up to 400 °C. No increase in N_2 and N_2O production was observed. The inert alumina in the second reactor was replaced by Pd/Al_2O_3 (a well known NOx reduction off-gas catalyst). Operating R2 at 220 °C the complexes were now completely converted to N_2 and N_2O as is illustrated in Fig.5, where the N_2 production is given as a function of the temperature of the first reactor with and without Pd-catalyst installed.

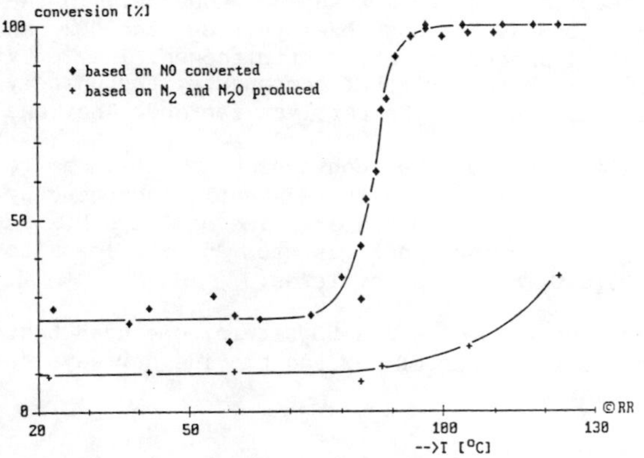

It is concluded that to perform a selective removal of NOx with urea a second reactor is necessary to decompose the complexes (easily) formed in the first one. Based on these results it was also concluded that at least one of the functions of the catalyst in the well known NOx removal process with NH_3 should be the decomposition of the complexes formed; it is not excluded that this will be the only function.

Though some other catalysts were tested and most promising results were obtained with a cheaper V_2O_5-catalyst, most of the experiments were performed with the Pd/Al_2O_3-catalyst installed in the second reactor and operated at 220 °C.

At temperatures in the first reactor higher than 110 °C some excess N_2 and N_2O was found due to the thermal decomposition of urea: supported urea can be decomposed at a lower temperature than the prilled one. However, the ammonia formed is oxidized in the second reactor.

Figure 4. NO_x-conversion over urea in the presence of O_2 as a function of temperature (τ = 8 s)

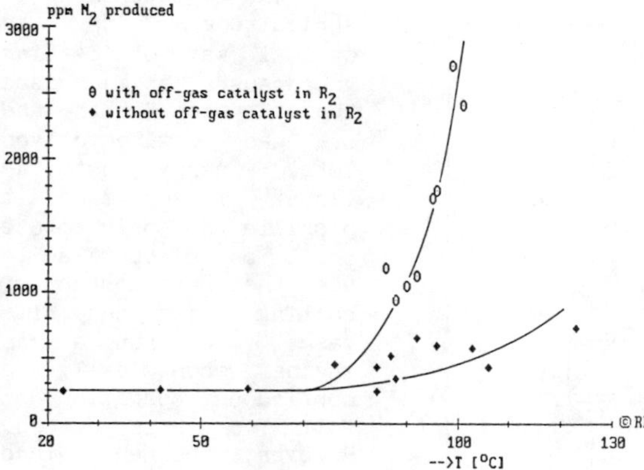

Figure 5. N_2 production over urea in the presence of O_2 as a function of temperature, with and without the Pd-catalyst installed

Some other N-containing species were tested for the escape route theory. Supported melamine (2.7 wt%) on alumina showed a similar S-shaped behaviour as urea,only at a higher temperature level. Complete conversion of NO was possible at about 260 °C, still using a 8 s space time. Even dried pelleted cow manure was tested. Though a selective removal of NOx is possible at very high space times, only 45 % conversion is reached at a space time of 8 s and a temperature of 140 °C. Higher temperatures were not allowed due to an awful smell!

The Use of Sulfur Dioxide.

With SO_2 solid complexes were formed in the presence of oxygen which proved to be $NOHSO_4$ and $(NO)_2S_2O_7$. No catalyst was used. This supports the suggestion as mentioned above that also in the NOx/NH_3-process the catalyst does not play a rôle in the reaction between NOx and NH_3, but only in the decomposition of the complexes formed. The complexes from SO_2 and NOx can of course not dissociate with simultaneous formation of nitrogen. They only dissociate at high temperatures into the original components. Fortunately these complexes exist as such, not only on the catalyst surface. They are solids and can be separated.
The formation of the complexes, hence the conversion of NOx, has to be carried out above 275 °C. The reaction velocity is greatly increased by pre-oxidizing SO_2 to SO_3. At 290 °C and a space time of 1.2 s NOx is converted to a degree of 80%. Cooling the gas to 45 °C leads to desublimation of $NOHSO_4$ and $(NO)_2S_2O_7$ as solid complexes.

Based on the results with N-containing species and SO_2 it is concluded that the use of the nitration/nitrosation power of NOx can provide new ways to selectively remove NOx from off-gases.

A PROCESS FOR THE USE OF OFF-GASES FROM TOTAL ENERGY INSTALLATIONS IN GREENHOUSES.

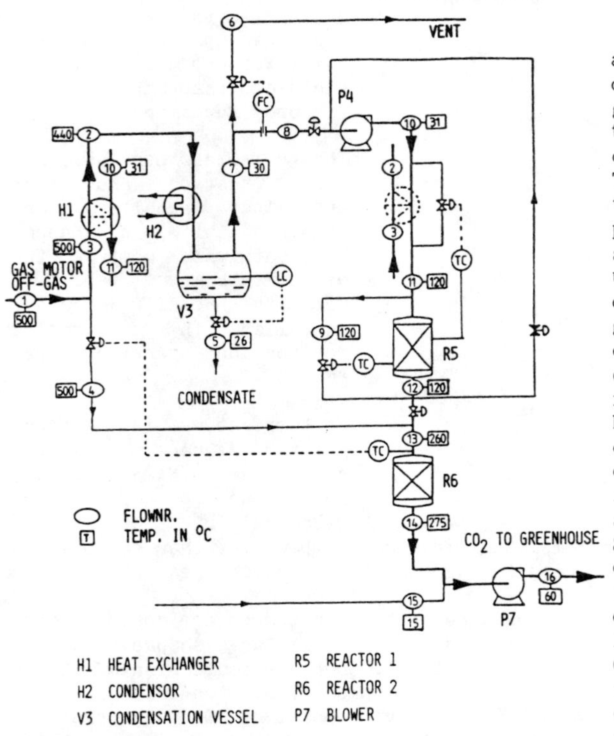

Figure 6. Process to utilize CO_2 from off-gases of gas engines

The interest in agriculture for an optimal way of heating greenhouses has increased the attention for the use of gas engine driven Total Energy installations. These make it possible not only to use shaft power but also to use the heat content of cooling water and off-gases. To run a gas engine economically a continuous operation at full load is needed. However, the heat demand of the greenhouse depends on the time of year and is strongly influenced by seasonal fluctuations. So, often a combination of a gas engine (operating continuously, delivering a basic heat load) and a gas-fired Central Heating installation (for maximum heat demand) is used. In those situations the CH-system is switched off for about four months a year.

For some years it is common practice to use carbon dioxide to stimulate plant growth in greenhouses. Instead of buying this separately, the CO_2 content of the off-gases can be applied to that purpose. However, the off-gases from the Central Heating installation have two disadvantages. First in an overall system as mentioned above it is switched off for about four months a year and secondly recent investigations [19] have made clear that even the low NOx concentrations resulting from the application of this off-gas may cause a latent damage (non-visible, but reducing growth) to the plants.

So the use of the CO_2-content of the gas engine off-gas would be a reasonable proposition if it would be possible to remove NOx (500 - 3500 ppm) and possible traces of CO and ethylene (both also harmful to plants). The use of the NOx/urea process as described above is now suggested and it was determined experimentally that, due to the overall oxidizing circumstances of this process, traces of CO and ethylene were almost completely converted to CO_2 in the second reactor. Instead of urea/Al_2O_3 (the alumina has to be reloaded for an economical operation) pure urea prills are suggested based upon preliminary experiments which show comparable results if operated at 120 -125 °C using water containing off-gases.

For a standard greenhouse of 10000 m^2 and an average height of 3.85 m a gas engine of 0.7 MW would take care of the basis heat load, the remainder being supplied by a 1.6 MW Central Heating system. Such a gas engine delivers 700 Nm^3/h off-gas with 8.2 % CO_2. The amount of NOx (700 ppm on an average) should be reduced to about 5 ppm. Figure 6 shows a process which still has to be tested, but is based on the laboratory experiments.

The off-gas at 500 °C is split into a flow which heats the first reactor (R5) and a secondary flow. In this reactor which operates at 120 °C NOx is converted. Together with the secondary flow the gas enters the second reactor (R6) where final conversion is reached at 265 °C on a V_2O_5-catalyst. A concentration of 4 ppm NOx is reached. With this set-up 1640 kg of urea per year is necessary.

A PROCESS TO COMBINE NOx-REMOVAL WITH SOx-REMOVAL FROM STACK GASES.

Figure 7 shows a concept for removing NOx in one stack gas simultaneously with removing the SOx- content from a second one.
The off-gas from a sulphuric acid plant with a capacity of 575 tons H_2SO_4 a day contains 0.2% SO_2 and 6.1% O_2. The flow is 34.5 Nm^3/sec. This gas is heated to 494 °C in heat exchangers. Then the SO_2 is converted in R_3 to SO_3 (conversion = 90%) on a V_2O_5 catalyst (12m^3). This gas is mixed with the off-gas from a nitric acid plant containing 0.1% NO, 0.1% NO_2, 3% O_2 and 1.3% H_2O. This flow is 19.1 Nm^3/s. The mixture has a temperature of 340 °C. In reactor R5 a homogeneous gasphase reaction occurs with a residence time of 1.5 s. $NOHSO_4$ is formed, together with some gaseous H_2SO_4. Cooling to 180 °C follows resulting in a H_2SO_4-mist in which $NOHSO_4$ is solved. A demister recovers the acid as a 67% solution of $NOHSO_4$ in 94% H_2SO_4. After cooling this is the final product which is an intermediate and can be used as such, e.g. in the production of caprolactam.

The final NOx concentration is neglegible. The SO_2-content has been reduced by 90 %. A more detailed description is published elsewhere [20].

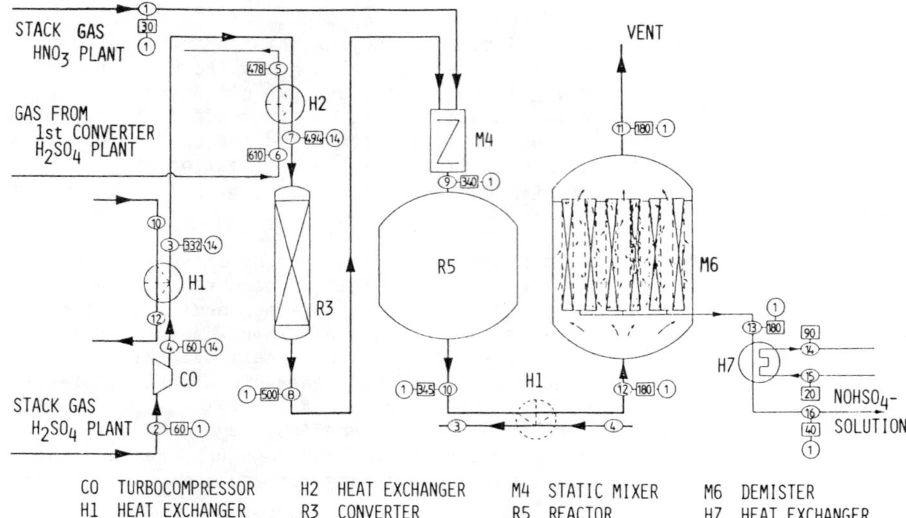

Figure 7. Process to combine NO_x-removal with SO_x-removal from stack-gases

CONCLUSIONS.

A better understanding of the selective removal of NOx from gases has been obtained. A model which originally was proposed to explain why first all of the óxygen must be converted before NOx is attacked, **unless** an escape route is made possible, has now been used to predict alternatives for a selective removal of NOx. This escape route is using the nitration/nitrosation power of NOx resulting in complexes. With nitrogen containing reagents (e.g. urea, melamine and even cow` manure) these complexes occur and finally dissociate into N_2 and N_2O. Other reagents are also possible, e.g. SO_2. Then the complexes can be removed and made useful.

Based upon this insight two technical processes are proposed : one using urea in Total Energy installations in greenhouses, making it possible to use CO_2 from gas engines to stimulate plant growth, the other to clean NOx- and SOx-containing off-gases simultaneously with a 67% $NOHSO_4$ solution in 94% H_2SO_4 as an attractive coproduct.

Finally it is made clear that a selective reduction of NOx does **not** exist. Even the so called selective reduction by NH_3 is actually only a nitrosation /nitration reaction resulting, more or less by coincidence in N_2 and N_2O. It is suggested that the rôle of the catalyst in this process is limited to the decomposition of the complexes formed.

ACKNOWLEDGEMENTS.

The authors like to thank P.E.M. van den Eijnden, P.E. Kleiborn, J.S. Hoornstra and M.A. Schwegler for their contributions during their MSc study in Chemical Engineering and A. Hexspoor, J.B. van Holst en A.A.M. Pruisken for their technical assistance.

REFERENCES.

1. NCI, 1986, (nr.22), 3 (editorial).

2. -Adrian, J.C., Vidon, B., Kuhlmann's New Nitric Acid Process, paper presented at the 170th Event of the European Federation of Chem.Eng., 1976, University of Salford.
   -NCI, 1985, (nr.22), 15 (editorial)

3. Lefers, J.B., De Boks, F.C., Van den Bleek, C.M., Van den Berg, P.J., Chem.Eng.Sci., 1980, 35, 145-53.

4. -Honda, 1973 Report by the Committee on Motor Vehicle Emissions, National Research Council, Washington DC.
   -NCI, 1985, (nr.20), 19 (editorial).

5. -Van der Sanden, A.M., Heidweiller, D.J., Poolman, P.J., Procestechn., 1985, (nr.9), 73
   -Flament, G., Phetan, W., Proc. US-Dutch Int.Symp., Maastricht, 24-28 May(1985), 603-621

6. Van den Bleek, C.M., Van den Berg, P.J., J.Chem.Tech.Biotechnol., 1980, 30, 467-475.
   A free copy without highly disturbing typing errors can be obtained from the authors.

7. Stedman, G., Adv.Inorg.Chem.Radiochem., 1979, 22, 113.

8. Lyn, D., Williams, H., Adv.Phys.Org.Chem., 1983, 19, 381.

9. Challis, B.C., Kyrtopoulos, S.A., J.Chem.Soc.Chem.Commun., 1976, 877.

10. Van den Bleek, C.M., Eek, J.W., Reedijk, J., Van den Berg, P.J., J.Chem.E.Symp.Series, 1979, 57, R1-R14.

11. Shelef, M., Kummer, J.T., Chem.Eng.Progr.Symp.Ser., 1971, 67(115), 74.

12. Ridd, J.H., Quart.Rev., 1961, 15, 418.

13. Stopperka, K., Wolf, F., Süss, G., Z.Anorg.Allg.Chem., 1968, 359, 14.

14. Fridman,A.L., Ioshin,V.P., Novihov,S.S., Russ.Chem.Rev., 1969, 38, 640.

15. Brønsted, J.N., Pedersen, K.J., Z.Phys.Chem., 1924, 108, 187, 216, 234.

16. Schwarz, R., Giese, H., Ber., 1934, 67, 1110, 1111, 1114.

17. Farber, M., Harrism, S.P., J.Phys.Chem., 1984, 88, 680.

18. Van den Eijnden, P.E.M., Delft Univ. of Technology, Dept of Chem.Eng., Internal Report 1980/12.

19. Wolting,H.G., Van Remortel,E.A.M.,Van Berkel,N., paper presented at the Symp.'Greenhouse Climate and its Control', May 19-24,1985, Wageningen.

20. Van den Eijnden, P.E.M., Delft Univ. of Technology, Dept of Chem.Eng., Internal Report 1981/9.

Wet oxidation of toxic sludges

J. Kálmán, Gy. Pálmai and I. Szebényi
Department of Chemical Technology
Technical University of Budapest

ABSTRACT

In the course of waste water treatment especially in case of industrial waste waters, concentrates containing chemicals and sludge-like hazardous wastes are produced in ever increasing quantities. They are disposed mainly by incineration after dewatering and drying. The disadvantage of this method is the high energy consumption necessary for the evaporation of the great amounts of water, that can be as high as 98 per cent.

In our paper the results of the experiments carried out for the disposal of sludge by means of wet oxidation are discussed. Sludge from the waste water treatment unit of a petrochemical plant containing hazardous organic compounds, ferric hydroxide, lime and excess sludge of a biological waste water treatment plant were tested in intermittent laboratory scale and continuous pilot-plant wet oxidation experiments. The optimum temperature and residence time were determined. Based on the results of the experiments an industrial scale unit was designed and the investment and operating costs were calculated. The paper proposes the application of the technology and discusses the conditions of the economic application of wet oxidation for the disposal of hazardous sludges.

KEYWORDS

Petrochemical industrial wastewater treatment, wet oxidation of toxic organic sludges, wet air regeneration of used powdered activated carbon.

INTRODUCTION

In the course of physico-chemical and biological treatment of industrial wastewaters a great amount of sludge is produced which can not be utilized for agricultural purposes because of its toxic heavy metal or organic content. These sludges are treated after thickening and dewatering by aerobic or anaerobic digestion, or they are disposed by oxidation /1/. Incineration is the wide-spread method of oxidative disposal of sludges, e. g. it is used in Hungary at Nitrokémia Industrial Works or abroad at BASF Co. or in the biggest municipal wastewater treatment plant of Tokyo. Recently, plants were built in several places for wet air oxidation of sludges in aqueous phase after thickening. This paper will review the preparations to build such a plant at one of the biggest chemical complexes in Hungary, at the Tisza Chemical Works /Tiszai Vegyi Kombinát/.

In the olefin plant of the works 900,000 tons of naphta are pyrolysed yearly to give ethylene, propylene and other products. From these feedstocks considerable amounts of polyethylene, polypropylene, other polymer products and different paints are manufactured. A quite great amount - 8000 m^3/d - of wastewater is accumulating from the different plants and administration buildings. This quantity of wastewater is treated at a central wastewater treating system.

The system works properly and the quality of the effluent fulfills legislative requirements /2/. The treated water, however, has a disagreeable odor even in the stabilization ponds following the secondary treatment system. The organic compounds causing the disagreeable taste and odor of the water often have carcinogenic effects even in small concentrations.

The presence of polycyclic aromatic hydrocarbons /PAHs/ in the waste water of the pyrolyse plant was probable, for this reason the change of PAH concentrations of wastewater was studied in the course of chemical, biological, activated carbon and stabilization pond treatment earlier /3/. It was stated that the PAHs are accumulated in sludges and they can be detected even in the internal organs of fishes living in the last unit of stabilization ponds, in the fish-pond /**Table 1**/.

A sludge amount of about 50 m^3/d originates from the primary settlers, from the treatment of paint-plant and olefin-plant wastewaters by lime and ferric chloride, respectively, and from excess sludges of biological treatment. In addition to these sludges, concentrates originate from settlers in smaller amounts because of rising and skimming of certain pollutants. Our aim is the joint treatment of them. The quantities and the main characteristics of the sludges originating from the wastewater treatment system are summarized in **Table 2** /3/. It can be stated that the COD and the dry matter content of the mixed sludge are about 16,000 mgO_2/l and 3 %, respectively.

TABLE 1

Polycyclic aromatic hydrocarbons in sludges and fish along the wastewater treatment system of Tisza Chemical Works

Sample	Fluoranthene	Indenopyrene	3,4-Benz--fluoranthene	Pyrene
	(μg/kg)	(μg/kg)	(μg/kg)	(μg/kg)
Activated sludge	16	8.8	11	0
Sludge at the end of the algal pond	2.2	2.2	0.2	0
Sludge at the end of the reed pond	5.5	0.3	0.3	0
Sludge at the end of the fish-pond	5.5	0.3	0.3	0
Fish, bowels	23	0.0	0.0	10
Fish, roe	11	0.4	0.2	0

TABLE 2

The quantities and main characteristics of the sludges originating from the wastewater treatment system of Tisza Chemical Works

Sludges from	Quantity	Percentage	COD	Cl^-	Dry matter at 105 ^{o}C	Ignition residue at 600 ^{o}C relating dry matter
	(m^3/week)	(%)	(mgO_2/l)	(mg/l)	(%)	(%)
Aerators	450	76	11,000	195	1.87	31.8
Longitudinal flow settler	100	17	37,000	248	7.97	69.3
Longitudinal flow settler, floating sludge	10	2	16,400	301	5.83	36.3
Dorr settler, floating sludge			20,400	212	3.39	34.1
Two-level settler	33	5	17,200	70.9	4.16	40.2
Quantities and weighted averages	593	100	15,878	199	3.07	38.6

EXPERIMENTAL EQUIPMENT AND METHOD

Laboratory-scale Batch Wet Air Oxidation Reactor

The scheme of the apparatus used for laboratory batch experiments is presented in **Figure 1.** The reactor is a LAM-202 type, 2 dm^3 stainless steel /18/8 CrNi/ autoclave equipped with a magnetic churning mixer. The reactor consists of two parts: reactor body and lid. The lid is sealed with a cone, and fixed to the lower section by screws. The mixer moving in a closed tube without mechanical sealing. Three needle valves with fine adjustment, a manometer and a safety valve with a rupture disc are screwed to the lid. The gas flows through one of the valves to the bottom of the reactor. The apparatus is equipped with an electrical heating jacket.

Regeneration was carried out as follows: 1-1,5 dm^3 excess sludge or an aqueous suspension of spent activated carbon was fed into the reactor, then the bomb was sealed. From a flask, air was pressed into the reactor that was heated to the reaction temperature required. The mixer was turned on and by adjustment of the valves controlling the flow of reactor influent and effluent gases, oxidation could be carried out in an appropriate air stream for the period of time required. In the course of the reaction, the amount and the composition of the outlet gas was determined. After completion of the reaction and appropriate cooling, the quality of the aqueous suspension and the activated carbon regenerated has been tested /4,5/.

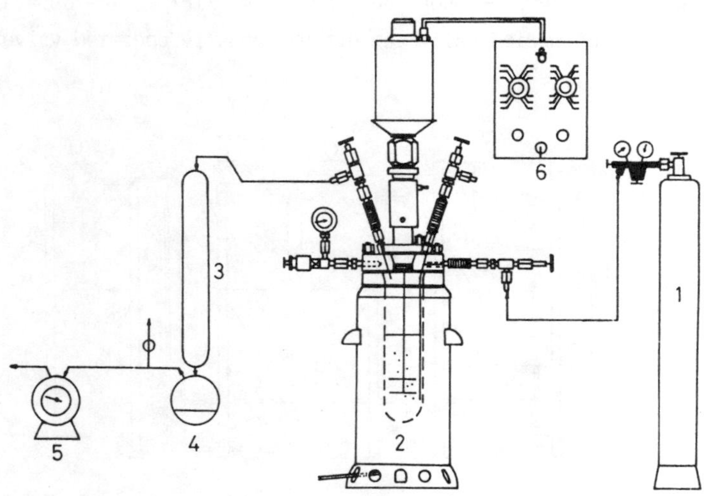

Figure 1. Laboratory-scale batch wet air oxidation system
1-Air cylinder, 2-Autoclave, 3-Cooler, 4-Trap, 5-Gas-meter, 6-Switch board

Pilot-plant Continuous Wet Air Oxidation Reactor

The schemes of the pilot-plant continuous activated sludge wastewater treatment and the wet air oxidation system are shown in **Figures 2 and 3.** The wastewater /volume flow of 100 to 1000 dm^3/h/ is led by pump /1/ into the top of activated sludge aeration tank /3/. The flow rate of water is measured by rotameter /2/. Air is led into the aeration tank by centrifugal blower /4/ through ceramic air distributors. The flow rate of air is controlled by rotameter /5/. The treated wastewater flows in overflow pipe /6/ from the aeration tank to settling tank /7/. The settled sludge is recirculated by mono pump /8/ into the aeration tank with a volume flow of 100 to 1000 dm^3/h and the excess sludge is periodically led into secondary settling tank /9/. The sludge is led by high-pressure pump /10/ from the secondary settling tank through heat exchanger /11/ and electrically heated preheater /13/ into wet oxidation reactor /12/. The reactor is supplied with electrical heating. The warmed and reacted sludge is recirculated from the reactor into the heat exchanger where its heat is utilized for heating of the inlet sludge then it is led through cooler /15/ into separator /16/. The gas phase is exhausted from the separator through membrane-type pressure control valve /17/. The liquid phase is run off by motor operated valve /19/ keeping the fluid level by level sensor and regulator /18/ in the separator. The sludge can also be periodically run off by manually operated valve.

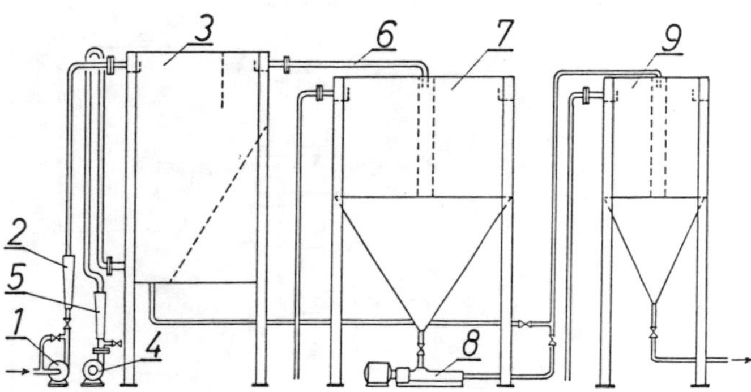

Figure 2. Pilot-plant continuous activated sludge wastewater treatment system

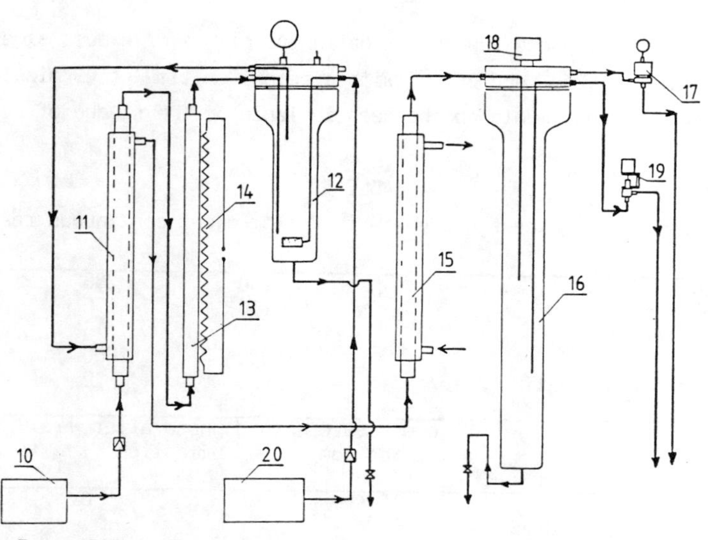

Figure 3. Pilot-plant continuous wet air oxidation system

Analytical Methods

The chemical oxygen demand of the samples was determined by the dichromate method. Other characteristics were determined according to the Hungarian standards.

EXPERIMENTAL RESULTS

The possibility of change-over to combined activated carbon-activated sludge wastewater treatment was taken into account in determation of suitable reaction circumstances of wet oxidation. For this reason the necessary minimum reaction temperature was determined at 40 minutes reaction time by using model compounds. On the basis of investigations on oxidation of glucose and acids formed by partial oxidation of it the reaction temperature of 240 to 250 $^{\circ}$C seems to be reasonable. The consumption of activated carbon is less than 5 % below 250 to 260 $^{\circ}$C which is still economically acceptable. The excess sludge oxidation versus reaction time function was investigated in this temperature range. It was stated that the necessary retention time is 30 minutes at a reaction temperature of 250 $^{\circ}$C for appropriate conversion.

In the continuous reactor oxidation reactions of excess sludge were carried out at 260 and 280 ^{0}C with a retention time of 20 minutes. The results of the experiments can be seen in **Table 3**. The sludge of

TABLE 3

Wet oxidation of excess sludge in the pilot-plant continuous reactor

Temperature /^{0}C/		260	280
Pressure /bar/		130	130
Fluid feed /l/h/		15	15
Air feed /l/h/		2000	2000
Calculated retention time /min/		20	20
	Sludge before wet oxidation	Fluid phase after wet oxidation	
Dry matter /g/l/	15.6	0	0
Ignition residue of dry matter /%/	37.5	0	0
COD /unfiltered/ /mg/l/	16,400	7200	6800
COD /filtered/ /mg/l/	900	6800	6400

dry matter content of treated sludge was 15.6 g/l which completely clarified after 20 minutes and at the same time the COD of the sludge reduced from a value of 16,400 mg/l to 7200 or 6800 mg/l. It can be seen that the COD of the untreated sludge originates practically from the high molecular weight water-insoluble sludge fraction /the dissolved COD is only 900 mg/l/. These sludge pieces are degraded and oxidized under the circumstances of wet oxidation in consequence of which partly carbon dioxide and water, partly water-soluble intermediates of low molecular weight are formed dependig on the circumstances of oxidation. The oxidation intermediates cause the remaining COD of the treated solution. Subsequent termical oxidation of intermediates to carbon dioxide and water is possible only by further increasing the temperature and the retention time.

On the basis of these experimental results the fundamental planning data of the pilot plant under construction are:

sludge flow	30 m^{3}/day
COD	17300 mgO$_2$/l
Cl$^-$ content	200 mg/l

dry matter	3 %
ignition residue	1 %
reaction temperature	260 °C
retention time	30 minutes.

Investment costs of the planned plant are 11 million Hungarian Forints. Estimated operating costs are 250 Ft/m³ of sludge. These costs are very close to the costs of other excess sludge treatment methods.

The wet oxidation technology is an alternative solution for excess sludge treatment even economically. Taking into account that at the change-over from activated sludge wastewater treatment to activated carbon-activated sludge wastewater treatment this system is suitable without any further investment and costs for regeneration of used activated carbon, this technology is considerably more economic than the before-mentioned other excess sludge treatment methods.

CONCLUSION

On the basis of laboratory and pilot-plant experiments for wet oxidation disposal of wastewater sludges as hazardous wastes it was stated that the organic components of sludges can be decomposed at a temperature of 260°C and at a pressure of 130 bars. Under these circumstances the used activated carbon can be regenerated and the carbon consumption is less than 5%. On the basis of investment and operating costs of the plant of this type the application of wet oxidation excess sludge treatment is feasible even economically.

ACKNOWLEDGEMENT

Thanks are due to R. Hajós and Z. Izsáki /analytical measurements/, L. Kovács /planning of pilot-plant/, I. Röhricht, G. Széplaki and L. Vörös /completion of pilot-plant/ for their valuable work.

REFERENCES

1. Hasit, Y. (1986). Sludge treatment, utilization and disposal. Journal WPCF, 58, No.6.
2. Kálmán, J. (1980). Polycyclic aromatic hydrocarbons in petrochemical

waste waters. <u>Presented at the Meeting of the Hungarian Hygienic Institute,</u> Budapest, Hungary.

3. Kálmán, J., Z. Izsáki, L. Kovács, L. Majerusz and I. Szebényi (1987). Upgrading activated sludge treatment using waste PAC of pharmaceutical factories. <u>Proceedings of the IAWPRC's "The Design and Operation of Large Wastewater Treatment Plants" International Conference,</u> Budapest, Hungary.

4. Donáth-Jobbágy, A., J. Kálmán and R. Hajós (1985). Intensification of activated sludge waste water treatment. <u>Proceedings of the IAWPRC's First Asian Conference on Treatment, Disposal and Management of Human Wastes,</u> Tokyo, Japan.

5. Kálmán, J. and I. Szebényi (1986). Waste water treatment with powdered carbon and wet oxidation regeneration. <u>Proceedings of the IAWPRC's 13th Biennal International Conference,</u> Rio de Janeiro, Brasil.

Chapter 6

THERMAL TREATMENT METHODS

HAZARDOUS WASTE TO ENERGY: A SUCCESSFUL CASE HISTORY

Thomas L. Rinker
General Manager
Waste Disposal Systems Group
Environmental Elements Corporation
Post Office Box 1318
Baltimore, Maryland 21203
USA

AN OVERVIEW

The creation of a new generation of waste management facilities in the United States has been a primary objective of federal and state environmental agencies since the passage of the nation's initial hazardous waste legislation in 1974, the Resource Conservation and Recovery Act (RCRA). The failure to achieve this objective was the principle reason for a significant strengthening of RCRA when it was reauthorized in 1984. Two of the most important concepts in the reauthorization legislation were the prohibition by statute of land disposal of organic chemical waste and the setting of stringent standards for the design and perpetual monitoring of land disposal facilities.

These provisions have been effective in achieving movement towards a new generation of waste management facilities as organic chemical producers and users throughout the United States now recognize that incineration is the only effective method to manage wastes in a manner that is environmentally appropriate and that ensures that there will be no lingering future liability. In short, the regulatory framework created by the passage and reauthorization of RCRA has produced a demand condition in the market for industrial waste incineration, particularly for commercial or "off-site" waste management companies.

In fact, this demand condition has resulted in a significant shortfall in adequate disposal capacity and has created a "sellers" market for off-site disposal capacity. In addition, the RCRA regulatory framework combined with heightened public concern over the potential adverse health and property value effects of hazardous waste management facilities has created an extremely difficult environment for the permitting and siting of new facilities. The results of this current capacity shortfall are the rapid escalation of off-site disposal prices (25 to 50 percent per year) and the allocation of capacity (miss the assigned slot and a generator has to store on-site the waste for several months). The difficulty in permitting new facilities insures that this condition will continue well into the future.

Relatively few generators were able to foresee this rapid change from a buyers to a sellers market and to move to put in place at the point of generation ("on-site") the incineration systems necessary to assure adequate capacity at well defined and controllable cost. Still fewer generators were able to foresee the benefit of aggregating waste from several manufacturing locations to provide a sufficient annual waste volume for economical recovery of the heat energy inherent in the waste.

In the United States, there is now one clear example of foresightedness in industrial waste management. That example is the Energy Recovery Unit of PPG Industries, Inc. located in Circleville, Ohio. This incineration based waste management facility is the first facility of its kind to be fully permitted under both federal and state RCRA regulations and fully sited under state facility siting regulations. The PPG Energy Recovery Unit (ERU) is now nearing completion of construction and the start of operation. This facility will destroy approximately 46,900 tons per year of waste, produce 42,000 pounds per hour of steam and avoid the use of 300,000 thousand cubic feet of natural gas per year for steam production in the host manufacturing plant.

THE NEED FOR THE FACILITY

The Coatings and Resins Group of PPG Industries, Inc. is a producer of a broad range of paints and coatings, with particular competence in the area of high performance automotive and speciality industrial coatings. Corporate and group management has consistently placed a high priority on quality of product and service and has used characteristics such as safety, housekeeping and environmental compliance as one set of measures of the total quality of the manufacturing operation.

With the passage of the initial RCRA legislation, the Coatings and Resins Group undertook in 1979 a detailed survey of the volume and characteristics of the waste being generated throughout the Group and a brief review of the applicability of alternative waste management strategies and disposal technologies. This initial study indicated the feasibility of a waste management strategy consisting of the following elements:

o   Further implementation of solvent recovery procedures and similar measures to reduce the volume of waste requiring disposal.

o   Aggregating all of the chemical wastes being generated at the twelve Group plants located in the United States and Canada.

o   Use of vehicles already making inter-plant hauls to transport the waste to a central point.

o   Siting of the incineration facility at a manufacturing location to provide a low cost source of process grade steam, to utilize existing plant infrastructure and to facilitate the community acceptance of the new facility.

o   Adding to the waste stream a variety of solid waste such as shipping bags and cartons to increase steam production to meet the host plant steam demand and enhance the cost effectiveness of the waste heat boiler.

o   Use of a rotary kiln based incineration system to allow for handling of the broadest possible spectrum of waste types and forms.

A second more detailed study by the Group in 1980 identified the most desirable plant site and provided more accurate economics and incineration system characteristics. The Circleville, Ohio manufacturing plant was chosen to host the incineration system based on the following considerations:

o   The Circleville plant is centrally located to the other plants in the Group, and manufactures intermediate products which are distributed to other plants for production of finished products. This characteristic allowed the use of established transport vehicles and procedures and provided a clear tie between the host plant and the plants which would send waste.

o   The Circleville plant generated the largest volume of bulk liquid waste.

o   The steam need of the Circleville plant closely matched the projected steam output of the incineration system and the construction schedule for the waste heat boiler matched the plant's intent to replace its aging natural gas fired steam production facility.

As a final step, the Group prepared a performance specification for the incineration facility, now christened the Energy Recovery Unit, and solicited proposals in 1981 from a variety of suppliers. The Environmental Elements Corporation proposal for a turnkey facility incorporating Von Roll, Ltd. technology was selected by the Group, and a contract was signed in May 1982. This selection was based in large measure on a desire by the Group to utilize well demonstrated, successfully applied technology.

THE WASTE STREAM

The waste stream processed at the ERU consists of a widely varying combination of solid, liquid and sludge wastes associated with the production of water based and solvent based paints and coatings and includes material such as: solvent reclamation sludges and residues; defective paints and resins; wash water and wastewater; wastewater treatment sludges; filter material; lab samples; cartridges and bags. The waste stream data used for the design as well as current projections are presented in Table I and II. Approximately sixty-five percent of the total waste stream comes from five facilities within 200 miles of the Circleville plant, with the Circleville plant and another facility within 60 miles contributing approximately fifty percent of the total waste stream.

TABLE I: WASTE PROFILE

Waste Component	Design Average	
	(pounds per hour - percent)	
Drums Containing Waste	2,865	22.6
Product In Cans	86	0.7
Baled Trash	937	7.4
Pumpable Waste		
Organic Bulk	1,885	14.8
Aqueous Bulk	4,621	36.5
Aqueous Liquid	2,277	18.0
Total	12,671	100.0

TABLE II: WASTE COMPOSITION

Waste Component	Design Average Composition
Carbon	26.9 percent
Hydrogen	3.5 percent
Oxygen	11.2 percent
Nitrogen	0.5 percent
Sulfur	0.1 percent
Chlorine	0.1 percent
Water	50.2 percent
Inerts	7.5 percent
High Heating Value	5,380 BTU per pound
Low Heating Value	4,519 BTU per pound
Total Heat Release	58.0 million BTU per hour

THE INCINERATION SYSTEM — SUMMARY

The rotary kiln based incineration system incorporated into the ERU consists of the following unit operations:

o   A waste feed system for pumpable waste and containerized and baled waste.

o   A rotary kiln complete with a front wall and designed for melting steel drums when in a slagging mode of operation.

o   A vertical secondary combustion chamber for conditioning the flue gas and completing the destruction process to an efficiency of 99.99 percent.

o   A waste heat recovery boiler for flue gas cooling and saturated steam production.

o   An electrostatic precipitator for particle removal with a 99.7 percent efficiency.

o   An induced draft fan to maintain a negative pressure in the combustion, boiler and precipitator components.

o   A packed bed scrubber for HCl removal with a 99 percent efficiency.

o   A flue gas monitoring system for continuous monitoring of stack emissions.

o   A forced draft air moving system for providing controlled quantities of combustion air.

o   A heavy duty slag and ash removal system.

o   A complete boiler make-up water, feed water and condensate treatment system.

o   A computer based distributed control system for automatic operation of the incineration train.

THE WASTE FEED AND COMBUSTION COMPONENTS

The destruction of the waste in the PPG incineration system is accomplished in the two distinct parts of the combustion process: the rotary kiln and the secondary combustion chamber. First, waste is fed into the rotary kiln through a refractory lined front wall. Drummed and baled waste is elevated from a ground level conveyor into an air lock chamber and then pushed into the kiln through a liquid cooled feed chute. Pumpable waste is injected using steam cooled lances designed to disperse the waste stream. Primary combustion air provided by a forced draft fan is added at the bottom of the front wall. A natural gas burner is also located on the front wall for start-up and shut-down of the incineration system and for maintaining a minimum flue gas temperature.

The drum and bale lift, air lock chamber and chute are designed to feed approximately twelve 55 and 85 gallon drums or 24 inches by 24 inches by 42 inches high bales of trash each hour. A total of five liquid waste injection lances are provided at the front wall as follows: one for aqueous waste, one for high BTU waste, two for sludge waste and one for special waste. In addition, a wastewater lance is provided at the secondary combustion chamber.

The rotary kiln itself is 13.0 feet in diameter inside the shell (11.5 feet inside the refractory) and 36.8 feet in length and rotates in the range of 0.05 to 0.50 rpm using a variable speed drive. A single layer of nine inch thick high alumina brick is used to protect the steel shell. The kiln is designed to operate in a "slagging" mode with a flue gas temperature at the discharge end of 2,400° F. A eutectic pool of molten salts is maintained in the tapered discharge end of the kiln in order to melt the steel drums that are fed to the kiln. A water spray system on the outside of the kiln shell provides the controlled shell cooling necessary to extend refractory life, and a fan provides cooling of the heat resistant castings at the kiln discharge end. Two self-adjusting, ceramic cloth based seals minimize the in-leakage of air at the feed and discharge ends of the kiln shell.

The discharge end of the kiln extends into the bottom of the vertical secondary combustion chamber. Molten slag from the kiln, as well as ash from the secondary combustion chamber, fall into a water filled quench tank and are removed by a steel belt conveyor. As the secondary combustion chamber structure extends into the water, the quench tank also provides a seal to prevent unwanted air from entering the system.

The secondary combustion chamber is approximately 15.1 feet by 14.8 feet in cross section inside the refractory and 34 feet in height above the centerline of the kiln. The secondary combustion chamber is lined with approximately 24 inches of refractory and insulation. Secondary combustion air is injected into the secondary combustion chamber approximately 13 feet above the centerline of the kiln through a series of blast nozzles. This secondary air provides the oxygen necessary to complete the burnout of the flue gas, reduce carbon monoxide and create a very turbulent regime. The secondary air also reduces the flue gas temperature to 1600° F. The total retention time of the flue gas above $2,200^\circ$ F, however, is 2.5 seconds. A stand-by natural gas fired burner is provided for start-up and shut-down of the incineration system.

Finally, flue gas recirculated from the exit of the electrostatic precipitator is injected at approximately 23 feet above the centerline of the kiln through a separate set of blast nozzles. This injection cools the flue gas to $1,350^\circ$ F to eliminate the carry over of plastic or semi-molten particulate material and insures a uniform profile of flue gas entering the boiler.

The upper portion of the secondary combustion chamber includes a "nose" or profile that insures that there is no short circuiting of the flue gas as it passes through the vertical chamber. This profile was developed and tested in a two dimensional flow study. The "nose" is constructed of refractory lined tubes that are part of the evaporation section of the boiler.

The overall design maximum continuous heat release in the combustion components is 58.0 million BTU per hour based on the lower heating value. A control loop monitors oxygen content and temperature in the secondary combustion chamber and then modulates the feed of high BTU liquid waste to maintain a relatively steady state heat release and reduce heat peaks to within 115 percent of the maximum continuous design.

THE BOILER COMPONENT

The waste heat boiler used at the ERU is a single drum, natural circulation, water tube design, with a straight horizontal gas flow. The top and sides of the boiler are constructed of water wall membranes. Four evaporator and one economizer sections are provided with the tube panels hung in vibratory registers that are mechanically rapped periodically to remove accumulated particles. A minimum center to center tube spacing of 4 inches is used to eliminate blockage due to accumulated particles.

An extra low gas velocity is used to eliminate erosion of the tubes, and the boiler is designed to maintain all tube wall temperatures in the range of 350° F to 700° F to avoid dew point and high temperature corrosion of the carbon steel tube surfaces.

The gross steam production rate of the boiler is 42,000 pounds per hour from a total heating surface of 15,700 square feet. The efficiency of the boiler is 68 percent based on a condensate return temperature of 220°F. A net of 36,500 pounds per hour of steam is available to the manufacturing plant and waste handling portion of the ERU. The average maximum steam demand is 28,000 pounds per hour with a peak demand of 37,000 pounds per hour.

Ash knocked from the boiler tubes falls into one of three hoppers and then into a dust tight drag chain conveyor through air lock valves. The ash is then emptied into a container truck for transport to a secure landfill together with slag from the kiln and residue from the scrubber blowdown evaporation system.

An air cooled condenser is provided to condense 35,000 pounds per hour of steam in the event that the manufacturing plant does not require the full steam capacity of the waste heat boiler. Boiler make-up water and feed water treatment equipment and returned condensate polishing equipment is provided. Also space has been allowed for the addition of a gas fired full capacity boiler and several small quick response boilers in the future as the manufacturing plant's existing boiler house is modernized.

THE GAS CLEANING COMPONENTS

Flue gas exits the waste heat boiler at 480° F and at a volume of 32,670 scfm. The first step in gas cleaning is the removal of particulate material in a conventional dry electrostatic precipitator. The ESP consists of a single chamber with three independent fields. The electrodes consist of a structural matrix of rigid 1.5 inch diameter tubes with 2.25 inch corona generating studs. The collection surfaces consist of 27 foot high roll formed steel plates with edge stiffeners to provide nineteen, 12 inch wide passages. Perforated plates in the entrance and exit nozzles insure an even distribution of the flue gas. An electrical impulse rapping system provides for the periodic removal of collected ash. The ESP provides 99.7 percent removal of particulate with a total collecting surface of 28,100 square feet. Ash is removed from the hopper of the ESP with a drag chain conveyor and is discharged into the boiler ash conveyor.

A portion of the particle free flue gas is recirculated back to the secondary combustion chamber for conditioning of the flue gas entering the boiler. The majority goes to an induced draft fan of a backward curved blade design. This fan produces the required negative pressure in the upstream components. As an example, the pressure in the secondary combustion chamber is controlled to a negative 0.5 inches. The fan is driven by a variable speed electric motor connected to a normal and emergency power bus. This arrangement allows operation of the fan at low speed during a power outage and associated system shut down.

The final gas cleaning component consists of a low pressure venturi quench and a packed bed scrubber. In the refractory lined quench, scrubber liquor is injected to saturate and cool the flue gas in order to protect the down stream fiberglass scrubbing tower. The tower consists of a single eight foot deep bed of saddle packing followed by a mesh pad type mist eliminator. Scrubber liquor is recirculated from a sump in the bottom of the tower. Caustic soda is injected into the recirculated tower scrubber liquor to maintain a constant pH of 6.5. Make-up water to account for water losses caused by evaporation and blowdown is added by a spray at the underside of the mesh pad. This arrangement keeps the mesh pad as clean as possible. During the time that make-up water is being added to the tower, scrubber liquor overflows into a recycle tank which serves as a sump for the quench scrubber liquor recirculation pump. Scrubber liquor is bled off from this sump to maintain a solids concentration of 11 to 12 percent. The quench and scrubber remove 99.0 percent of the HCl in the flue gas at a pressure drop of 2 inches and a scrubbing rate of 26 gallons per 1,000 scfm.

The liquid in the blowdown from the scrubber, in the range of 1 to 3 gallons per minute, is evaporated in an indirect steam drying system designed by PPG Industries, Inc., and the dried residue is mixed with the boiler ash in a pug mill prior to disposal in a secure landfill.

The characteristics of the flue gas leaving the stack of the incineration system are as presented in Table III. The top of the fiberglass stack is 120 feet above grade and the flue gas exits the stack at a velocity of 37 feet per second.

TABLE III: FLUE GAS COMPOSITION

Flue Gas Component	Design Average Composition
CO_2	6.0 percent by volume
H_2O	25.3 percent by volume
O_2	8.6 percent by volume
N_2	60.1 percent by volume
HCl	31 ppmv dry
SO_2	110 ppmv dry
NO_x	119 ppmv dry
CO	< 100 ppmv dry
VOC	68 ppmv dry
Particulate Material	0.05 grains per sdcf at 12 percent CO_2
Volume	30,320 standard cubic feet per minute
Temperature	140 ° F

CONTROL COMPONENT

The entire ERU is operated through a computer based distributed control system. The control room contains two color CRT control stations and associated keyboards and printers to display, track and record data, to alert operators of process variables or equipment status outside of the normal operating range, and to execute operational decisions. The control system is configured to provide automatic operation of the incineration system, with operator manual adjustment to fine tune performance based on waste feed composition. The control system allows the operator to view current operating variables and the trend of selected process variables over a period ranging from 2 minutes to 72 hours. The system also provides a visual display and hard copy record of variations from set point, and shuts the entire incineration system down in a predetermined sequence should certain variables exceed predetermined values. The control stations in the control room are connected using a data link to process controllers located in the motor control centers of the incineration system and the material handling system of the ERU. A separate keyboard, CRT and computer in the control room allows the controls engineer to make changes and improvements to the control configuration. Periodically, data is transmitted from the control stations to the ERU's data management computer for final long term record keeping and monthly and yearly report preparation.

OPERATIONAL FACTORS

The incineration system of the ERU is staffed with a crew consisting of a supervisor and four operators: one to work in the control room with the supervisor, one to work in the incineration area, one to work in the waste preparation and feed area and one for general maintenance. Four crews for a total of twenty persons provide continuous coverage. In addition, the ERU employs one superintendent, three persons in the laboratory and five persons in the waste reception and storage area.

Basic utility information for the incineration system is as follows:

Water Consumption	75	gallons per hour, maximum
Electric Power Consumption	540	KW
Natural Gas Consumption	3,500	thousand cubic feet per year
Caustic Soda Consumption	100	pounds per hour
Boiler Chemical Consumption	46	pounds per hour
Slag and Ash Production	950	pounds per hour
Boiler Blowdown	70	gallons per hour

THE PERMITTING AND SITING PROCESS

For an organization seeking to build and operate a hazardous waste management facility in Ohio, there are three separate regulatory paths to be followed:

o   A federal RCRA permit ("Part B" permit) from US EPA Region V.

o   Conventional permits to install (and, separately, to operate) air and water pollution control devices from Ohio EPA.

o   A state siting permit from the Ohio Hazardous Waste Facility Board (HWFAB) based on state RCRA criteria similar to the federal and on specific state siting criteria.

The HWFAB proceeding is the most complex and incorporates the technical and procedural requirements of the U.S. and Ohio EPA.

HWFAB is a regulatory body created by the Ohio Legislature in 1980 to oversee the development of new hazardous waste management facilities. To insure a state-wide perspective, the Board was granted preemptive authority. However, to ensure that full consideration be given to local concerns, local governmental representatives were by statute made a party to all application proceedings before the Board. The Board consisted, at the time of the PPG proceedings, of five persons. By statute, the Board includes as chairperson, the Director of the Ohio EPA, and as members, the Director of the Ohio Department of Natural Resources and the Chairperson of the Ohio Water Quality Department Authority. The other two members are university professors of geology and chemistry appointed by the Governor and with expertise appropriate to the duties of the Board. The Board is supported by legal and technical staff, but dependent on Ohio EPA for preliminary technical review of applications and for enforcement of permit conditions.

In summary, the statute creating HWFAB requires that the Board make the following determinations regarding any proposed facility:

o   The proposed facility represents the minimum adverse environmental impact considering the state of available technology and the nature and economics of various alternatives.

o   The proposed facility represents the minimum risk of:

-   contamination of groundwater and surface water
-   fires and explosions
-   accidents during transportation

o   The proposed facility complies with the regulations of the OEPA.

Circleville, Ohio where the ERU is located, is a relatively rural community approximately thirty miles south of Columbus. PPG has been a part of the community since 1962 and currently employs approximately 210 people. Employment will increase by approximately 30 people with the start-up of the new waste management facility. The Circleville plant is an excellent location for the ERU in that a trained labor pool exists, contingency and emergency response programs and facilities are in force, and transportation routes are established.

In October, 1982, PPG submitted its application to Ohio EPA for the Circleville Energy Recovery Unit. This submittal was preceded by a briefing on the ERU by plant management to, first, all employees of the Circleville plant, and, second, to all local government officials. To aid in these discussions,a model of the facility and large flow diagrams of the incineration system were prepared. In March 1983, Ohio EPA, acting on behalf of HWFAB, determined that the application was complete and met Ohio RCRA requirements. Concurrently with the HWFAB permit application, PPG applied to the Ohio EPA for a permit to install air pollution control components of the incineration system (PTI). In June 1983, Ohio EPA issued as final the permit to install. This permit included detailed operating conditions and limitations, and was immediately appealed by the county government. After receipt of the PPG RCRA application from Ohio EPA, HWFAB conducted an open public meeting to receive comments from all parts of the local community. Speakers at the meeting included several individuals concerned about potential negative impacts of the ERU, although the great majority of the speakers offered support. HWFAB then scheduled an adjudication hearing to consider whether or not the PPG application meet the statutory requirements for the siting of the facility.

Parallel with the HWFAB activity, PPG initiated discussions with local government officials aimed at reaching agreement on the application and on permit conditions as defined by the final OEPA permit to install. These officials, acting on behalf of the citizens in the Circleville area, successfully negotiated several air emission requirements more stringent then those recommended by Ohio EPA. They also negotiated ambient air and groundwater monitoring requirements to be implemented before and during operation of the ERU. These requirements were incorporated into a written agreement in July 1983 whereby the Board of County Commissioners of Pickaway County agreed not to oppose the PPG application before HWFAB or appeal the pending U.S. EPA RCRA permit, and agreed to withdraw their appeal to the Ohio EPA permit to install. In addition, as a result of the agreement, a local group organized to oppose the project elected to withdraw as a party to the HWFAB adjudication process. The negotiated agreement was accepted by HWFAB as part of the written record during a very brief (two hour) adjudication hearing. A permit was then granted in final form by HWFAB in February 1984.

Subsequent to the regulatory process, a small group of persons acting as individuals filed suit in February 1984 in the local court (Pickaway County) against PPG and several state officials, alleging that the HWFAB process was unconstitutional and that the ERU constituted a public nuisance. Many of these persons had belonged to the local opposition group that had supported the negotiated agreement. PPG moved for a summary judgment dismissing the suit. In July 1985, sixteen months after the suit was filed, the court rejected the constitutional and public nuisance arguments of the opponents, but allowed the suit to continue based on a private nuisance argument. No date was set by the court for further proceedings.

Following the local court ruling, the PPG Board of Directors voted in April 1986 to authorize the remainder of the project, and, after consultation with local government officials, construction was begun in May 1986.

The final stage in the permitting process is a performance evaluation or "trial burn" to determine the actual efficiency of the system and set the minimum operating conditions. This trial burn is scheduled to be conducted in August 1987 in accordance with a trial burn plan submitted with the permit application.

FINAL WORDS

During the summer of 1987, Environmental Elements Corporation will have the pleasure of turning over to PPG Industries, Inc. an operating incineration and energy recovery system at Circleville, Ohio. This will be the first major new incineration facility fully permitted and sited under RCRA. It is the kind of facility that is critically needed throughout North America. As professionals in the field of hazardous waste management we all need to work to ensure that, after thorough but reasonable regulatory and public review of specific projects, we enter the 1990's with a broad array of similar facilities under construction.

619

PYROLYSIS OF HAZARDOUS WASTES
WITH A MOBILE PLASMA ARC SYSTEM

Michael F. Joseph
Westinghouse Plasma Systems Canada Inc.
Niagara Falls, Ontario, Canada L2H 1J6

and

Thomas G. Barton
Royal Military College of Canada
Kingston, Ontario, Canada K7K 5L0

ABSTRACT

During a 20 month period ending in April 1986, a completely mobile system for the destruction of hazardous or highly toxic liquid organic waste had completed its initial testing in Canada. Upon completion of this testing, this unit was exported to the Love Canal waste site in New York State to undergo further demonstrations. This technology has been under development for the past 10 years and gives extremely high destruction efficiencies and low emissions for typically hard to destroy toxic wastes. A joint venture company, Westinghouse Plasma Systems Inc., was formed by Westinghouse Electric Corporation and Pyrolysis Systems Inc. This joint venture company now markets this technology which has been trademarked Pyroplasma.

The Pyroplasma unit with a throughput rate of approximately 12 kg/min is contained in a single 15 meter long tractor trailer and requires connections to power, water and sanitary sewer to become completely operational. Mobilization requires only 1 to 2 days in order to be operational at a waste site.

The heart of the technology lies within the plasma torch. This electrical devices efficiently converts electrical energy into thermal energy in the form of a heated gas with temperatures in excess of 10,000°C. The entire system in computer controlled with numerous safety, operational, and environmental failsafes as well as continuous process gas monitoring. Destruction efficiencies in excess of 99.9999% with low emissions of product of incomplete combustion make this technology both commercially and environmentally attractive.

INTRODUCTION

Pollution stems from overtaxing the natural purification processes offered by the environment. Presently, the large volumes and many types of hazardous industrial wastes currently produced pose a serious threat to both the public and the environment. It is estimated that over 300 million metric tons of hazardous and/or toxic wastes are generated on an annual basis [1,2]. This waste must be treated appropriately if man is to prevent irreparable damage to this biosphere.

In 1976, the United States Environmental Protection Agency (EPA) responded to this threat with the enactment of the Resource Conservation and Recovery Act (RCRA) and the Toxic Substance Control Act (TSCA). As a result of these two acts, a large increase in research activity was commenced to seek methods to destroy organic liquid and solid wastes. This was again accelerated in 1984 by the passing of the Hazardous and Solid Wastes Amendments (HSWA) to RCRA which requires a series of land fill ban restrictions on eventually all hazardous wastes by 1990. Obviously, the implementation of these amendments are dependent upon the availability of successfully demonstrated alternative technologies to replace the existing land disposal practice.

Prior to new technology being demonstrated, it must be able to comply with all the necessary criteria established by RCRA and TSCA. These regulations generally apply to the efficiency of destruction for the waste stream in question and certain restrictions as to the quality of the emissions that must be achieved and maintained. Efficiencies are measured as a destruction and removal efficiency (DRE) and expressed in percent as calculated by the following formula:

$$DRE = \frac{Waste\ In - Waste\ Out}{Waste\ In} \times 100$$

The waste is measured by the mass feed rate of the Principal Organic Hazardous Constituent (POHC). Under RCRA, hazardous waste facilities must maintain at least a 99.99% removal efficiency while technologies processing PCBs which are considered toxic and regulated under TSCA must operate at DREs in excess of 99.9999%. In addition, particulate and HCL emission restrictions also apply. Associated emissions of products of incomplete combustion are also regulated and standards set. Products of incomplete combustion are generally compounds which are not present in the actual feed stream but may be produced as a by-product of the destruction itself. Of greatest concern are the highly toxic chlorinated dioxins and furans.

Although incineration has been recognized as one of the best demonstrated available technologies (BDAT) there has been a rapid increase in the number of innovative chemical and thermal processes for the ultimate destruction or reduction of hazardous wastes. The Pyroplasma technology represents one of these developments. It was dependant upon several factors directly related to the required operational criteria. The plasma system offers an alternative to combustion as the prime source of a thermal energy for the destruction of organic compounds. Using plasma as the thermal source, a directly heated pyrolytic system could readily be developed. This technology had to be competitive from both a cost and performance standpoint and should have an appropriate commercial scale and be marketable. The present Pyroplasma units meet these operational criteria.

APPROACH

Work on a pilot laboratory scale prototype over a five year period at the Royal Military College of Canada yielded preliminary results which indicated that the Pyroplasma technology could indeed be a viable thermal destruction technique.

A thermal plasma, properly applied to toxic waste destruction, provides a pyrolytic or reducing environment with several distinct advantages over conventional combustion technologies [3]. The extremely high temperatures achieved assure high DREs and the reducing atmosphere reduces the production of oxygen containing PICs such as dioxins and furans. In addition, since the process is pyrolytic, the scale of the equipment is small and can be easily made mobile while still allowing for throughput rates as high as 12 kilograms per minute (Figure 1).

In 1982, Pyrolysis Systems Incorporated (PSI) of Canada signed a co-operative agreement with the New York State Department of Environmental Conservation (NYS DEC) providing funding for the design, construction, testing and operation at Love Canal of a four kilogram per minute commercial scale Pyroplasma unit. The first three phases of work are now complete and the last phase is awaiting the necessary permitting. In 1985, PSI signed a joint venture partnership with Westinghouse Electric Company forming Westinghouse Plasma Systems to commercialize this technology.

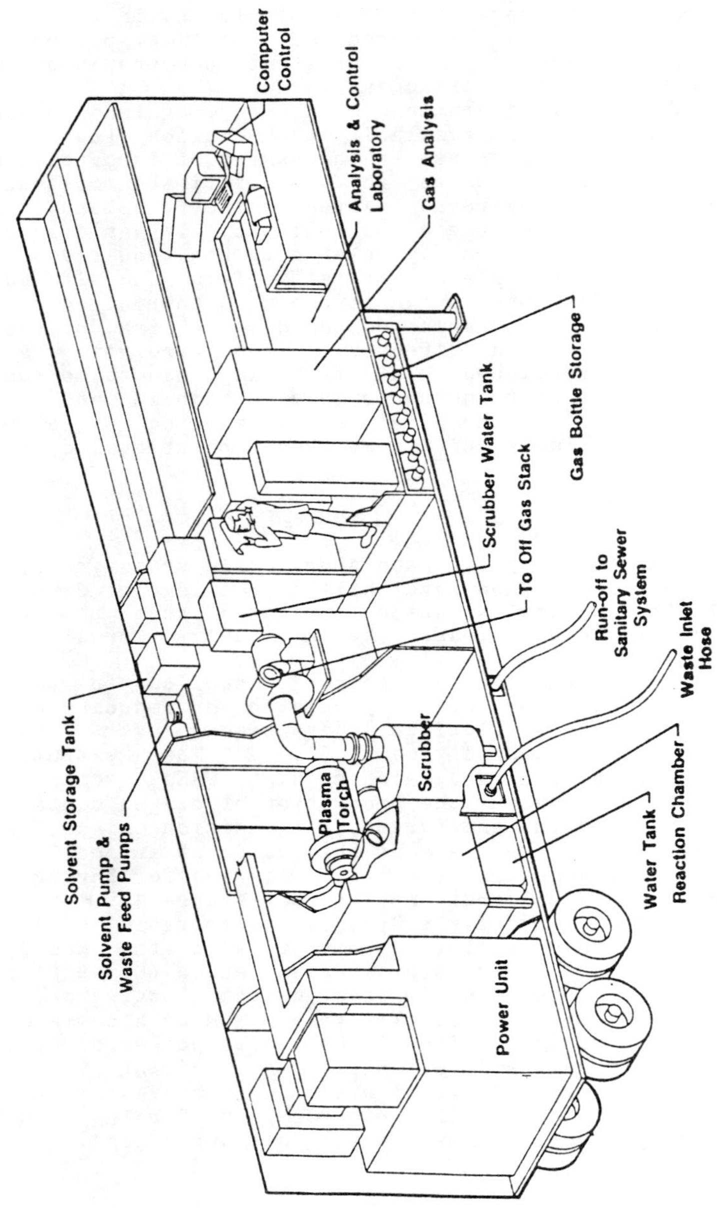

Figure 1:   Pyroplasma Trailer Schematic

Process Description

The heart of the system is the plasma torch (see Figure 2). A co-linear electrode arrangement and about 750 kilowatts (kW) of power are used to create an electric arc in the presence of a medium of low pressure air. Electrical energy absorbed by the air molecules raises them to highly excited states. During relaxation these molecules release thermal energy producing a plasma. When a process gas such as air is passed through this thermal plasma the result is a hot gas with temperatures in the $10,000^{o}C$ range. Waste fluids are injected directly into the plasma tailflame. The extremely high temperatures associated with the plasma torch provide enough energy for completed atomization of the waste molecules thus providing very high DREs with relatively low energy requirements, typically less than one kilowatt-hour per kilogram of waste.

The atoms then react in a recombination chamber according to chemical kinetic equilibriums and the minimization of Gibb's free energy, to form non-toxic type gases. The product gas typically comprises about 50% hydrogen, 30% carbon monoxide, 15% nitrogen and the remainder lower hydrocarbons such as methane and ethylene. Acid gas (HCl) formed from the destruction of chlorinated wastes is neutralized by caustic soda in a wet scrubber to give a slightly salted scrubber effluent which is discharged along with any particulate carbon that may be formed. The product gas is drawn off the scrubber by an induction fan which maintains a negative system pressure and the gas is flared. Fuel values of the product gas are between 2 - 3 times the actual power required for the torch. The composition of the gas may be altered by careful blending of the waste feed prior to destruction or by the addition of other constituents such as steam or air to the recombination chamber.

Operation begins with a non-toxic type solvent such as methanol and continues for about five to ten minutes until the recombination chamber stabilizes at a predetermined temperature, typically about $1200-1400^{o}C$. Stable chamber temperatures indicate adequate quenching of the torch power thus ensuring efficient destruction of the incoming waste. Since the destruction process is pyrolytic, occurring in the plasma field, system destruction efficiencies do not depend upon reactor chamber temperatures or residences times. Chamber temperature is controlled by adjustments in torch power, torch process gas flow or waste feed flow rate. The entire system is computer monitored through a series of temperature, pressure, flow and electrical sensors so that operational parameters and system efficiencies are maintained. There are a number of failsafes built into the system to minimize the risk to the environment even in the event of power failure, a worst case situation. Carbon monoxide, carbon dioxide, hydrogen, oxygen and total hydrocarbons of the bulk gas are measured on line for comparison to theoretical models.

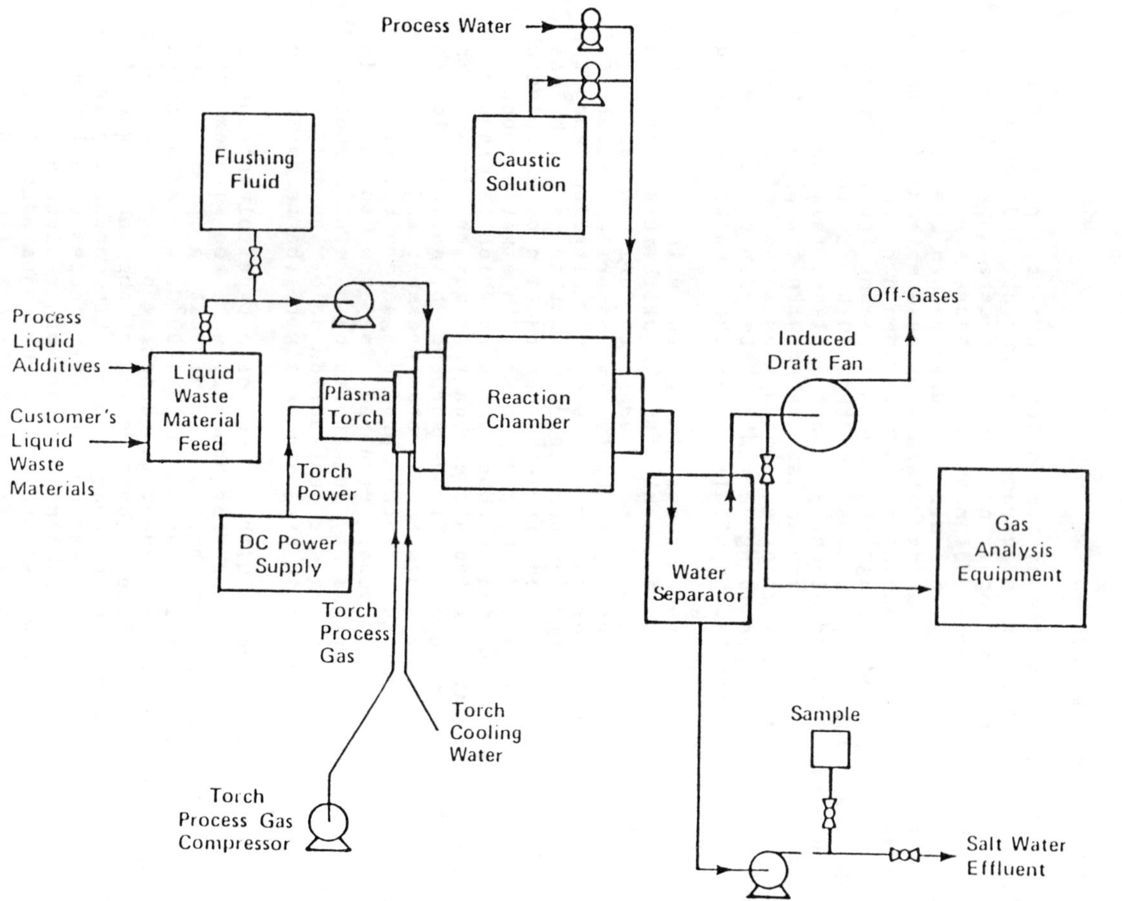

Figure 2: Pyroplasma Process Schematic

<u>Testing</u>

The Pyroplasma process is capable of destroying a wide spectrum of liquid organic wastes. Materials were selected for testing purposes to demonstrate the ability of the plasma to process difficult to destroy, highly chlorinated wastes. In 1985 and early 1986, a rigorous series of tests were conducted with methyl ethyl ketone (MEK), ethanol (EtOH), methanol (MeOH), carbon tetrachloride (CCl_4) and transformer Askarel fluid (PCBs and trichlorobenzene). All tests were observed by both federal and state authorities from the United States, as well as federal, provincial and local Canadian authorities. All sampling and analyses were performed by external agencies as part of the New York State contract with PSI and the data has been reviewed and approved by the EPA.

RESULTS

Mechanical operation of the mobile plasma pyrolysis system was first verified by processing a non-chlorinated blend of MEK/MeOH (1:1 volume). Following this, three one-hour CCl_4 runs were conducted to demonstrate (i) the destruction efficiency with a hard to destroy chlorinated compound, (ii) the stability of the system upon exposure to high concentrations of chlorine, and (iii) the effectiveness of the scrubber for HCl removal. The CCl_4 was blended with $MEK/MeOH/H_2O$ and fed at a rate of one kilogram of CCl_4 per minute. DREs were greater than 99.99995% for all three runs based on measured residuals in all effluent streams. Stack emissions for CCl_4 were analytically non-detectable (see Table 1).

TABLE 1
CCl_4 test results

Test (#)	Mass (%)	Input (kg/hr)	Output Air (mg/hr)	Output Water (mg/hr)	Total (mg/hr)	DE (%)
1	35.0	64.2	29.27*	2.51	31.78	99.99995
2	40.0	60.6	22.79*	9.85	32.64	99.99995
3	35.0	64.2	22.89*	6.26	29.15	99.99995

* number reported is analytical detection limit

Following this stage of testing seven tests were performed using a blend of Askarel (Aroclors 1254/1260 in trichlorobenzene) in MEK/MeOH to determine the PCB destruction capability of the process. Three one-hour, one two-hour, and three five-hour tests with feedstocks approximately 30-35% by mass Askarel gave DREs consistently greater than 99.99999% (see Table 2).

TABLE 2
PCB Test Results

Test (#)	Mass (%)	Input (kg/hr)	Output Air (mg/hr)	Output Water (mg/hr)	Output Total (mg/hr)	DRE (%)
3	14.1	23.9	0.65*	4.09	4.73	99.999997
4	14.1	18.8	2.00	10.16	12.16	99.999994
5	11.5	19.8	6.96	18.76	25.72	99.99996
6	14.3	18.0	0.43*	0.41	0.84	99.999998
7	12.5	17.5	0.02*	93.12	93.14	99.9999999
8	12.8	16.9	0.03*	19.56	19.59	99.9999999
9	17.5	26.7	0.33	0.75	1.08	99.999999

* number reported is analytical detection limit

Table 3 summarizes the HCl, particulate, NO_x, O_2, CO and CO_2 emissions for the flared product gas. Emissions of HCl, particulate and NO_x were below accepted Canadian and American regulatory agency guidelines.

TABLE 3
Flared product gas parameters@

Test (#)	O_2 (%)	CO (%)	CO_2 (%)	HCl (mg/M^3)	NO_x (ppm)	particulate (mg/M^3)
3	14.0	0.01	5.5	NA	117	10
4	14.5	0.01	5.0	43	NA	180
5	16.5	0.01	3.0	68	139	12.5
6	15.8	0.00	3.8	1.07	96	15.84
7	14.0	0.00	5.1	2.68	115	7.60
8	15.3	0.00	4.3	1.30	108	10.96
9	13.3	0.00	5.2	26.8	81	11.9

@ refers to PCB tests #3-#9 of Table 2

In addition to the high destruction efficiencies, emissions of products of incomplete combustion (PICs) were usually below detection limits in both the gas and water effluents. For example, no chlorinated dioxins were detected in either the gas or water effluents in PCB test #9, the highest PCB concentration test. These results are even more meaningful if you consider the relatively low gas and water volumes associated with this pyrolytic process in combination with the emission concentrations expressed in mass per unit volume.

PROGNOSIS

At present, two Pyroplasma trailers are operational and permit applications have been filed where applicable. Four more units are under construction to meet the anticipated growth in the need for their application. Additional research is being conducted to apply similar technology to the problem of solid hazardous and/or toxic waste.

ACKNOWLEDGMENTS

The co-operation and assistance of the Royal Military College of Canada, the Environmental Protection Service and the Ministry of the Environment of Canada, and the Environmental Protection Agency and New York State Department of Environmental Conservation are much appreciated.

REFERENCES

1. Matey, J.S., and L.F. Tischler, 1984 Hazardous Waste Survey, _Journal of the Air Pollution Control Assoc._, 1984, _36_(6), pp.737-740.

2. Rich, L.A., Hazardous Waste Management, _Chemical Week_, August 20, pp.26-64.

3. Barton, T.G., Mobile Plasma Pyrolysis, _Hazardous Waste_, 1984, _1_(2), pp.237-247.

THERMAL TREATMENT OF HAZARDOUS AND TOXIC WASTES FROM THE
PHARMACEUTICAL AND RESIN INDUSTRIES

J. A. J. Clark,
Marketing & Business Development Executive,
Robert Jenkins Systems Ltd.,
Wortley Road, Rotherham,
South Yorkshire. S61 1LT

ABSTRACT

This paper deals with the problems involved in the incineration of process effluent streams containing organic and inorganic components. In particular the practical aspects and problems are considered together with environmental considerations in respect of gaseous emissions.

The use of thermal oxidation commonly known as incineration has long been practiced but there are drawbacks, particularly when burning wastes which contain halogens and inorganic salts. Modern chemical processes produce wastes which are increasingly difficult to burn without some form of pretreatment prior to incineration and flue gas treatment thereafter.

Given some thought and consideration during the planning stage the disposal costs can be reduced considerably by making use of recovered heat for use in the production or treatment process.

At the same time care in selecting the treatment route can significantly improve plant operation, reliability and maintenance costs.

TEXT

The principal effluent streams produced by the Resin and Pharmaceutical industries can present considerable problems if they are to be rendered harmless to the environment.

Whilst certain types of effluent can be easily and adequately treated by processes such as filtration, anaerobic digestion and other biological means there are still those wastes for which the only disposal means is by incineration or high temperature thermal oxidation.

The three effluent streams from these industries which are the most common cases for thermal treatment are:

Aqueous Liquid Wastes : These are typically waste water streams contaminated with organics such as phenol, formaldethyde and alcohols. The level of organics is usually less than 5 to 10%. There may also be dissolved inorganic salts.

Organic Liquid Wastes : In the main these are solvents which are not capable of being recycled. The composition is usually a wide mixture of chlorinated and non chlorinated compounds and can also be highly volatile or extremely stable.

Gaseous Streams : These can vary considerably from largely inert gas streams with small amounts of highly toxic or objectionable organics, such as Methyl Mercaptan in Nitrogen, or on the other extreme relatively high amounts of organic vapour in air or inerts.

These three different types of streams have their own distinctive problems when it comes to their disposal by incineration. To deal with each in turn together with the difficulties encountered is perhaps the best way of describing the requirements.

Aqueous Wastes

There are three principal problems which have to be recognised and overcome with aqueous streams. The predominantly water based waste requires a considerable amount of heat to transpose it from the liquid phase to what is an essentially superheated vapour. Added to this heat it is a necessity to provide sufficient free oxygen in the combustion air, generally about 10% by volume in the final combustion products – this is required for essentially complete 'burnout' of pollutants in the waste stream.

The second complication may be varying waste composition. The levels of organics will inevitably vary and it follows that the calorific value also changes, often significantly. If the composition changes relatively slowly a control system can be designed which will adequately control incinerator auxiliary firing equipment to give a stable temperature level.

Possibly the most serious drawback to the incineration of aqueous wastes is the presence of dissolved alkaline salts. The action of sodium and potassium salts in particular are likely to cause serious damage to refractory linings.

As an illustration it has been observed that whilst levels of sodium of 100 mg/litre and below can in some circumstances give acceptable life, the increasing of the sodium concentration by a factor of 10 can result in fluxing of the lining at the rate of 2 to 3 cm per week on certain plants. Above temperatures of 800 °C, particularly with potassium, attack on refractories increases rapidly.

Incineration temperatures have therefore to be a compromise on being sufficiently high to ensure destruction of organics yet low enough to be below the temperature of which fluxing attack of the lining starts to occur. Also with lower temperatures fuel usage is reduced.

If very high levels of dissolved salts are present and removal at source is not possible the only way to incinerate may be to vapourise in an evaporator/crystallizer and pass the steam and organic vapours released through a fume incineration system. The solid or sludge residue can be removed for other means of disposal.

Incinerators of this type are often difficult to fit waste heat recovery equipment to if dissolved solids are present.

If inorganics are not removed prior to incineration it may be necessary, depending on the atmospheric pollution levels applicable to the site, to fit gas cleaning plant to remove offensive particulate matter.

Organic Wastes

The majority of organic liquid wastes are used solvents which are incapable of recovery for either reuse in the process or cannot be economically refined. The solvents may have heavy organics dissolved in them such as tars or resins. As a further complication the solvents may be halogenated.

The incineration of non halogenated solvents generally presents few problems provided care is taken to ensure correct atomisation quality, spray pattern, mixing or turbulence in the combustion chamber and residence time. Temperatures are generally in the range of 750 °C to 800 °C.

The addition of dissolved heavy organics residues can present problems if they are not adequately dispersed giving smoke emissions and unburned hydrocarbons in the combustion products. If halogenated solvents are present in the effluent it is almost certain that gas cleaning equipment will be needed to reduce Hydrogen Chloride, Chlorine and any other halogens present to statutory levels. Gas cleaning systems usually feature a water quench tower to cool the off gases from the incinerator and remove hydrogen chloride followed by a caustic scrubber to remove chlorine.

It is the nature of Hydrochloric acid that it is one of the most corrosive agents encountered in industry. This is perhaps the single most difficult obstacle to overcome in the design and construction of a gas cleaning plant. Virtually all common corrosion resistant materials have been used in an attempt to give acceptable life. Despite many claims of success few plants will last longer than two years without major renewals of components in the hot gas ductwork and gas cleaning system, indeed some components last a matter of months.

A further highly important factor when dealing with the incineration of chlorinated hydrocarbons is the potential for formation of dioxins and furans. To minimise the risk of formation of these complex compounds it is necessary that the incineration temperature is at least 1100 C and also the residence time at this temperature is sufficiently long.

This high temperature requirement also complicates and increases the severity and potential for corrosion of downstream gas cleaning plant.

If at all possible it is desirable to use non halogenated solvents to minimise problems with waste disposal, often however there is no alternative.

Fumes and Gaseous Effluent

Both the resin and pharmaceutical industry use purged vessels for production. The off gases from these reactors inevitably needs treatment to remove contaminants before release to the atmosphere. Extract air from certain plant areas also can contain organic vapours which need removal.

Typical fume streams may be Nitrogen or steam based with contaminants such as mercaptans, hydrogen sulphide, hydrogen cyanide, ammonia and more complicated organics. If the fume is mainly inert gas it is necessary to introduce the fume into an oxygen rich environment in the incinerator. The means of achieving this are shown later.

Extract air streams, provided they contain organic vapour concentrations less than 25% of the lower explosive limit, can often be used directly as combustion air.

If the fume stream contains chemically combined nitrogen such as ammonia or Hydrogen Cyanide in significant quantities then problems may arise with formation of large quantities of oxides of nitrogen (NOx) which are highly undesirable. If these nitrogenous compounds are present steps can be taken to drastically reduce the formation of NOx by staged combustion whereby the fume stream is subjected to high temperature in a sub stoichiometric oxygen atmosphere at 1100 °C followed by quenching with excess air to the lower temperature of 800 °C.

Depending on the atmospheric pollution legislation in force in the country where the plant is sited it may be necessary to install gas cleaning equipment to remove objectionable pollutants such as oxides of sulphur and nitrogen.

Multipurpose Incineration Plants

It is seldom the case that a plant in the industry at which this paper is aimed has only one effluent stream to dispose of. From a cost point of view it is inevitably desirable that all mobile waste streams, that is to say in the liquid and gaseous phases, are all disposed of in one plant.

This was not always a simple matter, burner designs and chamber arrangements often made more than a simple fume and single liquid stream a difficult problem.

The development of the HYGROTHERM multijet multistream matrix burner made the difficulty of dealing with a multiplicity of streams much less of a problem. The burner, as illustrated in figure 1 enabled simultaneous combustion of contaminated airstreams, inert fume streams, aqueous and organic liquid streams. A number of these plants have been commissioned in the past few years and examples can be seen operating on sites throughout the world.

The HYGROTHERM incineration unit comprises distinct steps. Referring to Figure 2 air and/or fume is introduced to the burner plenum. The air then flows through the venturi matrix. Behind each venturi is a gas spud to introduce auxiliary fuel. The air/fume and fuel then burn in the burner tunnel with a short highly intense flame across the whole area of the matrix. Temperatures up to 1600 °C can be reached in this initial reaction.

As a substitute for gas it is possible to use oil as an auxiliary fuel, the oil being atomised to give a suitable flame profile.

The remainder of the air flows around the front of the burner and then mixes with the stream of hot products from the burner. The combined mixture dropping in temperature then expands through the orifice ring entering the combustion chamber in a highly turbulent manner.

Following the orifice ring a baffle is then used to impart further turbulence. The baffle position has been optimised as a result of test work.

The function of the chamber inlet orifice and baffle ensure that the mean incineration temperature is reached in a very short space of time with the result that coring and wall slippage are effectively eliminated.

To introduce the liquid waste streams the burner design readily lends itself to the fitting of waste injection guns. The gun ports replace ports in the burner tile matrix.

The liquid wastes are atomised using compressed air or steam. The nozzle design is specifically designed for the particular application. It can be seen that the liquid waste is introduced into the zone of highest temperature to aid rapid vaporisation and thermal cracking of atomised droplets prior to mixing with additional air.

FIG. 1 : MULTI JET MULTI STREAM INCINERATOR BURNER

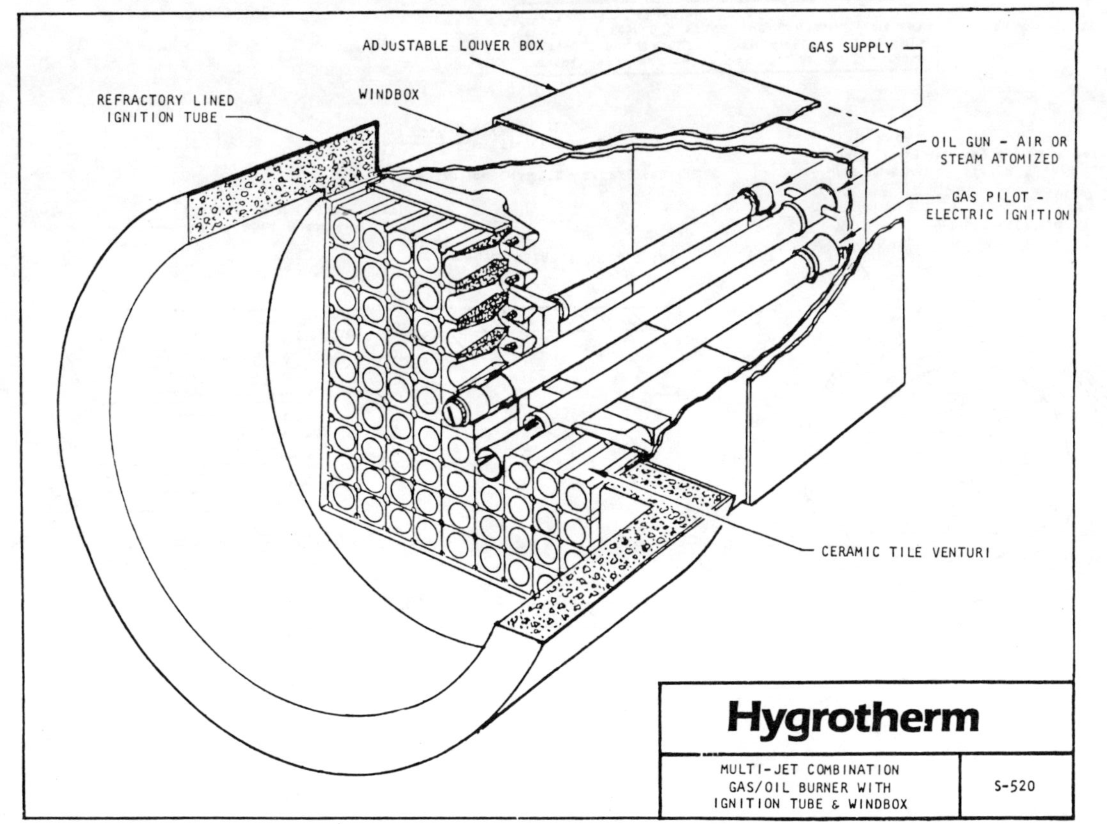

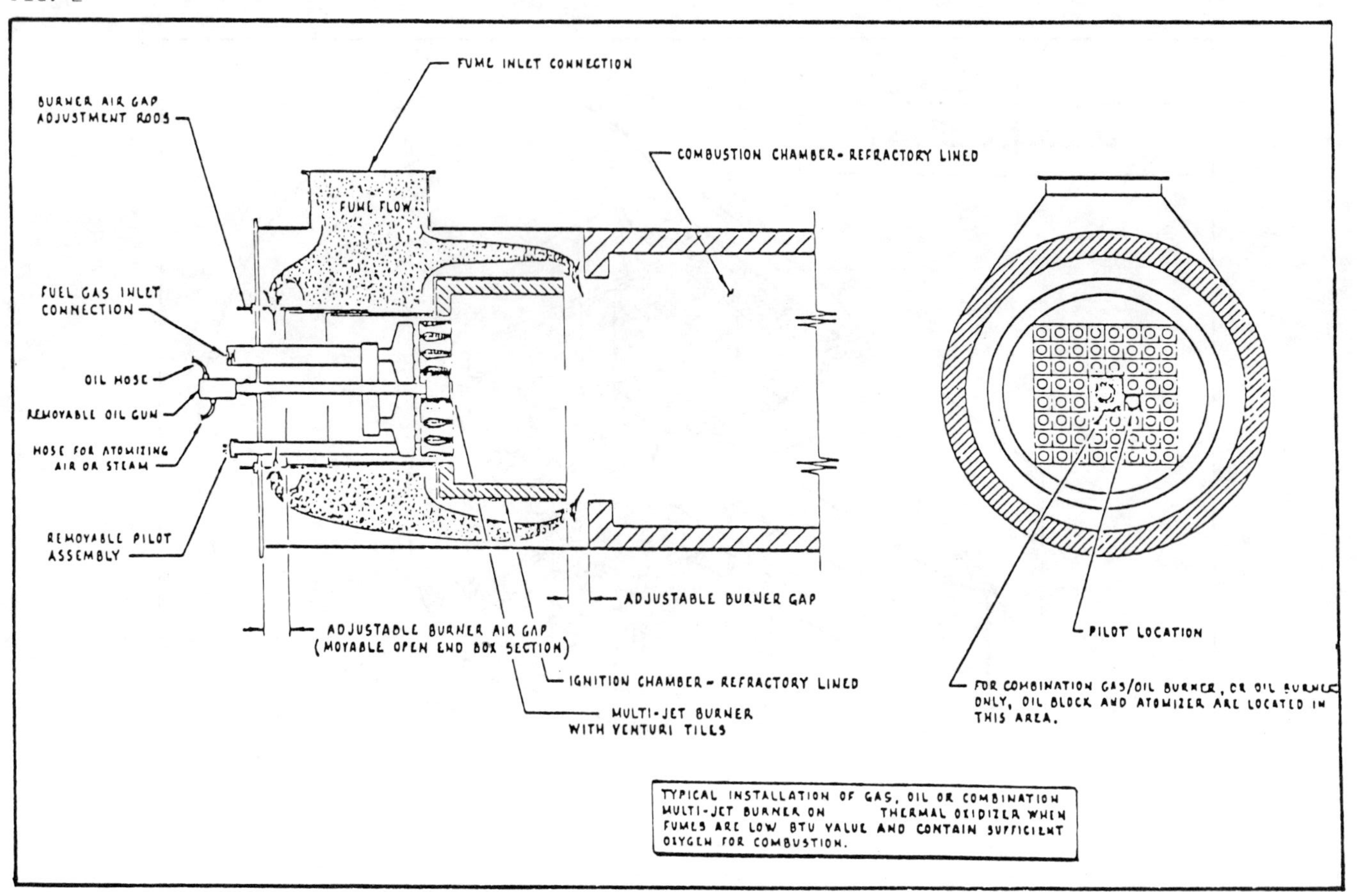

FIG. 2
BURNER AIR GAP ADJUSTMENT RODS
FUME INLET CONNECTION
COMBUSTION CHAMBER - REFRACTORY LINED
FUEL GAS INLET CONNECTION
FUME FLOW
OIL HOSE
REMOVABLE OIL GUN
HOSE FOR ATOMIZING AIR OR STEAM
REMOVABLE PILOT ASSEMBLY
ADJUSTABLE BURNER AIR GAP (MOVABLE OPEN END BOX SECTION)
IGNITION CHAMBER - REFRACTORY LINED
MULTI-JET BURNER WITH VENTURI TILES
ADJUSTABLE BURNER GAP
PILOT LOCATION
FOR COMBINATION GAS/OIL BURNER, OR OIL BURNER ONLY, OIL BLOCK AND ATOMIZER ARE LOCATED IN THIS AREA.
TYPICAL INSTALLATION OF GAS, OIL OR COMBINATION MULTI-JET BURNER ON THERMAL OXIDIZER WHEN FUMES ARE LOW BTU VALUE AND CONTAIN SUFFICIENT OXYGEN FOR COMBUSTION.

The control system is fully automatic and will adjust the auxiliary fuel firing rate to give a preset temperature as the effluent composition varies.

The problems of burning particular fume, solvent and aqueous streams are as described previously and the design of a multisteam incinerator clearly has to be assessed very carefully to foresee the consequences of all combinations of waste flow and conditions. The cost of operation also needs to be fully understood together with the maintenance requirements. An incineration facility is part of the site waste disposal function and it tends to be the universal case that whilst many companies spend and put considerable effort into a production plant they are not inclined to spend money on waste disposal or effluent disposal equipment or its maintenance.

Although it is accepted that waste disposal is a high cost operation it should be realised and recognised that to cut costs to the bone on equipment and its maintenance can ultimately result in prosecution and closure.

Some of the steps to observe when preparing a specification or study for a new facility are:

- Define exactly or as closely as possible what waste streams are to be processed by the incinerator. The quantities, forms, variations, available pressure, storage and handling methods should all be specified.

- The standards of the flue gas emissions which will be needed to satisfy air pollution legislation should be determined together with any forthcoming amendments to legislation.

- Take a realistic view of operating costs and what will be required on maintenance levels. This is particularly important where wet gas cleaning systems are used because many parts can be regarded as consumable items.

Appendix

OVERVIEW OF WASTE MANAGEMENT PRACTICES IN MAJOR INDUSTRIAL GROUPS

PETROCHEMICAL INDUSTRY
MANUFACTURING AND WASTE MANAGEMENT OVERVIEW

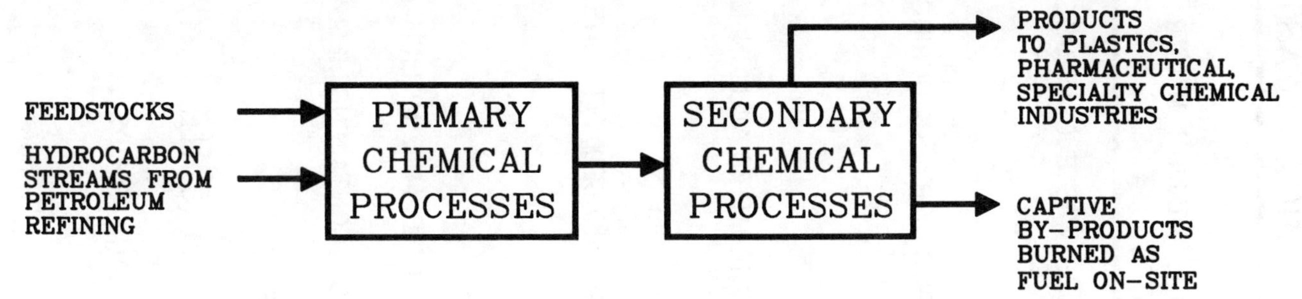

WASTE MANAGEMENT

96% OF WASTE GENERATED MANAGED ON-SITE

4% OF WASTE GENERATED MANAGED OFF-SITE

PETROCHEMICAL WASTE OVERVIEW

● INDUSTRY GENERATES 180 MILLION MT/YR RCRA HAZARDOUS WASTE (1984 EPA SURVEY)

- 68% OF ALL HAZARDOUS WASTE GENERATED IN U.S.

- (CMA SURVEY)

- GENERATION RATE REDUCED BY 8% DURING 1981-82

● WASTE MANAGEMENT

- 96% OF WASTE GENERATED (173 MILLION MT/YR) MANAGED

  ON-SITE

- 4% (7 MILLION MT/YR) MANAGED OFF-SITE

● WASTE CHARACTERISTICS

- 97% OF WASTES GENERATED (175 MILLION MT/YR) ARE

  DILUTE AQUEOUS LIQUIDS CONTAINING LESS THAN 1%

  HAZARDOUS CONSTITUENTS

- 3% OF WASTES GENERATED (5 MILLION MT/YR) ARE

  NON-AQUEOUS LIQUIDS AND SOLIDS

DILUTE AQUEOUS WASTES

SOURCES (EXAMPLES)

- PROCESS WASTEWATER
- COOLING WATER

- CONTAMINATED STORM WATER
- STEAM CONDENSATE

- MAINTENANCE WASH WATER

<u>DILUTE AQUEOUS WASTES</u>

TYPICAL COMPONENTS AND CHARACTERISTICS

● <u>**WATER-SOLUBLE PROCESS CHEMICALS**</u>

ORGANICS (OXIDES, ALCOHOLS, ESTERS)

BOD   100–10,000 mg/L

COD   500–100,000 mg/L

INORGANICS (CHLORIDES, SULFATES)

TDS   100–50,000 mg/L

METAL IONS 1–100 mg/L

CHROMIUM, NICKEL, COPPER

COBALT, MOLYBDENUM, ALUMINUM

SODIUM, CALCIUM

● <u>**EMULSIFIED ORGANICS**</u>

AROMATICS

HIGHER MOLECULAR WEIGHT HYDROCARBONS

● <u>**CAN BE EXTREMELY ACIDIC OR ALKALINE**</u>

NON–AQUEOUS LIQUID AND SOLID WASTES

● LIQUIDS (EXAMPLES)

- BY–PRODUCTS NOT BURNED ON–SITE

- SPENT PROCESS SOLVENTS, EXTRACTANTS

- SKIMMINGS FROM WASTEWATER SEPARATORS

● ORGANIC SOLIDS (EXAMPLES)

- BY–PRODUCT POLYMERS, DISTILLATION BOTTOMS

- SOIL CONTAMINATED FROM CHEMICAL LEAKS AND SPILLS

- EXCESS SLUDGE FROM WASTEWATER BIOTREATMENT

● INORGANIC SOLIDS (EXAMPLES)

- PROCESS FILTER CAKE AND PRECOAT

- WASTEWATER AND COOLING WATER TREATMENT SLUDGES

- SPENT CATALYSTS

ANALYSIS OF ON–SITE MANAGEMENT PRACTICES

96% OF HAZARDOUS WASTE GENERATED

– 173 MILLION MT/YR

– ALMOST ALL DILUTE AQUEOUS LIQUIDS

MANAGEMENT PRACTICE	PERCENTAGE	TYPICAL COST=$/MT)
CHEM/PHYS/BIO TREATMENT	72%	1–5
INJECTION WELL	15%	1–3
IMPOUNDMENT/LANDFARM	10%	N/A
RECYCLE/REUSE	3%	N/A

ANALYSIS OF OFF-SITE MANAGEMENT PRACTICES

4% OF HAZARDOUS WASTE GENERATED

— 7 MILLION MT/YR

AQUEOUS LIQUIDS	2 MILLION MT/YR
NON-AQUEOUS LIQUIDS AND SOLIDS	5 MILLION MT/YR

MANAGEMENT PRACTICE	PERCENTAGE	TYPICAL COST=$/MT)
LANDFILL	33%	100–300
INJECTION WELL	25%	50–150
RECLAIMERS	24%	N/A
INCINERATORS	18%	600–1,000

<u>PLANT MODIFICATIONS</u>

- REDUCE WATER USAGE (PROCESS AND COOLING)

- SEGREGATION OF AQUEOUS WASTE STREAMS

- MULTIPLE USE OF AQUEOUS REAGENTS (ACID, CAUSTC)

- BACK-EXTRACTION OF WASH WATER CONTAMINANTS

TYPICALLY EXPENSIVE, REQUIRES LEAD TIME FOR R&D, PILOT TESTING

<u>OPERATING CHANGES</u>

- PROCESS AUTOMATION/CONTROLS

- EQUIPMENT MONITORING

- PREVENTIVE MAINTENANCE

- EQUIPMENT CLEANING

REQUIRES LESS CAPITAL AND LEAD TIME,

FOCUS ON CONTROLS AND SPILL PREVENTION

PHARMACEUTICAL INDUSTRY
PRODUCTION & WASTE MANAGEMENT OVERVIEW

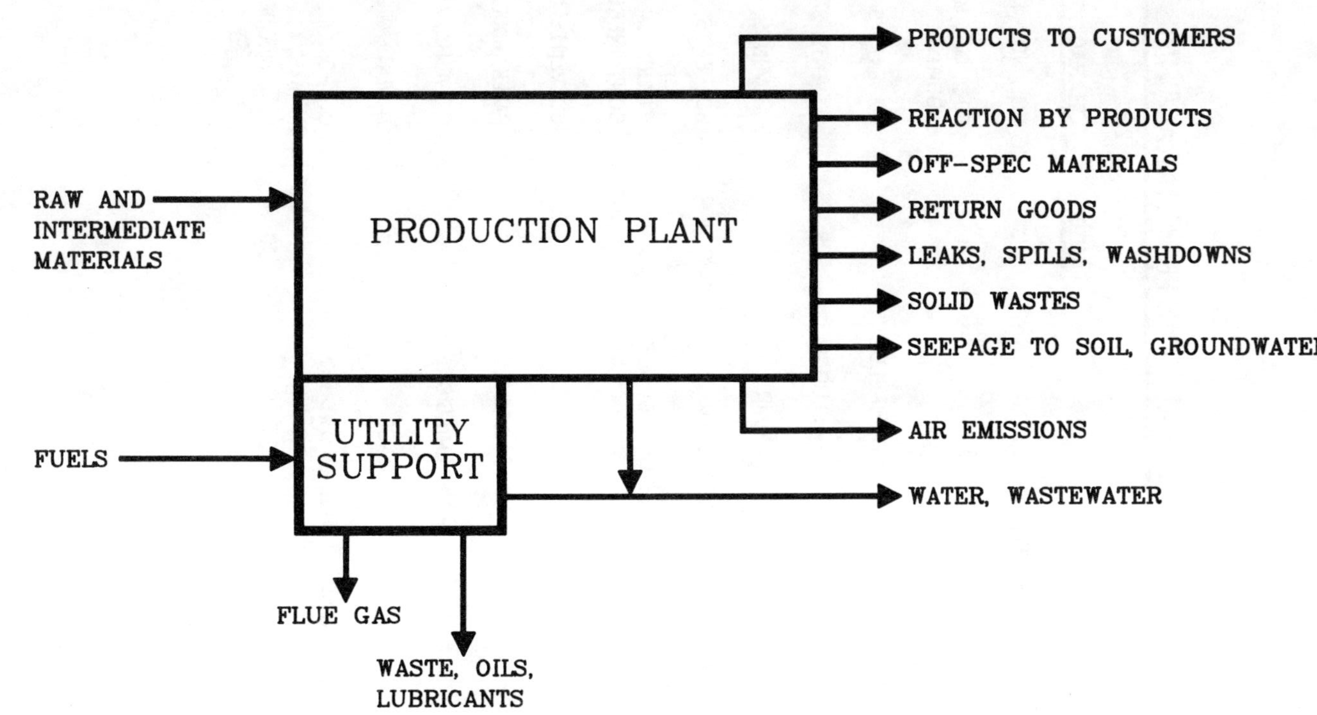

GENERAL WASTE CHARACTERISTICS

SOURCE	TYPE	CHARACTERISTICS
● BOILERS—STACK	G	SO_x, NO_x, TSP, THERMAL
—BLOWDOWN	L	PO_4, Cl^-, METALS, ALKALINITY, ADDITIVES
● WATER/WASTEWATER—EFFLUENT	L	SOLVENTS, SALTS, METALS, BRINES
—SLUDGE	S	SOLVENTS, SALTS, METALS, BRINES
—AIR	G	SOLVENTS, BACTERIA, ODORS
● PRODUCTION—OFF—SPEC	S, L	SOLVENTS, SALTS, ORGANICS
—RETURN GOODS	S, L	SOLVENTS, SALTS, ORGANICS, POWDERS, CONTAINERS
—LEAKS/SPILLS	S, L	SOLVENTS, SALTS, ORGANICS
—SOLID WASTES	S	PAPER, WOOD, CONTAINERS, TRASH
—AIR EMISSIONS	G	SOLVENTS, ORGANICS, POWDERS
—SEEPAGE	S, L	SOLVENTS, ORGANICS, SALTS, METALS
● COOLING TOWER—BLOWDOWN	L	Cl^-, BIOCIDES, SALTS, ADDITIVES
—DRIFT	G	Cl^-, BIOCIDES, SALTS, ADDITIVES

CURRENT WASTE MANAGEMENT PRACTICES

SOURCE	PRACTICES
● BOILERS—STACK	COMBUSTION CONTROLS, LOW—S FUELS, BABRIC FILTER, ESP
—BLOWDOWN	TO WASTEWATER TREATMENT
● WATER/WASTEWATER—EFFLUENT	ON—SITE (PRE)TREATMENT, REUSE, RECYCLE, DISCHARGE
—SLUDGE	DEWATERING, LANDFILL DISPOSAL, INCINERATION
—AIR	SELECTIVE ODOR CONTROLS
● PRODUCTION—OFF—SPEC	REWORKED INTO PRODUCT, DISPOSAL
—RETURN GOODS	CRUSH/SHRED, LANDFILL DISPOSAL, INCINERATION
—LEAKS/SPILLS	CONTROL AREAS, SPCC PLANS, TRAINING, MAINTENANCE
—SOLID WASTES	SEGREGATION, RESALE VALUE, DISPOSAL
—AIR EMISSIONS	SCRUBBER, ACTIVATED CARBON, RECOVERY/REUSE
—SEEPAGE	SPCC PLANS, TRAINING, CLEAN—UP
● COOLING TOWER—BLOWDOWN	TO WASTEWATER TREATMENT
—DRIFT	DRIFT ELIMINATORS, REDUCE DRAFT CONTROL, TOWER LOCATION

<u>BEST WASTE MANAGEMENT PRACTICES</u>

<u>SOURCE</u>	<u>BEST PRACTICES</u>
●BOILERS—STACK	SCRUBBER, ESP, COMB. CONTROL, FUEL MIX
—BLOWDOWN	NON—CR TREATMENT, CHELATING AGENTS
●WATER/WASTEWATER—EFFLUENT	SECONDARY/TERTIARY TREATMENT, SEGREGATION
—SLUDGE	INCINERATION, LAND DISPOSAL
—AIR	COVERED/ENCLOSED TREATMENT
●PRODUCTION—OFF—SPEC	INCINERATION, LANDFILL DISPOSAL
—RETURN GOODS	INCINERATION, LANDFILL DISPOSAL
—LEAKS/SPILLS	SPCC/PM PROGRAMS, TRAINING, RESPONSE
—SOLID WASTES	SEGREGATION, INCINERATION
—AIR EMISSIONS	SCRUBBER, ACTIVATED CARBON
—SEEPAGE	SPCC/PM PROGRAMS, TRAINING, RESPONSE
●COOLING TOWER—BLOWDOWN	NON—CR TREATMENT, INCREASE OF CONC.
—DRIFT	HIGH—EFFICIENCY DRIFT ELIMINATORS

<u>WASTE MINIMIZATION OPPORTUNITIES</u>

<u>SOURCE</u>	<u>WASTE MINIMIZATION</u>
●BOILERS−STACK	REDUCED NEEDS, INCREASED EFFICIENCY
−BLOWDOWN	REDUCED NEEDS, INCREASED EFFICIENCY
●WATER/WASTEWATER−EFFLUENT	MATCH WATER QUALITY WITH NEEDS, WATER/ WASTE REDUCTION AND REUSE
−SLUDGE	DEWATERING, SEGREGATION
−AIR	USE CONTAMINATED AIR FOR COMBUSTION AIR
●PRODUCTION−OFF−SPEC	INCREASE EFFICIENCY AND QA/QC
−RETURN GOODS	MATCH PRODUCTION TO DEMAND
−LEAKS/SPILLS	INCREASE TRAINING, PM PRACTICES
−SOLID WASTES	INCINERATION, WITH ENERGY RECOVERY
−AIR EMISSIONS	REUSE/RECOVERY OF PRODUCTS
−SEEPAGE	INCREASE TRAINING, PM PRACTICES
●COOLING TOWER−BLOWDOWN	REDUCE ENERGY NEEDS/COOLING USE
−DRIFT	PM PRACTICES

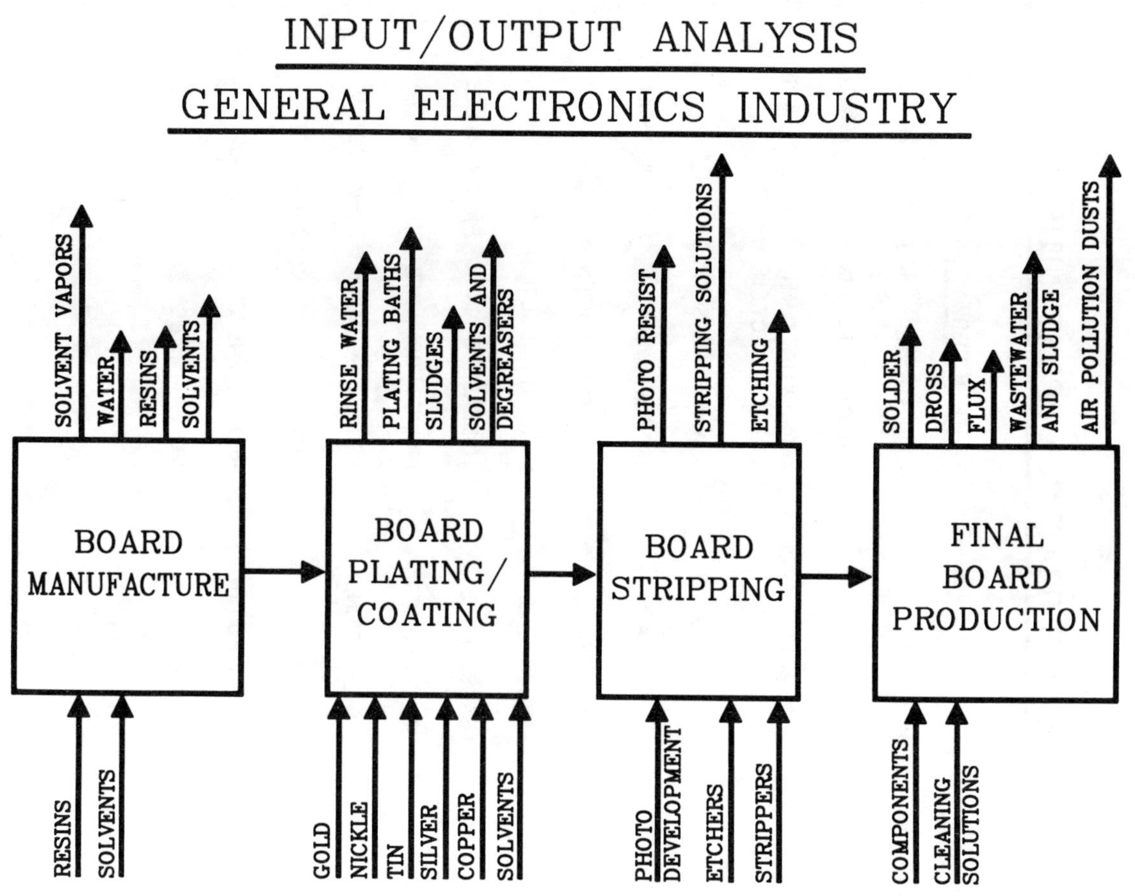

INPUT/OUTPUT ANALYSIS
GENERAL ELECTRONICS INDUSTRY
BOARD MANUFACTURE
BOARD PLATING/ COATING
BOARD STRIPPING
FINAL BOARD PRODUCTION
SOLVENT VAPORS
WATER
RESINS
SOLVENTS
RINSE WATER
PLATING BATHS
SLUDGES
SOLVENTS AND DEGREASERS
PHOTO RESIST
STRIPPING SOLUTIONS
ETCHING
SOLDER
DROSS
FLUX
WASTEWATER
AND SLUDGE
AIR POLLUTION DUSTS
RESINS
SOLVENTS
GOLD
NICKLE
TIN
SILVER
COPPER
SOLVENTS
PHOTO DEVELOPMENT
ETCHERS
STRIPPERS
COMPONENTS
CLEANING SOLUTIONS

CURRENT MANAGEMENT PRACTICES

● **ON—SITE LANDFILLING**

  - OFF—SPEC BOARDS

  - AIR POLLUTION CONTROL DUSTS

  - TREATMENT SLUDGES

● **ON—SITE TREATMENT**

  - PLATING SOLUTIONS

  - SOLVENT FUMES

  - WASTE WATER

● **ON—SITE RECOVERY**

  - NOBLE METAL SOLUTIONS

  - SOLVENTS

  - STRIPPING AND ETCHING WASTES

● **OFF—SITE DISPOSAL**

  - SOLVENTS

  - SLUDGES

  - DUSTS

  - PLATING BATHS

  - SPENT STRIPPING AND ETCHING

  - SOLDERING DROSS AND FLUX

ELECTRONICS INDUSTRY

BEST MANAGEMENT PRACTICES

SOLVENTS	RECOVERY
PLATING BATHS	RECOVERY/DETOXIFICATION
STRIPPERS/ETCHERS SOLUTIONS	RECOVERY/DETOXIFICATION

ELECTRONICS INDUSTRY

WASTE MINIMIZATION PRACTICE

SOLVENT	SEPARATE MIXTURES/SUBSTITUTE AQUEOUS SOLUTIONS/RECYCLE
PLATING BATHS	RECYCLE/RECOVERY SYSTEMS
RINSE WATERS	RECYCLE SYSTEM/REGENERATION RECOVERY SYSTEMS
STRIPPERS/ETCHERS	RECYCLE/REUSE

ELECTRONICS INDUSTRY

INPUT/OUTPUT ANALYSIS

BASIC STEEL INDUSTRY

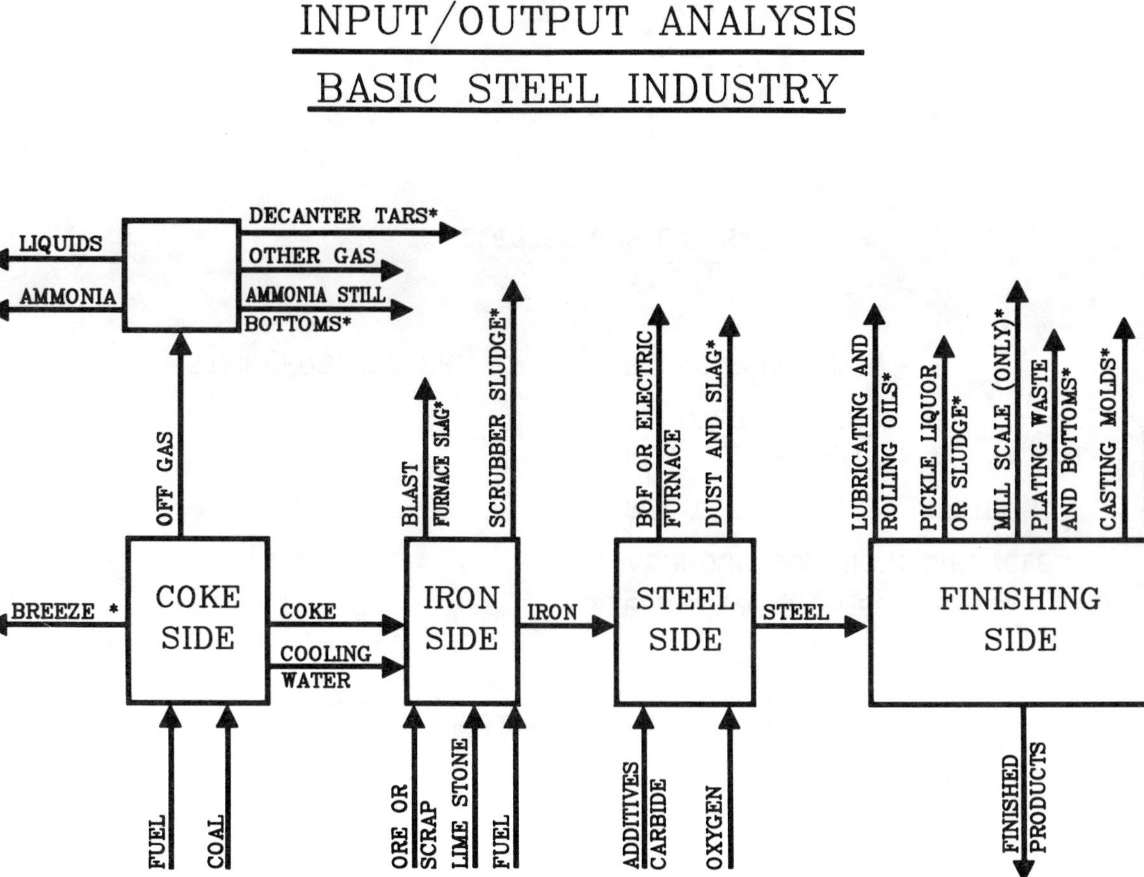

CURRENT WASTE MANAGEMENT PRACTICES

● **ON—SITE LANDFILL**

  - SLAGS
  - SCALES
  - BREEZE
  - OTHER SOLID TRASH
  - TREATMENT SLUDGES

● **WASTE FUEL IN BOILERS**

  - OILS AND GREASERS
  - SOLVENTS

● **RECLAMATION**

  - SLAGS
  - SOLVENTS
  - SCALES
  - PICKLE LIQUOR

● **OFF—SITE DISPOSAL**

  - TARS
  - PLATING SLUDGES
  - SOLVENTS/DEGREASERS

● **ON—SITE TREATMENT**

  - PICKLE LIQUOR
  - RINSE WATER
  - COOLING WATER

BASIC IRON AND STEEL INDUSTRY

BEST MANAGEMENT PRACTICES

WASTE	TECHNOLOGY
AMMONIA STILL BOTTOMS	SOLIDIFICATION/ENCAPSULATION
DECANTER TAR	INCINERATION/FUEL BLENDING
SLAGS	CONSTRUCTION MATERIALS
BAGHOUSE/SCRUBBER SLUDGE	FERROUS/METAL RECOVERY
OILS/GREASES	RECOVERY/FUEL
SOLVENTS	RECOVERY/FUELS
MILL SCALE	OIL RECOVERY/FERROUS RECOVERY (SINTER PLANT)
PLATING BOTTOMS	RECOVERY
PICKLE LIQUOR	REGENERATION

BASIC IRON AND STEEL INDUSTRY

WASTE MINIMIZATION PRACTICES

WASTE	MINIMIZATION METHOD
DECANTER TAR	FUEL BLENDING
DUSTS AND SLAGS	USE ELECTRIC AND/OR INDUCTION FURNACE
OILS/GREASES	RECYCLE/REUSE
SOLVENTS	ELIMINATE HALOGENATED/RECOVERY/ ELIMINATE MIXTURES
MILL SCALE	RECYCLE/THERMAL OR SOLVENT STRIPPING
PICKLE LIQUOR	REGENERATION/RECOVERY/ELIMINATE MIXTURES
PLATING BOTTOMS	RECOVERY OF METALS/ELIMINATE CYANIDE

BASIC IRON AND STEEL INDUSTRY

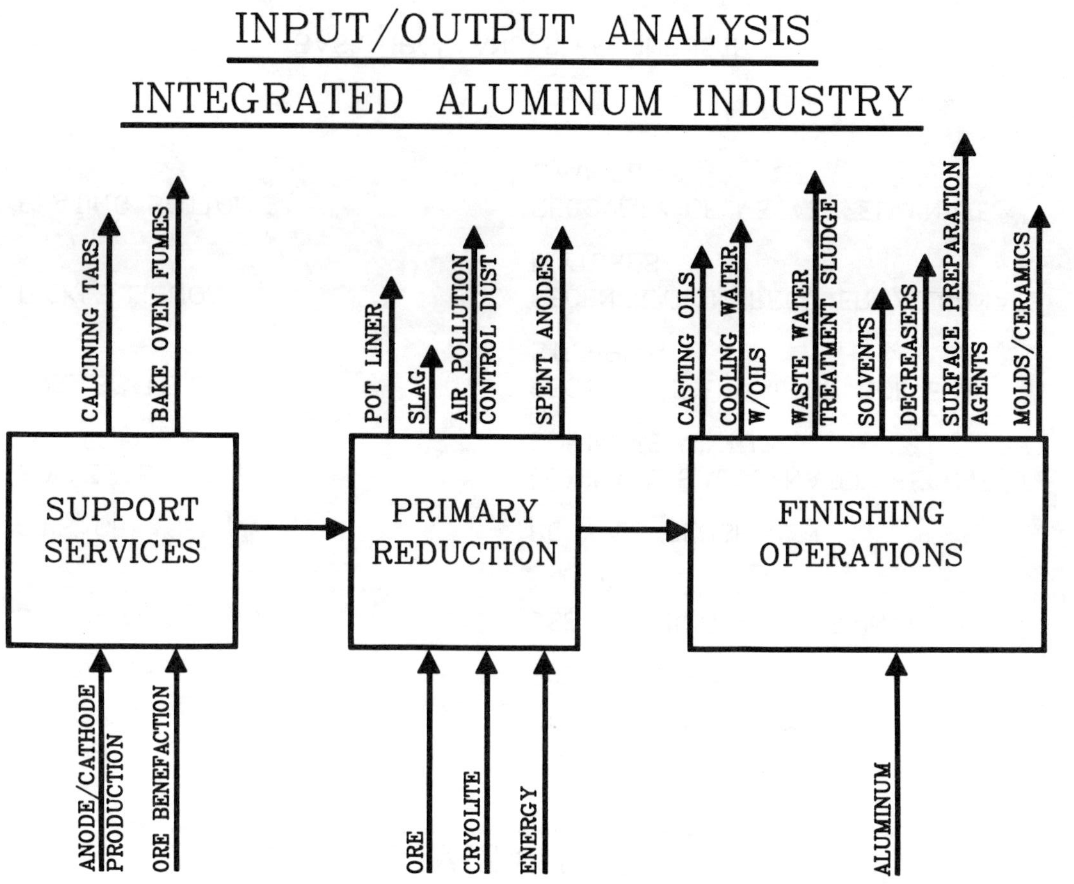

INPUT/OUTPUT ANALYSIS
INTEGRATED ALUMINUM INDUSTRY
SUPPORT SERVICES
PRIMARY REDUCTION
FINISHING OPERATIONS
CALCINING TARS
BAKE OVEN FUMES
ANODE/CATHODE PRODUCTION
ORE BENEFACTION
POT LINER
SLAG
AIR POLLUTION CONTROL DUST
SPENT ANODES
ORE
CRYOLITE
ENERGY
CASTING OILS
COOLING WATER
W/OILS
WASTE WATER TREATMENT SLUDGE
SOLVENTS
DEGREASERS
SURFACE PREPARATION AGENTS
MOLDS/CERAMICS
ALUMINUM

<u>CURRENT MANAGEMENT PRACTICES</u>

- **<u>ON-SITE LANDFILL</u>**

    - TRASH

    - DUSTS

    - MOLDS/CERAMICS

- **<u>ON-SITE STORAGE</u>**

    - POT LINER

- **<u>ON-SITE TREATMENT/RECOVERY</u>**

    - CASTING OILS

    - SURFACE TREATING CHEMICALS

    - DEGREASERS

- **<u>OFF-SITE DISPOSAL</u>**

    - OIL FILTRATION SLUDGE

    - ANODE/CATHODE TARS

    - DUSTS

ALUMINUM INDUSTRY

BEST MANAGEMENT PRACTICES

POT LINER WASTE	SECURE LANDFILL/ENERGY RECOVERY
OILY SLUDGE	LAND DISPOSAL/ENCAPSULATION
ANODES	RECYCLE/REUSE
DEGREASERS	WASTE WATER TREATMENT
SOLVENTS	RECYCLE
SURFACE ETCHING	

PRIMARY ALUMINUM

WASTE MINIMIZATION PRACTICES

POT LINER WASTES	ENERGY RECOVERY/REUSE
OILY SLUDGE	RECOVERY/ENERGY RECOVERY
ANODES	RECYCLE/REUSE
DEGREASERS	REGENERATE/REUSE
SOLVENTS	RECOVERY/ELIMINATE MIXTURES
SURFACE ETCHING	RECYCLE/REUSE

PRIMARY ALUMINUM